Reiner Horst · Hoang Tuy

Global Optimization

Deterministic Approaches

With 55 Figures

Springer-Verlag
Berlin Heidelberg New York
London Paris Tokyo
Hong Kong

Professor Dr. Reiner Horst
University of Trier
Department of Mathematics
P.O.Box 3825
D-5500 Trier, FRG

Professor Dr. Hoang Tuy
Vien Toan Hoc
Institute of Mathematics
P.O.Box 631, BO-HO
10000 Hanoi, Vietnam

ISBN 3-540-52368-5 Springer-Verlag Berlin Heidelberg New York Tokyo
ISBN 0-387-52368-5 Springer-Verlag New York Heidelberg Berlin Tokyo

Printing: Weihert-Druck GmbH, Darmstadt
Bookbinding: T. Gansert GmbH, Weinheim-Sulzbach
2142/7130-543210

PREFACE

The enormous practical need for solving global optimization problems coupled with a rapidly advancing computer technology has allowed one to consider problems which a few years ago would have been considered computationally intractable. As a consequence, we are seeing the creation of a large and increasing number of diverse algorithms for solving a wide variety of multiextremal global optimization problems.

The goal of this book is to systematically clarify and unify these diverse approaches in order to provide insight into the underlying concepts and their properties. Aside from a coherent view of the field much new material is presented.

By definition, a multiextremal global optimization problem seeks at least one global minimizer of a real–valued objective function that possesses different local minimizers. The feasible set of points in $\mathbb{R}^n$ is usually determined by a system of inequalities. It is well known that in practically all disciplines where mathematical models are used there are many real–world problems which can be formulated as multiextremal global optimization problems.

Standard nonlinear programming techniques have not been successful for solving these problems. Their deficiency is due to the intrinsic multiextremality of the formulation and not to the lack of smoothness or continuity, for often the latter properties are present. One can observe that local tools such as gradients, subgradients, and second order constructions such as Hessians, cannot be expected to yield more than local solutions. One finds, for example, that a stationary point is often detected for which there is even no guarantee of local minimality. Moreover, determining the local minimality of such a point is known to be NP–hard in the sense of computational complexity even in relatively simple cases. Apart from this deficiency in the local situation, classical methods do not recognize conditions for global optimality.

For these reasons global solution methods must be significantly different from standard nonlinear programming techniques, and they can be expected to be – and are – much more expensive computationally. Throughout this book our focus will be on typical procedures that respond to the inherent difficulty of multiextremality and which take advantage of helpful specific features of the problem structure. In certain sections, methods are presented for solving very general and difficult global problems, but the reader should be aware that difficult large scale global optimization problems cannot be solved with sufficient accuracy on currently available computers. For these very general cases our exposition is intended to provide useful tools for transcending local optimality restrictions, in the sense of providing valuable information about the global quality of a given feasible point. Typically, such information will give upper and lower bounds for the optimal objective function value and indicate parts of the feasible set where further investigations of global optimality will not be worthwhile.

On the other hand, in many practical global optimizations, the multiextremal feature involves only a small number of variables. Moreover, many problems have additional structure that is amenable to large scale solutions.

Many global optimization problems encountered in the decision sciences, engineering and operations research have at least the following closely related key properties:

(i) convexity is present in a limited and often unusual sense;
(ii) a global optimum occurs within a subset of the boundary of the feasible set.

With the current state of the art, these properties are best exploited by deterministic methods that combine analytical and combinatorial tools in an effective way. We find that typical approaches use techniques such as branch and bound, relaxation, outer approximation, and valid cutting planes, whose basic principles have long appeared in the related fields of integer and combinatorial optimization as well as convex minimization. We have found, however, that application of these fruitful ideas to global optimization is raising many new interesting theoretical and computational questions whose answers cannot be inferred from previous successes. For example, branch and bound methods applied to global optimization problems generate infinite processes, and hence their own convergence theory must be developed. In contrast, in integer programming these are finite procedures, and so their convergence properties do not directly apply. Other examples involve important results in convex minimization that reflect the coincidence of local and global solutions. Here also one cannot expect a direct application to multiextremal global minimization.

In an abundant class of global optimizations, convexity is present in a reverse sense. In this direction we focus our exposition on the following main topics:

(a) minimization of concave functions subject to linear and convex constraints (i.e., "concave minimization");
(b) convex minimization over the intersection of convex sets and complements of convex sets (i.e., "reverse convex programming"); and
(c) global optimization of functions that can be expressed as a difference of two convex functions (i.e., "d.c.–programming").

Another large class of global optimization that we shall discuss in some detail has been termed "*Lipschitz Programming*", where now the functions in the formulation are assumed to be Lipschitz continuous on certain subsets of their domains. Although neither of the aforementioned properties (i)–(ii) is necessarily satisfied in Lipschitz problems, much can be done here by applying the basic ideas and techniques which we shall develop for the problem classes (a), (b), and (c) mentioned above.
Finally, we also demonstrate how global optimization problems are related to *solving systems* of *equations and/or inequalities.* As a by–product, then, we shall present some new solution methods for solving such systems.

The underlying purpose of this book is to present general methods in such a way as to enhance the derivation of special techniques that exploit frequently encountered additional problem structure. The multifaceted approach is manifested occasionally in some computational results for these special but abundant problems. However, at the present stage, these computational results should be considered as preliminary.

The book is divided into three main parts.

Part A introduces the main global optimization problem classes we study, and develops some of their basic properties and applications. It then discusses the fundamental concepts that unify the various general methods of solution, such as outer approximation, concavity cuts, and branch and bound.

Part B treats concave minimization and reverse convex programming subject to linear and reverse convex constraints. In this part we present additional detail on specially structured problems. Examples include decomposition, projection, separability, and parametric approaches.

In Part C we consider rather general global optimization problems. We study d.c.–programming and Lipschitz optimization, and present our most recent attempts at solving more general global optimization problems. In this part, the specializations most naturally include biconvex programming, indefinite "all–quadratic" optimization, and design centering as encountered in engineering design.

Each chapter begins with a summary of its contents.

The technical prerequisites for this book are rather modest, and are within reach of most advanced undergraduate university programs. They include a sound knowledge of elementary real analysis, linear algebra, and convexity theory. No familiarity with any other branch of mathematics is required.

In preparing this book, we have received encouragement, advice, and suggestions from a large group of individuals. For this we are grateful to Faiz Al–Khayyal, Harold P. Benson, Neil Koblitz, Ken Kortanek, Christian Larsen, Janos Pintér, Phan Thien Thach, Nguyen van Thoai, Jakob de Vries, Graham Wood, and to several other friends, colleagues and students. We are indebted to Michael Knuth for drawing the figures.

We acknowledge hospitality and / or financial support given by the following organizations and institutions: the German Research Association (DFG), the Alexander von Humboldt Foundation, the Minister of Science and Art of the State Lower Saxony, the National Scientific Research Center in Hanoi, the Universities of Oldenburg and Trier.

Our special thanks are due to Frau Rita Feiden for the efficient typing and patient retyping of the many drafts of the manuscript.

Finally, we thank our families for their patience and understanding.

December 1989 Reiner Horst Hoang Tuy

CONTENTS

PART A

INTRODUCTION AND BASIC TECHNIQUES

Part A introduces the main global optimization problem classes we study, and develops some of their basic properties and applications. It then discusses some fundamental concepts that unify the various general methods of solution, such as outer approximation, concavity cuts, and branch and bound.

CHAPTER I

SOME IMPORTANT CLASSES OF GLOBAL OPTIMIZATION PROBLEMS

In Chapter I, we introduce the main classes of global optimization problems that we study: concave minimization, reverse convex constraints, d.–c. programming, and Lipschitz optimization. Some basic properties of these problems and various applications are discussed. It is also shown that very general systems of equalities and (or) inequalities can be formulated as global optimization problems.

1. GLOBAL OPTIMIZATION

We define a **standard global optimization** problem as follows.

Given a nonempty, closed set $D \subset \mathbb{R}^n$ and a continuous function $f: A \longrightarrow \mathbb{R}$, where $A \subset \mathbb{R}^n$ is a suitable set containing D, find at least one point $x^* \in D$ satisfying $f(x^*) \le f(x)$ for all $x \in D$ or show that such a point does not exist.

For the sake of simplicity of presentation it will sometimes be assumed that a solution $x^* \in D$ exists. In many cases, D will be compact and f will be continuous on an open set $A \supset D$. Then, clearly, the existence of x^* is assured by the well–known

Theorem of Weierstraß. In other important cases, one encounters compact feasible sets D and objective functions f that are continuous in the (relative) interior of D but have discontinuities on the boundary of D.

Throughout this book, a global optimization problem will be denoted by

$$\begin{array}{ll} \text{minimize} & f(x) \\ \text{s.t.} & x \in D \end{array} . \tag{1}$$

A point $x^* \in D$ satisfying $f(x^*) \leq f(x) \ \ \forall x \in D$ is called a *global minimizer* of f over D. The corresponding value of f is called the *global minimum* of f over D and is denoted by min f(D). The set of all solutions of problem (1) will be denoted by argmin f(D).

Note that since

$$\max f(D) = -\min(-f(D)) ,$$

global maximization problems are included in (1).

Sometimes we shall be able to find all solutions. On the other hand, we frequently have to require additional properties of f and D, one of them being **robustness** of D.

Definition I.1. *A closed subset $D \subset \mathbb{R}^n$ is called* ***robust*** *if it is the closure of an open set.*

Note that a convex set $D \subset \mathbb{R}^n$ with nonempty interior is robust (cf., e.g., Rockafellar (1970), Theorem 6.3).

Let $\|\cdot\|$ denote the Euclidean norm in $\mathbb{R}^n$ and let $\varepsilon > 0$ be a real number. Then an (open) ε–neighbourhood of a point $x^* \in \mathbb{R}^n$ is defined as the open ball

$$N(x^*, \varepsilon) := \{x \in \mathbb{R}^n : \|x - x^*\| < \varepsilon\}$$

centered at x^* with radius ε.

A point $x^* \in D$ is called a ***local minimizer*** *of f over D* if there is an $\varepsilon > 0$ such that

$$f(x^*) \leq f(x) \quad \forall x \in N(x^*, \varepsilon) \cap D$$

holds.

In order to understand the enormous difficulties inherent in global optimization problems and the computational cost of solving them, it is important to notice that *all standard techniques in nonlinear optimization can at most locate local minima.* Moreover, there is no local criterion for deciding whether a local solution is global. Therefore, conventional methods of optimization using such tools as derivatives, gradients, subgradients and the like, are, in general, not capable of locating or identifying a global optimum.

Remark I.1. Several *global* criteria for a global minimizer have been proposed. Let D be bounded and robust and let f be continuous on D. Denote by $\mu(M)$ a measure of a subset $M \subset \mathbb{R}^n$. Furthermore, let $x^* \in D$ and

$$S := \{x \in D: f(x) \leq f(x^*)\}.$$

Then, obviously, $\mu(S) = 0$ implies that x^* is a global minimizer of f over D. But, apart from very special cases, there is no numerically feasible method for computing $\mu(S)$. A theoretical iterative scheme for solving global optimization problems that is based on this criterion was proposed by Galperin and Zheng (1987) (for a thorough exposition of the related so–called integral methods, see also Chew and Zheng (1988)).

Another global criterion is given by Falk (1973a). Let D have the above property and suppose that $f(x) > 0 \ \ \forall x \in D$. Let $\bar{x} \in D$. Define for $k \in \mathbb{N}$

$$r(\bar{x}, k) := \int_D \left[\frac{f(x)}{f(\bar{x})}\right]^k dx .$$

Then, it is shown in Falk (1973) that $\bar{x}$ is a global maximizer of f over D if and only if the sequence $\{r(\bar{x}, k)\}_{k \in \mathbb{N}}$ is bounded.

Although it is not very practical for solving the general global optimization problem, Falk's criterion has been used by Duong (1987) to propose an algorithm for finding an approximation of the global minimum of a polynomial over a polytope. A somewhat similar idea was suggested in Zhirov (1985) for globally minimizing a polynomial over a parallelepiped.

Note that certain important classes of optimization problems have the property that every local minimum is a global minimum. A well–known example is convex minimization where the objective function is a convex function and where the feasible set is a convex set. Since, in these classes, standard optimization procedures for finding local solutions will yield the global optimum, considerable effort has gone into characterizing families of functions and feasible sets having the property that every local minimum is a global solution. For a number of results on this question see Martos (1967), Mangasarian (1969), Martos (1975), Zang and Avriel (1975), Netzer and Passy (1975), Zang et. al. (1976), Avriel (1976), Avriel and Zang (1981), Horst (1982 and 1984a,b), Gasanov and Rikun (1984 and 1985), Horst and Thach (1988).

Throughout this volume global optimization problems are considered ***where standard optimization techniques fail because of the existence of local minima that are not global.*** These global optimization problems will be called ***multiextremal global optimization*** problems.

Due to the inherent difficulties mentioned above, the methods devised for analysing multiextremal global optimization problems are quite diverse and significantly different from the standard tools referred to above.

Though several general theoretical concepts exist for solving problem (1), in order to build a numerically promising implementation, additional properties of the problem's data usually have to be exploited.
Convexity, for example, will often be present. Many problems have linear constraints. Other problems involve Lipschitzian functions with known Lipschitz constants.

In recent years a rapidly growing number of proposals has been published for solving specific classes of multiextremal global optimization problems. It seems to be impossible to present a thorough treatment of all these methods in one volume. However, many of them can be interpreted as applications and combinations of certain recent basic approaches. Knowledge of these approaches not only leads to a deeper understanding of various techniques designed for solving specific problems, but also serves as a guideline in the development of new procedures.

This book presents the state of the art in certain deterministic concepts used in most of the methods for solving multiextremal global optimization problems that we believe to be promising for further research. These concepts will be applied to derive algorithms for solving broad classes of multiextremal global optimization problems that are frequently encountered in applications.

In order to describe these classes, the following definitions are introduced. We assume the reader to be familiar with basic notions and results on convexity.

Definition I.2. *Let $C \subset \mathbb{R}^n$ be convex. A function $h: C \longrightarrow \mathbb{R}$ is called* ***d.c.*** *on C if there are two convex functions $p: C \longrightarrow \mathbb{R}$, $q: C \longrightarrow \mathbb{R}$ such that*

$$h(x) = p(x) - q(x) \quad \forall x \in C.$$

A function that is d.c. on $\mathbb{R}^n$ will be called d.c..

An inequality $h(x) \le 0$ is called a ***d.c. inequality*** *whenever h is d.c..*

In Definition I.2, d. c. is an abbreviation for "difference of two convex functions".

Definition I.3. *Let $M \subset \mathbb{R}^n$. A function $h: \mathbb{R}^n \longrightarrow \mathbb{R}$ is called* ***Lipschitzian on M*** *if there is a real constant $L = L(h,M) > 0$ such that*

$$|h(x) - h(y)| \le L \, \|x - y\| \quad \forall x,y \in M.$$

An inequality $h(x) \le 0$ is called a ***Lipschitzian inequality*** *(on M) whenever h is Lipschitzian on M.*

Definition I.4. *Let $h: \mathbb{R}^n \longrightarrow \mathbb{R}$ be a convex function. Then the inequality $h(x) \le 0$ is called* **convex** *whereas the inequality $h(x) \ge 0$ is called* **reverse convex**.

Sometimes one encounters quasiconvex functions. Recall that $h: C \longrightarrow \mathbb{R}$ with C convex, is quasiconvex if and only if, for all $\alpha \in \mathbb{R}$, the level sets $\{x \in C: h(x) \le \alpha\}$ are convex. An alternative characterization of quasiconvexity is given by

$$h(\lambda x + (1-\lambda)y) \le \max \{h(x), h(y)\} \quad \forall x,y \in C, 0 \le \lambda \le 1 .$$

The concepts to be described in the subsequent chapters will be used to develop algorithms for solving broad classes of multiextremal global optimization problems. The feasible set D and the objective function f can belong to one of the following classes.

Feasible Set D:

- *convex and defined by finitely many convex inequalities,*
- *intersection of a convex set with finitely many complements of a convex set and defined by*

  finitely many convex and finitely many reverse convex inequalities,
- *defined by finitely many Lipschitzian inequalities.*

Objective Function f:

- *convex,*
- *concave,*
- *d.c.,*
- *Lipschitzian,*
- *certain generalizations of these four classes.*

In addition, some classes of problems will be considered in which D is defined by a finite number of *d.c. inequalities*. Finally, it will be shown that various systems of *equalities and inequalities* can be solved by the algorithmic concepts to be presented.

The sections that follow contain an introduction to some basic properties of the problems introduced above. Many applications will be described, and various connections between these classes will be revealed.

2. CONCAVE MINIMIZATION

2.1. Definition and Basic Properties

One of the most important global optimization problems is that of ***minimizing a concave function over a convex set*** (or – equivalently – of maximizing a convex function over a convex set):

$$\begin{array}{ll} \text{minimize } f(x) \\ \text{s.t. } x \in D \end{array}, \tag{2}$$

where $D \subset \mathbb{R}^n$ is nonempty, *closed* and *convex* and where $f: A \longrightarrow \mathbb{R}$ is *concave on a suitable set* $A \subset \mathbb{R}^n$ *containing* D.

The concave minimization problem (2) is a multiextremal global optimization problem: it is easy to construct, for example, concave functions f and polytopes D having the property that every vertex of D is a local minimizer of f over D.

Example I.1. Let $f(x) = -\|x\|^2$, where $\|\cdot\|$ denotes the Euclidean norm, and let $D = \{x \in \mathbb{R}^n: a \le x \le b\}$ with $a, b \in \mathbb{R}^n$, $a < 0$, $b > 0$ (all inequalities are understood with respect to the componentwise order of $\mathbb{R}^n$). It is easily seen that every vertex of D is a local minimizer of f over D: Let $v = (v_1,\dots,v_n)^T$ be a vertex of D. Then there are two index sets $I_1, I_2 \subset \{1,\dots,n\}$ satisfying $I_1 \cup I_2 = \{1,\dots,n\}$, $I_1 \cap I_2 = \emptyset$ such that we have

$$v_i = a_i \ (i \in I_1),\ v_i = b_i \ \ (i \in I_2).$$

Let

$$0 < \varepsilon < \min \{-a_i, b_i: i = 1,\ldots,n\}$$

and consider the cube

$$B(v,\varepsilon) := \{y \in \mathbb{R}^n: \max_i |y_i - v_i| \le \varepsilon\}.$$

Clearly, we have

$$N(v,\varepsilon) = \{y \in \mathbb{R}^n: \|y - v\| < \varepsilon\} \subset B(v,\varepsilon).$$

All $x \in B(v,\varepsilon) \cap D$ satisfy

$$x_i = a_i + y_i \quad (i \in I_1), \quad x_i = b_i - y_i \quad (i \in I_2),$$

where $0 \le y_i \le \varepsilon$.

By the definition of ε, it follows that we have

$$(x_i)^2 \le (a_i)^2 \quad (i \in I_1) \ , \ (x_i)^2 \le (b_i)^2 \quad (i \in I_2),$$

hence $f(v) \le f(x) \ \forall x \in N(v,\varepsilon) \cap D$.

Some basic properties of problem (2), however, make concave programming problems easier to handle than general multiextremal global optimization problems.

Besides convexity of the feasible set D which will be heavily exploited in the design of algorithms for solving (2), the most interesting property is that a concave function f attains its global minimum over D at an extreme point of D.

Theorem I.1. *Let $f: D \longrightarrow \mathbb{R}$ be concave and let $D \subset \mathbb{R}^n$ be nonempty, compact and convex. Then the global minimum of f over D is attained at an extreme point of D.*

Proof. The global minimum of f over the compact set D exists by the well-known Theorem of Weierstraß since a concave function defined on $\mathbb{R}^n$ is continuous everywhere. It suffices to show that for every $x \in D$ there is an extreme point v of D such that $f(x) \geq f(v)$ holds.

By the Theorems of Krein–Milman / Caratheodory, there is a natural number $k \leq n+1$ such that

$$x = \sum_{i=1}^{k} \lambda_i v^i, \quad \sum_{i=1}^{k} \lambda_i = 1, \; \lambda_i \geq 0 \quad (i=1,\dots,k) , \tag{3}$$

where v^i $(i=1,\dots,k)$ are extreme points of D. Let v satisfy $f(v) = \min \{f(v^i): i=1,\dots,k\}$. Then we see from the concavity of f and from (3) that we have

$$f(x) \geq \sum_{i=1}^{k} \lambda_i f(v^i) \geq f(v) \cdot (\sum_{i=1}^{k} \lambda_i) = f(v) . \qquad \blacksquare$$

From the viewpoint of computational complexity, a concave minimization problem is *NP–hard*, even in such special cases as that of minimizing a quadratic concave function over very simple polytopes such as hypercubes (e.g., Pardalos and Schnitger (1987)). One exception is the negative Euclidean norm that can be minimized over a hyperrectangle by an $0(n)$ algorithm (for related work see Gritzmann and Klee (1988), Bodlaender et al. (1989)). Among the few practical instances of concave minimization problems for which polynomial time algorithms have been constructed are certain production scheduling and inventory models that can be regarded as special network flow problems (e.g., Zangwill (1968 and 1985), Zangwill (1969), Love (1973), Konno (1973 and 1988), Afentakis et al. (1984), Bitran et al. (1984)).

The range of practical applications of concave minimization problems is very broad. Large classes of decision models arising from operations research and mathematical economics and many engineering problems lead to formulations such as problem (2). Furthermore, many models which originally are not concave can be transformed into equivalent concave minimization problems. We briefly discuss some important examples and relationships in the next sections.

Surveys of concave programming are given in Horst (1984), Pardalos and Rosen (1986 and 1987), and Horst (1990).

2.2. Brief Survey of Direct Applications

Many problems consist in choosing the levels x_i of n activities i=1,...,n restricted to $a_i \leq x_i \leq b_i$, $a_i, b_i \in \mathbb{R}_+$, $a_i < b_i$, producing independent costs f_i: $[a_i, b_i] \longrightarrow \mathbb{R}_+$, subject to additional convex or (in most applications) linear inequality constraints. The objective function $f(x) = \sum_{i=1}^{n} f_i(x_i)$ is separable, i.e., the sum of n functions $f_i(x_i)$. Each f_i typically reflects the fact that the activities incur a fixed setup cost when the activity is started (positive jump at $x_i = 0$) as well as a variable cost related to the level of the activity.

If the variable cost is linear and the setup cost is positive, then f_i is nonlinear and concave and we have the objective functions of the classical fixed charge problems (e.g., Murty (1969), Bod (1970), Steinberg (1970), Cabot (1974)). Frequently encountered special examples include fixed charge transportation problems and capacitated as well as uncapacitated plant location (or site selection) problems (e.g., Manne (1964), Gray (1971), Dutton et al. (1974) Barr et al. (1981)). Multilevel fixed charge problems have also been discussed by Jones and Soland (1969). Interactive fixed charge problems were investigated by Kao (1979), Erenguc and Benson (1986), Benson and Erenguc (1988) and references therein.

Often, price breaks and setup costs yield concave functions f_i having piecewise linear variable costs. An early problem of this type was the bid evaluation problem (e.g., Bracken and McCormick (1968), and Horst (1980a)). Moreover, piecewise linear concave functions frequently arise in inventory models and in connection with constraints that form so–called Leontiev substitution systems (e.g., Zangwill (1966), Veinott (1969), and Koehler et al. (1975)).

Nonlinear concave variable costs occur whenever it is assumed that as the number of units of a product increases, the unit cost strictly decreases (*economies of scale*). Different concave cost functions that might be used in practice are discussed, by, e.g., Zwart (1974). Often the concave functions f_i are assumed to be quadratic (or approximated by quadratic functions; the most well–known examples are quadratic transportation and related network–flow problems (e.g., Koopmans and Beckman (1957), Cabot and Francis (1974), Bhatia (1981), Bazaraa and Sherali (1982), Florian (1986)).

Many other situations arising in practice lead to minimum *concave objective network problems*. Examples include problems in communications network planning, hydraulic or sewage network planning, plant location problems, inventory and production planning etc. (e.g., Afentakis et al. (1984), Bitran et al. (1984), Ciaponi and Papiri (1976), Dajani and Hasit (1974), Dutton et al. (1974), Elshafei (1975), Florian (1986), Florian et al. (1971), Gallo and Sodini (1979 and 1979 a), Gray (1971), Jensen and Barnes (1980), Kleitman and Claus (1972), Konno (1973 and 1988), Love (1973), Mahjoub (1983), Manne (1964), Minoux (1986), Minoux and Serreault (1981), Soland (1974), Taha (1973), Yaged (1971), and Zangwill (1966, 1968, 1969 and 1985)).

Global optimization of the general (possibly indefinite) quadratic case is discussed by many authors, e.g., Ritter (1965 and 1966), Cabot and Francis (1970), Balas (1975a), Tammer (1976), Konno (1976a and 1980), Gupta and Sharma (1983), Rosen (1983, 1984 and 1984a), Aneja et al. (1984), Kalantari (1984), Schoch (1984), Thoai (1984), Pardalos (1985, 1987, 1988 and 1988a), Benacer and Tao (1986), Kalantari and Rosen (1986 and 1987), Pardalos and Rosen (1986 and 1987), Rosen and Pardalos (1986), Thoai (1987), Tuy (1987) Pardalos and Gupta (1988).

The solution of nonconvex quadratic programming is the main topic of the book of Pardalos and Rosen (1987).

Often, nonseparable concave objective functions occur in the context of models related to economies of scale or the maximization of utility functions having increasing marginal values. Some models yield quasiconcave rather than concave objective functions that in many respects behave like concave functions in minimization problems. Moreover, quasiconcave functions are often concave in the region of interest (e.g., Rössler (1971) and references therein). Sometimes these functions can be suitably transformed into concave functions that yield equivalent minimization problems (e.g., Horst (1984)).

A model arising from production planning is discussed in Rössler (1971), and a model for national development is treated in Brotchi (1971).

Many other specific problems lead directly to concave minimization; examples are discussed in, e.g., Grotte (1975), and McCormick (1973 and 1983). Of particular interest are certain engineering design problems. Many of them can be formulated as suitable concave objective network problems (see above). Other examples arise from VLSI chip design (Watanabe (1984)), in the fabrication of integrated circuits (Vidigal and Director (1982)), and in diamond cutting (Nguyen et al. (1985)). The last two examples are so–called design centering problems that turn out to be linear, concave or d.c. programming problems (cf. Vidigal and Director (1982), Thach et al. (1987), Thach (1988), see also Section I.3.).

2.3. Integer Programming and Concave Minimization

One of the most challenging classes of optimization problems with a wide range of applications is integer programming. These are extremum problems with a discrete feasible set.

In this section, it will be shown that concave programming is a sort of bridge between integer and nonlinear programming. Broad classes of integer programming problems can be formulated as equivalent concave programming problems, where equivalence is understood in the sense that the sets of optimal solutions coincide.

This equivalence is well–known for the quadratic assignment problem (e.g., Bazaraa and Sherali (1982), Lawler (1963)) and the 3–dimensional assignment problem (e.g., Frieze (1974)). The zero–one integer linear programming problem was reduced to a quadratic concave programming problem subject to linear constraints by Raghavachari (1969). More general results on the connections between integer and nonlinear programming are given in Giannessi and Niccolucci (1976), where the Theorem I.2 below is presented.

Let $C \subseteq \mathbb{R}^n$, $B = \{0,1\}$, $B^n = B \times \ldots \times B$ (n times), $f: \mathbb{R}^n \longrightarrow \mathbb{R}$. Consider the integer programming problem

$$\begin{array}{ll} \text{minimize} & f(x) \\ \text{s.t.} & x \in C \cap B^n \end{array} \qquad (4)$$

Define $e := (1,1\ldots,1)^T \in \mathbb{R}^n$, $E := \{x \in \mathbb{R}^n : 0 \leq x \leq e\}$ and associate with (4) the nonlinear problem

$$\begin{array}{ll} \text{minimize} & [f(x) + \mu x(e-x)] \\ \text{s.t.} & x \in C \cap E \end{array} \qquad (5)$$

which depends on the real number μ. Then the following connection between (4) and (5) holds.

Theorem I.2. *Let C be a closed subset of $\mathbb{R}^n$ satisfying $C \cap B^n \neq \emptyset$ and suppose that $f: \mathbb{R}^n \longrightarrow \mathbb{R}$ is Lipschitzian on an open set $A \supset E$ and twice continuously differentiable on E. Then there exists a $\mu_0 \in \mathbb{R}$ such that for all $\mu > \mu_0$ we have*

(i) *(4) and (5) are equivalent,*

(ii) *$f(x) + \mu x(e - x)$ is concave on E.*

Proof. (i): Set $\varphi(x) := x(e - x)$. The function $\varphi(x)$ is continuous on E, and obviously we have

$$\varphi(x) = 0 \ \ \forall x \in B^n, \ \varphi(x) > 0 \ \ \forall x \in E \setminus B^n . \qquad (6)$$

We show that there is a $\mu_0 \in \mathbb{R}$ such that, whenever we have $\mu > \mu_0$, then the global minimum of $f(x) + \mu\varphi(x)$ over $C \cap E$ is attained on B^n.

First, note that, for all $y \in B^n$, there exists an open neighbourhood $N(y, \varepsilon) = \{x \in \mathbb{R}^n: \|x-y\| < \varepsilon\}$ such that for all $x \in N(y,\varepsilon) \cap (E \setminus B^n)$ we have

$$\varphi(x) \geq (1-\varepsilon)\, \|x-y\| \tag{7}$$

($\|\cdot\|$ denotes the Euclidean norm).

To see this, let $\varepsilon < 1$, $r := \|x-y\| < \varepsilon$, and $u := \frac{1}{r}(x-y)$. Then, as $x = y + ru \in E$, we have

$$\varphi(x) = \sum_{j=1}^{n} (ru_j + y_j)\,(1 - y_j - ru_j)\,.$$

Since $y_j = 1$ implies $u_j \leq 0$, and $y_j = 0$ implies $u_j \geq 0$, we may express $\varphi(x)$ in the following way:

$$\varphi(x) = \sum_{u_j>0} ru_j(1-ru_j) + \sum_{u_j<0} (1+ru_j)\,(-ru_j) = \sum_{j=1}^{n} r|u_j|\,(1-r|u_j|)$$

$$= r\sum_{j=1}^{n} |u_j| - r^2\sum_{j=1}^{n} |u_j|^2\,.$$

Using $0 < r < \varepsilon < 1$, $\sum_{j=1}^{n} |u_j|^2 = \frac{1}{r^2}\|x-y\|^2 = 1$ and $\sum_{j=1}^{n} |u_j| \geq \|u\| = 1$, we finally see that

$$\varphi(x) \geq r\,(1-\varepsilon) = (1-\varepsilon)\,\|x-y\|$$

holds.

Now let $C_1 := C \cap B^n$, $C_2 := C \cap E$ and define

$$F_y := \frac{f(y) - f(x)}{\varphi(x)} \quad \forall x \in C_2 \setminus C_1,\ y \in C_1\,.$$

We prove that $F_y(x)$ is always bounded in some neighbourhood of y. To see this, consider $A(y, \varepsilon) := A \cap N(y,\varepsilon)$, where A is the open set containing E introduced in the assumptions of Theorem I.2. Then, by (7), it follows that

$$\varphi(x) \geq (1-\varepsilon)\|x-y\| \quad \forall x \in A(y, \varepsilon) \cap (C_2 \setminus C_1)$$

holds. Moreover, by assumption, f is Lipschitzian on A, i.e. there is a constant $L > 0$ such that we have

$$|f(x) - f(y)| \leq L\|x - y\| \quad \forall x,y \in A .$$

The last two inequalities yield

$$|F_y(x)| \leq \frac{L}{1-\epsilon} < +\infty \quad \forall x \in A(y, \varepsilon) \cap (C_2 \setminus C_1), y \in C_1 .$$

The family of sets $\{A(y^i,\varepsilon): y^i \in B^n\}$ is a finite cover of C_1. Let $k = 2^n$ and consider

$$C_3: = (\bigcup_{i=1}^{k} A(y^i, \varepsilon)) \cap C_2.$$

Clearly, $C_1 \subset C_3$, and $\mu > \frac{L}{1-\epsilon}$ implies that

$$f(x) + \mu\varphi(x) > f(y^i) \quad \forall x \in C_3 \setminus C_1 , y^i \in C_1 \tag{8}$$

holds.

Finally, consider the compact set

$$C_4: = C_2 \setminus C_3 = C_2 \cap (\mathbb{R}^n \setminus (\bigcup_{i=1}^{k} A(y^i, \varepsilon))) .$$

By the definition of the sets involved, the following relations hold:

$$C_2 \setminus C_1 = C_4 \cup (C_3 \setminus C_1) , \quad C_4 \cap C_1 = \emptyset ,$$

$$C_4 \cap (C_3 \setminus C_1) = \emptyset , \quad C_1 \cap (C_3 \setminus C_1) = \emptyset .$$

Since f is continuous on E and C_2 is a compact set contained in E, the quantities $m_f := \min f(C_2)$ and $M_f := \max f(C_2)$ exist. By a similar argument, we see that also $m_\varphi := \min \varphi(C_4)$ exists. Note that we have $\varphi(x) > 0$ on C_4, hence $m_\varphi > 0$ and

$$\mu_1 := \frac{M_f - m_f}{m_\varphi} \geq 0 .$$

It follows that

$$f(x) + \mu\varphi(x) > m_f \tag{9}$$

holds for all $x \in C_4$ whenever $\mu > \mu_1$.

Choose $\mu > \max \{\frac{L}{1-\epsilon}, \mu_1\}$. Then, both inequalities (8) and (9) are fulfilled and the global minimum m of $f(x) + \mu\varphi(x)$ over C_2 cannot be attained on $C_4 \cup (C_3 \setminus C_1) = C_2 \setminus C_1$. It follows that m has to be attained on C_1. But $\varphi(x)$ vanishes on C_1. In other words, the problems (4) and (5) are equivalent.

(ii): Let by $\nabla^2 f$ denote the Hessian matrix of f. The Hessian $\nabla^2 f$ exists on E and its elements are bounded there. Then, by a well known criterion on definiteness of symmetric matrices and diagonal dominance, there is a $\mu_2 > 0$ such that, whenever $\mu > \mu_2$, the Hessian $\nabla^2 f - \text{diag}(2\mu)$ of $f(x) + \mu x^T(e-x)$ is negative semidefinite on E, and this implies concavity of $f(x) + \mu x^T(e-x)$.

From the above, Theorem I.2 follows for $\mu_0 = \max \{\frac{L}{1-\epsilon}, \mu_1, \mu_2\}$. ■

Now let C be a *convex* set. Then problem (5) is a concave minimization problem, whenever $\mu > \mu_0$, and Theorem I.2 shows that *large classes of integer programming problems* are equivalent to *concave minimization problems.*

Important special classes of integer programming problems are the *integer linear programming problem*

$$\begin{array}{ll} \text{minimize} & cx \\ \text{s.t.} & Ax \leq b,\ x \in B^n , \end{array} \tag{10}$$

where $c \in \mathbb{R}^n$, $b \in \mathbb{R}^m$, $A \in \mathbb{R}^{m\times n}$, and the *integer quadratic programming problem*

$$\begin{array}{ll} \text{minimize} & cx + \frac{1}{2}\, x(Cx) \\ \text{s.t.} & Ax \leq b ,\quad x \in B^n \end{array} \tag{11}$$

which adds to the objective function of (10) a quadratic term with matrix $C \in \mathbb{R}^{n\times n}$ (assumed symmetric without loss of generality).

Suppose that the feasible sets of problems (10) and (11) are not empty. Then, obviously, the assumptions of Theorem I.2. are satisfied for problems (10) and (11).

Note that problem (4) also covers the cases where $x \in B^n$ is replaced by $x_j \in \mathbb{N} \cup \{0\}$, and the variables x_j are bounded. A simple representation of x_j by (0–1)–variables is then

$$x_j = \sum_{i=0}^{K} y_{ij} 2^i \ , \quad y_{ij} \in \{0,1\} \ ,$$

where K is an integer upper bound on $\log_2 x_j$.

In practice, when dealing with integer programming problems, a transformation into an equivalent concave minimization problem may be of benefit only for special cases (e.g., Bazaraa and Sherali (1982), Frieze (1974)). The connections discussed above, however, may lead to an adaptation of typical ideas used in concave minimization to approaches for solving certain integer problems (e.g., Adams and Sherali (1986), Beale and Forrest (1978), Erenguc and Benson (1987), and Glover and Klingman (1973)).

2.4. Bilinear Programming and Concave Minimization

One of the most often encountered difficult multiextremal global problems in mathematical programming is the bilinear programming problem, whose general form is

$$\begin{aligned} &\text{minimize } f(x,y) = px + x(Cy) + qy \\ &\text{s.t. } x \in X,\ y \in Y \end{aligned} \tag{12}$$

where X, Y are given closed convex polyhedral sets in $\mathbb{R}^n$, $\mathbb{R}^m$ respectively, and

$p \in \mathbb{R}^n$, $q \in \mathbb{R}^m$, $C \in \mathbb{R}^{n \times m}$. Problem (12) was studied in the bimatrix game context by, e.g., Mangasarian (1964), Mangasarian and Stone (1964), Altman (1968). Further applications include dynamic Markovian assignment problems, multicommodity network flow problems and certain dynamic production problems. An extensive discussion of applied problems which can be formulated as bilinear programming problems is given by Konno (1971a).

The first solution procedures were either locally convergent (e.g., Altman (1968), Cabot and Francis (1970)) or completely enumerative (e.g., Mangasarian and Stone (1964)).

Cabot and Francis (1970) proposed an extreme point ranking procedure. Subsequent solution methods included various relaxation and cutting plane techniques (e.g., Gallo and Ülkucü (1977), Konno (1971, 1971a and 1976), Vaish and Shetty (1976 and 1977), Sherali and Shetty (1980a), or branch and bound approaches (e.g., Falk (1973)). A general treatment of cutting–plane methods and branch and bound techniques will be given in subsequent chapters. Related approaches to bilinear programming can be found in Al–Khayyal (1977), Czochralska (1982 and 1982 a) and Thieu (1986).

An extension to *biconvex programming* problems, where the objective function in (12) is *biconvex*, i.e.,

$$f(x,y) = f_1(x) + x(Cy) + f_2(y)$$

with f_1, f_2 convex, and where the constraints form a convex set in $\mathbb{R}^{n+m}$, is given in Al–Khayyal and Falk (1983).

Most of the methods mentioned above use in an explicit or implicit way the close relationship between problem (12) and concave minimization. Formulations of bilinear problems as concave minimization problems are discussed, e.g., in Altman (1968), Konno (1976), Gallo and Ülkucü (1977). In the converse direction, Altman (1968) and Konno (1976a) have shown that the minimization of a concave quadratic function under linear constraints can be reduced to a bilinear programming problem.

Frieze (1974) has reduced the 3–dimensional assignment problem to a bilinear programming problem and then to a special concave programming problem. A general discussion of the relationship between bilinear and concave programming is given by Thieu (1980).

Theorem I.3. *In problem (12) assume that Y has at least one vertex and that for every $x \in X$*

$$\begin{aligned} &\min f(x,y) \\ &s.t. \ \ y \in Y \end{aligned} \tag{13}$$

has a solution. Then problem (12) can be reduced to a concave minimization problem with piecewise linear objective function and linear constraints.

Proof. Note first that the assumptions of Theorem I.3 are both satisfied if Y is nonempty and compact. Denote by V(Y) the set of vertices of Y. It is known from the theory of linear programming that for every $x \in X$ the solution of (13) is attained at least at one vertex of Y. Problem (12) can be restated as

$$\min_{x \in X, y \in Y} f(x,y) = \min_{x \in X} \{\min_{y \in Y} f(x,y)\}$$

$$= \min_{x \in X} \{\min_{y \in V(Y)} f(x,y)\} = \min_{x \in X} f(x), \text{ where}$$

$$f(x) := \min_{y \in V(Y)} f(x,y) = \min_{y \in Y} f(x,y).$$

The set V(Y) is finite, and for each $y \in V(Y)$, f(x,y) is an affine function of x. Thus, f(x) is the pointwise minimum of a finite family of affine functions, and hence is concave and piecewise linear. ■

In order to obtain a converse result, let $f(x) := 2px + x(Cx)$, where $p \in \mathbb{R}^n$ and $C \in \mathbb{R}^{n \times n}$ is symmetric and negative semi–definite. Consider the quadratic concave programming problem

$$\begin{array}{ll} \text{minimize } f(x) \\ \text{s.t. } x \in X \end{array} \tag{14}$$

where X is a closed convex polyhedral set in $\mathbb{R}^n$.

Define $f(x,y) := px + py + x(Cy)$, let $Y = X$, and consider the bilinear programming problem

$$\begin{array}{ll} \text{minimize } f(x,y) \\ \text{s.t. } x \in X,\ y \in Y \end{array} \tag{15}$$

Theorem I.4. *Under the assumptions above the following equivalence holds: If x^* solves (14), then (x^*, x^*) solves (15). If $(\bar{x}, \bar{y})$ solves (15), then $\bar{x}$ and $\bar{y}$ solve (14).*

Proof. By definition, $f(x^*) \le f(x)\ \ \forall x \in X$. In particular,

$$\begin{aligned} f(x^*) &\le f(\bar{x}) = f(\bar{x},\bar{x}) , \\ f(x^*) &\le f(\bar{y}) = f(\bar{y}, \bar{y}) . \end{aligned} \tag{16}$$

On the other hand, we obviously have

$$f(\bar{x}, \bar{y}) = \min_{x \in X, y \in Y} f(x, y) \le \min_{x \in X} f(x, x) = \min_{x \in X} f(x) = f(x^*) . \tag{17}$$

Combining (16) and (17), we obtain

$$f(\bar{x}, \bar{y}) \le f(x^*) \le \min \{f(\bar{x}), f(\bar{y})\} ,$$

and so Theorem I.4 is established if we prove that

$$f(\bar{x}, \bar{y}) = f(\bar{x}) = f(\bar{y}) . \tag{18}$$

Obviously, we have

$$\begin{aligned} 0 &\le f(\bar{x}, \bar{x}) - f(\bar{x}, \bar{y}) = p(\bar{x} - \bar{y}) + \bar{x}(C(\bar{x} - \bar{y})) , \\ 0 &\le f(\bar{y}, \bar{y}) - f(\bar{x}, \bar{y}) = p(\bar{y} - \bar{x}) + \bar{y}(C(\bar{y} - \bar{x})) . \end{aligned} \tag{19}$$

Adding these inequalities yields

$$(\bar{x} - \bar{y})(C(\bar{x} - \bar{y})) \geq 0$$

which, by the negative semi–definiteness of C, implies $C(\bar{x} - \bar{y}) = 0$, i.e.

$$C\bar{x} = C\bar{y} . \tag{20}$$

Inserting (20) into (19), we obtain $p(\bar{x} - \bar{y}) = 0$, i.e.

$$p\bar{x} = p\bar{y} . \tag{21}$$

Thus, by (20) and (21),

$$f(\bar{x}, \bar{y}) = p\bar{x} + p\bar{y} + \bar{x}(C\bar{y}) = 2p\bar{x} + \bar{x}(C\bar{x}) = 2p\bar{y} + \bar{y}(C\bar{y}) ,$$

hence (18) follows. ■

Note that the proof of Theorem I.4 did not make use of the special form of the feasible set X in (14). Hence, Theorem I.4 can be extended to more general feasible sets.

Note as well that, by Theorem I.2 and Theorem I.4, the integer linear programming problem (10) can be reduced to a bilinear problem.
Bilinear programming problems will be treated in Chapter IX.

2.5. Complementarity Problems and Concave Minimization

Let $D \subseteq \mathbb{R}^n$, $g,h: \mathbb{R}^n \longrightarrow \mathbb{R}^m$. The problem of finding $x \in D$ such that

$$g(x) \geq 0,\ h(x) \geq 0,\ g(x)h(x) = 0 \tag{22}$$

is called a ***complementarity problem.***

Complementarity problems play a very important role in the decision sciences and in applied mathematics. There are close relations to fixed–point problems and variational inequalities. The literature on complementarity problems contains an enormous number of papers and some excellent textbooks. We only cite here the monographs of Lüthi (1976), Garcia and Zangwill (1981), Gould and Tolle (1983), Murty (1988), and the report Stone (1981). See also the survey of Al–Khayyal (1986a) and references therein.

A frequently studied special case is the ***linear complementarity problem*** which is obtained from (22) if we set $D = \mathbb{R}^n$, $n = m$, $g(x) = x$ and $h(x) = Mx+q$, $M \in \mathbb{R}^{n \times n}$, $q \in \mathbb{R}^n$.

It has been shown in Mangasarian (1978) that a linear complementarity problem can be converted into an equivalent concave minimization problem. Certain linear complementarity problems are even solvable as linear programs (e.g., Cottle and Pang (1978), Mangasarian (1976, 1978 and 1979), Al–Khayyal (1985 and 1986a)).

Thieu (1980) proved that certain classes of nonlinear complementarity problems are equivalent to concave minimization problems.

Theorem I.5. *In (22) let D be a convex set and let g and h be concave mappings on D. Suppose that (22) has a solution. Then (22) is equivalent to the concave minimization problem*

$$\begin{aligned} &\text{minimize } f(x) := \sum_{i=1}^{m} \min \{g_i(x), h_i(x)\} \\ &\text{s.t. } g(x) \geq 0,\ h(x) \geq 0,\ x \in D. \end{aligned} \tag{23}$$

Proof. Let x^* be a solution of (22). Then, obviously, x^* is feasible for (23). From $g(x^*) \geq 0$, $h(x^*) \geq 0$, and $g(x^*)h(x^*) = 0$ it follows that $f(x^*) = 0$. But all feasible points x of (23) satisfy $f(x) \geq 0$; hence x^* is an optimal solution of (23).

Conversely, since (22) is assumed to have a solution, every optimal solution x^* of (23) has to satisfy $f(x^*) = 0$. Thus we have $g(x^*) \geq 0$, $h(x^*) \geq 0$, $\min \{g_i(x^*), h_i(x^*)\} = 0$, $(i=1,\dots,m)$, hence $g(x^*)h(x^*) = 0$. ■

Again, since the convexity of D and the concavity of g and h were not used in the proof, the result can be generalized to arbitrary D, g, and h. However, in that case, (23) does not necessarily constitute a concave programming problem.

In the linear case, when D is a convex polyhedral set, and h and g are affine functions, then f(x) is concave and piecewise linear.

An algorithm for solving linear complementarity problems that is based on the above relationship with concave minimization is proposed by Thoai and Tuy (1983). For related approaches, see also Al–Khayyal (1986a and 1987), Pardalos and Rosen (1988) and references therein. These procedures will be treated in Chapter IX.

2.6. Max–Min Problems and Concave Minimization

A linear max–min problem can be stated as

$$\begin{aligned} &\underset{x}{\text{maximize}}\ \underset{y}{\min}\ \{cx + dy\} \\ &\text{s.t. } Ax + By \leq b,\ x, y \geq 0 \end{aligned} \tag{24}$$

where $x, c \in \mathbb{R}^n$, $y, d \in \mathbb{R}^m$, $A \in \mathbb{R}^{s\times n}$, $B \in \mathbb{R}^{s\times m}$, and $b \in \mathbb{R}^s$.

The relationship between problem (24) and concave minimization was discussed and exploited by Falk (1973).

Theorem I.6. *Let the feasible set of (24) be nonempty and compact. The problem (24) is equivalent to a concave minimization problem with piecewise linear objective function and linear constraints.*

Proof. Let $D = \{(x, y) \in \mathbb{R}^{n+m}: Ax + By \leq b,\ x, y \geq 0\}$ denote the feasible set of (24). Consider the orthogonal projection P of D onto the space $\mathbb{R}^n$ of the variable x, i.e., $P := \{x \in \mathbb{R}^n: x \geq 0,\ Ax + By \leq b \text{ for at least one } y \geq 0\}$. Further, for $x \in P$, let

$D_x := \{y \in \mathbb{R}^m, By \le b - Ax, y \ge 0\}$. Then (24) may be rewritten as

$$\underset{x \in P}{\text{maximize}} \{cx + \min_{y \in D_x} dy\}.$$

It is well known from the theory of parametric linear programming that the function

$$f(x) := \min_y \{dy: By \le b - Ax, y \ge 0\}$$

is convex and piecewise linear on P. Therefore, (24) is equivalent to the convex maximization (or, equivalently, concave minimization) problem

$$\text{maximize } (f(x) + cx). \quad \text{s.t. } x \in P$$ ■

3. D.C. PROGRAMMING AND REVERSE CONVEX CONSTRAINTS

3.1. D.C. Programming: Basic Properties

Recall from Section I.1 that a real–valued function f defined on a convex set $C \subset \mathbb{R}^n$ is called **d.c. on C** if, for all $x \in C$, f can be expressed in the form

$$f(x) = p(x) - q(x) \tag{25}$$

where p and q are convex functions on C.
The function f is called d.c. if f is d.c. on $\mathbb{R}^n$. The representation (25) is said to be a **d.c. decomposition** of f.

A global optimization problem is called a **d.c. programming problem** if it has the form

$$\begin{array}{ll} \text{minimize } f(x) & \\ \text{s.t. } x \in C, g_j(x) \le 0 \ (j=1,...,m) \end{array} \tag{26}$$

where $C \subset \mathbb{R}^n$ is convex and all functions f and g_j are d.c. on C.

Note that C is usually given by a system of convex inequalities, so that, for the sake of simplicity of notation, we will sometimes set $C = \mathbb{R}^n$ without loss of generality.

Clearly, every concave minimization problem is also a d.c. programming problem, and it follows from Example I.1 (Section I.2.1) that (26) is a multi–extremal optimization problem.

The following results show that the class of d.c. functions is very rich and, moreover, that it enjoys a remarkable stability with respect to operations frequently encountered in optimization.

Theorem I.7. *Let f, f_i (i=1,...,m) be d.c.. Then the following assertions hold.*

(i) *Every linear combination of the f_i (i=1,...,m) is d.c..*

(ii) $\max_{i=1,..,m} f_i$ *and* $\min_{i=1,..,m} f_i$ *are d.c..*

(iii) $|f(x)|$, $f^+(x) := \max\{0, f(x)\}$ *and* $f^-(x) := \min\{0, f(x)\}$ *represent d.c. functions.*

Proof. Let $f = p - q$, $f_i = p_i - q_i$ (i=1,...,m) be d.c. decompositions of f and the f_i, respectively.

(i): $f_i + f_j = (p_i + p_j) - (q_i + q_j)$ is d.c. since the sum of two convex functions is convex. Obviously, $(-f_i) = q_i - p_i$ is d.c. and $\lambda_i f_i = \lambda_i p_i - \lambda_i q_i$ is d.c. for $\lambda_i \in \mathbb{R}$. It follows that a linear combination of d.c. functions is d.c.

(ii): $f_i = p_i - q_i = p_i + \sum_{\substack{j=1 \\ j \neq i}}^{m} q_j - \sum_{j=1}^{m} q_j$. The last sum does not depend on i.

Therefore, we have

$$\max_{i=1,..,m} f_i = \max_{i=1,..,m} \{p_i - q_i\} = \max_{i=1,..,m} \{p_i + \sum_{\substack{j=1 \\ j \neq i}}^{m} q_j\} - \sum_{j=1}^{m} q_j .$$

This is a d.c. decomposition, since the sum and the maximum of finitely many convex functions are convex.

Similarly, we see that

$$\min_{i=1,..,m} f_i = \sum_{j=1}^{m} p_j + \min_{i=1,..,m} \{-(\sum_{\substack{j=1 \\ j \neq i}}^{m} p_j) - q_i\} = \sum_{j=1}^{m} p_j - \max_{i=1,..,m} \{(\sum_{\substack{j=1 \\ j \neq i}}^{m} p_j) + q_i\}$$

is d.c.

(iii): Suppose that we have $p(x) \geq q(x)$. Then $|f(x)| = p(x) - q(x) = 2p(x) - (p(x) + q(x))$ holds.

Now let $p(x) < q(x)$. Then we have $|f(x)| = q(x) - p(x) = 2q(x) - (p(x) + q(x))$.

Hence, it follows that

$$|f| = 2\max\{p,q\} - (p + q)$$

which is a d.c. decomposition of $|f|$.

Similarly, we see that

$$f^+ = \max\{p, q\} - q ,$$

$$f^- = p - \max\{p, q\}$$

are d.c. decompositions of f^+ and f^-, respectively. ■

It can be shown that the d.c. property is also preserved under other operations which will not be used in this book. An example is the product of two d.c. functions (cf. Hiriart–Urruty (1985) and references given there).

A main result concerning the recognition of d.c. functions goes back to Hartman (1959). Before stating it, let us agree to call a function $f: \mathbb{R}^n \longrightarrow \mathbb{R}$ **locally d.c.** *if, for* *every $x^0 \in \mathbb{R}^n$, there exists a neighbourhood $N = N(x^0,\varepsilon)$ of x^0 and convex functions p_N, q_N such that*

$$f(x) = p_N(x) - q_N(x) \quad \text{for all } x \in N. \tag{27}$$

Theorem I.8. *Every locally d.c. function is d.c.*

The proof requires some extension techniques that are beyond the scope of this book and is omitted (cf. Hartman (1959), Ellaia (1984), Hiriart–Urruty (1985)).

Denote by C^2 the class of functions $\mathbb{R}^n \to \mathbb{R}$ whose second partial derivatives are continuous everywhere.

Corollary I.1. *Every function $f \in C^2$ is d.c.*

Proof. Corollary I.1 is merely a consequence of Theorem I.8. If $f \in C^2$, then we see that, for all $x^0 \in \mathbb{R}^n$, the Hessian $\nabla^2 f$ of f is bounded on the closed neighbourhood $\overline{N}(x^0,\varepsilon) := \{x \in \mathbb{R}^n : \|x - x^0\| \le \varepsilon\}$, $\varepsilon > 0$. As in the proof of Theorem I.2, part (ii), it then follows that there is a real number $\mu > 0$ such that $f(x) + \mu\|x\|^2$ is convex on $\overline{N}(x^0,\varepsilon)$. Due to this simple fact, one can find a d.c. decomposition

$$f(x) = (f(x) + \mu\|x\|^2) - \mu\|x\|^2$$

of f on $\overline{N}(x^0,\varepsilon)$, i.e., f is locally d.c., and hence d.c. (cf. Theorem I.8). ∎

Furthermore, it turns out that any problem of minimizing a continuous real function over a compact subset D of $\mathbb{R}^n$ can, in principle, be approximated as closely as desired by a problem of minimizing a d.c. function over D.

Corollary I.2. *A real valued continuous function on a compact (convex) subset D of $\mathbb{R}^n$ is the limit of a sequence of d.c. functions on D which converges uniformly in D.*

Proof. Corollary I.2 is a consequence of the famous Stone–Weierstraß Theorem or of the original Theorem of Weierstraß which states that every continuous function on D is the limit of a uniformly convergent sequence of polynomials on D. Corollary I.2 follows from Corollary I.1, since every polynomial is C^2. ■

Of course, the main concern when using Corollary I.2 is how to construct an appropriate approximation by d.c. functions for a given continuous function on D. However, in many special cases of interest, the exact d.c. decomposition is already given or easily found (cf. Theorem I.7, Bittner (1970), Hiriart–Urruty (1985) and the brief discussion of direct applications below).

Example I.2. Let f(x) be separable, i.e., we have $f(x) = \sum_{i=1}^{n} f_i(x_i)$, where each f_i is a real–valued function of one real variable on a given (possibly unbounded) interval. In many economic applications where the functions f_i represent, e.g., utility functions or production functions, each f_i is differentiable and has the property that there is a point $\bar{x}_i$ such that $f_i(x_i)$ is concave for $x_i < \bar{x}_i$ and convex for $x_i \geq \bar{x}_i$ (Fig. I.1). Likewise, we often encounter the case where $f_i(x)$ is convex for $x_i < \bar{x}_i$ and concave for $x_i \geq \bar{x}_i$.

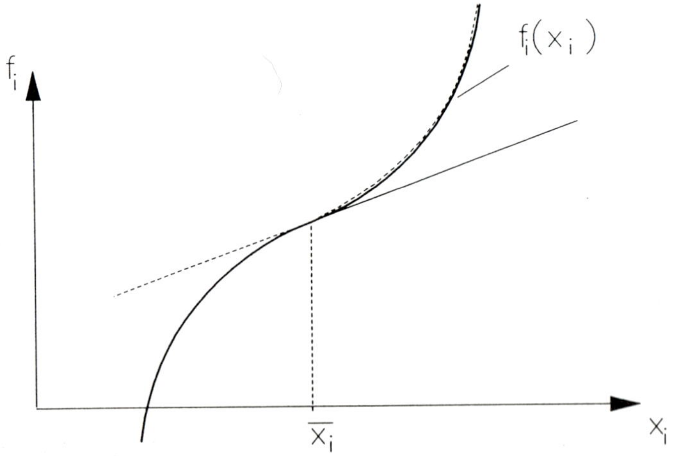

Fig. I.1

In the case corresponding to Fig. I.1 where f_i is concave–convex, we obviously have the d.c. decomposition $f_i = p_i - q_i$ given by

$$p_i(x_i) := \begin{cases} (f_i'(\bar{x}_i)\,(x_i - \bar{x}_i) + f_i(\bar{x}_i))\,/2 & (x_i < \bar{x}_i) \\ f_i(x_i) - (f_i'(\bar{x}_i)\,(x_i - \bar{x}_i) + f_i(\bar{x}_i))\,/2 & (x_i \geq \bar{x}_i) \end{cases},$$

$$q_i(x_i) := \begin{cases} (f_i'(\bar{x}_i)\,(x_i - \bar{x}_i) + f_i(\bar{x}_i))\,/2 - f_i(x_i) & (x_i < \bar{x}_i) \\ -(f_i'(\bar{x}_i)\,(x_i - \bar{x}_i) + f_i(\bar{x}_i))\,/2 & (x_i \geq \bar{x}_i) \end{cases}.$$

Let $f(x)$ be a real–valued function defined on a compact interval $I = \{x: a \leq x \leq b\}$, $b > a$, of the real line. Then, f is said to be a **piecewise linear function** if I can be partitioned into a finite number of subintervals such that in each subinterval $f(x)$ is affine. That is, there exist values $x_1, x_2, \ldots, x_r$ such that

$$f(x) = \alpha_i x + \beta_i \ \text{ for } x_{i-1} \leq x < x_i\,, \quad i=1,\ldots,r+1\,,$$

$$f(x_{r+1}) = \alpha_{r+1} x_{r+1} + \beta_{r+1}\,,$$

where $x_0 := a$, $x_{r+1} := b$ and α_i, β_i are known real constants ($i=1,\ldots,r+1$). A similar definition holds for noncompact intervals I.

A piecewise linear function is continuous if $\alpha_i x_i + \beta_i = \alpha_{i+1} x_i + \beta_{i+1}$ for $i=1,\ldots,r$.

Corollary I.3. *A continuous piecewise–linear function on an interval I is d.c. on I.*

Proof. Let $f: I \longrightarrow \mathbb{R}$ be continuous and piecewise linear. By setting $f(x) = \alpha_1 x + \beta_1$ for $-\infty < x \leq x_0$, and $f(x) = \alpha_{r+1} x + \beta_{r+1}$ for $x_{r+1} \leq x < \infty$, we see that f can be extended to a continuous piecewise–linear function $\bar{f}$ on $\mathbb{R}$. From the definition of a continuous piecewise–linear function it follows that, for each point $x \in \mathbb{R}$, there is a neighbourhood $N(x,\varepsilon)$ such that $\bar{f}$ can be expressed in $N(x,\varepsilon)$ as the pointwise maximum or minimum of at most two affine functions. Corollary I.3 is then established via Theorem I.7 (ii) and Theorem I.8. ■

Corollary I.3 holds also for continuous piecewise–linear functions on $\mathbb{R}^n$, $n > 1$.

3.2. D.C. Programming: Applications

Example I.3. Minimizing an indefinite quadratic form is a d.c. programming problem. Let A be a symmetric n×n–matrix and let

$$f(x) = \frac{1}{2} x(Cx) .$$

Then f(x) is obviously d.c. by to Corollary I.1 and there are several ways of finding positive semi– definite matrices P and Q such that f(x) can be expressed in the form

$$f(x) = \frac{1}{2} x(Px) - \frac{1}{2} x(Qx) .$$

Example I.4. Let M be an arbitrary nonempty closed subset of $\mathbb{R}^n$ and let $d_M^2(x)$ denote the square of the distance from a point $x \in \mathbb{R}^n$ to M. Then it can be shown that $\|x\|^2 - d_M^2(x)$ is convex (cf. Asplund (1973)). Consequently,

$$d_M^2(x) = \|x\|^2 - (\|x\|^2 - d_M^2(x))$$

is a d.c. decomposition of $d_M^2(x)$. Note that $d_M^2(x)$ is convex whenever M is convex.

In many econometric applications, we encounter a situation where the objective function is of the form $p(y) - q(z)$ where $p: \mathbb{R}^{n_1} \longrightarrow \mathbb{R}$, and $q: \mathbb{R}^{n_2} \longrightarrow \mathbb{R}$ are convex. Extending the considerations of Section I.2.2, we see that such a cost function reflects the fact that in some activities the unit cost increases when the scale of activity is enlarged (diseconomies of scale), whereas in other activities the unit cost decreases when the scale of activity is enlarged. Likewise, in optimal investment planning, one might have to find an investment program with maximal profit, where the profit function is $(-p(y)) + q(z)$ with $(-p(y))$ concave (according to a law of diminishing marginal returns) and with $q(z)$ convex (according to a law of increasing marginal returns).

Certain general economic models also give rise to d.c. functions in the constraints. Suppose that a vector of activities x has to be selected from a certain convex set $C \subset \mathbb{R}^n$, which is often called the set of all technologically feasible activities. Suppose also that the selection has to be made so as to minimize the cost f(x), under the condition that some kind of "utility" has to be kept above a certain level. This leads to constraints of the form

$$u_i(x) \geq c_i ,$$

where $u_i(x)$ represents some utility depending on the activity level x, and where c_i is a minimal level of utility required. In many cases, the $u_i(x)$ can be assumed to be separable, i.e., of the form

$$u_i(x) = \sum_{j=1}^{n} u_{ij}(x_j)$$

(cf. Example I.2) with u_{ij} either convex or concave or of the form discussed in Example I.2. For a thorough discussion of the underlying economic models, see, e.g., Hillestad (1975), Zaleesky (1980 and 1981).

D.C. programming problems are also frequently encountered in engineering and physics (cf., e.g., Giannessi et al. (1979), Heron and Sermange (1982), Mahjoub (1983), Vidigal and Director (1982), Strodiot et al. (1985 and 1988), Polak and Vincentelli (1979), Toland (1978), Tuy (1986 and 1987)).

Moreover, for certain d.c. programming problems, there is a nice **duality theory** developed by Toland (1978 and 1979), that leads to several applications in engineering, mathematics and physics (cf., e.g., Toland (1978 and 1979), Hiriart–Urruty (1985)).

In **engineering design**, we often encounter constraints of the form

$$g(x):= \max_{s \in S} G(x,s) \leq 0 \tag{28}$$

where $G: \mathbb{R}^n \times \mathbb{R}^s \longrightarrow \mathbb{R}$ and $S \subset \mathbb{R}^s$ is compact. Here, x represents the *design vector* and (28) is used to express bounds on time and frequency responses of a dynamical system as well as tolerancing or uncertainty conditions in worst case design (cf. Polak (1987)). The function g(x) belongs to the class of so–called lower C^2–functions if G is a function which has partial derivatives up to order 2 with respect to x and which, along with all these derivatives, is jointly continuous in $(x,s) \in \mathbb{R}^n \times S$. But lower C^2–functions are d.c.. For a detailed discussion of lower C^2–functions see Rockafellar (1981) and Hiriart–Urruty (1985).

Example I.5. Let $D \subset \mathbb{R}^n$ be a *compact* set, and let $K \subset \mathbb{R}^n$ be a *convex compact* set. Assume that the origin 0 is an interior point of K. Define the function $r_D: D \longrightarrow \mathbb{R}$ by

$$r_D(x):= \max \{r: x + rK \subset D\} . \tag{29}$$

Then the problem

$$\begin{aligned} &\text{maximize } r_D(x) \\ &\text{s.t. } x \in D \end{aligned} \tag{30}$$

is called a **design centering** problem.

An application of the design centering problem is the optimum shape design problem from the diamond industry where one wants to know how to cut the largest diamond of shape K that can be cut from a rough stone $D \supset K$ (Nguyen et al. (1985)). Other applications are discussed by Tuy (1986) and Polak and Vincentelli (1979).

In a more general context consider any fabrication process where random variations may result in a very low production yield. A method to minimize the influence of these random variations consists of centering the nominal value of the designable parameters in the so-called region of acceptability. This leads to a problem of the form (29), (30) (cf. Vidigal and Director (1982)).

In many cases, D can be described by finitely many d.c. inequalities. Moreover, it has been shown in Thach (1988) that in these cases (30) is actually a d.c. programming problem. A somewhat different approach for polyhedral sets D is described in Thoai (1987). Generalizations are discussed in Thach et al. (1988). We shall come back to this problem in Chapter X.

Example I.6. *The jointly constrained biconvex programming problem* is of the form

$$\text{minimize } f(x) + xy + h(y)$$
$$\text{s.t. } (x,y) \in D \subset \mathbb{R}^n \times \mathbb{R}^n$$

where D is a closed convex subset of $\mathbb{R}^n \times \mathbb{R}^n$, and f and h are real-valued convex functions on D. The objective function is a d.c. function, since $xy = \frac{1}{4}(\|x+y\|^2 - \|x-y\|^2)$.

The intuition of many people working in global optimization is that most of the "reasonable" optimization problems actually involve d.c. functions. It is easy to see, for example, that a problem of the form

$$\begin{array}{l}\text{minimize } f(x) \\ \text{s.t. } x \in D\end{array}, \tag{31}$$

where D is a closed subset of $\mathbb{R}^n$ and $f: \mathbb{R}^n \to \mathbb{R}$ is continuous, can be converted into a d.c. program (e.g., Tuy (1985)). Introducing the additional real variable t, we can obviously write problem (31) in the equivalent form

$$\begin{array}{l}\text{minimize } t \\ \text{s.t. } x \in D, f(x) \le t\end{array}. \tag{32}$$

The feasible set $M:= \{(x,t) \in \mathbb{R}^{n+1}: x \in D, f(x) \le t\}$ in (32) is closed and the condition $(x,t) \in M$ is equivalent to the constraint $d_M^2(x,t) \le 0$, where $d_M(x,t)$ denotes the distance from (x,t) to M. This is a d.c. inequality (cf. Example I.4).

3.3. Reverse Convex Constraints

Recall from Section I.1 that a constraint $h(x) \ge 0$ is called **reverse convex** whenever $h: \mathbb{R}^n \longrightarrow \mathbb{R}$ is convex. Obviously, every optimization problem with concave, convex or d.c. objective function and a combination of convex, reverse convex and d.c. constraints is a d.c. programming problem. However, it is worthwhile to consider reverse convex constraints separately. One reason for doing so is that we often encounter convex or even linear problems having only one or very few additional reverse convex constraints. Usually, these problems can be solved more efficiently by taking into account their specific structure instead of using the more general approaches designed for solving d.c. problems (e.g., Tuy (1987), Horst (1988), Thoai (1988)). Another reason is that every d.c. programming problem can be converted into an equivalent relatively simple problem having only one reverse convex constraint. This so–called **canonical d.c. programming problem** will be derived below.

Example I.7. Given two mappings $h: \mathbb{R}^n \longrightarrow \mathbb{R}^m$, and $g: \mathbb{R}^n \longrightarrow \mathbb{R}^m$, the condition $h(x)\, g(x) = 0$ is often called a **complementarity condition** (cf. Section I.2.5). Several applications yield optimization problems where an additional simple linear complementarity condition of the form $xy = 0$ $(x,y \in \mathbb{R}^n)$ has to be fulfilled.

An example is given by Schoch (1984), where it is shown that the indefinite quadratic programming problem is equivalent to a parametric linear program subject to a complementarity condition.

Another example has been discussed in Giannessi et al. (1979). In offshore technology, a submarine pipeline is usually laid so that it rests freely on the sea bottom.

Since, however, the sea bed profile is usually irregularly hilly, it is often regularized by means of trench excavation in order to bury the pipe for protection and to avoid excessive bending moments on the pipe. The optimization problem which arises is to minimize the total cost of the excavation, under the condition that the free contact equilibrium configuration of the pipe nowhere implies excessive bending. It has been shown by Giannessi et al. (1979) that the resulting optimization problem is a *linear program with one additional complementarity condition* of the form

$$x \geq 0,\ y \geq 0,\ xy = 0 \quad (x,y \in \mathbb{R}^n)\ .$$

Obviously, this condition is equivalent to

$$x \geq 0,\ y \geq 0,\ \sum_{i=1}^{n} \min \{x_i, y_i\} \leq 0\ ,$$

the last inequality being a reverse convex constraint, since the function $\sum_{i=1}^{n} \min \{x_i, y_i\}$ is concave.

Example I.8. A (0–1) restriction can be cast into the reverse convex form. For instance, the constraint

$$x_i = 0 \text{ or } x_i = 1$$

can be rewritten as

$$-x_i + (x_i)^2 \geq 0\ ,\quad 0 \leq x_i \leq 1\ .$$

Example I.9. Let G be an open convex set defined by the convex inequality $g(x) < 0$, $g\colon \mathbb{R}^n \to \mathbb{R}$ convex. Let K be a compact, convex set contained in G. Then the problem of minimizing the distance d(x) from K to $\mathbb{R}^n \setminus G$ is a *convex minimization problem* with the additional reverse convex constraint $g(x) \geq 0$ (Fig. I.2).

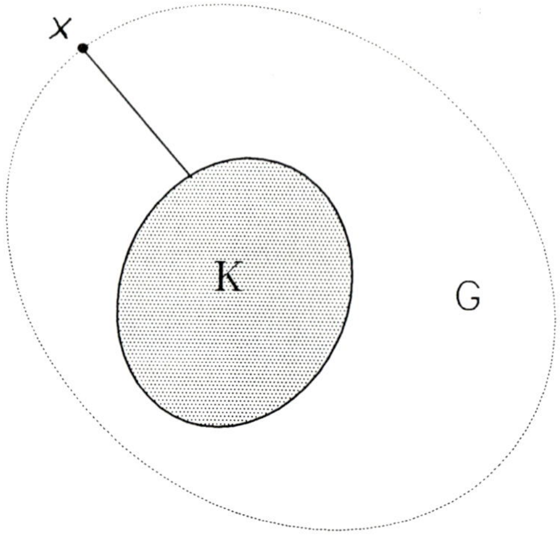

Fig. I.2

Certain classes of optimization problems involving reverse convex constraints were studied by Rosen (1966), Avriel and Williams (1970), Meyer (1970), Ueing (1972), Bansal and Jacobsen (1975), Hillestad and Jacobsen (1980 and 1980 a), Tuy (1983 and 1987), Thuong and Tuy (1984), Horst and Dien (1987), Horst (1988). Avriel and Williams (1970) showed that reverse convex constraints may occur in certain engineering design problems. Hillestad (1975) and Zaleesky (1980 and 1981) discussed economic models yielding reverse convex constraints (cf. Section I.3.2). In an abstract setting, Singer (1980) related reverse convex constraints to certain problems in approximation theory, where the set of approximating functions is the complement of a convex set.

3.4. Canonical D.C. Programming Problems

A striking feature of d.c. programming problems is that any d.c. problem can always be reduced to a canonical form which has a linear objective function and only two constraints, one of them being a convex inequality, the other being reverse convex (cf. Tuy (1986)).

Definition I.5. *A **canonical d.c. program** is an optimization problem of the form*

$$\begin{aligned} &\textit{minimize } cx \\ &\textit{s.t. } h(x) \le 0,\ g(x) \ge 0 \end{aligned} \tag{33}$$

where $c \in \mathbb{R}^n$, $h, g: \mathbb{R}^n \to \mathbb{R}$ *are convex.*

Theorem I.9. *Every d.c. programming problem of the form*

$$\begin{aligned} &\textit{minimize } f(x) \\ &\textit{s.t. } x \in C,\ g_j(x) \le 0\ (j=1,\dots,m) \end{aligned} \tag{34}$$

where C is defined by a finite system of convex inequalities $h_k(x) \le 0$, $k \in I \subset \mathbb{N}$, *and where* f, g_j *are d.c. functions, can be converted into an equivalent canonical d.c. program.*

Proof. By introducing an additional variable t, we see that problem (34) is equivalent to the following one:

$$\begin{aligned} &\text{minimize } t \\ &\text{s.t. } h_k(x) \le 0\ (k \in I),\ g_j(x) \le 0\ (j=1,\dots,m),\ f(x)-t \le 0 \end{aligned}$$

Therefore, by changing the notation, we have obtained a linear objective function. Furthermore, Theorem I.7 shows that the finite set of d.c. inequalities $f(x) - t \le 0$, $g_j(x) \le 0$ $(j=1,\dots,m)$ can be replaced by a single *d.c. inequality*

$$r(x,t) := \max \{f(x)-t,\ g_j(x):\ j = 1,\dots,m\} \le 0 .$$

Moreover, the d.c. inequality

$$r(x,t) = p(x,t) - q(x,t) \leq 0$$

with p and q convex is equivalent to the system

$$p(x,t) - z \leq 0 \;, \quad z - q(x,t) \leq 0 \tag{35}$$

involving the additional real variable z. The first inequality in (35) is convex and the second is reverse convex.

Finally, setting $h(x,t,z) := \max \{p(x,t) - z, h_j(x) \colon j \in I\}$ and $g(x,t,z) := q(x,t) - z$, we see that problem (34) is transformed into an equivalent canonical d.c. program. ■

Along the same lines, various other transformations can be carried out that transform, e.g., a d.c. problem into a program having convex or concave objective functions, etc. Obviously, since a canonical d.c. problem has simpler structure than the general d.c. problem, transformations of the above type will be useful for the development of certain algorithms for solving d.c. problems. Note, however, that these transformations increase the number of variables of the original problem. Because the numerical effort required to solve these kinds of difficult problems generally grows exponentially with the number of variables, we will also attempt to solve d.c. problems without prior transformations (cf. Chapter X).

We now consider the canonical d.c. problem (33), and we let ∂H and ∂G denote the boundaries of the sets $H := \{x \colon h(x) \leq 0\}$ and $G := \{x \colon g(x) \geq 0\}$, respectively. For the sake of simplicity we shall assume that H is bounded and that $H \cap G$ is not empty, so that an optimal solution of problem (33) exists.

Definition I.6. *The reverse convex constraint $g(x) \geq 0$ is called* ***essential*** *in the canonical d.c. program (33) if the inequality*

$$\min \{cx \colon x \in H\} < \min \{cx \colon x \in H \cap G\}$$

holds.

It can be assumed that $g(x) \geq 0$ is always essential because otherwise, problem (33) would be equivalent to the convex minimization problem

$$\begin{aligned} &\text{minimize } cx \\ &\text{s.t. } x \in H \end{aligned} \tag{36}$$

which can be solved by standard optimization techniques.

Theorem I.10. *Consider the canonical d.c. program. Suppose that H is bounded, $H \cap G$ is nonempty and the reverse convex constraint is essential. Then the canonical d.c. program (33) always has a solution lying on $\partial H \cap \partial G$.*

Proof. Since the reverse convex constraint is essential, there must be a point $w \in H$ satisfying

$$g(w) < 0, \; cw < \min \{cx: x \in H \cap G\} . \tag{37}$$

As $\mathbb{R}^n \setminus G$ is convex, we know that for every $x \in G$ there is a point

$$\pi(x) = \lambda x + (1-\lambda)w, \quad \lambda \in (0,1] ,$$

where the line segment $[w,x]$ intersects the boundary ∂G. The number λ is uniquely determined by the equation

$$g(\lambda x + (1-\lambda)w) = 0 .$$

Now let $x \in H \cap G$ satisfy $g(x) > 0$. Then we have

$$\pi(x) = \lambda x + (1-\lambda)w, \quad 0 < \lambda < 1 ,$$

and it follows from (37) that

$$c\pi(x) = \lambda cx + (1-\lambda)cw < \lambda cx + (1-\lambda)cx = cx .$$

Therefore, problem (33) is equivalent to

$$\begin{aligned} &\text{minimize } cx \\ &\text{s.t. } x \in H \cap \partial G \end{aligned} \tag{38}$$

Now consider an optimal solution $\bar{x}$ of (38). Denote by int M the interior of a set M. Then the closed convex set $(\mathbb{R}^n \setminus G) \cup \partial G$ has a *supporting hyperplane* at $\bar{x}$. The intersection of this hyperplane with the compact convex set H is a compact convex set K. It is well known from linear programming theory that cx attains its minimum over K at an extreme point u of K (cf. also Theorem I.1).

Clearly, we have $u \in \partial H$, and since $K \subset H \setminus \text{int } (\mathbb{R}^n \setminus G)$, it follows that u is a feasible point of problem (33). Finally, we see by construction that $\bar{x} \in K$, hence $cu \le c\bar{x} = \min \{cx : x \in H \cap G\}$. This implies that u solves (33). From (38) it then follows that $u \in \partial G$. ■

In certain applications we have that the function h(x) is the maximum of a finite number of affine functions, i.e., the canonical d.c. program reduces to linear programs with one additional reverse convex constraint. It will be shown in Part B of this volume, that in this case the global minimum is usually attained on an edge of H.

Problems with reverse convex constraints will be discussed in Chapter IX and Chapter X.

4. LIPSCHITZIAN OPTIMIZATION AND SYSTEMS OF EQUATIONS AND INEQUALITIES

4.1. Lipschitzian Optimization

Recall from Definition I.3 that a real-valued function f is called **Lipschitzian** on a set $M \subset \mathbb{R}^n$ if there is a constant $L = L(f,M) > 0$ such that

$$|f(x) - f(y)| \leq L \|x - y\| \quad \forall x,y \in M . \tag{39}$$

It is well known that all *continuously differentiable functions f with bounded gradients* on M are Lipschitzian on M, where

$$L = \sup \{\|\nabla f(x)\| : x \in M\} \tag{40}$$

is a Lipschitz constant ($\|\cdot\|$ again denotes the Euclidean norm).

Obviously, if f is Lipschitzian with constant L, then f is also Lipschitzian with all constants $L' > L$.

The value of knowing a Lipschitz constant L arises from the following simple observation. Suppose that the diameter $d(M) := \sup \{\|x - y\| : x, y \in M\} < \infty$ of M is known. Then we easily see from (39) that

$$f(x) \geq f(y) - L\|x - y\| \geq f(y) - Ld(M) \tag{41}$$

holds. Let $S \subset M$ denotes a finite sample of points in M where the function values have been calculated. Then it follows from (41) that we have

$$\min f(S) \geq \inf f(M) \geq \max f(S) - Ld(M) , \tag{42}$$

i.e., knowledge of L and d(M) leads to computable bounds for inf (M).

Note that the following well–known approach for solving the problem

$$\begin{array}{ll} \text{minimize} & f(x) \\ \text{s.t.} & x \in D \end{array} , \tag{43}$$

with $D \subset \mathbb{R}^n$ nonempty and compact and f Lipschitzian on D, is based directly on (41):

Start with an arbitrary point $x^0 \in D$, define the first approximating function by

$$F_0(x) := -L\|x - x^0\| + f(x^0)$$

and the next iteration point by

$$x^1 \in \operatorname{argmin} F_0(D) .$$

In Step k, the approximating function is

$$F_k(x) := \max_{0 \le i \le k} \{f(x^i) - L\|x - x^i\|\} \tag{44}$$

and the next iteration point is

$$x^{k+1} \in \operatorname{argmin} F_k(D) . \tag{45}$$

It is easy to show that any accumulation point of the sequence $\{x^k\}$ solves problem (43). This algorithm was proposed by Piyavskii (1967 and 1972) and Shubert (1972) for one–dimensional intervals D. A straightforward extension to the n–dimensional case was proposed by Mladineo (1986). Alternative aspects were considered in Mayne and Polak (1984), Horst and Tuy (1987), and Pinter (1983 and 1986). See also Bulatov (1977), Evtushenko (1985). The crucial part of this method is the minimization in (45). Unfortunately, since F_k is the pointwise maximum of a finite family of concave functions, this minimization problem is, except in the one–dimensional case, a very difficult one. Actually $F_k(x)$ is a d.c. function (cf. Theorem I.7), and applying the above approach results in solving a sequence of increasingly complicated d.c. programming subproblems. In the chapters that follow, some different and, hopefully, more practical approaches will be presented.

Lipschitzian optimization problems of the form (43), where D is either a convex set or the intersection of a convex set with finitely many complements of convex sets or a set defined by a finite number of Lipschitzian inequalities, are encountered in many economic and engineering applications. Examples are discussed, e.g., in Dixon and Szego (1975 and 1978), Strongin (1978), Zielinski and Neumann (1983), Fedorov (1985), Zilinskas (1982 and 1986), Pinter et al. (1986), Tuy (1986). Algorithms such as those proposed in Horst (1987 and 1988), Pinter (1988), Thach and Tuy (1987) will be discussed in Chapter XI.

Apparently problem (43) is a very general problem. Most problems discussed in the previous sections are actually also Lipschitzian. This follows from the fact that a convex function is Lipschitzian on a compact subset of the relative interior of its effective domain (cf. Rockafellar (1970), Theorem 10.4), and, moreover, the Lipschitzian property is preserved under operations such as forming linear combinations and maxima or minima of finitely many Lipschitzian functions (cf., also Section I.4.2).

However, we point out that all methods for solving Lipschitzian optimization problems to be presented in this book require ***knowledge of a Lipschitzian constant*** *for some or all of the functions involved.*

Though such a constant can often be estimated, this requirement sets limits on the application of Lipschitzian optimization techniques, since – in general – finding a good estimate for L (using, e.g., (40)) can be almost as difficult as solving the original problem. In most applications where Lipschitzian techniques have been proposed, the sets M are successively refined in such a way that one can use adaptive approximation of L (cf. Strongin (1973) and Pinter (1986)). Another means of calculating suitable approximations of L is by interval analysis (cf., e.g., Ratscheck and Rokne (1984 and 1988), and Ratscheck (1985)).

Example I.10. Many practical problems may involve indefinite, separable quadratic objective functions and / or constraints (cf., e.g., Pardalos et al (1987), Pardalos and Rosen (1986 and 1987)). To solve some of these problems, Al–Khayyal et al. (1989) proposed several variants of a branch and bound scheme that require a Lipschitz constant L for

$$f(x) := \sum_{k=1}^{n} (1/2\, p_k x_k^2 + q_k x_k + r_k)$$

on a rectangle $M = \{x\colon a_k \le x_k \le b_k,\ k=1,\dots,n\}$. In this case, the relation (40) yields

$$L := \max_{y \in M} \Big(\sum_{k=1}^{n} (p_k y_k + q_k)^2 \Big)^{1/2} .$$

Using monotonicity and separability we see that

$$\max_{y \in M} \left(\sum_{k=1}^{n} (p_k y_k + q_k)^2 \right)^{1/2}$$

and

$$\sum_{k=1}^{n} \max_{a_k \le y_k \le b_k} |p_k y_k + q_k|$$

have the same optimal solutions. Hence we have

$$L = \left[\sum_{k \in I_1} (p_k a_k + q_k)^2 + \sum_{k \in I_2} (p_k b_k + q_k)^2 \right]^{1/2},$$

where

$$I_1 = \left\{ k: -\frac{q_k}{p_k} \ge \frac{a_k + b_k}{2} \right\},$$

$$I_2 = \left\{ k: -\frac{q_k}{p_k} < \frac{a_k + b_k}{2} \right\}.$$

4.2. Systems of Equations and Inequalities

Solving systems of nonlinear equations and / or inequalities is a basic activity in numerical analysis. It is beyond the scope of our book to present here an overview of applications and methodology in that field (see, e.g., Forster (1980), Allgower and Georg (1980 and 1983) Dennis and Schnabel (1983)). It is the purpose of this section to show that unconventional global optimization methods designed for solving Lipschitzian optimization problems or d.c. programs can readily be applied to treat systems of equations and / or inequalities. Consider the system

$$f_i(x) = 0 \quad (i \in I_1) \tag{46}$$

$$f_i(x) \le 0 \quad (i \in I_2) \tag{47}$$

subject to $x \in D \subset \mathbb{R}^n$, where I_1, I_2 are finite index sets satisfying $I_1 \cap I_2 = \emptyset$. Suppose that D is nonempty and compact and that all functions f_i are continuous on D.

The system (46), (47) can be transformed into an equivalent global optimization problem.

Suppose first that we have $I_2 = \emptyset$ in (46), (47), i.e., we consider the **system of equations** (46).
Let $F: \mathbb{R}^n \longrightarrow \mathbb{R}^{|I_1|}$ be the mapping that associates to $x \in \mathbb{R}^n$ the vector with components $f_i(x)$ $(i \in I_1)$, and let $\|\cdot\|_N$ denote any norm on the image space of F. Then we have

Lemma I.1. *$x^* \in D$ is a solution of the system of equations (46) if and only if*

$$0 = \|F(x^*)\|_N = \min \{\|F(x)\|_N : x \in D\} \tag{48}$$

holds.

Proof. Lemma I.1 is an immediate consequence of the norm properties $\|z\|_N \geq 0$ $\forall z$ and $(\|z\|_N = 0 \Longleftrightarrow z = 0)$ ■

Let $f(x) = \|F(x)\|_N$. Then by virtue of Lemma I.1, the optimization problem

$$\begin{array}{l} \text{minimize } f(x) \\ \text{s.t. } x \in D \end{array} \tag{49}$$

contains all of the information on (46) that is usually of interest. We see that min $f(D) > 0$ holds if and only if (46) has no solution, and in the case min $f(D) = 0$ the set of solutions of (49) coincides with the set of solutions of (46).

Suppose now that we have $I_1 = \emptyset$ in (46), (47), i.e., we consider the **system of inequalities** (47). The following Lemma is obvious.

Lemma I.2. *$x^* \in D$ solves the system of inequalities (47) if and only if*

$$\max \{f_i(x^*): i \in I_2\} \leq 0 \tag{50}$$

holds.

Lemma I.2 suggests that we consider the optimization problem

$$\begin{array}{ll} \text{minimize} & \bar{f}(x) \\ \text{s.t.} & x \in D \end{array} , \tag{51}$$

where $\bar{f}(x) = \max \{f_i(x): i \in I_2\}$. Whenever a procedure for solving (51) detects a point $x^* \in D$ satisfying $\bar{f}(x^*) \leq 0$, then a solution of (47) has been found. The system (47) has no solution if and only if $\min \bar{f}(D) > 0$.

The case $I_1 \neq \emptyset$ and $I_2 \neq \emptyset$ can be treated in a similar way. As above, we see that $x^* \in D$ is a solution of (46) and (47) if and only if for

$$\tilde{f}(x) = \max \{\|F(x)\|_N, f_i(x) \ (i \in I_2)\} \tag{52}$$

we have

$$0 = \tilde{f}(x^*) = \min \tilde{f}(D) . \tag{53}$$

Now let all the functions in (46) and (47) be d.c., and let

$$\|F(x)\|_N = \|F(x)\|_1 = \sum_{i \in I_1} |f_i| \text{ or } \|F(x)\|_N = \|F(x)\|_\infty = \max_{i \in I_1} |f_i(x)| .$$

Then, by virtue of Theorem I.7 (Section I.3.1), each of the objective functions f, $\bar{f}$, $\tilde{f}$ in (49), (51) and (53) is d.c.. In other words, **whenever the functions f_i involved in a system of equations and / or inequalities of the form (46) (47) are d.c., then this system can be solved by d.c. programming techniques.**

Now suppose that all of the functions f_i in (46), (47) are Lipschitzian on $M \supset D$ with known Lipschitz constants L_i. Consider for $z = (z_1,...,z_m)^T \in \mathbb{R}^m$ the p–norms

$$\|z\|_p = \left(\sum_{i=1}^{m} |z_i|^p\right)^{\frac{1}{p}},\ 1 \le p < \infty, \text{ and } \|z\|_\infty = \max_{i=1,..,m} |z_i|.$$

Lemma I.3. *Let f_i be Lipschitzian on $M \subset \mathbb{R}^n$ with Lipschitz constants L_i (i=1,...,m) and let $F = (f_1,...,f_m)^T$. Then $\|F(x)\|_p$ $(1 \le p \le \infty)$ and $\max_{i=1,..,m} f_i(x)$ define Lipschitzian functions on M with Lipschitz constants $\sum_{i=1}^{m} L_i$ and $\max_{i=1,..,m} L_i$, respectively.*

Proof. We use two well–known properties of the norms involved. First, it follows from the triangle inequality that for any norm $\|\cdot\|_N$ in $\mathbb{R}^m$ and any z^1, $z^1 \in \mathbb{R}^m$ we have

$$|\ \|z^1\|_N - \|z^2\|_N\ | \le \|z^1 - z^2\|_N \,. \tag{54}$$

Furthermore, for $1 \le p \le \infty$ and $z \in \mathbb{R}^m$, the inequality

$$\|z\|_p \le \|z\|_1 \tag{55}$$

is satisfied (Jensen's inequality).

Using (54), (55) and the Lipschitz constants L_i of f_i (i=1,...,m), we obtain the following relations:

$$|\ \|F(x)\|_p - \|F(y)\|_p| \le \|F(x) - F(y)\|_p \le \|F(x) - F(y)\|_1$$

$$= \sum_{i=1}^{m} |f_i(x) - f_i(y)| \le \sum_{i=1}^{m} L_i \|x - y\| = \|x - y\|\left(\sum_{i=1}^{m} L_i\right),$$

where $\|x - y\|$ denotes the Euclidean norm of $(x - y)$ in $\mathbb{R}^n$.

Therefore, for all $1 \le p \le \infty$, $\|F(x)\|_p$ defines a Lipschitzian function on M with $L = \sum_{i=1}^{m} L_i$ being a Lipschitz constant which is independent of p.

In a similar way, since

$$|\max_i f_i(x) - \max_i f_i(y)| \le \max_i |f_i(x) - f_i(y)| \le \max_i \{L_i\|x-y\|\} = (\max_i L_i)\ \|x-y\|,$$

we see that $\max_{i=1,..m} f_i(x)$ defines a Lipschitz function with Lipschitz constant $\max_{i=1,..,m} L_i$. ■

Lemma I.3 provides Lipschitz constants of the objective functions in (49), (51) and (53); and in a manner similar to the d.c. case we have:

Whenever the functions f_i involved in a system of equations and / or inequalities of the form (46) and (47) are Lipschitzian with known Lipschitz constants L_i, then the system can be solved by Lipschitzian optimization techniques.

The above transformations have been applied by Horst and Thoai (1988) in order to solve systems of equations and inequalities. Another method for solving systems of nonlinear equations via d.c.–programming was proposed by Thach (1987).

CHAPTER II

OUTER APPROXIMATION

Outer approximation of the feasible set by a sequence of simpler relaxed sets is a basic method in many fields of optimization. In this technique, the current approximating set is improved by a suitable additional constraint (a cut).

In this chapter, outer approximation is developed with regard to the specific needs of global optimization. First, a simple general convergence principle is presented which permits nonlinear cuts and unbounded feasible sets. Specialization to outer approximation by polyhedral convex sets enables us to derive a large class of "cutting–plane" methods. Then, some "constraint dropping" strategies and several computational issues are addressed. These include the calculation of new vertices and extreme directions generated from a polyhedral convex set by a linear cut and the identification of redundant constraints.

1. BASIC OUTER APPROXIMATION METHOD

In this section we present a class of methods which are among the basic tools in many fields of optimization and which have been used in many forms and variants. The feasible set is relaxed to a simpler set D_1 containing D, and the original objective function f is minimized over the relaxed set. If the solution of this relaxed problem is in D, then we are done; otherwise an appropriate portion of $D_1 \setminus D$ is cut off by an additional constraint, yielding a new relaxed set D_2 that is a better approximation of D than D_1. Then, D_1 is replaced by D_2, and the procedure is repeated. These methods are frequently called **outer approximation** or **relaxation** methods.

Since the pioneering papers of Gomory (1958 and 1960), Cheney and Goldstein (1959) and Kelley (1960), outer approximation in this sense has developed into a basic tool in combinatorial optimization and (nonsmooth) convex programming. In global optimization, where certain theoretical and computational questions arise that cannot be inferred from previous applications in other fields, outer approximation has been applied in various forms for solving most of the problem classes that we introduced in the preceding chapter. Examples include concave minimization (e.g., Hoffman (1981), Thieu, Tam and Ban (1983), Tuy (1983), Thoai (1984)), problems having reverse convex constraints (e.g., Tuy (1987)), d.c. programming (e.g., Tuy (1986), Thoai (1988)) and Lipschitzian optimization (e.g., Thach and Tuy (1987)).

We follow the lines of a simple and general treatment given by Horst, Thoai and Tuy (1987 and 1989).

Consider the global optimization problem

$$\text{(P)} \qquad \begin{array}{l} \text{minimize } f(x), \\ \text{s.t. } x \in D \end{array} \qquad (1)$$

where $f: \mathbb{R}^n \longrightarrow \mathbb{R}$ is **continuous** and $D \subset \mathbb{R}^n$ is **closed**.

We shall suppose throughout this chapter that min f(D) exists.

A widely used outer approximation method for solving (P) is obtained by replacing it by a sequence of simpler "relaxed" problems

$$(Q_k) \qquad \begin{array}{l} \text{minimize } f(x), \\ \text{s.t. } x \in D_k \end{array} \qquad (2)$$

where $\mathbb{R}^n \supset D_1 \supset D_2 \supset \dots \supset D$ and

$$\min f(D_k) \xrightarrow[(k\to\infty)]{} \min f(D).$$

Usually, the sets D_k belong to a family $\underline{F}$ with the following properties:

(a) The sets $D_k \subset \mathbb{R}^n$ are closed, and any problem (Q_k) with $D_k \in \underline{F}$ has a solution and can be solved by available algorithms;

(b) for any $D_k \in \underline{F}$ containing D and any point $x^k \in D_k \backslash D$ one can define a constraint function $\ell_k\colon \mathbb{R}^n \longrightarrow \mathbb{R}$ satisfying

$$\ell_k(x) \le 0 \quad \forall x \in D\,, \tag{3}$$

$$\ell_k(x^k) > 0\,, \tag{4}$$

$$\{x \in D_k\colon \ell_k(x) \le 0\} \in \underline{F}\,. \tag{5}$$

Under these conditions the following solution method suggests itself:

Outer Approximation Method:

Choose $D_1 \in \underline{F}$ such that $D_1 \supset D$. Set $k \leftarrow 1$.

Iteration k (k = 1,2,...):

Solve the relaxed problem (Q_k) obtaining a solution $x^k \in \operatorname{argmin} f(D_k)$.
If $x^k \in D$, then stop: x^k solves (P).
Otherwise construct a constraint function $\ell_k\colon \mathbb{R}^n \longrightarrow \mathbb{R}$ satisfying (3), (4), (5) and set

$$D_{k+1} = D_k \cap \{x\colon \ell_k(x) \le 0\},\ k \leftarrow k+1. \tag{6}$$

Go to the next iteration.

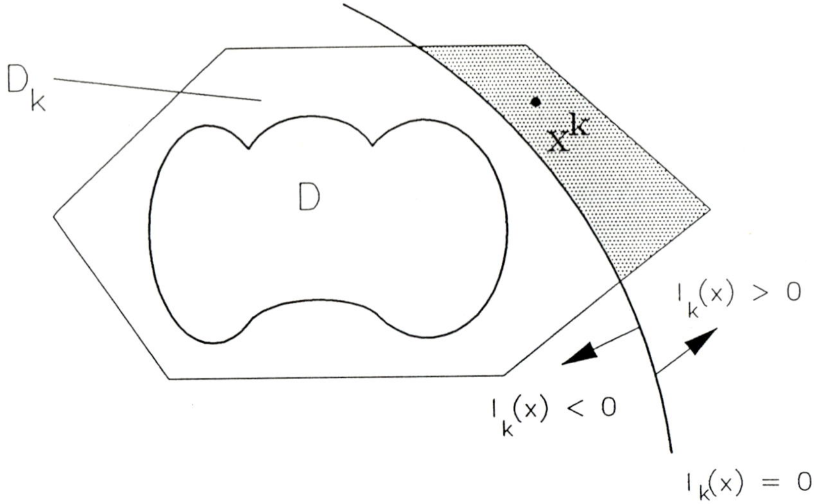

Fig. II.1. Portion of $D_k \setminus D$ containing x^k is cut off.

Conditions (3) and (4) imply that the set $\{x \in \mathbb{R}^n: \ell_k(x) = 0\}$ strictly separates $x^k \in D_k \setminus D$ from D. The additional constraint $\ell_k(x) \leq 0$ cuts off a subset of D_k. However, since we have $D \subset D_k$ for all k (no part of D is cut off), each D_k constitutes an outer approximation of D.

Note that, since $D_k \supset D_{k+1} \supset D$, we have

$$\min f(D_k) \leq \min f(D_{k+1}) \leq \min f(D) , \tag{7}$$

and $x^k \in D$ implies $x^k \in \text{argmin } f(D)$.

The following convergence results require, apart from the conditions stated in the theorems, only that the objective function is continuous and the feasible set is closed (cf. the assumptions in (1)).

Theorem II.1. *In the context of the outer approximation method above, assume that*

(i) ℓ_k *is continuous for each* $k = 1,2,\ldots$ *and for each convergent subsequence* $\{x^{k_q}\} \subset \{x^k\}$ *satisfying* $x^{k_q} \xrightarrow[(q\to\infty)]{} \bar{x}$ *we have*

$$\ell_{k_q}(x^{k_q}) \xrightarrow[(q\to\infty)]{} \ell(\bar{x}), \text{ where } \ell\colon \mathbb{R}^n \longrightarrow \mathbb{R},$$

and

(ii) $\ell(\bar{x}) = 0$ *implies* $\bar{x} \in D$.

Then every accumulation point of the sequence $\{x^k\}$ *belongs to D and solves (P).*

Proof. Let $\bar{x}$ be an accumulation point of $\{x^k\}$ and let $\{x^{k_q}\}$ be a corresponding subsequence such that $x^{k_q} \xrightarrow[(q\to\infty)]{} \bar{x}$. By (6), we have $\ell_{k_q}(x^{k_{q+1}}) \le 0 \ \forall q$. By assumption (i), letting $q \to \infty$ it follows that

$$\lim_{q\to\infty} \ell_{k_q}(x^{k_{q+1}}) = \lim_{q\to\infty} \ell_{k_q}(x^{k_q}) = \ell(\bar{x}) \le 0 . \tag{8}$$

On the other hand, by (4), we have $\ell_{k_q}(x^{k_q}) > 0 \ \forall q$, which similarly implies

$$\ell(\bar{x}) \ge 0 . \tag{9}$$

From (8) and (9) it follows that $\ell(\bar{x}) = 0$; and hence by (ii) we have $\bar{x} \in D$.
But since $f(x^k) \le f(x) \ \forall x \in D_k \supset D$, it follows, by continuity of f, that $f(\bar{x}) \le f(x)$ $\forall x \in D$, i.e., $\bar{x}$ solves (P). ■

Assumption (i) is fullfilled, e.g., when $\ell_{k_q}(x)$ converges uniformly to $\ell(x)$ or if it converges continuously to $\ell(x)$ (cf. Kall (1986)).

Note that, for an outer approximation method, convergence in the sense of Theorem II.1 is, of course, assured if every accumulation point $\bar{x}$ of $\{x^{k_q}\}$ is in D.

But the proof of Theorem II.1 shows that, by (3), (4) and the assumption (i) in Theorem II.1, for every convergent subsequence $\{x^{k_q}\}$ we have $\ell_{k_q}(x^{k_q}) \xrightarrow[(q\to\infty)]{} 0$. It turns out that this fact and also the introduction of the function $\ell(x)$ (which is obviously not necessary in Theorem II.1) can be used to establish convergence of several algorithmic realizations in an easy way; some of these will be discussed in the next section. Another version of the convergence theorem is

Theorem II.1'. *In the context of the outer approximation method above, assume that*

(i) ℓ_k *is lower semi-continuous for each* $k = 1,2,\ldots$,

(ii) *for each convergent subsequence* $x^{k_q} \xrightarrow[(q\to\infty)]{} \bar{x}$ *we have*

$$\lim_q \ell_{k_q}(x^{k_q}) = \lim_q \ell_{k_q}(\bar{x}) \le 0 \text{ , and}$$

$$\lim_q \ell_{k_q}(\bar{x}) = 0 \text{ implies } \bar{x} \in D.$$

Then every accumulation point of the sequence $\{x^k\}$ *belongs to D and solves (P).*

2. OUTER APPROXIMATION BY CONVEX POLYHEDRAL SETS

In most realizations of outer approximation methods the sets D_k are convex polyhedral sets, and the constraints $\ell_k(x)$ are affine functions such that $\{x: \ell_k(x) = 0\}$ define hyperplanes strictly separating x^k and D. The set D is usually assumed to be a closed or even compact convex set and the procedures are often called **cutting plane methods**.

Let

$$D = \{x \in \mathbb{R}^n: g(x) \le 0\} , \tag{10}$$

where $g: \mathbb{R}^n \longrightarrow \mathbb{R}$ is convex. Clearly, D is a closed convex set. Note that in the case of several convex constraints $g_i(x) \leq 0$ $(i \in I)$, we set

$$g(x) = \max_{i \in I} g_i(x) \tag{11}$$

(g is, of course, also convex).

Suppose that we are given a convex polyhedral set $D_1 \supset D$ (for the construction of D_1, see Section II.4). Denote by xy the inner product of x, $y \in \mathbb{R}^n$. Let at each iteration k the constraint function ℓ_k be defined by

$$\ell_k(x) = p^k(x - y^k) + \beta_k \tag{12}$$

where $p^k \in \mathbb{R}^n$, $y^k \in D_1$, $\beta_k \in \mathbb{R}$ are suitably chosen.

In (12) we now set

$$p^k \in \partial g(y^k),\ \beta_k = g(y^k)\ , \tag{13}$$

where $\partial g(y^k)$ denotes the **subdifferential** of g at y^k.

If in addition we choose $y^k = x^k$, then (12) and (13) yield the so–called **KCG–algorithm**, essentially proposed by Cheney and Goldstein (1959) and by Kelley (1960) for convex programming problems.

Moreover, if a point w satisfying $g(w) < 0$, is given and if we choose y^k as the unique point where the line segment $[w,x^k]$ meets the boundary ∂D of D, i.e., $\{y^k\} = [w,x^k] \cap \partial D$, then (12), (13) yield the so–called **supporting hyperplane** approach of Veinott (1967) (cf. Fig. II.2).

Note that, by the convexity of g, we have

$$y^k = \lambda_k w + (1 - \lambda_k)x^k\ , \tag{14}$$

where λ_k is the unique solution of the univariate convex programming problem

$$\min\ \{\lambda \in [0,1]: \lambda w + (1-\lambda)x^k \in D\}\ , \tag{15}$$

or equivalently, of

$$g(\lambda w + (1-\lambda)x^k) = 0, \quad \lambda \geq 0 . \tag{15'}$$

When applying these methods to solve convex programming problems

$$\min \{f(x): \tilde{g}(x) \leq 0\} , \tag{16}$$

where f, $\tilde{g}$: $\mathbb{R}^n \longrightarrow \mathbb{R}$ are convex, one usually transforms (16) into the equivalent problem

$$\min \{t: g(x,t) \leq 0\} , \tag{16'}$$

where $t \in \mathbb{R}$ and $g(x,t) = \max \{\tilde{g}(x), f(x) - t\}$. The relaxed problems (Q_k) are then **linear programming** problems.

Instead of subgradients $p^k \in \partial g(y^k)$ one can as well use ε–subgradients (cf., e.g., Parikh (1976)).

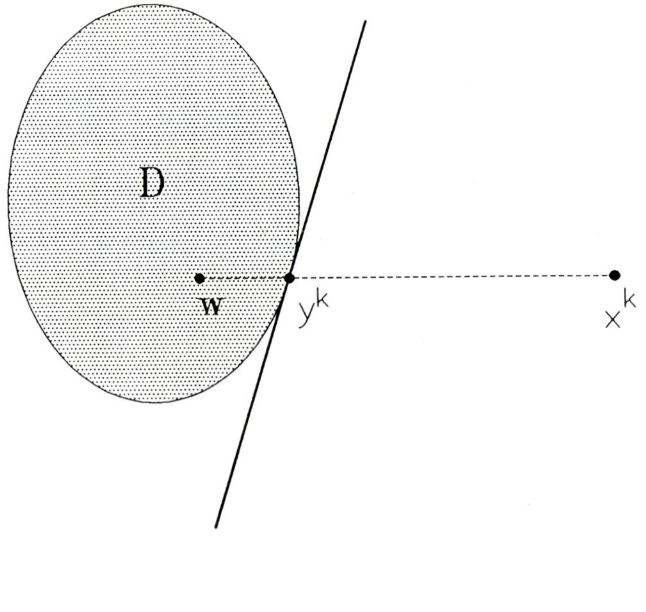

Fig. II.2 The supporting hyperplane method.

In a modified form (essentially relating to the solution of the relaxed problems (Q_k)), both classical methods have been applied for solving concave minimization problems: the KCG approach by Thieu, Tam and Ban (1983) and the supporting hyperplane method by Hoffman (1981). Furthermore, the general incorporation of outer approximation by cutting planes into procedures for solving global optimization problems was discussed in Tuy (1983), Horst (1986) and Tuy and Horst (1988). We will return to these discussions later.

Another approach to outer approximation by convex polyhedral sets uses duality and decomposition (see, e.g., Dantzig and Wolfe (1960), Mayne and Polak (1984) and references given therein). In several of these dual approaches, the important question of constraint dropping strategies is addressed. This question will be discussed in the next section, where we will also mention some additional important approaches to outer approximation by cutting planes.

The following class of outer approximation by polyhedral convex sets – introduced in Horst, Thoai and Tuy (1987 and 1989) – provides a large variety of methods that include as very special cases the KCG and supporting hyperplane methods. Consider problem (P) with feasible set D as defined by (10). Assume that an initial polyhedral convex set $D_1 \supset D$ is given and let the constraints ℓ_k be affine, i.e., of the form (12).

Denote

$$D^0 := \{x \in \mathbb{R}^n : g(x) < 0\} . \tag{17}$$

We assume that $\partial g(x) \setminus \{0\} \neq \emptyset \ \ \forall x \in \mathbb{R}^n \setminus D^0$, where $\partial g(x)$ denotes (as above) the subdifferential of g at x. This assumption is certainly fulfilled if $D^0 \neq \emptyset$ (Slater condition): let $z \in D^0$ and $y \in \mathbb{R}^n \setminus D^0$, $p \in \partial g(y)$. Then, by the definitions of $\partial g(y)$ and D^0, we have

$$0 > g(z) \geq p(z-y) + g(y) \geq p(z-y),$$

which implies $p \neq 0$, i.e., $0 \notin \partial g(y)$.

Theorem II.2. *Let K be any compact convex subset of D^0. In (12) for each $k = 1,2,...$ choose*

$$y^k \in conv\,(K \cup \{x^k\}) \setminus D^0,\ p^k \in \partial g(y^k) \setminus \{0\},\ \beta_k = g(y^k)\,, \tag{18}$$

where $x^k \in argmin\, f(D_k)$.

Then the conditions (3), (4) of the outer approximation method and the assumptions (i), (ii) of Theorem II.1 are fulfilled.

The set $Y^k := conv\,(K \cup \{x^k\}) \setminus D^0$ is displayed in Fig. II.3.

Proof. **(3):** Since $p^k \in \partial g(y^k)$, we have $g(x) \geq p^k(x - y^k) + g(y^k) = \ell_k(x)$ $\forall x \in \mathbb{R}^n$. Hence using $x \in D \Longleftrightarrow g(x) \leq 0$, we see that $\ell_k(x) \leq 0\ \ \forall x \in D$.

(4): If $D^0 = \emptyset$, then $K = \emptyset$; hence $y^k = x^k$ and $\ell_k(x^k) = g(x^k) > 0$, since $x^k \notin D$. Let $D^0 \neq \emptyset$, $K \neq \emptyset$. Then every point $y^k \in conv\,(K \cup \{x^k\}) \setminus D^0$ can be expressed in the form

$$y^k = \lambda_k z^k + (1 - \lambda_k)x^k\,, \tag{19}$$

where $z^k \in K$, $0 \leq \lambda_k < 1$. Then

$$x^k - y^k = \alpha_k(y^k - z^k),\ \text{where}\ \ \alpha_k = \frac{\lambda_k}{1-\lambda_k} \geq 0\,. \tag{20}$$

Since K is compact and is contained in D^0, there exists a number $\delta > 0$ such that $g(x) \leq -\delta < 0\ \ \forall x \in K$.

Using $p^k \in \partial g(y^k)$, $g(y^k) \geq 0$, we obtain

$$0 > -\delta \geq g(z^k) \geq p^k(z^k - y^k) + g(y^k) \geq p^k(z^k - y^k),$$

$$\text{hence}\ \ p^k(y^k - z^k) \geq \delta > 0.$$

From this and (20) it follows that

$$\ell_k(x^k) = p^k(x^k - y^k) + g(y^k) = \alpha_k p^k(y^k - z^k) + g(y^k) \geq \alpha_k\delta + g(y^k) > 0$$

whenever $\alpha_k > 0$ (i.e., $x^k \neq y^k$). However, for $\alpha_k = 0$, i.e., $y^k = x^k$, we have $\ell_k(x^k) = g(x^k) > 0$, since $x^k \notin D$.

(i): For each $k = 1,2,\ldots$, the affine function ℓ_k defined by (12) and (18) is obviously continuous.

Let $\{x^q\}$ be any convergent subsequence of $\{x^k\}$, and let $x^q \to \bar{x}$. Then there is a $q_0 \in \mathbb{N}$ sufficiently large such that $x^q \in B := \{x \in \mathbb{R}^n: \|x - \bar{x}\| \leq 1\}$ $\forall q > q_0$. Since K and $\{\partial g(y): y \in B\}$ are compact sets (cf., e.g., Rockafellar (1970), Chapter 24) and $\lambda_q \in [0,1]$, we can assume (by passing to a suitable subsequence if necessary) that

$$p^q \to \bar{p}, \lambda_q \to \bar{\lambda} \in [0,1), z^q \to \bar{z} \text{ as } q \to \infty . \tag{21}$$

Then it follows that

$$y^q = \lambda_q z^q + (1-\lambda_q)x^q \to \bar{y} = \bar{\lambda}\bar{z} + (1-\bar{\lambda})\bar{x}, \beta_q = g(y^q) \to g(\bar{y}) = \bar{\beta}. \tag{22}$$

Observe that we cannot have $\bar{\lambda} = 1$, because $\bar{\lambda} = 1$ would imply $\bar{y} = \bar{z}$, which is impossible, since $\bar{z} \in K \subset \text{int } D$. Note that, since the set valued mapping defining $\partial g(y)$ is closed (cf., e.g., Rockafellar (1970), Chapter 24), we even have $\bar{p} \in \partial g(\bar{y})$. Let $\ell(x) = \bar{p}(x-\bar{y}) + \bar{\beta}$. Then we see that

$$\ell_q(x^q) = p^q(x^q - y^q) + \beta_q \to \bar{p}(\bar{x} - \bar{y}) + \bar{\beta} = \ell(\bar{x}) . \tag{23}$$

(ii): Let $x^q \to \bar{x}$, $\ell_q(x^q) \to \ell(\bar{x}) = 0$.

By (20) – (23), we have

$$0 = \ell(\bar{x}) = \bar{p}(\bar{x} - \bar{y}) + \bar{\beta} = \frac{\bar{\lambda}}{1-\bar{\lambda}}\bar{p}(\bar{y} - \bar{z}) + \bar{\beta} . \tag{24}$$

But since $y^q \notin D^0$, we have $\beta_q = g(y^q) \geq 0$; hence, by continuity of the convex function g, we have $\bar{\beta} = g(\bar{y}) \geq 0$. Moreover, while verifying (4), we saw that $p^q(y^q - z^q) \geq \delta > 0$, and hence $\bar{p}(\bar{y} - \bar{z}) \geq \delta > 0$. From (24) it then follows that $\bar{\lambda} = 0$; and hence $\bar{y} = \bar{x}$ (cf. (22)), and also that $\bar{\beta} = g(\bar{y}) = 0$, hence $\bar{x} \in D$. ■

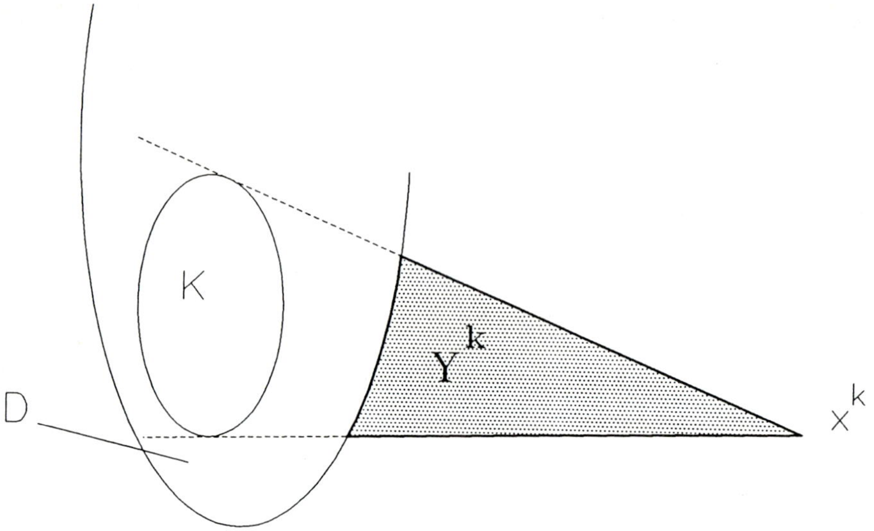

Fig. II.3 Choice of y^k in Theorem II.2.

For any compact, convex set $K \subset D^0$ we obviously may always choose $y^k = x^k$. The resulting algorithm is the KCG–method. If we fix $K = \{w\}$, i.e., $z^k = w$, where $g(w) < 0$, and determine y^k as the unique point where the line segment $[w,x^k]$ meets the boundary of D, then we obtain the supporting hyperplane algorithm. The KCG and the supporting hyperplane methods, which in their basic form were originally designed for convex programming problems, are useful tools in that field if the problems are of moderate size and if the accuracy of the approximate solution obtained when the algorithm is stopped, does not need to be very high. The rate of convergence, however, generally cannot be expected to be faster than linear (cf., e.g., Levitin and Polyak (1966) and Wolfe (1970)).

Moreover, in the case of cutting planes, it has been observed that near a solution, the vectors p^k of successive hyperplanes may tend to become linearly dependent, causing numerical instabilities.

In order to solve convex programming problems and also to obtain local solutions in nonlinear programming, recently several faster cutting plane methods have been proposed that combine the idea of outer approximation methods with typical local elements as line searches etc.. We will return briefly to these methods when discussing constraint dropping strategies in the next section. In global optimization, where rapid local methods fail and where – because of the inherent difficulties of global (multiextremal) optimization – only problems of moderate size can be treated, the class of relaxation methods by polyhedral convex sets that was introduced in Theorem II.2 constitutes a basic tool, especially when combined with branch and bound techniques (cf. Chapter IV). However, the great variety of concrete methods included in Theorem II.2 needs further investigation in order to find the most promising realizations.

Another way of constructing supporting hyperplanes in the outer approximation approach is by projecting the current iteration point x^k onto the feasible set D. Denote by $\pi(x)$ the projection of $x \in \mathbb{R}^n$ onto the closed convex set D, i.e.,

$$\|x - \pi(x)\| := \min \{\|x - y\| : y \in D\} . \tag{25}$$

It is well known that $\pi(x)$ exists, is unique, and is continuous everywhere; moreover, we have

$$(x - \pi(x))\,(y - \pi(x)) \le 0 \qquad \forall y \in D \tag{26}$$

(cf., e.g., Luenberger (1969)).

Theorem II.3. *If in (12) for each $k = 1,2,...$ we choose*

$$p^k = (x^k - \pi^k),\ y^k = \pi^k \text{ and } \beta_k = 0, \text{ i.e., } \ell_k(x) = (x^k - \pi^k)\,(x - \pi^k)\,, \tag{27}$$

where $x^k \in \operatorname{argmin} f(D_k)$, $\pi^k = \pi(x^k)$, then the outer approximation conditions (3), (4) and the assumptions (i), (ii) of Theorem II.1 are satisfied.

Proof: **(3)**: By (27) and (26), we have $\ell_k(x) \le 0 \ \forall x \in D$.

(4): $\ell_k(x^k) = (x^k - \pi^k)\,(x^k - \pi^k) = \|x^k - \pi^k\|^2 > 0$, since $x^k \notin D$.

(i): The affine function ℓ_k defined by (27) is continuous for each k. Let $\{x^q\}$ be a subsequence of $\{x^k\}$ satisfying $x^q \rightarrow \bar{x}$. Since the function π is continuous, we have $\pi^q \rightarrow \pi(\bar{x})$.

Let $\ell(x) = (\bar{x} - \pi(\bar{x}))(x-\pi(\bar{x}))$. Then $\ell_q(x^q) = (x^q - \pi^q)(x^q - \pi^q) \rightarrow (\bar{x} - \pi(\bar{x}))(\bar{x} - \pi(\bar{x})) = \ell(\bar{x})$.

(ii): Let $x^q \rightarrow \bar{x}$, $\ell_q(x^q) \rightarrow \ell(\bar{x}) = 0$.

Then it follows that $\ell(\bar{x}) = (\bar{x} - \pi(\bar{x}))(\bar{x} - \pi(\bar{x})) = \|\bar{x} - \pi(\bar{x})\|^2 = 0$. This implies $\bar{x} = \pi(\bar{x}) \in D$. ■

Note that Theorem II.3 cannot be subsumed under Theorem II.2. As an example, consider the case where the ray from x^k through π^k does not meet the set D^0 (Fig. II.4).

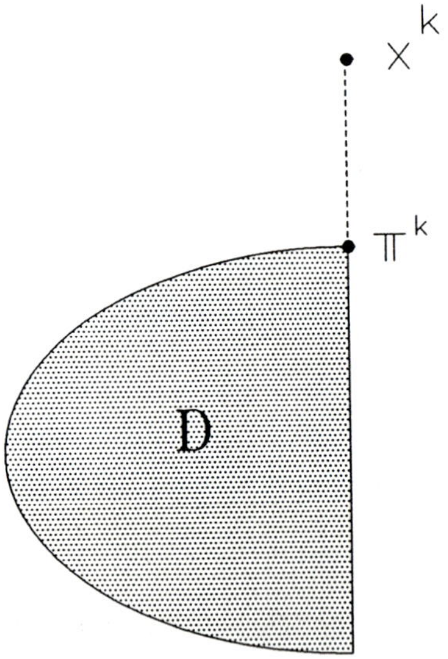

Fig. II.4. Theorem II.3 is not included in Theorem II.2.

3. CONSTRAINT DROPPING STRATEGIES

One clear disadvantage of the outer approximation methods discussed so far is that the size of the subproblems (Q_k) increases from iteration to iteration, since in each step a new constraint $\ell_k(x) \le 0$ is added to the existing set of constraints, but no constraint is ever deleted. Most of the theoretical work on outer approximation by polyhedral convex sets has centered on the crucial question of dropping (inactive) constraints. Topkis (1970) proposed a cutting plane method where certain constraints could be dropped. Eaves and Zangwill (1971) – by introducing the notions of a cut map and a separator – gave a general and abstract theory of outer approximation by cutting planes. This theory was generalized further by Hogan (1973) to cover certain dual approaches. Examples of dual approaches that include constraint dropping strategies are given in Gonzaga and Polak (1979) and in Mayne and Polak (1984).

Though convex programming is not within the scope of our treatment, we would like to mention that for (possibly nonsmooth) convex problems several cutting plane approaches have been proposed that converge linearly or faster; moreover, they enable one to drop constraints in such a manner that the number of constraints used in each step is bounded. Most of these algorithms do not fit into our basic approach, since in each step a quadratic term is added to the objective function of the original problem, while the outer approximation methods presented here use the original objective function throughout the iterations. For details, we refer to Fukushima (1984) and to the book of Kiwiel (1985). However, these results on local optimization and convex programming, respectively, could not yet be carried over to the **global** optimization of multiextremal problems.

Below, we present a theorem on constraint dropping strategies in global optimization that can be applied to all algorithms satisfying the assumptions of Theorem II.1, hence, e.g., also to the classes of procedures discussed in the preceding section. Note that these results apply also to **nonlinear cuts**.

We shall need separators as defined in Eaves and Zangwill (1971).

Definition II.1. *A function* $\delta: D_1 \setminus D \longrightarrow \mathbb{R}_+$ *is called a separator if for all sequences* $\{x^k\} \subset D_1 \setminus D$ *we have that* $x^k \longrightarrow \bar{x}$, $\delta(x^k) \longrightarrow 0$ *imply* $\bar{x} \in D$.

Note that for $x \in D_1 \setminus D$ we must have $\delta(x) > 0$. Each function $\delta: D_1 \setminus D \longrightarrow \mathbb{R}$ that is lower semicontinuous and satisfies $\delta(x) > 0 \quad \forall x \in D_1 \setminus D$ is a separator. Another example is given by the distance

$\delta(x) = d_D(x)$ of $x \in D_1 \setminus D$ from D.

For practical purposes, when separators are used to drop certain old cuts, a straightforward choice of separator is any penalty function, as is well known from standard local nonlinear programming techniques. For example, let

$$D = \{x \in \mathbb{R}^n : g_i(x) \leq 0 \quad (i = 1,...,m)\}$$

with continuous $g_i: \mathbb{R}^n \longrightarrow \mathbb{R} \quad (i = 1,...,m)$. Then all functions of the form

$$\delta(x) = \sum_{i=1}^{m} [\max \{0, g_i(x)\}]^{\beta}, \ \beta \geq 1 \tag{28}$$

are examples of separators.

Let

$$L_k := \{x \in \mathbb{R}^n : \ell_k(x) \leq 0\} . \tag{29}$$

Then we would like be able to say that

$$D_{k+1} = D_1 \cap \{\bigcap_{i \in I_k} L_i\} , \tag{30}$$

where I_k is a subset of $\{1,...,k\}$, $|I_k| < k$, defines a convergent outer approximation method. Let $\uparrow, \downarrow$ denote monotonically increasing or decreasing convergence, respectively.

Theorem II.4. *Let δ be a separator. Let $\{\varepsilon_{ij}\}$, $j \ge i$, be a double-indexed sequence of nonnegative real numbers such that*

$$\varepsilon_{ii} = 0, \qquad \varepsilon_{ij} > 0 \quad \text{for } j > i \tag{31}$$

$$\varepsilon_{ij} \uparrow \bar{\varepsilon}_i \text{ as } j \to \infty, \qquad \bar{\varepsilon}_i \downarrow 0 \text{ as } i \to \infty . \tag{32}$$

Assume that the sequence $\{\ell_k(x)\}$ in the outer approximation method satisfies (3), (4) and the requirements (i), (ii) of Theorem II.1. Then

$$D_{k+1} = D_1 \cap (\bigcap_{i \in I_k} L_i) \quad k = 1,2,\ldots \tag{33}$$

with

$$I_k = \{i\colon 1 \le i \le k,\ \delta(x^i) \ge \varepsilon_{ik}\} \tag{34}$$

defines a convergent outer approximation method, i.e., every accumulation point of $\{x^k\}$ solves (P).

Proof. Let $\{x^k\} \subset D_1 \setminus D$ be the sequence of iteration points defined by an outer approximation method satisfying (31), (32), (33), (34). Let $\{x^q\}$ be a subsequence of $\{x^k\}$ converging to $\bar{x}$.

If $\delta(x^q) \longrightarrow 0$, then $\bar{x} \in D$ by Definition II.1.

Suppose that $\delta(x^q) \not\longrightarrow 0$. By passing to a suitable subsequence if necessary, we may assume that $\delta(x^q) \ge \varepsilon > 0$ $\forall q$. Since $\bar{\varepsilon}_i \downarrow 0$ we may also assume that $\varepsilon \ge \bar{\varepsilon}_i$ for all i considered. Then using (32), we have $\delta(x^i) \ge \varepsilon \ge \bar{\varepsilon}_i \ge \varepsilon_{i,q-1}$ $\forall i < q-1$. By (34), this implies $i \in I_{q-1}$, hence $x^q \in L_i$, i.e., $\ell_i(x^q) \le 0$ $\forall i < q$. Letting $i \to \infty$, we see from (i) of Theorem II.1 that $\ell(\bar{x}) \le 0$.

But by (4) we have $\ell_q(x^q) > 0$ $\forall q$; hence by (i), $\ell(\bar{x}) \ge 0$. By (ii) of Theorem II.1, then it follows that $\bar{x} \in D$ which implies that $\bar{x}$ solves (P). ∎

Many double-indexed sequences satisfying the above requirements exist. For example, let, $\{\eta_i\}$ be any monotonically decreasing sequence of positive real numbers that converge to 0. Then $\varepsilon_{ij} = \eta_i - \eta_j$ $(i \leq j)$ satisfies (31), (32).

Note that, since $\varepsilon_{kk} = 0$, we always have $k \in I_k$. Moreover, because ε_{ik} is monotonically increasing with k for fixed i, we conclude that $i \notin I_k$ implies $i \notin I_s$ for all $s > k$: a dropped cut remains dropped forever. Likewise, a cut retained at some iteration k may be dropped at a subsequent iteration $s > k$. On the other hand, for $i,k \longrightarrow \infty$, we have $\varepsilon_{ik} \longrightarrow 0$, which at first glance seems to imply that few "late" constraints can be dropped for large i,k, by (34). However, note that we also have $\delta(x^i) \longrightarrow 0$, so that the number of constraints to be dropped for large i,k depends on the relative speed of convergence of $\delta(x^i)$ and ε_{ik}. This limit behaviour can be influenced by the choices of δ and ε_{ik}, and should be investigated further.

Another constraint dropping strategy that is useful when outer approximation methods are applied to solve concave minimization will be presented in Chapter VI.

4. ON SOLVING THE SUBPROBLEMS (Q_k)

We now discuss the question of how to solve the subproblems (Q_k) in outer approximation methods that use affine cuts. In global optimization, these algorithms have often been applied to problems satisfying

$$\min f(D_k) = \min f(V(D_k)) ,$$

where D_k is a polytope and $V_k = V(D_k)$ denotes the vertex set of D_k. The most well–known examples are concave minimization and stable d.c. programming that can be reduced to parametric concave minimization (cf. Chapters I, IX and X).

In this section we shall address three questions:

– *How do we determine the initial polytope D_1 and its vertex set V_1?*

– *How do we find the vertex set V_{k+1} of D_{k+1} in each step, if D_k, V_k and an affine cut are given?*

– *How do we identify redundant constraints?*

The problem of finding all vertices and redundant constraints of a polytope given by a system of linear inequalities has been treated often (e.g., Manas and Nedoma (1968), Matheis (1973), Gal (1975), Dyer and Proll (1977 and 1982), Matheis and Rubin (1980), Dyer (1983), Khang and Fujiwara (1989)). Since, however, in our case V_k and the new constraint are given, we are interested in developing methods that take into account our specific setting rather than applying one of the standard methods referred to above. We will briefly also treat the case of unbounded polyhedral convex sets D_k, where vertices and extreme directions are of interest.

4.1 Finding an Initial Polytope D_1 and its Vertex Set V_1

Let D be a nonempty, convex and compact set of dimension n. D_1 should be a simple polytope tightly enclosing D and having a small number of vertices.

Denote

$$\alpha_j := \min \{x_j : x \in D\} \quad (j = 1,...,n) \tag{35}$$

and

$$\alpha := \max \{ \sum_{j=1}^{n} x_j : x \in D\} . \tag{36}$$

Then it is easily seen that

$$D_1 := \{x \in \mathbb{R}^n : \alpha_j - x_j \le 0 \ (j=1,...,n), \ \sum_{j=1}^{n} x_j - \alpha \le 0\} \tag{37}$$

is a simplex containing D. The n+1 facets of D_1 are defined by the n+1 hyperplanes

$\{x \in \mathbb{R}^n: x_j = \alpha_j\}$ (j=1,...,n), and $\{x \in \mathbb{R}^n: \sum_{i=1}^{n} x_i = \alpha\}$. Each of these hyperplanes is a supporting hyperplane of D. The set of vertices of D is

$$V_1 = \{v^0, v^1, \dots, v^n\}$$

where

$$v^0 = (\alpha_1, \dots, \alpha_n)^T \tag{38}$$

and

$$v^j = (\alpha_1, \dots, \alpha_{j-1}, \beta_j, \alpha_{j+1}, \dots, \alpha_n) \ (j=1,\dots,n) \tag{39}$$

with

$$\beta_j = \alpha - \sum_{i \neq j} \alpha_i .$$

Note that (35), (36) define n + 1 convex optimization problems with linear objective functions. For this reason their solution can be efficiently computed using standard optimization algorithms. If D is contained in the orthant $\mathbb{R}^n_+$ (e.g., if the constraints include the inequalities $x_j \geq 0$ $(j = 1,\dots,n)$), it suffices to compute α according to (36), and then

$$D_1 = \{x \in \mathbb{R}^n: x_j \geq 0 \ (j=1,\dots,n), \ \sum_{j=1}^{n} x_j \leq \alpha\} \tag{40}$$

is a simplex containing D with vertices $v^0 = 0$, $v^j = \alpha e^j$ (j=1,..,n), where e^j is the j–th unit vector in $\mathbb{R}^n$.

If the constraints comprise the inequalities $\alpha_j \leq x_j \leq \beta_j$ (j=1,...,n) with given lower and upper bounds α_j and β_j, then obviously D_1 may be the rectangular set

$$D_1 = \{x \in \mathbb{R}^n: \alpha_j \leq x_j \leq \beta_j \ \ (j=1,\dots,n)\} . \tag{41}$$

The vertex set is then

$$V_0 = \{v^1, \dots, v^{2^n}\}$$

with

$$v^k = (\tau_1^k,...,\tau_n^k)^T , \tag{42}$$

where

$$\tau_i^k = \alpha_i \text{ or } \beta_i \quad (i = 1,...,n) . \tag{43}$$

and k ranges over the 2^n possible different combinations in (42), (43).

Note that more sophisticated constructions of D_1 are available in the case of linear constraints (cf. Part B).

4.2. Computing New Vertices and New Extreme Directions

Let the current polyhedral convex set constituting the feasible set of the relaxed problem (Q_k) at iteration k be defined as

$$D_k = \{x \in \mathbb{R}^n: \ell_i(x) \leq 0, i \in K\} , \tag{44}$$

where $K \subset \mathbb{N}$ is a finite set of indices and $k \notin K$.

Let $\ell_k(x) \leq 0$ be the new constraint defining

$$D_{k+1} = \{x \in \mathbb{R}^n: \ell_i(x) \leq 0, i \in K \cup \{k\} \} . \tag{45}$$

Denote by V_k, V_{k+1} the vertex sets of D_k and D_{k+1}, respectively, and let U_k, U_{k+1} denote the sets of extreme directions of D_k and D_{k+1}, respectively.
The following lemma characterizes the new vertices and extreme directions of D_{k+1}.

Lemma II.1. *Let $P \subseteq \mathbb{R}^n$ be an n-dimensional convex polyhedral set with vertex set V and set U of extreme directions. Let $\ell(x) = ax + \beta \leq 0$ ($a \in \mathbb{R}^n$, $\beta \in \mathbb{R}$) be an additional linear constraint, and let V', U' be the vertex set and set of extreme directions of*

$$P' = P \cap \{x: \ell(x) \leq 0\}.$$

Then we have

a) $w \in V' \setminus V$ *if and only if* w *is the intersection of the hyperplane* $\{x: \ell(x) = 0\}$ *with an edge* $[v^-, v^+]$ *of* P *satisfying* $\ell(v^-) < 0$, $\ell(v^+) > 0$ *or with an unbounded edge emanating from a vertex* $v \in V$ *in a direction* $u \in U$ *satisfying either* $\ell(v) < 0$ *and* $au > 0$ *or* $\ell(v) > 0$ *and* $au < 0$.

b) $u \in U' \setminus U$ *if and only if* u *satisfies* $au = 0$ *and is of the form* $u = \lambda u^- + \mu u^+$ *with* $\lambda, \mu > 0$, $au^- < 0$, $au^+ > 0$, *which defines a two-dimensional face of the recession cone of* P.

Proof. Suppose that $P = \{x \in \mathbb{R}^n: \ell_i(x) \leq 0, i \in K\}$, where $\ell_i(x) = a^i x + \beta_i$ $(i \in K)$.

a): The "if" part of the assertion is obvious.

Now let $w \in V' \setminus V$. Since $w \in V'$, among the linear constraints defining P' there are n linearly independent constraints which are active at w. One of these constraints must be $\ell(x) \leq 0$, i.e., we have $\ell(w) = 0$, since otherwise $w \in V$.

Let $\ell_i(x) \leq 0$, $i \in J$, $|J| = n - 1$, be the $n - 1$ constraints that remain active at w if we delete $\ell(x) \leq 0$. Then

$$F := \{x \in P: \ell_i(x) = 0, i \in J\}$$

is a face of P containing w. We have $\dim F = 1$ since $\ell_i(x) = 0$, $i \in J$, are linearly independent constraints and $|J| = n-1$. It follows that F is an edge of P.

If F is bounded, then we have $F = [v^-, v^+]$, $v^- \in V$, $v^+ \in V$ and $w \in [v^-, v^+]$ but $w \neq v^-$, $w \neq v^+$. This implies $\ell(v^-) \neq 0$, $\ell(v^+) \neq 0$. Since $\ell(w) = 0$, we must have $\ell(v^-) \cdot \ell(v^+) < 0$.

If F is an unbounded one-dimensional face (unbounded edge) of P, then $F = \{v + \alpha u: \alpha \geq 0\}$, $v \in V$, $u \in U$ and $w \neq v$, i.e., $w = v + \alpha_0 u$, $\alpha_0 > 0$. Since

$\ell(w) = 0$, this implies that $\ell(v) \neq 0$, $au \neq 0$ and $\ell(v) \cdot au < 0$.

b) The "if" part of the assertion is again obvious.

Now let C and C' denote the recession cones of P and P', respectively.

We have

$$C = \text{cone } U = \{y \in \mathbb{R}^n : a^i y \leq 0 \ (i \in K)\},$$

$$C' = \text{cone } U' = C \cap \{y \in \mathbb{R}^n : ay \leq 0\}.$$

Let $u \in U' \setminus U$. Then, as in the proof of part a) we conclude that among the constraints defining C' there are (n–1) linearly independent constraints active at u. Since $u \notin U$, one of these active constraints has to be $ax \leq 0$, i.e., we have $au = 0$. Let J denote the index set of the (n–2) linearly independent constraints that remain if we delete $ax \leq 0$, $|J| = n-2$. Then

$$G := \{y \in C : a^i y = 0, \ i \in J\}$$

is a smallest face of C containing the ray $\{\alpha u : \alpha \geq 0\}$. Certainly, $G \neq \{\alpha u : \alpha \geq 0\}$, since otherwise, u would be an extreme direction of P, i.e., $u \in U$. Therefore, $\dim G = 2$, and G is a two–dimensional cone generated by two extreme directions $u^-, u^+ \in U$, $u^- \neq u$, $u^+ \neq u$.

Thus we have

$$u = \lambda u^- + \mu u^+ \text{ with } \lambda, \mu > 0.$$

This implies $au^- \neq 0$, $au^+ \neq 0$ and $(au^-)(au^+) < 0$, since $0 = au = \lambda au^- + \mu au^+$. ∎

Note that part a) of Lemma II.1 requires that P possesses edges and part b) requires the existence of at least one two–dimensional face of the recession cone of P. It is easy to see that otherwise we cannot have new vertices or new extreme directions, respectively.

A straightforward application of this result is the following procedure for calculating V_{k+1} and U_{k+1} from V_k and U_k (cf. Thieu, Tam and Ban (1983), Thieu (1984)).

Let $D_k = \{x \in \mathbb{R}^n : \ell_i(x) = a^i x + \beta_i \le 0 \ (i \in K)\}$, where K is a finite set of indices satisfying $k \notin K$.

Method I:

Let $\ell_k(x) = a^k x + \beta_k$, and define

$$V_k^+ := \{v \in V_k : \ell_k(v) > 0\},\ V_k^- := \{v \in V_k : \ell_k(v) < 0\} , \tag{46}$$

$$U_k^+ := \{u \in U_k : a^k u > 0\},\ U_k^- := \{u \in U_k : a^k u < 0\} . \tag{47}$$

α) Finding the vertices of $V_{k+1} \setminus V_k$ that are points where $\{x \in \mathbb{R}^n : \ell_k(x) = 0\}$ meets a bounded edge of D_k:

For any pair $(v^-, v^+) \in V_k^- \times V_k^+$ let $w = \alpha v^- + (1-\alpha)v^+$, where

$$\alpha = \ell_k(v^+) \,/\, (\ell_k(v^+) - \ell_k(v^-)) \quad (\text{i.e., } \ell_k(w) = 0) . \tag{48}$$

Compute $I(w) = \{i \in K : \ell_i(w) = 0\}$. (49)

If the rank of the matrix A(w) having the rows a^i, where $\ell_i(x) = a^i x + \beta_i$, $a^i \in \mathbb{R}^n$, $\beta_i \in \mathbb{R}$, $i \in I(w)$, is less than n–1, then w cannot be in $V_{k+1} \setminus V_k$. Otherwise, w is in $V_{k+1} \setminus V_k$.

If V_k is bounded, then $V_{k+1} \setminus V_k$ is determined in this way.

Note that, by the proof of Lemma II.1, we have

$$I(w) = \{i \in K : \ell_i(v^-) = \ell_i(v^+) = 0\} . \tag{50}$$

Since the calculation of the rank of A(w) may be time–consuming, one may simply calculate I(w) via (50) and consider all points w defined by (48), (49) which satisfy $|I(w)| = n-1$. Then, in general we obtain a set V'_{k+1} larger than V_{k+1}. But since $V_{k+1} \subseteq V'_{k+1} \subset D_{k+1}$, and since we assume that $\operatorname{argmin} f(D_{k+1}) = \operatorname{argmin} f(V_{k+1})$, we have $\operatorname{argmin} f(D_{k+1}) = \operatorname{argmin} f(V'_{k+1})$. The computational effort for the whole procedure may be less if we use V'_{k+1} instead of V_{k+1}.

β) Finding the vertices of $V_{k+1} \setminus V_k$ that are points where $\{x \in \mathbb{R}^n: \ell_k(x) = 0\}$ meets an unbounded edge of D_k:

For any pair $(u, v) \in \{U_k^- \times V_k^+\} \cup \{U_k^+ \times V_k^-\}$, determine $w = v + \alpha u$, where $\alpha = -\ell_k(v) / a^k u$ (i.e., $\ell_k(w) = 0$). Compute $I(w) = \{i: \ell_i(w) = 0, i \in K\}$ and as in the bounded case decide from the rank of $A(w)$ whether $w \in V_{k+1} \setminus V_k$. Note that in this case we have $I(w) = \{i: \ell_i(v) = 0, a^i u = 0, i \in K\}$.

γ) Finding the new extreme directions:

For any pair $(u^-,u^+) \in U_k^- \times U_k^+$ determine $u = (a^k u^+)u^- - (a^k u^-)u^+$.

It is easily seen that $u \in$ cone U_k and $a^k u = 0$. Let $J(u) = \{j: a^j u^- = a^j u^+ = 0, j \in K\}$.

If the rank of the system of equations

$$a^j x = 0 \quad j \in J(u)$$

is less than $n - 2$, then u cannot be an extreme direction of D_{k+1}, otherwise $u \in U_{k+1} \setminus U_k$.

Method II.

In the most interesting case in practice when D and D_k are bounded, the vertices of $V_{k+1} \setminus V_k$ may be calculated by linear programming techniques (cf., e.g., Falk and Hoffman (1976) and Hoffman (1981), Horst, Thoai and de Vries (1988)).

We say that a vertex v of a polytope P is a neighbour in P of a vertex w of P if $[v, w]$ is an edge of P.

Let $\ell_k(x) \leq 0$ be the new constraint and consider the polytope

$$\tilde{D}_{k+1} := D_k \cap \{x \in \mathbb{R}^n: \ell_k(x) \leq 0\}.$$

Then, from Lemma II.1, it follows that $w \in V_{k+1} \setminus V_k$ if and only if w is a vertex of

$\tilde{D}_{k+1}$ satisfying $w \notin V_k$, $\ell_k(w) = 0$ and having a neighbour v in $\tilde{D}_{k+1}$, $v \in V_k^-$.

Falk and Hoffman (1976) represent each vertex v in V_k^- by the **entire** system of inequalities defining $\tilde{D}_{k+1}$. This system must be transformed by a suitable pivoting such that each $v \in V_k^-$ is represented in the form

$$s + Tt = b,$$

where s and t are the vectors of basic and nonbasic variables, respectively, corresponding to v. By performing dual pivoting on all current nonbasic variables in the row corresponding to $\ell_k(x) \geq 0$, one obtains the neighbour vertices of v in the new cutting hyperplane or detects that such a neighbour vertex does not exist. The case of degenerate vertices is not considered in Falk and Hoffman (1976) and Hoffman (1981).

The proposal of Horst, Thoai and de Vries (1988) is based on the following consideration.

For each $v \in V_k^-$, denote by E(v) the set of halflines (directions) emanating from v, each of which contains an edge of D_k, and denote by I(v) the index set of all constraints that are active (binding) at v. By Lemma II.1, the set of new vertices coincides with those intersection points of the halflines $e \in E(v)$ with the hyperplane $H_k := \{x: \ell_k(x) = a^k x + \beta_k = 0\}$ that belong to D_k.

Suppose first that v is nondegenerate. Then I(v) contains exactly n linearly independent constraints, and the set of inequalities

$$a^{i_k} x + \beta_{i_k} \leq 0,\ i_k \in I(v) \quad (k = 1,...,n)$$

defines a polyhedral cone with vertex at v and is generated by the halflines $e \in E(v)$.

To carry out the calculations, in a basic form of the procedure, introduce slack variables $y_1,...,y_{n+1}$ and construct a simplex tableau

x_1	x_2	...	x_n	y_1	y_2	...	y_n	y_{n+1}	RHS
$a_1^{i_1}$	$a_2^{i_2}$	...	$a_n^{i_1}$	1	0	...	0	0	$-\beta_{i_1}$
$\vdots$				$\vdots$	$\vdots$		$\vdots$	$\vdots$	$\vdots$
$a_1^{i_n}$	$a_2^{i_n}$	...	$a_n^{i_n}$	0	0	...	1	0	$-\beta_{i_n}$
α_1	α_2	...	α_n	0	0	...	0	1	$-\beta$

where the last row corresponds to the inequality $\ell_k(x) \le 0$.

Perform pivot operations in such a way that all variables x_i $(i=1,...,n)$ become basic variables. In the last tableau on hand after these operations, let s_j $(j=1,...,2n+1)$ be the first $2n+1$ elements of the last row (i.e., the elements on the left in the row corresponding to the cut $\ell_k(x) \le 0$). For each j satisfying $s_j \neq 0$, a point $w \in H$ is obtained by a pivot operation with pivot element s_j. If we have $w \in D_k$, i.e., $a^i w + \beta_i \le 0$ $(i \in K)$, then w is a new vertex of D_{k+1}.

In the case of a degenerate vertex $v \in V_k^-$, the operation presented above must be considered for each system of linearly independent constraints. This can be done in a similar way by considering the system of all equations binding at v. This system then admits several basic forms (simplex tableaus) corresponding to v, for each of which the above calculation can be carried out. The transition from one of these basic forms to another can be done by pivoting with pivot rows corresponding to a basic variable with value 0. For an efficient bookkeeping that prevents a new vertex w from being determined twice, one can apply of the devices used with algorithms for finding all vertices of a given polytope (e.g., Matheis and Rubin (1980); Dyer and Proll (1977 and 1983); Horst, Thoai and de Vries (1988)). A comprehensive treatment of the determination of the neighbouring vertices of a given degenerate vertex in polytopes is given in Kruse (1986) (cf. also Gal et al. (1988)).

In the procedure above, the new vertices are calculated from V_k^-. Obviously, instead of V_k^- one can likewise use V_k^+. Since the number of elements of these two sets may differ considerably, one should always use the set with the smaller number of vertices.

We give some remarks on a comparison of these methods.

Note first that the calculation of the new vertices w from one of the known sets V_k^- or V_k^+ is justified by the observation that in almost all of the examples where we applied a cutting plane procedure to solve concave minimization problems, the number of elements in V_k^- (or in V_k^+, respectively) was considerably smaller than the number of new vertices created by the cut (for a related conjecture see e.g., Matheis and Rubin (1980) and references given there).

Table II.1 below shows some related results of 20 randomly generated examples.

Problem No.	n	m	V_k	$sign$	V_1	V_2	V_{k+1}
1	5	8	12	V^+	3	9	18
2	—	14	18	V^-	6	12	18
3	—	19	18	V^+	6	12	24
4	—	23	24	V^+	6	12	30
5	10	13	27	V^+	1	10	36
6	—	18	122	V^-	28	84	112
7	—	22	112	V^+	52	120	180
8	—	25	360	V^+	120	304	544
9	—	28	680	V^-	304	336	640
10	—	35	688	V^+	192	360	856
11	20	23	57	V^-	18	54	72
12	—	25	192	V^+	4	168	256
13	—	26	256	V^+	48	624	832
14	—	29	1008	V^+	432	1152	1728
15	—	31	2376	V^-	780	1728	2508
16	—	32	2508	V^+	598	1474	3884
17	50	51	51	V^+	1	5	100
18	—	54	675	V^-	48	1248	1296
19	—	56	2484	V^+	108	2376	4752
20	—	58	9113	V^+	351	8718	16580

Table II.1: number of vertices in a cutting procedure

In Table II.1, we have used the following headings:

n: dimension of D_k, m: number of constraints that defines D_k;

$|V_k|$ and $|V_{k+1}|$: number of vertices of D_k and D_{k+1}, respectively;

sign: indicates which set of vertices is taken for calculation the new vertices $(V^+ = V_k^+, V^- = V_k^-)$;

V_1: number of elements of the set indicated in sign;

V_2: number of newly generated vertices.

For a comparison of the procedure of Horst, Thoai and de Vries (HTV) with the method of Falk–Hoffman, note that in an outer approximation algorithm using cutting planes one usually starts with a simplex or an n–rectangle defined by $(n + 1)$ or 2n constraints, respectively. Then in each step an additional constraint is added. Therefore, even when applying certain constraint dropping strategies, the number of constraints defining the polytope $D_k = P$ is always greater than n, and it increases from step to step in the master cutting plane algorithm. In typical examples one usually has to expect that the number of constraints defining D_k will be considerably larger than the dimension n of the variable space.

The Falk–Hoffman approach has to handle all these constraints for each vertex $v^+ \in V_k^+$, whereas the number of rows in the tableau of the HTV–procedure never exceeds $n + 1$. Moreover, the case of degenerate vertices $v^+ \in V_k^+$ is easier to handle in the HTV–approaches than in the method of Falk–Hoffman.

Thieu–Tam–Ban's method, because of its simplicity, operates well for very small problems. However,since, in this approach one must investigate the intersection of all line segments $[v^-,v^+]$, $v^- \in V_k^-$, $v^+ \in V_k^+$ with the hyperplane H_k, the computational effort is prohibitive even for problems of medium size. Note that, for $|V_k^-| = 200$, $|V_k^+| = 100$, we already have 2.10^4 line segments $[v^-,v^+]$, $v^- \in V_k^-$, $v^+ \in V_k^+$.

A numerical comparison is given in Horst, Thoai and de Vries (1988). Several hundred randomly generated test problems were run on an IBM–PC/AT using the original FORTRAN code of Thieu–Tam–Ban for their relaxed version and our

FORTRAN code for the procedure presented in the preceding section. In the relaxed version of Thieu–Tam–Ban, the decision on $w \in V_{k+1}$ or $w \notin V_{k+1}$ is based only on the number of constraints that are binding at w and omits the relatively expensive determination of their rank (relaxed procedure). The results show that the Thieu–Tam–Ban algorithm should not be used if $|V|$ exceeds 100 (cf. Table II.2 below).

Problem No.	n	m	$\mid V_k \mid$	$\mid V_{k+1} \mid$	$T1$ $[Min]$	$T2$ $[Min]$
1	3	7	9	10	0.008	0.004
2	—	16	10	10	0.018	0.010
3	5	28	34	42	0.390	0.300
4	10	14	27	32	0.420	0.170
5	—	17	48	48	0.720	0.670
6	—	12	32	56	0.120	0.110
7	—	18	56	80	0.580	1.040
8	—	19	80	80	1.12	2.18
9	—	16	160	256	3.16	20.610
10	—	18	288	256	3.55	27.310
11	—	27	400	520	2.65	53.650
12	—	30	672	736	1.91	81.850
13	20	23	57	72	0.96	1.980
14	—	24	72	220	1.21	8.250
15	—	25	220	388	7.54	81.640
16	—	26	388	1233	7.71	303.360

Table II.2: A comparison of the algorithm of the HTV method with the Thieu–Tam–Ban procedure

In Table II.2, we have used the following headings:

n: dimension of D_k ,

m: number of constraints defining D_k ,

$|V_k|$ and $|V_{k+1}|$: number of vertices of D_k and D_{k+1}, respectively,

T1: CPU – time for the method of Horst, Thoai and de Vries,

T2: CPU – time for Thieu–Tam–Ban's relaxed procedure.

Another way of treating the unbounded case is as follows (cf. Tuy (1983)). Assume $D_1 \subset \mathbb{R}^n_+$. Recall first that, under the natural correspondence between points of the hyperplane $\{(1,x): x \in \mathbb{R}^n\}$ in $\mathbb{R}^{n+1}$ and points $x \in \mathbb{R}^n$, a point $x \in D_k$ can be represented by the ray $\rho(x) = \{\alpha(1,x): \alpha \geq 0\}$, and a direction of recession y can be represented by $\bar{\rho}(y) = \{\alpha(0,y): \alpha \geq 0\}$. In $\mathbb{R}^{n+1}_+$ define the n–simplex

$$S = \{(t,\tilde{x}) \in \mathbb{R}^{n+1}_+ : \sum_{i=1}^{n} \tilde{x}_i + t = 1\} .$$

Associate to each point $x \in D_k$ the point s(x), where the ray $\rho(x)$ meets S, and to each direction y of recession y of D_k associate the point s(y) where the ray $\bar{\rho}(y)$ meets S. In this way there is a one–to–one correspondence between the points and directions of recession of D_k and the points of the corresponding polytope $s(D_k) \subset S$.

A vertex $(t,\tilde{x})$ of $s(D_k)$ corresponds to an (ordinary) vertex $x = \frac{\tilde{x}}{t}$ of D_k if $t > 0$, or to a vertex at infinity (an extreme direction $\tilde{x}$) of D_k if $t = 0$. Furthermore, since the corresponding point of D is simply $x = \frac{\tilde{x}}{t}$ for $(t,\tilde{x}) \in s(D_k)$, $t > 0$, we conclude that if

$$D_k := \{x \in D_1 : \ell_i(x) = a^i x + \beta_i \leq 0, (i = 1,..., k-1)\} ,$$

then

$$s(D_k) = \{(t, \tilde{x}) \in s(D_1): \tilde{\ell}_i(t, \tilde{x}) := a^i \tilde{x} + \beta_i t \leq 0, \ (i=1,...,k-1)\} .$$

Denote the vertex set of $s(D_k)$ by W_k. Then, it follows from these observations, that W_{k+1} can be derived from W_k by one of the methods presented above. Therefore, the set of vertices and extreme directions of D_{k+1} can be computed whenever the vertices and extreme directions of D_k and the new constraint are known.

4.3. Identifying Redundant Constraints

Redundancy of constraints in linear programming has been discussed by many authors (cf., e.g., Matheis and Rubin (1980) and references given there). However, in contrast to the common linear programming setting, in our context all vertices of a given polytope are known; hence we present a characterization of redundant constraints by means of the vertex set of a given polytope.

Let $P := D_k = \{x \in \mathbb{R}^n: \ell_i(x) \leq 0, i \in K\}$, $k \notin K$, and $P' := D_{k+1} = P \cap \{x \in \mathbb{R}^n: \ell_k(x) \leq 0\}$ be polytopes defined as above. A constraint $\ell_j(x) \leq 0$, $j \in K$ is called redundant for P if its removal does not change P.

Definition II.2. *A constraint $\ell_j(x) \leq 0$, $j \in K$, is redundant for a polytope $P = \{x \in \mathbb{R}^n: \ell_i(x) \leq 0, i \in K\}$ if there is an $i_0 \in K \setminus \{j\}$ such that we have*

$$F_j := P_{i_0} \cap \{x \in \mathbb{R}^n: \ell_j(x) = 0\} \subset \{x \in \mathbb{R}^n: \ell_{i_0}(x) \geq 0\} , \tag{51}$$

where

$$P_{i_0} := \{x \in \mathbb{R}^n: \ell_i(x) \leq 0, i \in K \setminus \{i_0\}\} . \tag{52}$$

We also say that the constraint $\ell_j(x) \leq 0$ is redundant for P relative to ℓ_{i_0}

In a cutting plane approach as described above, the first polytope can usually be assumed to have no redundant constraints. Let P' be a polytope generated from a given polytope P by a cut $\ell_k(x) \leq 0$. Then, by the definition of a cut, $\ell_k(x) \leq 0$ cannot be redundant for P'. If we assume that redundancy is checked and redundant constraints are eliminated in each iteration of the cutting–plane algorithm (i.e., P has no redundant constraints), then P' can only possess redundant constraints relative to the cut ℓ_k. These redundant constraints can be eliminated by the following assertion. Again denote by V(P) the vertex set of a polytope P.

Theorem II.6. *Assume that* $V^-(P) := \{v \in V(P): \ell_k(v) < 0\} \neq \emptyset$. *Then a constraint* $\ell_j(x) \le 0$, $j \in K$, *is redundant for P' relative to* ℓ_k *if and only if we have*

$$\ell_j(v) < 0 \ \forall v \in V^-(P) . \tag{53}$$

Proof. Since F_j is a polytope, we have $F_j = \text{conv } V(F_j)$, where conv $V(F_j)$ denotes the convex hull of the vertex set $V(F_j)$ of F_j. Then, by Definition II.2, $\ell_j(x) \le 0$ is redundant for P' relative to ℓ_k if and only if conv $V(F_j) \subset \{x: \ell_k(x) \ge 0\}$. Obviously, because of the convexity of the halfspace $\{x: \ell_k(x) \ge 0\}$, this inclusion is equivalent to the condition $V(F_j) \cap V^-(P) = \emptyset$, which in turn holds if and only (53) is satisfied.

■

If one does not want to investigate redundancy at each step of the cutting plane method, then one can choose to identify only the so-called strictly redundant constraints. As we shall see, this can be done at an arbitrarily chosen iteration.

Definition II.3. *A constraint* $\ell_j(x) \le 0$, $j \in K$, *is strictly redundant for a polytope* $P = \{x \in \mathbb{R}^n: \ell_i(x) \le 0,\ i \in K\}$ *if there is an* $i_0 \in K \setminus \{j\}$ *such that we have*

$$F_j := P_{i_0} \cap \{x \in \mathbb{R}^n: \ell_j(x) = 0\} \subset \{x \in \mathbb{R}^n: \ell_{i_0}(x) > 0\} . \tag{54}$$

where P_{i_0} *is defined by (52).*

We also say that the constraint $\ell_j(x) \le 0$ *is strictly redundant for P relative to* ℓ_{i_0}.

The following assertion shows that, whenever a constraint is strictly redundant for P, it is strictly redundant for P' relative to the new constraint $\ell_k(x) \le 0$.

Theorem II.7. a) *A constraint* $\ell_j(x) \le 0$, $j \in K$, *is strictly redundant for P relative to* $\ell_k(x) \le 0$ *if and only if we have*

$$\ell_j(v) < 0 \ \forall v \in V(P) \setminus V^+(P) . \tag{55}$$

b) Every constraint that is strictly redundant for P is strictly redundant for P' relative to $\ell_k(x) \le 0$.

Proof. a) The proof of (55) is similar to the proof of (53) (replace $V^-(P)$ by $V(P) \setminus V^+(P)$).

b) Let $\ell_j(x) \le 0$ be strictly redundant for P and let ℓ_{i_0} satisfy $F_j \subset \{x: \ell_{i_0}(x) > 0\}$. Then we have $P_{i_0} \cap \{x: \ell_{i_0}(x) \le 0\} = P \subset \{x: \ell_j(x) < 0\}$. It follows that $\ell_j(v) < 0$ $\forall v \in V(P)$. Hence, we see that $\ell_j(v) < 0 \ \ \forall v \in V(P) \setminus V^+(P)$ holds, and, by part a) of Theorem II.7, $\ell_j(x) \le 0$ is strictly redundant for P' relative to $\ell_k(x) \le 0$. ■

Remark II.1. Let the feasible set $D := \{x \in \mathbb{R}^n : g(x) \le 0\}$ with strictly convex $g: \mathbb{R}^n \to \mathbb{R}$ and assume that int $D \neq \emptyset$. If all of the facets of the initial polytope D_1 and all hyperplanes generated by an outer approximation method support D, then it is easily seen that redundant constraints cannot occur.

CHAPTER III

CONCAVITY CUTS

In Chapter II we discussed the general concept of a cut and the use of cuts in the basic technique of outer approximation. There, we were mainly concerned with using cuts in a "**conjunctive**" manner: typically, cuts were generated in such a way that no feasible point of the problem is excluded and the **intersection** of all the cuts contains the whole feasible region. This technique is most successful when the feasible region is a convex set, so that supporting hyperplanes can easily be constructed to separate this convex set from a point lying outside.

We now discuss cuts that are used in a "**disjunctive**" manner: each cut taken separately may exclude certain points of the current region of interest, but the **union** of all cuts constructed at a given stage covers this region entirely. This technique is often used for approximating a certain nonconvex set and separating it from a point lying outside.

1. CONCEPT OF A VALID CUT

Consider the problem of minimizing a continuous function $f: \mathbb{R}^n \longrightarrow \mathbb{R}$ over a polyhedron D of full dimension in $\mathbb{R}^n$. Suppose that γ is the smallest value of f(x) taken over all feasible points where an evaluation has been made up to a given stage in the process of solving the problem. Then we may restrict our further search to the subset of D consisting only of points x satisfying $f(x) < \gamma$. In other words we may discard all points $x \in D$ in the set

$$G = \{x \in \mathbb{R}^n: f(x) \ge \gamma\} \tag{1}$$

from further consideration. Of course we could do this by adjoining the constraint $f(x) < \gamma$ to the feasible set D. However, this would lead us to consider the new feasible set

$$D^*(\gamma) = D \cap \{x: f(x) < \gamma\} ,$$

which except in certain simple cases is no longer polyhedral, and so would be difficult to handle. Therefore, in order to preserve the polyhedral structure of the constraint set, we have to consider instead an affine function $\ell(x)$ such that the constraint $\ell(x) \ge 0$ does not exclude any feasible point x with $f(x) < \gamma$, i.e., such that

$$D^*(\gamma) \subset \{x \in D: \ell(x) \ge 0\}\ ^{1)} . \tag{2}$$

Definition III.1 *A linear inequality $\ell(x) \ge 0$ satisfying (2) is called a γ–**valid cut** for (f,D).*

Typically, we have a point $z \in D$ such that $f(z) > \gamma$ and we would like to have the cut eliminate it, i e., we require that

$$\ell(z) < 0 .$$

When f(x) is a convex function, so that the set $D^*(\gamma)$ is convex, any hyperplane $\ell(x) = 0$ strictly separating z from $D^*(\gamma)$ is a γ–valid cut that excludes z. For example, if $p \in \partial f(z)$, then $f(z) + p(x-z) \le f(x)$, and the inequality

$$f(z) + p(x-z) \le \gamma \quad (\text{i.e., } \ell(x) = \gamma - f(z) - p(x-z) \ge 0)$$

---

1) In this chapter it will be more convenient to denote a cut by $\ell(x) \ge 0$ rather than by $\ell($ in Chapter II.

defines a γ–valid cut that excludes z (see Chapter II).

A more complicated situation typical of global optimization occurs when f(x) is a **concave function**. Then the set $D^*(\gamma)$ is nonconvex, and the subgradient concept can no longer be used. However, since the complement of $D^*(\gamma)$ (relative to D), i.e., the set $D \cap G$, is convex, we may try to exploit this convexity in order to construct a hyperplane strictly separating z from $D^*(\gamma)$.

Assume that a vertex x^0 of D is available which satisfies $x^0 \in \text{int } G$ (i.e., $f(x^0) > \gamma$) and is nondegenerate, i.e., there are exactly n edges of D emanating from x^0. Let $u^1,u^2,...,u^n$ denote the directions of these edges. If the i–th edge is a line segment joining x^0 to an adjacent vertex y^i, then one can take $u^i = y^i - x^0$. However, note that certain edges of D may be unbounded (the corresponding vectors u^i are extreme directions of D). Since D has full dimension (dim D = n) the vectors $u^1,u^2,...,u^n$ are linearly independent and the cone vertexed at x^0 and generated by the halflines emanating from x^0 in the directions $u^1,u^2,...,u^n$ is an n–dimensional cone K with exactly n edges such that $D \subset K$ (in fact K is the smallest cone with vertex at x^0 that contains D).

Now for each i=1,...,n let us take a point $z^i \neq x^0$ on the i–th edge of K such that $f(z^i) \geq \gamma$ (see Fig. III.1). These points exist by virtue of the assumption $x^0 \in \text{int } G$. The n×n–matrix

$$Q = (z^1-x^0, z^2-x^0,...,z^n-x^0)$$

with z^i-x^0 as its i–th column, is nonsingular because its columns are linearly independent. Let $e = (1,1,...,1)$ be the row vector of n ones.

Theorem III.1. *If $z^i = x^0 + \theta_i u^i$ with $\theta_i > 0$ and $f(z^i) \geq \gamma$, then the linear inequality*

$$eQ^{-1}(x - x^0) \geq 1 \tag{3}$$

defines a γ–valid cut for (f,D).

Proof. Since $z^1-x^0, z^2-x^0, \ldots, z^n-x^0$ are linearly independent, there is a unique hyperplane H passing through the n points $z^1, z^2, \ldots, z^n$. The equation of this hyperplane can be written in the form $\pi(x-x^0) = 1$, where π is some n-vector. Since this equation is satisfied by each z^i, we can wirte

$$\pi(z^i-x^0) = 1 \quad (i=1,2,\ldots,n) , \tag{4}$$

i.e.,

$$\pi Q = e ,$$

hence $\pi = eQ^{-1}$. Thus, the equation of the hyperplane H through $z^1, z^2, \ldots, z^n$ is $eQ^{-1}(x-x^0) = 1$.

Now let $S = [x^0, z^1, \ldots, z^n] := \text{conv}\,\{x^0, z^1, \ldots, z^n\}$ denote the simplex spanned by $x^0, z^1, \ldots, z^n$. Clearly $S = K \cap \{x: eQ^{-1}(x-x^0) \leq 1\}$. Since $x^0, z^i \in G$ $(i=1,2,\ldots,n)$ and the set G is convex, it follows that $S \subset G$, in other words, $f(x) \geq \gamma$ $\forall x \in S$. Then, if $x \in D$ satisfies $f(x) < \gamma$, we must have $x \notin S$; and since D is contained in the cone K, this implies $eQ^{-1}(x-x^0) > 1$. Therefore,

$$\{x \in D: f(x) \leq \gamma\} \subset \{x \in D: eQ^{-1}(x-x^0) \geq 1\} ,$$

proving that the cut (3) is γ-valid for (f,D). ■

Let $U = (u^1, u^2, \ldots, u^n)$ be the nonsingular $n \times n$-matrix with columns $u^1, u^2, \ldots, u^n$. Then $x \in K$ if and only if $x = x^0 + \sum_{i=1}^{n} t_i u^i$, with $t = (t_1, \ldots, t_n)^T \geq 0$, i.e.,

$$K = \{x: x = x^0 + Ut,\ t \geq 0\} . \tag{5}$$

Clearly, $t = U^{-1}(x-x^0)$. Using (5), from Theorem III.1 we can derive the following form of a valid cut which in computations is more often used than the form (3).

Corollary III.1. *For any $\theta = (\theta_1, \theta_2, \ldots, \theta_n) > 0$ such that $f(x^0 + \theta_i u^i) \geq \gamma$ the linear inequality*

$$\sum_{i=1}^{n} \frac{t_i}{\theta_i} \geq 1 , \tag{6}$$

where $t = U^{-1}(x-x^0)$, *defines a* γ*–valid cut for (f,D).*

Proof. If $z^i = x^0 + \theta_i u^i$ (i=1,2,...,n) and we denote $Q = (z^1-x^0, z^2-x^0, \ldots, z^n-x^0)$, then $Q = U \text{ diag } (\theta_1, \ldots, \theta_n)$. Hence,

$$Q^{-1} = \text{diag } (\frac{1}{\theta_1}, \ldots, \frac{1}{\theta_n}) \, U^{-1} ,$$

$$eQ^{-1}(x - x^0) = (\frac{1}{\theta_1}, \ldots, \frac{1}{\theta_n}) \, U^{-1}(x - x^0) = (\frac{1}{\theta_1}, \ldots, \frac{1}{\theta_n}) \, (t_1, \ldots, t_n)$$

$$= \sum_{i=1}^{n} \frac{t_i}{\theta_i} ,$$

and so (3) is equivalent to (6). ■

Definition III.2. *A* γ*–valid cut which excludes a larger portion of the set* $D \cap G = \{x \in D: f(x) \geq \gamma\}$ *than another one, is said to* ***dominate*** *the latter, or to be* ***stronger (deeper)*** *than the latter.*

It is straightforward to see that, if $\theta_i \geq \theta_i'$ ($\forall$i), then the cut $\sum_{i=1}^{n} \frac{t_i}{\theta_i'} \geq 1$ cannot dominate the cut (6). Therefore, the strongest γ–valid cut of type (6) corresponds to $\theta_i = \alpha_i$ (i=1,2,...,n), where

$$\alpha_i = \sup \{\theta \in \mathbb{R}_+ : x^0 + \theta u^i \in G\} = \sup \{\theta \in \mathbb{R}_+ : f(x^0 + \theta u^i) \geq \gamma\} . \tag{7}$$

Since this cut originated in concave programming, and since the set $D^*(\gamma)$ that it separates from x^0 is "concave", we shall refer to this strongest cut, i.e., to the cut

$$\sum_{i=1}^{n} \frac{t_i}{\alpha_i} \geq 1$$

as a **concavity cut**.

Definition III.3. *A cut of the form (6) with $\theta_i = \alpha_i$ satisfying (7) is called a* ***γ-valid concavity cut for (f,D)), constructed at the point x^0.***

This cut was first introduced by Tuy (1964).

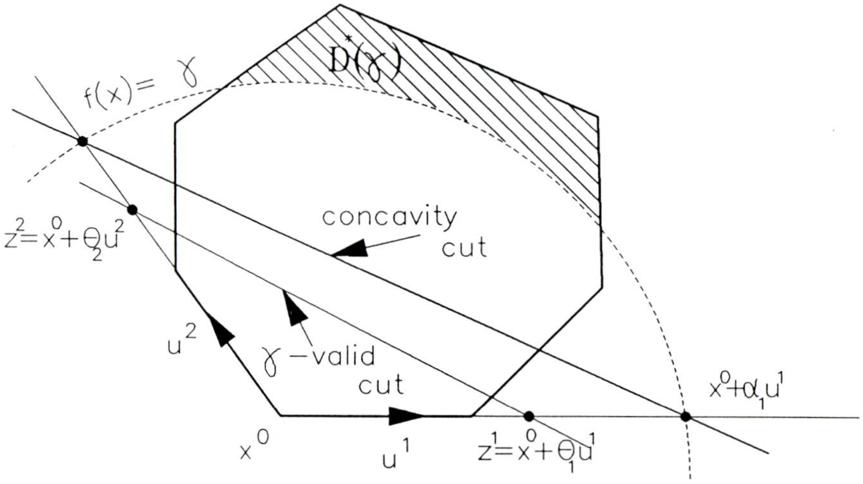

Fig. III.1 A γ-valid cut

Note that some of the values α_i defined by (7) may be equal to $+\infty$. This occurs when u^i is a recession direction of the convex set G. Setting $\frac{1}{\alpha_i} = 0$ when $\alpha_i = +\infty$, we see that if $I = \{i: \alpha_i < +\infty\}$, then the concavity cut is given by

$$\sum_{i \in I} \frac{t_i}{\alpha_i} \geq 1 . \tag{8}$$

Its normal vector is $\pi = (\pi_1, \pi_2, \ldots, \pi_n)$ with $\pi_i = \frac{1}{\alpha_i}$ $(i \in I)$, $\pi_j = 0$ $(j \notin I)$, i.e., in this case the hyperplane of the concavity cut is parallel to each direction u^j with $j \notin I$ (see Fig. III.2). Of course, the cut (8) will exist only if I is nonempty. Otherwise, the cone K will have all its edges contained in G, in which case $K \subset G$

and there is no point $x \in D$ with $f(x) < \gamma$ (in this situation one can say that the cut is "infinite").

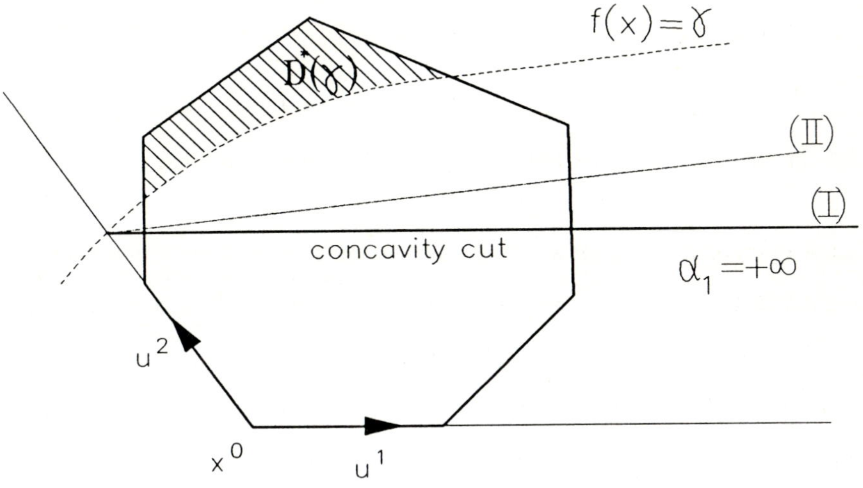

Fig. III.2 The case $\alpha_i = +\infty$

2. VALID CUTS IN THE DEGENERATE CASE

The construction of the cut (3) or (6) relies heavily on the assumption that x^0 is a nondegenerate vertex of D. When this assumption fails to hold, there are more than n edges of D emanating from x^0. Then the smallest cone K vertexed at x^0 and containing D has $s > n$ edges. If, as before, we take a point $z^i \neq x^0$ satisfying $f(x^i) \geq \gamma$ on each edge of K, then there may not exist any hyperplane passing through all these z^i.

Several methods can be used to deal with this case.

The first method, based on linear programming theory, is simply to avoid degeneracy. In fact, it is well–known that by a slight perturbation of the vector b one can always make a given polyhedron

$$Ax \le b$$

nondegenerate (i.e., have only nondegenerate vertices). However, despite its conceptual simplicity, this perturbation method is rarely practical.

The second method, suggested by Balas (1971), is to replace D by a slightly larger polyhedron D' that admits x^0 as a nondegenerate vertex. In fact, since x^0 is a vertex, among the constraints that define D, one can always find n linearly independent constraints that are binding for x^0. Let D' denote the polyhedron obtained from D by omitting all the other binding constraints for x^0. Then x^0 is a nondegenerate vertex of D', and a γ–valid cut for (f,D') can be constructed at x^0. Since $D' \supset D$, this will also be a γ–valid cut for (f,D).

In practical computations, the most convenient way to carry out this method is as follows. Using standard linear programming techniques, write the constraints defining D in the form

$$x_B = x_B^0 + Ux_N \,, \tag{9}$$

$$x = (x_B, x_N) \ge 0 \,, \tag{10}$$

where $x_B = (x_i \,, i \in B)$ is the vector of basic variables corresponding to the basic solution $x^0 = (x_B^0, 0)$, $x_N = (x_j, j \in N)$, $|N| = n$, is the vector of the corresponding nonbasic variables, and U is a matrix given by the corresponding simplex tableau. Setting $x_N = t$, we can represent the polyhedron D by the system in t:

$$x_B^0 + Ut \ge 0 \,, \quad t \ge 0 \,. \tag{11}$$

The vertex x^0 now corresponds to $t^0 = 0 \in \mathbb{R}^n$, while the concave function f(x) becomes a certain concave function $\varphi(t)$ such that $\varphi(0) = f(x^0) > \gamma$ (hence, the convex set $\{x: f(x) > \gamma\}$ becomes the convex set $\{t: \varphi(t) > \gamma\}$ in t–space). The polyhedron D is now contained in the orthant $t \geq 0$ which is a cone vertexed at the origin $t^0 = 0$ and having exactly n edges (though some of these edges may not be edges of D, but rather lie outside D). Therefore, all the constructions discussed in the previous section apply. In particular, if e^i denotes the i–th unit vector in t–space, and if $\varphi(\theta_i e^i) \geq \gamma$, then a γ–valid concavity cut is given by the same inequality (6) of Corollary III.1. Note that when the vertex x^0 is nondegenerate (so that D' = D), the variables t_i (i=1,...,n) and the matrix U in (11) can be identified with the variables t_i (i=1,...,n) and the matrix U in (5).

The third method, proposed by Carvajal–Moreno (1972), is to determine a normal vector π of a γ–valid cut not as a solution of the system of equations (4) (as we did in the nondegenerate case), but, instead, as a basic solution of the system of inequalities

$$\pi u^i \geq \frac{1}{\alpha_i} \quad (i=1,2...,s) , \tag{12}$$

where $u^1,u^2,...,u^s$ are directions of the edges of D emanating from x^0 ($s > n$), and α_i is defined by (7) (with the usual convention $\frac{1}{\alpha_i} = 0$ if $\alpha_i = +\infty$).

Lemma III.1. *The system (12) is consistent and has at least one basic solution.*

Proof. Recall that a basic solution of a linear system such as in (12) is a solution which corresponds to an extreme point (a vertex) of the polyhedron described by this system. Since dim K = n and K has a vertex (i.e., the lineality of K = 0), the polar K_0 of $-K + x^0$ must have full dimension: dim $K_0 = n$, so that K_0 contains an interior point, i.e., a point t satisfying $tu^i > 0$ $\forall i=1,2,...,s$ (cf. Rockafellar (1970), Corollary 14.6.1). Then the vector $\pi = \lambda t^T$, with $\lambda \in \mathbb{R}$, $\lambda > 0$ sufficiently large, will satisfy the system (12). But it is well known that the recession cone of the polyhedron (12) is the cone

$$\pi u^i \geq 0 \quad (i=1,2,\ldots,s)$$

(cf. Rockafellar (1970), page 62), and since this also describes the polar K_0, it follows that K_0 must have lineality 0, i.e., it must be pointed (cf. Rockafellar (1970), Corollary 14.6.1). This is equivalent to saying that the polyhedron (12) has at least one extreme point (cf. Rockafellar (1970), Corollary 18.5.3). Hence, the system (12) has at least one basic solution. ■

Proposition III.1. *Any solution π of (12) provides a γ–valid cut for (f,D) by means of the inequality*

$$\pi(x - x^0) \geq 1 .$$

Proof. Denote $M = \{x \in K: \pi(x - x^0) \leq 1\}$, where, we recall, K is the cone vertexed at x^0 with s edges of direction $u^1,u^2,\ldots,u^s$. Since $D \subset K$, it suffices to show that

$$M \subset G = \{x: f(x) \geq \gamma\} \tag{13}$$

In fact, consider any point $x \in M$. Since $x \in K$, we have

$$x = x^0 + \sum_{i=1}^{s} \lambda_i u^i \ , \ \lambda_i \geq 0 .$$

Furthermore, since $\pi(x - x^0) \leq 1$, it follows that

$$1 \geq \sum_{i=1}^{s} \lambda_i(\pi u^i) \geq \sum_{i=1}^{s} \frac{\lambda_i}{\alpha_i} .$$

Hence, assuming $\alpha_i < +\infty \ \forall i$ and setting $\mu = \sum_{i=1}^{s} \frac{\lambda_i}{\alpha_i}$ we can write

$$x = \sum_{i=1}^{s} \frac{\lambda_i}{\alpha_i}(x^0 + \alpha_i u^i) + (1-\mu)x^0 , \tag{14}$$

which means that x is a convex combination of the points x^0 and $x^0+\alpha_i u^i$ $(i=1,\ldots,s)$. Since $f(x^0) \geq \gamma$ and $f(x^0+\alpha_i u^i) \geq \gamma$, it follows that $f(x) \geq \gamma$, proving (13).

In the case where some of the α_i may be $+\infty$, the argument is similar, but instead of (14) we write

$$x = \sum_{i\in I} \frac{\lambda_i}{\alpha_i}(x^0 + \alpha_i u^i) + (1-\mu)x^0 ,$$

where $I = \{i: \alpha_i < +\infty\}$, $\mu = \sum_{i\in I} \frac{\lambda_i}{\alpha_i}$. ■

If π is a basic solution of (12), then it satisfies n of the inequalities of (12) as equalities, i.e., the corresponding cutting hyperplane passes through n points $z^i = x^0 + \alpha_i u^i$ (with the understanding that if $\alpha_i = +\infty$ for some i, this means that the hyperplane is parallel to u^i).

A simple way to obtain a basic solution of the system (12) is to solve a linear program such as

$$\text{minimize} \sum_{i\in I} \alpha_i \pi u^i \quad \text{s.t.} \quad \pi u^i \geq \frac{1}{\alpha_i} \quad (i=1,\ldots,s) .$$

When $s = n$ (i.e., the vertex x^0 is nondegenerate), the system (12) has a unique basic solution which corresponds exactly to the concavity cut.

3. CONVERGENCE OF CUTTING PROCEDURES

The cuts which are used in the outer approximation method (Chapter II) are **conjunctive** in the following sense: at each step k=1,2,..., of the procedure we construct a cut $L_k = \{x: \ell_k(x) \geq 0\}$ such that the set to be considered in the next step is

$$D_{k+1} = \bigcap_{h\leq k} L_h \supset D ,$$

where D is the feasible set. Since each L_k is a halfspace, it is clear that this method can be applied only if the feasible set D is convex.

By constrast, the valid cuts developed above may be used in **disjunctive** form.

To be specific, let us again consider the problem of minimizing a continuous function f over a polyhedron D in $\mathbb{R}^n$. If γ is the best feasible value of f(x) known up to a given stage, we would like to find a feasible point x with $f(x) < \gamma$ or else to be sure that no such point exists (i.e., γ is the global optimal value). We construct a γ–valid cut $L_0 = \{x: \ell_0(x) \geq 0\} \supset D^*(\gamma)$ and find an element $x^1 \in D \cap L_0$. If we are lucky and $f(x^1) < \gamma$, then we are done. Otherwise, we try to construct a γ–valid cut $L_1 = \{x: \ell_1(x) \geq 0\}$ to exclude x^1. However, sometimes this may not be possible; for example, if the situation is as depicted in Fig. III.3, then there is no way to exclude x^1 by a **single** γ–valid cut. Therefore, in such cases we construct a system of cuts $L_{1,1} = \{x: \ell_{1,1}(x) \geq 0\}, \ldots, L_{1,N_1} = \{x: \ell_{1,N_1}(x) \geq 0\}$ such that

$$L_1 = \bigcup_{j=1}^{N_1} L_{1,j} \supset D^*(\gamma) ,$$

i.e., the union (**disjunction**) of all these cuts together forms a set L_1 which plays the role of a γ–valid cut for (f,D), namely it excludes x^1 without excluding any point of $D^*(\gamma)$.

Thus, a general cutting procedure works as follows. Given a target set Ω which is a subset of some set D (for instance $\Omega = D^*(\gamma)$), we start with a set $L_0 \supset \Omega$. At step $k = 1,2,\ldots$, we find a point $x^k \in D \cap L_{k-1}$. If x^k happens to belong to Ω, then we are done. Otherwise, construct a system of cuts $L_{k,j}$ $(j=1,\ldots,N_k)$ such that

$$L_k = \bigcup_{j=1}^{N_k} L_{k,j} \supset \Omega , \tag{15}$$

$$D \cap L_k \subset D \cap L_{k-1} ; \quad x^k \in (D \cap L_{k-1}) \setminus L_k , \tag{16}$$

and go to the next step.

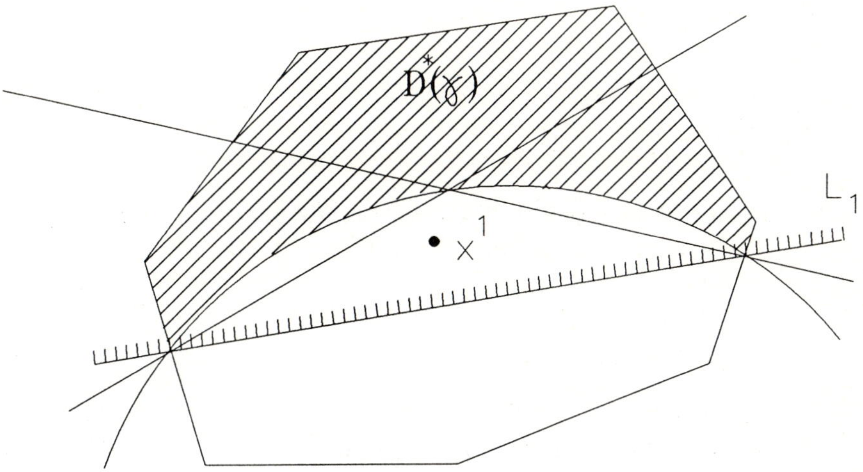

Fig. III.3. Disjunctive cuts

Since the aim of a cutting procedure is eventually to obtain an element of Ω, the procedure will be successful either if $x^k \in \Omega$ at some step k (in which case we say that the procedure is **finite** or **finitely convergent**), or if the sequence $\{x^k\}$ is bounded and any cluster point of it belongs to Ω (in which case we say that the procedure is **convergent**, or **infinitely convergent**).

Note that, by (16), $D \cap L_k \subset D \cap L_h$ for all $h \leq k$, and therefore

$$x^k \in \left[\bigcap_{h<k} L_h\right] \setminus L_k . \tag{17}$$

Many convergence proofs for cutting plane procedures are based on the following, simple, but useful proposition.

Lemma III.2 *(Bounded Convergence Principle). Let L_k, $k = 1,2,...$, be a sequence of arbitrary sets in $\mathbb{R}^n$. If $\{x^k\}$ is a bounded sequence of points satisfying*

$$x^k \in \bigcap_{h<k} L_h , \tag{18}$$

then

$$d(x^k, L_k) \longrightarrow 0 \quad (k \longrightarrow \infty)$$

(where d is the distance function in $\mathbb{R}^n$*). The same conclusion holds, if instead of (18) we have*

$$x^k \in \bigcap_{h>k} L_h . \tag{19}$$

Proof. Suppose that (18) holds, but $d(x^k, L_k)$ does not tend to 0. Then there exists a positive number ε and an infinite sequence $\{k_\nu\} \subset \{1,2,...\}$ satisfying $d(x^{k_\nu}, L_{k_\nu}) \geq \varepsilon$ for all ν. Since $\{x^k\}$ is bounded, we may assume, by passing to subsequences if necessary, that x^{k_ν} converges, so that $\|x^{k_\mu} - x^{k_\nu}\| < \varepsilon$ for all μ,ν sufficiently large. But, by (18), for all $\mu > \nu$ we have $x^{k_\mu} \in L_{k_\nu}$, and hence

$$\|x^{k_\mu} - x^{k_\nu}\| \geq d(x^{k_\nu}, L_{k_\nu}) \geq \varepsilon .$$

This contradiction proves the lemma.

The case (19) is analogous, except that for $\mu > \nu$ we have $x^{k_\nu} \in L_{k_\mu}$, and hence $\|x^{k_\mu} - x^{k_\nu}\| \geq d(x^{k_\mu}, L_{k_\mu}) \geq \varepsilon$. ■

Now consider a cutting procedure in which in (15) $N_k = 1$ (k=1,2,...), so that each L_k consists of a single cut:

$$L_k = \{x\colon \ell_k(x) \geq 0\} .$$

Then

$$\ell_k(x^k) < 0 , \tag{20}$$

while

$$\ell_h(x^k) \geq 0 \quad (h = 0,1,\dots,k-1). \tag{21}$$

Assume that the sequence $\{x^k\}$ is bounded.

Theorem III.2. *Let $H_k = \{x\colon \ell_k(x) = 0\}$. If there exists an $\varepsilon > 0$ such that $d(x^k,H_k) \geq \varepsilon$ for all k, then the procedure is finite.*

Proof. It suffices to apply Lemma III.2 to $\{x^k\}$ and $L_k = \{x\colon \ell_k(x) \geq 0\}$. In view of (21) we have (18). But from (21) $d(x^k,L_k) = d(x^k,H_k)$, and since $d(x^k,H_k) \geq \varepsilon$ for all k, it follows from Lemma III.2 that the procedure must be finite. ■

From this Theorem we can derive the following results (see Bulatov (1975, 1977)).

Corollary III.2. *Suppose that every cut is of the form*

$$\ell_k(x) := \pi^k(x-x^k) - 1 \,.$$

If the sequence $\{\pi^k\}$ is bounded, then the procedure is finite.

Proof. Clearly, $d(x^k,H_k) = \|\pi^k\|^{-1}$. Therefore, if $\|\pi_k\| \leq c$ for some $c \in \mathbb{R}$, $c > 0$, then $d(x^k,H_k) \geq 1/c$, and the conclusion follows from Theorem III.2. ■

Corollary III.3. *Suppose that every cut is of the form (3), i.e.,*

$$\ell_k(x) = eQ_k^{-1}(x-x^k) - 1 \,.$$

If the sequence $\|Q_k^{-1}\|$, $k = 1,2,\dots$, is bounded, then the procedure is finite.

Proof. This follows from Corollary III.2 for $\pi^k = eQ_k^{-1}$. ■

Corollary III.4. *Suppose that every cut is of the form (6), i.e.,*

$$\ell_k(x) = \sum_{i=1}^{n} t_i/\theta_{ik} - 1 ,$$

where $t \in \mathbb{R}^n$ is related to $x \in \mathbb{R}^n$ by

$$x = x^k + U_k t , \; t \geq 0 .$$

If the sequence $\|U_k^{-1}\|$ is bounded and there exists $\theta > 0$ such that $\theta_{ik} \geq \theta$ for all $i = 1,2,\ldots,n$ and all k, then the procedure is finite.

Proof. As seen from the proof of Corollary III.1, we have:

$$\pi^k = (1/\theta_{1k}, 1/\theta_{2k}, \ldots, 1/\theta_{nk}) U_k^{-1} ,$$

from which the result follows. ■

The conditions indicated in Theorem III.2 and its corollaries are not easy to enforce in practical implementations. Nevertheless, in Part B we shall see how they can sometimes be used to establish the finiteness or convergence of a given cutting procedure.

4. CONCAVITY CUTS FOR HANDLING REVERSE CONVEX CONSTRAINTS

Cuts of the type (3) or (6), which were originally devised for concave programming, have proved to be useful in some other problems as well, especially in integer programming (see Glover (1972, 1973 and 1973a), Balas (1971, 1972, 1975, 1975a and 1979), Young (1971)). In fact, Corollary III.1 can be restated in the following form (see Glover (1972)), which may be more convenient for use in certain questions not directly related to the minimization of a concave function f(x).

Proposition III.2. *Let G be a convex set in* $\mathbb{R}^n$ *whose interior contains a point* x^0 *but does not contain any point of a given set S. Let U be a nonsingular n×n matrix with columns* $u^1,u^2,\ldots,u^n$. *Then for any constants* $\theta_i > 0$ *such that* $x^0 + \theta_i u^i \in G$ $(i=1,\ldots,n)$, *the cut*

$$\sum_{i=1}^{n} \frac{t_i}{\theta_i} \geq 1 ,$$

where $t = U^{-1}(x-x^0)$, *excludes* x^0 *without excluding any point* $x \in S$ *satisfying* $U^{-1}(x-x^0) \geq 0$.

Proof. Clearly $K = \{x: U^{-1}(x-x^0) \geq 0\}$ is nothing but the cone generated by n halflines emanating from x^0 in the direction $u^1,u^2,\ldots,u^n$. It can easily be verified that the argument used to prove Theorem III.1 and its corollary carries over to the case when G is an arbitrary convex set with $x^0 \in \text{int } G$, and the set $D^*(\gamma)$ is replaced by $S \cap K$, where S has no common point with int G.
The proposition can also be derived directly from Corollary III.1 by setting $D = K$, $f(x) = -p(x-x^0)$, $\gamma = -1$, where $p(x) = \inf \{\lambda > 0 : \frac{1}{\lambda} x \in G\}$ is the gauge of G, so that $G = \{x: p(x-x^0) \leq 1\}$. ■

To emphasize the role of the convex set G in the construction of the cut, Glover (1972) has called this cut a **convexity cut**. In addition, the term "**intersection cut**" has been used by Balas (1970a), referring to the fact that the cut is constructed by taking the intersection of the edges of the cone K with the boundary of G.

In the sequel, we shall refer to the cut obtained in Proposition III.2 as a **concavity cut relative to (G,K)**. Such a cut is valid for K\G, in the sense that it does not exclude any point of this set.

In Section III.1 concavity cuts were introduced as a tool for handling nonconvex objective functions. We now show how these cuts can also be used to handle nonconvex constraints.

As a typical example let us consider a problem

(P) $\quad$ minimize $f(x)$ s.t. $x \in D, g(x) \geq 0$

such that by omitting the last constraint $g(x) \geq 0$ we obtain a relatively easy problem

(Q) $\quad$ minimize $f(x)$ s.t. $x \in D$.

This is the case, for example, if D is a convex set, while both functions f(x) and g(x) are convex. Setting

$$G = \{x: g(x) \leq 0\} ,$$

we can rewrite the last constraint of (P) as

$$x \notin \text{int } G . \tag{22}$$

In order to solve the problem (P) by the outer approximation method discussed in Chapter II, we can proceed as follows.

Start with $D_0 = D$.

At iteration k=0,1,2,... find an optimal solution x^k of the relaxed problem

(Q_k) $\quad$ minimize $f(x)$ s.t. $x \in D_k$.

If $x^k \notin$ int G, then stop (x^k solves (P)). Otherwise, construct a cut $\ell_k(x) \geq 0$ to exclude x^k without excluding any feasible point of (P). Form the new relaxed problem (Q_{k+1}) with constraint set

$$D_{k+1} = D_k \cap \{x: \ell_k(x) \geq 0\} ,$$

and go to iteration k+1.

Obviously, a crucial point in this approach is the construction of the cuts. If x^0 denotes an optimal solution of (Q_0), then we can assume that $x^0 \in$ int G (otherwise x^0 would solve (P)). Furthermore, in most cases we can find a cone K vertexed at x^0

having exactly n edges (of directions $u^1,u^2,...,u^n$), and such that $D \subset K$. Then all the conditions of Proposition III.2 are fulfilled (with $S = D \setminus \text{int } G$), and a cut of type (6) can be constructed that excludes x^0 without excluding any feasible point of (P). The cut at any iteration k=1,2,... can be constructed in a similar way.

There are cases where a cone K with the above mentioned properties does not exist, or is not efficient because the cut constructed using this cone would be too shallow (see, for example, the situation with x^1 depicted in Fig. III.3). However, in these cases, it is always possible to find a number of cones $K_{0,1}, K_{0,2}, ..., K_{0,N_0}$ covering D such that the concavity cuts $\ell_{0,i}(x) \geq 0$ relative to $(G,K_{0,i})$ $(i=1,...,N_0)$ form a disjunctive system which does not exclude any point of $K \setminus G$. (We shall see in Part B how these cones and the corresponding cuts can be constructed effectively.)

Thus, concavity cuts offer a tool for handling reverse convex constraints such as (22).

In integer programming these cuts are useful, too, because an integrality condition like $x_j \in \{0,1\}$ $(j \in J)$ can be rewritten as $x_j^2 \geq x_j$, $0 \leq x_j \leq 1$ $(j \in J)$; and hence it implies $\sum_{j \in J} h_j(x_j^2 - x_j) \geq 0$ for arbitrary nonnegative numbers h_j $(j \in J)$. Note that the convex set

$$G = \{x: \sum_{j \in J} h_j(x_j^2 - x_j) \leq 0\}$$

does not contain in its interior any point x satisfying $x_j \in \{0,1\}$ $(j \in J)$. Therefore, if $0 \leq x_j^0 \leq 1$ for all $j \in J$ and $0 < x_j^0 < 1$ for some $j \in J$ with $h_j > 0$, then $x^0 \in \text{int } G$, and a concavity cut can be constructed to exclude x^0 without excluding any x satisfying $x_j \in \{0,1\}$ $(j \in J)$.

The reader interested in a more detailed discussion of suitable choices of the h_j or, more generally, of the use of concavity cuts in integer programming is referred to the cited papers of Glover, Balas, and Young. For a detailed treatment of disjunctive cuts, see Jeroslow (1976 and 1977) and also Sherali and Shetty (1980).

5. A CLASS OF GENERALIZED CONCAVITY CUTS

The cuts (6) (or (3)) do not exhaust the class of γ–valid cuts for (f,D). For example, the cut (II) in Fig. III.2, which is stronger than the concavity cut (I), cannot be obtained by Theorem III.1 or its corollary (to see this, it suffices to observe that by definition, the concavity cut is the strongest of all cuts obtained by Corollary III.1; moreover, a cut obtained by Corollary III.1 never meets an edge of K in its negative extension, as does the cut (II).

In this section we describe a general class of cuts that usually includes stronger valid cuts than the concavity cut.

Definition III.4. *Let G be a convex subset of* $\mathbb{R}^n$ *with an interior point* x^0*, and let K be a full dimensional polyhedral cone with vertex at* x^0*. A cut that does not exclude any point of* $K \setminus G$ *is called a* ***(G,K)–cut*** *(cf. Fig. III.4, page 105).*

When $G = \{x: f(x) \geq \gamma\}$ with f(x) a concave function, we recover the concept of a γ–valid cut.

Assume that the cone K has s edges ($s \geq n$) and that for each $i \in I \subset \{1,2,...,s\}$ the i–th edge meets the boundary of G at a point $z^i \neq x^0$, while for $i \notin I$ the i–th edge is entirely contained in G, i.e., its direction belongs to the recession cone R(G) of G (cf. Rockafellar (1970)). Denote by u^i the direction of the i–th edge and define

$$\alpha_i = \max \{\alpha \geq 0: x^0 + \alpha u^i \in G\} \quad (i \in I) , \tag{23}$$

$$\beta_{ij} = \inf \{\beta \geq 0: \alpha_i u^i + \beta u^j \in R(G)\} \quad (i \in I, j \notin I) . \tag{24}$$

Clearly, $\alpha_i > 0$ because $z^i \neq x^0$, $\beta_{ij} > 0$ (otherwise u^i would belong to R(G), because the latter cone is convex and closed and β_{ij} may be $+\infty$ for certain (i,j), which means that $\alpha_i u^i + \beta u^j \notin R(G)$ for any $\beta \geq 0$.

Since $u^j \in R(G)$ for $j \notin I$, it follows from (24) that for $0 < \alpha \leq \alpha_i$ and $\beta > 0$ we have

$$\alpha u^i + \beta u^j \in R(G) \Longleftrightarrow \alpha/\beta \leq \alpha_i/\beta_{ij} \ (i \in I, j \notin I) . \tag{25}$$

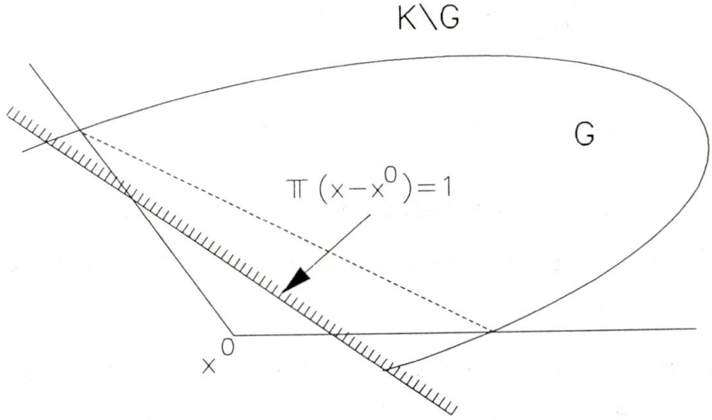

Fig. III.4. (G,K)–cut

Theorem III.3. *The inequality*

$$\pi(x-x^0) \geq 1 \tag{26}$$

defines a (G,K)–cut if and only if π satisfies the following relations:

$$\alpha_i \pi u^i \geq 1 \quad (i \in I) , \tag{27}$$

$$\pi(\alpha_i u^i + \beta_{ij} u^j) \geq 0 \quad (i \in I, j \notin I) . \tag{28}$$

(It is understood that if $\beta_{ij} = +\infty$, then the condition $\pi(\alpha_i u^i + \beta_{ij} u^j) \geq 0$, i.e., $\pi(\frac{\alpha_i}{\beta_{ij}} u^i + u^j) \geq 0$ means that $\pi u^j \geq 0$).

Proof. First observe that the inequality (26) defines a valid cut for $K \setminus G$ if and only if the polyhedron

$$M = \{x \in K: \pi(x-x^0) \leq 1\}$$

is contained in G. Because of the convexity of all the sets involved, the latter condition in turn is equivalent to requiring that

(i) each vertex of M belong to G,

while

(ii) all extreme directions of M belong to the cone R(G).

(i): Let $I^+ = \{i: \pi u^i > 0\}$. We must have $I \subset I^+$, since if $\pi u^i \leq 0$ for some $i \in I$. Then for $\theta > 0$ sufficiently large, we would have $x = x^0 + \theta u^i \in K \setminus G$, whereas $\pi(x-x^0) = \theta\pi u^i \leq 0 < 1$.

It is easily seen that the extreme points of M consist of x^0 and the points where the edges of K intersect the hyperplane $\pi(x-x^0) = 1$, i.e., the points

$$x^0 + \theta_i u^i \quad (i \in I^+), \text{ with } \theta_i = 1/(\pi u^i) > 0 . \tag{29}$$

According to (23), for $i \in I$ these points belong to G if and only if $\theta_i \leq \alpha_i$, i.e., if and only if (27) holds. On the other hand, for $i \notin I$, since u^i is a direction of recession for G, we have $x^0 + \theta_i u^i \in G$ for all $\theta_i > 0$.

(ii) Let us now show that any extreme direction of M is either a vector u^j $(j \notin I^+)$ or a vector of the form

$$\theta_i u^i + \xi_j u^j \quad (i \in I^+, j \notin I^-) , \tag{30}$$

where θ_i is defined by (29), $I^- = \{j: \pi u^j < 0\}$, and

$$\xi_j = -1/(\pi u^j) > 0 \quad (j \notin I) . \tag{31}$$

To see this, consider the cone $K_0 = K - x^0$. Then it is easily seen that K_0 and the origin can be separated by a hyperplane $tx = 0$, $t \in \mathbb{R}^n \setminus \{0\}$ satisfying $tu^i > 0$ for all $i=1,2,\ldots,s$. Since the recession cone $R(M)$ of M is

$$R(M) = \{x \in K_0 \colon \pi x \leq 0\} ,$$

the extreme directions of M are given by the vertices of the polytope

$$P = \{x \in R(M) \colon tx = 1\}$$

$$= \{x \in K_0 \colon tx = 1, \pi x \leq 0\} = S \cap \{x \colon \pi x \leq 0\} ,$$

with $S = \{x \in K_0 \colon tx = 1\}$. But the vertices of S are obviously $v^i = \lambda_i u^i$ $(i=1,2,\ldots,s)$ with $\lambda_i = 1/(tu^i) > 0$. Therefore, a vertex of P must be either a v^j such that $\pi v^j \leq 0$ (i.e., $j \notin I^+$) or the intersection of the hyperplane $\pi x = 0$ with an edge of S, i.e., a line segment joining two vertices v^i, v^j such that $\pi v^i > 0$ (i.e., $\pi u^i > 0$) and $\pi v^j < 0$ (i.e., $\pi u^j < 0$). In the latter case, let u^{ij} denote the intersection of the line segment $[v^i, v^j]$ with the hyperplane $\pi x = 0$. Since the hyperplane $\pi x = 1$ meets the halfline from 0 through v^i at $\theta_i u^i$ and the halfline from 0 through v^j at $-\xi_j u^j$ (Fig. III.5), it is clear that u^{ij} is parallel to $\theta_i u^i + \xi_j u^j$. We have thus proved that the extreme

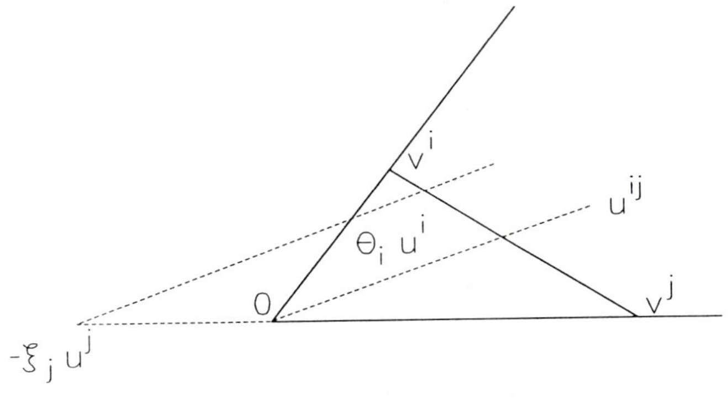

Fig. III.5

directions of M consist of the vectors u^j ($j \notin I^+$) and a set of vectors of the form (30).

Since $j \notin I^+$ implies $j \notin I$, any vector u^j ($j \notin I^+$) belongs to R(G). Moreover, according to (25), for $i \in I$, $j \in I^-$(which implies $j \notin I$) a vector of the form (30) belongs to the recession cone of G if and only if

$$\alpha_i \xi_j \geq \theta_i \beta_{ij}$$

and hence, from (29) and (31), if and only if

$$\alpha_i \pi u^i \geq -\beta_{ij} \pi u^j .$$

The proof of Theorem III.3 is completed by noting that for $i \in I$, $j \notin I \cup I^-$ we have $\pi u^i > 0$, $\pi u^j \geq 0$, so that (28) trivially holds (by convention the inequality $\pi(\alpha_i u^i + \beta_{ij} u^j) \geq 0$ means $\pi u^j \geq 0$ if $\beta_{ij} = +\infty$) while for $i \in I^+ \setminus I$, $j \in I^-$ we have u^i, $u^j \in R(G)$ (because $i \notin I$, $j \notin I$), so that $\theta_i u^i + \xi_j u^j \in R(G)$ for all $\theta_i > 0$, $\xi_i > 0$. ■

6. CUTS USING NEGATIVE EDGE EXTENSIONS

Theorem III.3 yields a very general class of cuts, since the inequalities (27) and (28) that characterize this class are linear in π and can be satisfied in many different ways.

The cut (6) in the nondegenerate case ($s = n$), the Carvajal–Moreno cut for the degenerate case (Proposition III.1) and the cut obtained by Proposition III.2 are special variants of this class, which share the common feature that their coefficient vectors π satisfy the inequalities $\pi u^i \geq 0$ ($i=1,2,\ldots,s$). Though these cuts are easy to construct, they are not always the most efficient ones. In fact, Glover (1973 and 1974) showed that if G is a polyhedron and some edges of K (say, the edges of direction u^j, $j \in J$) strictly recede from all boundary hyperplanes of G, then stronger cuts than the concavity cut can be constructed which satisfy $\pi u^j < 0$ for $j \in J$ (thus, these cuts use negative extensions of the edges rather than the usual positive extensions: an example is furnished by the cut (II) in Fig. III.2, already mentioned in the previous section).

We close this chapter by showing that these cuts can also be derived from Theorem III.3.

As before, let I denote the index set of those edges of K which meet the boundary of G.

Theorem III.4. *The system*

$$\alpha_i \pi u^i \geq 1 \quad (i \in I), \tag{32}$$

$$(\max_{i \in I} \beta_{ij}) \pi u^j \geq -1 \quad (j \notin I), \tag{33}$$

where β_{ij} *is defined by (23), (24), has at least one basic solution and any basic solution is a (G,K) cut.*

Proof. Since (12) implies (32) and (33), the consistency of the system (12) implies that of the system (32) and (33). Obviously, any solution of the latter satisfies the conditions (27), (28). Hence, by Theorem III.3, it generates a (G,K)–cut. It remains to show that the polyhedron (32) and (33) has at least one extreme point. But its recession cone is defined by the system

$$\pi u^i \geq 0 \quad (i = 1,2,\dots,s)$$

and is pointed, as was shown in the proof of Lemma III.1. This implies the existence of at least one basic solution to the system (32) and (33) (cf. Rockafellar (1970), Corollary 18.5.3). ∎

Remark III.1. To obtain a cut of the type indicated in Theorem III.4, it suffices to solve, for example, the linear program

$$\text{minimize} \sum_{i \in I} \alpha_i \pi u^i \quad \text{subject to (32) and (33)} .$$

This cut is obviously stronger than the Carvajal–Moreno cut obtained by minimizing the same linear function over the smaller polyhedron (12). In particular, if s=n (nondegenerate case), then this cut is determined by the system of equations

$$\alpha_i \pi u^i = 1 \qquad (i \in I) ,$$

$$(\max_{i \in I} \beta_{ij}) \pi u^j = -1 \qquad (j \notin I) .$$

From the latter equation it follows that $\pi u^j < 0$ for $j \notin I$, provided that $\max_{i \in I} \beta_{ij} < +\infty$.

When G is a polyhedron, the recession cone R(G) of G is itself polyhedral and readily available from the constraints determining G. In this case, the values β_{ij} can be easily computed from (24), and we necessarily have $\max_{i \in I} \beta_{ij} < +\infty$ for every j such that u^j is strictly interior to R(G).

CHAPTER IV

BRANCH AND BOUND

A widely used method to solve various kinds of difficult optimization problems is called branch and bound. In this technique, the feasible set is relaxed and subsequently split into parts (branching) over which lower (and often also upper) bounds of the objective function value can be determined (bounding).

In this chapter, branch and bound is developed with regard to the specific needs of global optimization. First, a general prototype method is presented which includes all the specific approaches that will be discussed in subsequent chapters. A general convergence theory is developed and applied to concave minimization, d.c.–programming and Lipschitzian optimization. The basic operations of branching and bounding are discussed in detail.

1. A PROTOTYPE BRANCH AND BOUND METHOD

Consider again the global optimization problem

$$(P) \qquad \begin{array}{l} \text{minimize } f(x) \\ \text{s.t. } x \in D \end{array}, \qquad (1)$$

where $f: A \longrightarrow \mathbb{R}$ and $D \subset A \subset \mathbb{R}^n$. The set A is assumed to contain all of the subsets that will be used below.

For the moment we only assume that **min f(D) exists**. Further assumptions will be needed later.

A basic idea that is widely used in many forms to handle difficult optimization problems is "branch and bound" in the sequel abbreviated BB. BB has been successfully applied to certain specific types of problem (P), such as minimizing a concave function over a convex set (see, e.g., Falk and Soland (1969), Soland (1971), Bulatov (1977), Horst (1976, 1980, 1984 and 1986), Benson (1982), Thoai and Tuy (1980), Tuy, Thieu and Thai (1985), Pardalos et al. (1987)), minimization of Lipschitz– functions (e.g., Pinter (1986), Horst and Tuy (1987), Horst (1987)), minimizing the difference of two convex functions (d.c.–optimization) (e.g., Horst (1988), Horst and Dien (1987)), minimization under reverse convex constraints (e.g., Muu (1985), Horst (1988)). The references listed here are not exhaustive; these classes of problems will be treated in some detail in the later parts of this book. General BB concepts in global optimization, sometimes in combination with other methods such as outer approximation and their applications to concave minimization, Lipschitzian optimization and d.c.–optimization, have been proposed in Horst (1976), Benson (1982), Tuy, Thieu and Thai (1985), Horst (1986, 1988 and 1989), Horst and Tuy (1987), Tuy and Horst (1988), Benson and Horst (1988), Horst, Thoai and Benson (1988).

The following presentation is based on Horst (1986 and 1988), Tuy and Horst (1988).

The idea of BB methods in global optimization is rather simple:

– Start with a relaxed feasible set $M_0 \supset D$ and split (partition) M_0 into finitely many subsets M_i, $i \in I$.
– For each subset M_i determine lower and (if possible) upper bounds $\beta(M_i)$, $\alpha(M_i)$ respectively, satisfying

$$\beta(M_i) \le \inf f(M_i \cap D) \le \alpha(M_i).$$

Then $\beta := \min_{i \in I} \beta(M_i)$, $\alpha := \min_{i \in I} \alpha(M_i)$ are "overall" bounds, i.e., we have $\beta \le \min f(D) \le \alpha$.

– If $\alpha = \beta$ (or $\alpha - \beta \le \varepsilon$, for prescribed $\varepsilon > 0$), then stop.

Otherwise select some of the subsets M_i and partition these chosen subsets in order to obtain a refined partition of M_0. Determine new, hopefully better bounds on the new partition elements, and repeat the process in this way.

An advantage of BB methods is that during the iteration process one can usually **delete** certain subsets of D, since one knows that min f(D) cannot be attained there. A typical disadvantage is that, as a rule, the accuracy of an approximate solution can only be measured by the difference $\alpha - \beta$ of the current bounds. Hence a "good" feasible point found early may be detected as "good" only much later after many further refinements.

Definition IV.1. *Let B be a subset of* $\mathbb{R}^n$ *and I be a* ***finite*** *set of indices. A set* $\{M_i : i \in I\}$ *of subsets of B is said to be a* ***partition*** *of B if*

$$B = \bigcup_{i \in I} M_i,$$

$$M_i \cap M_j = \partial M_i \cap \partial M_j \quad \forall i, j \in I, i \neq j,$$

where ∂M_i *denotes the (relative) boundary of* M_i.

Let M denote an element of a current partition used in a BB procedure as indicated above. For M_0 and all partition sets M, it is natural to use most simple polytopes or convex polyhedral sets, such as simplices, rectangles and polyhedral cones. In this context, a polytope M is often given by its vertex set, a polyhedral cone by its generating vectors.

An important question arising from BB procedures will be that of properly deleting partition sets satisfying $M \cap D = \emptyset$. Clearly, for many feasible sets a decision on whether we have $M \cap D = \emptyset$ or $M \cap D \neq \emptyset$ will be difficult based on the information at hand (cf. Section IV.5).

Definition IV.2. *A partition set M satisfying $M \cap D = \emptyset$ is called* ***infeasible****; a partition set M satisfying $M \cap D \neq \emptyset$ is called* ***feasible****. A partition set M is said to be* ***uncertain*** *if we do not know whether M is feasible or infeasible.*

In the following description of a general BB method we adopt the usual convention that infima and minima taken over an empty set equal $+\infty$.

Prototype BB procedure:

Step 0 (Initialization):

Choose $M_0 \supseteq D$, $S_{M_0} \subset D$, $-\infty < \beta_0 \leq \min f(D)$.

Set $\mathcal{M}_0 = \{M_0\}$, $\alpha_0 = \min f(S_{M_0})$, $\beta(M_0) = \beta_0$.

If $\alpha_0 < \infty$, then choose $x^0 \in \operatorname{argmin} f(S_{M_0})$ (i.e., $f(x^0) = \alpha_0$).

If $\alpha_0 - \beta_0 = 0$, then stop: $\alpha_0 = \beta_0 = \min f(D)$, x^0 is an optimal solution. Otherwise, go to Step 1.

Step k (k = 1,2,...):

At the beginning of Step k we have the current partition $\mathcal{M}_{k-1}$ of a subset of M_0 still of interest. Furthermore, for every $M \in \mathcal{M}_{k-1}$ we have $S_M \subseteq M \cap D$ and bounds $\beta(M)$, $\alpha(M)$ satisfying

$$\begin{aligned} &\beta(M) \leq \inf f(M \cap D) \leq \alpha(M) && \text{if M is known to be feasible,} \\ &\beta(M) \leq \inf f(M) && \text{if M is uncertain.} \end{aligned} \tag{2}$$

Moreover, we have the current lower and upper bounds β_{k-1}, α_{k-1} satisfying

$$\beta_{k-1} \le \min f(D) \le \alpha_{k-1}.$$

Finally, if $\alpha_{k-1} < \infty$, then we have a point $x^{k-1} \in D$ satisfying $f(x^{k-1}) = \alpha_{k-1}$ (the best feasible point obtained so far).

k.1. Delete all $M \in \mathcal{M}_{k-1}$ satisfying

$$\beta(M) \ge \alpha_{k-1}.$$

Let $\mathcal{R}_k$ be the collection of the remaining sets in the partition $\mathcal{M}_{k-1}$.

k.2. Select a nonempty collection of sets $\mathcal{P}_k \subset \mathcal{R}_k$ and construct a partition of every member of $\mathcal{P}_k$. Let $\mathcal{P}'_k$ be the collection of all new partition elements.

k.3. Delete each $M \in \mathcal{P}'_k$ for which it is known that $M \cap D = \emptyset$ or where it is otherwise known that min f(D) cannot occur. Let $\mathcal{M}'_k$ be the collection of all remaining members of $\mathcal{P}'_k$.

k.4. Assign to each $M \in \mathcal{M}'_k$ a set S_M and a quantity $\beta(M)$ satisfying

$$S_M \subseteq M \cap D,$$

$$\beta(M) \le \inf f(M \cap D) \quad \text{if } M \text{ is known to be feasible,}$$

$$\beta(M) \le \inf f(M) \quad \text{if } M \text{ is uncertain.}$$

Furthermore, it is required that we have

$$S_M \supseteq M \cap S_{M'}, \quad \beta(M) \ge \beta(M') \text{ whenever } M \subset M' \in \mathcal{M}_{k-1}.$$

Set $\alpha(M) = \min f(S_M)$.

k.5. Set $\mathcal{M}_k = (\mathcal{R}_k \setminus \mathcal{P}_k) \cup \mathcal{M}'_k$.

Compute

$$\alpha_k = \inf \{\alpha(M) : M \in \mathcal{M}_k\}$$

and

$$\beta_k = \min \{\beta(M): M \in \mathscr{M}_k\}.$$

If $\alpha_k < \infty$, then let $x^k \in D$ such that $f(x^k) = \alpha_k$.

k.6. If $\alpha_k - \beta_k = 0$, then stop: $\alpha_k = \beta_k = \min f(D)$, x^k is an optimal solution. Otherwise, go to Step k+1.

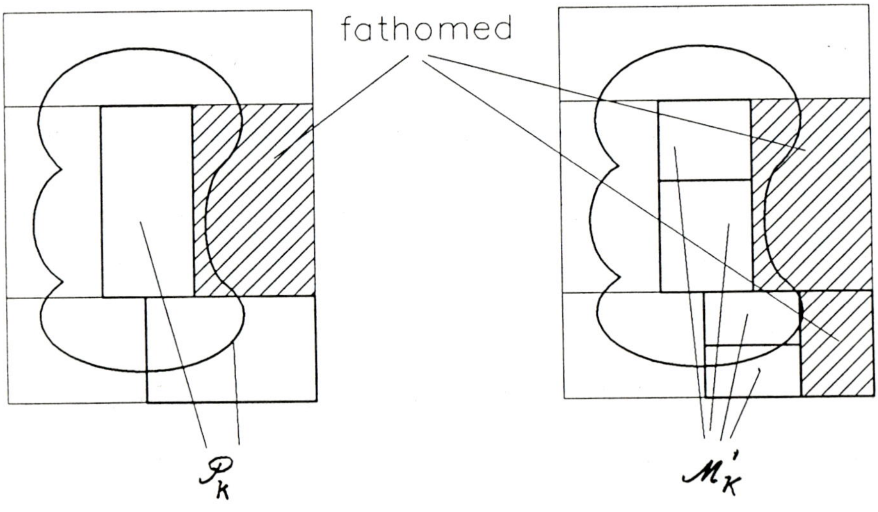

Fig. IV.1. A BB step

Remarks IV.1. **(i)** We say that a partition element $M \in \mathscr{M}_k$ is **fathomed** if $\beta(M) \geq \alpha_{k-1}$. Such a partition element will be deleted in Step k. The stopping criterion $\alpha_k = \beta_k$ means that all partition elements are fathomed.

(ii) Step k is to be performed only if there remain unfathomed partition elements, i.e., if $\mathscr{R}_k \neq \emptyset$. In general, the partition operation is defined only on a certain family $\mathscr{F}$ of subsets of $\mathbb{R}^n$ (e.g., simplices, or rectangles, or cones of a certain type). In order that Step k can actually be carried out, one must require that $\mathscr{R}_k \subset \mathscr{F}$, i.e., every unfathomed partition element should be capable of further refinement.

(iii) In Step k.4 one can obviously replace any $M \in \mathcal{M}_k'$ by a smaller set $\tilde{M} \subset M$ such that $\tilde{M} \in \mathcal{I}$ and $\tilde{M} \cap D = M \cap D$.

(iv) For each partition set M, $S_M \subset M \cap D$ is the collection of feasible points in M that can be found by reasonable effort when a BB algorithm is performed. We assume that min $f(S_M)$ exists or $S_M = \emptyset$. In the applications, the sets S_M are usually finite. However, the above procedure is also defined in the case where many or even all sets S_M are empty and we may even have $\alpha_k = \infty$ for all k. We will return to this point later, but we first treat the case where "enough" feasible points are available. Then the conditions imposed on S_M and $\beta(M)$ ensure that $\{\alpha_k\} = \{f(x^k)\}$ is a nonincreasing sequence, $\{\beta_k\}$ is a nondecreasing sequence, and $\alpha_k \geq \min f(D) \geq \beta_k$. Thus, the difference $\alpha_k - \beta_k$ measures the proximity of the current best solution x^k to the optimum. For a given tolerance $\varepsilon > 0$, the algorithm can be stopped as soon as $\alpha_k - \beta_k \leq \varepsilon$.

Since $\{\alpha_k\}$ and $\{\beta_k\}$ are monotonically nonincreasing and nondecreasing, respectively, the limits $\alpha = \lim_{k\to\infty} \alpha_k$ and $\beta = \lim_{k\to\infty} \beta_k$ necessarily exist, and, by construction, they satisfy $\alpha \geq \min f(D) \geq \beta$. The algorithm is said to be **finite** if $\alpha_k = \beta_k$ occurs at some step k, while it is **convergent** if $\alpha_k - \beta_k \longrightarrow 0$, i.e., $\alpha = \lim_{k\to\infty} f(x^k) = \beta = \min f(D)$.

It is clear that the concrete choice of the following three basic operations is crucially important for the convergence and efficiency of the prototype branch and bound procedure:

Bounding (how to determine $\alpha(M)$, $\beta(M)$?),

Selection (how to choose $\mathcal{P}_k$?)

and

Refining (how to construct the partitions?).

Example IV.1:

Consider the concave minimization problem:

$$\text{minimize } f(x_1, x_2) := -(x_1 - 20)^2 - (x_2 - 10)^2$$

$$\text{s.t.} \qquad x_2 - \tfrac{1}{2}x_1 \leq 10 \,,$$

$$(x_2 - 10)^2 + (x_1)^2 \leq 500 \,,$$

$$x_1 \geq 0, \ x_2 \geq 0.$$

The procedure that follows is one of several possibilities for solving this problem by a BB algorithm.

Step 0: M_0 is chosen to be the simplex conv $\{(0,0), (0,40), (40,0)\}$ having vertices $(0,0)$, $(0,40)$, $(40,0)$.

To obtain lower bounds, we construct the affine function $\varphi(x_1, x_2) = a_1x_1 + a_2x_2 + a_3$ that coincides with f at the vertices of M_0. Solving the corresponding system of linear equations

$$a_3 = -500$$

$$40a_2 + a_3 = -1300$$

$$40a_1 + a_3 = -500$$

yields $\varphi(x_1, x_2) = -20x_2 - 500$. Because of the concavity of $f(x_1, x_2)$, $\varphi(x_1, x_2)$ is underestimating $f(x_1, x_2)$, i.e., we have $\varphi(x_1,x_2) \leq f(x_1,x_2) \quad \forall (x_1,x_2) \in M_0$. A lower bound β_0 can be found by solving the convex optimization problem (with linear objective function)

$$\min \varphi(x_1, x_2), \text{ s.t. } (x_1, x_2) \in M_0 \cap D = D.$$

We obtain $\beta_0 = -900$ (attained at $(20,20)$).

The two feasible points $(0,0)$ and $(20,20)$ are at hand, and we set

$S_{M_0} = \{(0,0), (20,20)\}$, $\alpha_0 = \min f(S_{M_0}) = f(0,0) = -500$ and $x^0 = (0,0)$.

Step 1: We partition M_0 into the two simplices $M_{1,1} = \text{conv}\,\{(0,0), (0,40), (20,20)\}$ and $M_{1,2} = \text{conv}\,\{(0,0), (20,20), (40,0)\}$.

As in Step 0, we construct lower bounds $\beta(M_{1,1})$ and $\beta(M_{1,2})$ by minimizing over $M_{1,1} \cap D$ the affine function $\varphi_{1,1}$ that coincides with f at the vertices of $M_{1,1}$ and by minimizing over $M_{1,2} \cap D$ the affine function $\varphi_{1,2}$ that coincides with f at the vertices of $M_{1,2}$, respectively.

One obtains

$$\varphi_{1,1}(x_1,x_2) = 40x_1 - 20x_2 - 500,\ \varphi_{1,2}(x_1,x_2) = 20x_2 - 500;\ \beta(M_{1,1}) = -700$$

$$\text{(attained at } (0,10)) \text{ and } \beta(M_{1,2}) = -500 \text{ (attained at } (0,0)),$$

which implies $\beta_1 = -700$. The sets of feasible points in $M_{1,1}$, $M_{1,2}$ which are known until now are

$$S_{M_{1,1}} = \{(0,0)(0,10)(20,20\},\ S_{M_{1,2}} = \{(0,0),(20,20)\}.$$

Hence $\alpha(M_{1,1}) = \alpha(M_{1,2}) = f(0,0) = -500$, $\alpha_1 = -500$ and $x^1 = (0,0)$.

Step 2. The set $M_{1,2}$ can be deleted since $\beta(M_{1,2}) = -500 = \alpha_1$. We partition $M_{1,1}$ into the two simplices

$$M_{2,1} = \text{conv}\{(0,0),(0,20),(20,20)\},\ \ M_{2,2} = \text{conv}\{(0,20),(0,40),(20,20)\}.$$

In a similar way to the above, we calculate the subfunctional $\varphi_{2,1} = 20x_1 - 500$. The subfunctional $\varphi_{2,2}$ does not need to be calculated, since $M_{2,2} \cap D = \{(20,20)\}$. We obtain

$$\beta(M_{2,1}) = -500 \text{ (attained at } (0,0)) \text{ and } \beta(M_{2,2}) = f(20,20) = -100,$$

$$\text{hence } \beta_2 = -500;$$

$$S_{M_{2,1}} = \{(0,0),(0,10),(0,20),(20,20)\},\ \alpha(M_{2,1}) = -500\ ;$$

$$S_{M_{2,2}} = \{(20,20)\},\ \alpha(M_{2,2}) = -100.$$

Hence $\alpha_2 = -500$ and $x^2 = (0,0)$. Since $\beta_2 = \alpha_2$, we stop the procedure; $x^2 = (0,0)$ is the optimal solution.

The particular way of subdividing simplices applied above will be justified in Section IV.3.1.

Another possibility for calculating the lower bounds would have been simply by minimizing f over the vertex set of the corresponding partition element M. Keeping the partitioning unchanged, we then would have obtained $\beta_0 = -1300$, $\beta_1 = -1300$, deletion of $M_{1,2}$ as above, deletion of $M_{2,2}$ (since $M_{2,2} \cap D \subset M_{2,1} \cap D$), $\beta_2 = -500 = \alpha_2$.

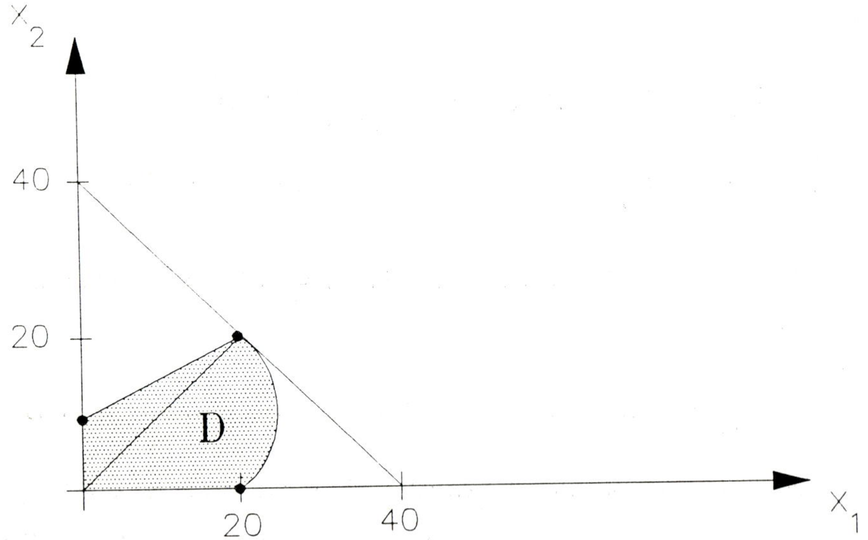

Fig. IV.2. Feasible set D and simplices for Example IV.1

The example shows both the mentioned disadvantage that for a solution found very early it may take a long time to verify optimality, and the typical advantage that parts of the feasible region may be deleted from further consideration.

The BB–procedure can be visualized by a graph hanging at M_0: the nodes correspond to successively generated partition sets, two nodes being connected by an arc if the second is obtained by a direct partition of the first (cf. the proof of the following Theorem IV.1).

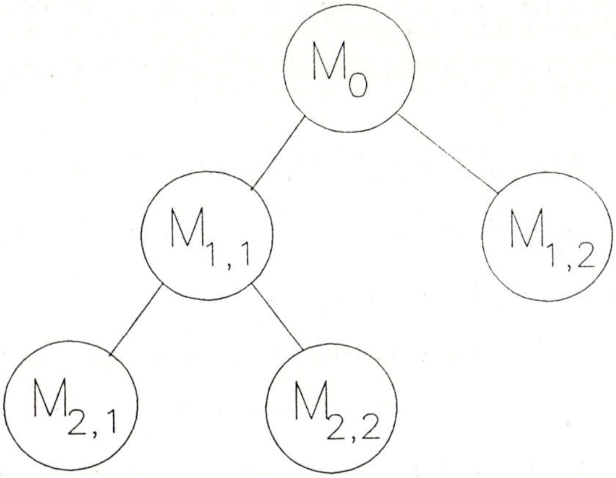

Fig. IV.3. The graph corresponding to Example IV.1.

2. FINITENESS AND CONVERGENCE CONDITIONS

Finiteness and convergence conditions depend on the limit behaviour of the difference $\alpha_k - \beta_k$. Therefore, in order to study convergence properties for realizations of the BB procedure, it is natural to consider decreasing (nested) sequences of successively refined partition elements, i.e., sequences $\{M_{k_q}\}$ such that

$$M_{k_q} \in \mathcal{P}_{k_q}, \; M_{k_{q+1}} \subset M_{k_q} .$$

Definition IV.3. *A bounding operation is called* ***finitely consistent*** *if, at every step, any unfathomed partition element can be further refined, and if any decreasing sequence* $\{M_{k_q}\}$ *of successively refined partition elements is finite.*

Theorem IV.1. *In a BB procedure, suppose that the bounding operation is finitely consistent. Then the procedure terminates after finitely many steps.*

Proof. Since any unfathomed partition element can be further refined, the procedure stops only when $\alpha_k = \beta_k$ and an optimal solution has been attained. A directed graph G can be associated to the procedure in a natural way. The nodes of G consist of M_0 and all partition elements generated by the algorithm. Two nodes are connected by an arc if and only if the first node represents an immediate ancestor of the second node, i.e., the second is obtained by a direct partition of the first.
Obviously, in terms of graph theory, G is a rooted tree with root M_0. A path in G corresponds to a decreasing sequence $\{M_{k_q}\}$ of successively refined partition elements, and the assumption of the theorem means that every path in G is finite.

On the other hand, by Definition IV.1, each partition consists of finitely many elements: hence, from each node of G only a finite number of arcs can emanate (the "out–degree" of each node is finite, the "in–degree" of each node different from M_0 is one, by construction).

Therefore, for each node M, the set of descendants of M, i.e., the set of nodes attainable (by a path) from M must be finite. In particular, the set of all descendants of M_0 is finite, i.e., the procedure itself is finite. ■

Remark IV.2. The type of tree that appears in the proof is discussed in Berge

1958) (the so-called "Γ-finite graphs", cf. especially Berge (1958, Chapter 3, Theorem 2).

In the sequel, convergence conditions for the infinite BB procedure are considered. It will turn out that convergence of an infinite BB procedure is guaranteed if an obvious requirement is imposed on the selection operator, and if a certain consistency property of the bounding operations is satisfied on nested sequences of successively refined partition elements. An immediate consequence of Theorem IV.1 is the following corollary.

Corollary IV.1. *If a BB procedure is infinite, then it generates at least one infinitely decreasing sequence* $\{M_{k_q}\}$ *of successively refined partition elements.*

Definition IV.4. *A bounding operation is called* ***consistent*** *if at every step any unfathomed partition element can be further refined, and if any infinitely decreasing sequence* $\{M_{k_q}\}$ *of successively refined partition elements satisfies*

$$\lim_{q\to\infty} (\alpha_{k_q} - \beta(M_{k_q})) = 0. \tag{3}$$

Remarks IV.3. (i) By construction, α_{k_q} is a nonincreasing sequence, $\beta(M_{k_q})$ is a nondecreasing sequence and $\alpha_{k_q} > \beta(M_{k_q})$, since otherwise M_{k_q} would be deleted in the next step, and so it could not be an element of a decreasing sequence of successively refined partition elements. However, since $\beta(M_{k_q}) \geq \beta_{k_q}$, (3) does not necessarily imply $\lim_{k\to\infty} \alpha_k = \lim_{k\to\infty} \beta_k = \min f(D)$ via $\lim_{q\to\infty} \alpha_{k_q} = \lim_{q\to\infty} \beta_{k_q}$.

In order to guarantee convergence an additional requirement must be imposed on the selection operation.

(ii) The relation (3) may be difficult to verify in practice, since α_{k_q} is not necessarily attained at M_{k_q}. In view of the inequality $\alpha(M_{k_q}) \geq \alpha_{k_q} \geq \beta(M_{k_q})$ and the properties just mentioned, (3) will be implied by the more practical requirement

$$\lim_{q\to\infty} (\alpha(M_{k_q}) - \beta(M_{k_q})) = 0 , \tag{4}$$

which simply says that, whenever a decreasing sequence of partition sets converges to a certain limit set, the bounds also must converge to the exact minimum of f over this limit set.

Definition IV.5. *A selection operation is called* ***complete*** *if for every* $M \in \bigcup_{p=1}^{\infty} \bigcap_{k=p}^{\infty} \mathcal{R}_k$ *we have*

$$inf\, f(M \cap D) \geq \alpha := \lim_{k\to\infty} \alpha_k .$$

Stated in words, this means that any portion of the feasible set which is left "unexplored forever" must (in the end) be not better than the fathomed portions.

Theorem IV.2. *In an infinite BB procedure suppose that the bounding operation is consistent and the selection operation is complete. Then*

$$\alpha := \lim_{k\to\infty} \alpha_k = \lim_{k\to\infty} f(x^k) = min\, f(D).$$

Proof. Since, by construction, we have $\alpha \geq \min f(D)$, it suffices to prove that $f(x) \geq \alpha \quad \forall x \in D$.

(i) If $x \in D$ belongs to a partition set M that is deleted at iteration k, then by step k.1 of the prototype BB procedure we have $f(x) \geq \beta(M) \geq \alpha_{k-1} \geq \alpha$.

(ii) If $x \in D$ belongs to a partition set $M \in \bigcup_{p=1}^{\infty} \bigcap_{k=p}^{\infty} \mathcal{R}_k$, then by the completeness property we have $\inf f(M \cap D) \geq \alpha$, and hence $f(x) \geq \alpha$.

(iii) If neither of the two previous cases holds for $x \in D$, then any partition set M containing x must be partitioned at some iteration, i.e., it must belong to $\mathcal{M}'_k$ for some k. Therefore, one can find a decreasing sequence $\{M_{k_q}\}$ of partition sets satisfying $M_{k_q} \in \mathcal{M}'_{k_q}$, $x \in M_{k_q}$. By consistency, it follows that $\lim_{q\to\infty} \alpha_{k_q} = \alpha = \lim_{q\to\infty} \beta(M_{k_q})$. Hence $f(x) \geq \alpha$, since $f(x) \geq \beta(M_{k_q})$ for every q. ■

Theorem IV.2 does not include results on the behaviour of the sequence $\{\beta_q\}$ of lower bounds. By construction (since $\{\beta_q\}$ is a nondecreasing sequence bounded from above by min f(D)), we only have the existence of

$$\beta := \lim_{k\to\infty} \beta_k \leq \inf f(D).$$

However, as already seen, the degree of approximation attained at Step k is measured by the difference of the bounds $\alpha_k - \beta_k$. It is therefore desirable to apply selection operations that improve lower bounds and imply that $\alpha = \beta$.

Definition IV.6. *A selection operation is said to be* ***bound improving*** *if, at least each time after a finite number of steps, $\mathcal{P}_k$ satisfies the relation*

$$\mathcal{P}_k \cap \operatorname{argmin} \{\beta(M): M \in \mathcal{R}_k\} \neq \emptyset, \tag{5}$$

i.e., at least one partition element where the actual lower bound is attained is selected for further partition in Step k of the prototype algorithm.

Several selection rules not explicitly using (5) are actually bound improving. For example, define for every partition set M let $\mathcal{T}(M)$ denote the index of the step where M is generated, and choose the **oldest** partition set, i.e., select

$$\mathcal{P}_k = \operatorname{argmin} \{\mathcal{T}(M): M \in \mathcal{R}_k\}.$$

Alternatively, for every partition set M we could define a quantity $\delta(M)$ closely related to the size of M, for example, $\delta(M)$ could be the diameter of M, the volume of M, or the length of the longest edge if M is a polytope. Let the refining operation be such that, for every compact M, given $\varepsilon > 0$, after a finite number of refinements of M we have $\delta(M_i) \le \varepsilon$ for all elements M_i of the current partition of M. Choose the largest partition element, i.e., select

$$\mathscr{P}_k = \text{argmax}\,\{\delta(M): M \in \mathscr{R}_k\}.$$

Both selections are bound improving simply because of the finiteness of the number of partition elements in each step which assures that any partition set M will be deleted or split again after finitely many steps.

Theorem IV.3. *In the infinite BB procedure, suppose that let the bounding operation is consistent and the selection operation is bound improving. Then the procedure is convergent:*

$$\alpha := \lim_{k\to\infty} \alpha_k = \lim_{k\to\infty} f(x^k) = \min f(D) = \lim_{k\to\infty} \beta_k = \beta.$$

Proof. Let us mark every partition element M satisfying $M \in \mathscr{P}_k \cap \text{argmin}\,\{\beta(M): M \in \mathscr{R}_k\}$, i.e., such that $\beta(M) = \beta_{k-1}$ at some Step k where (5) occurs. Since (5) occurs after finitely many steps, it follows that, if the procedure is infinite, it must generate infinitely many marked partition elements.

Among the members of the finite partition of M_0 there exists one, say M_{k_0}, with infinitely many marked descendants; then, among the members of the partition of M_{k_0} there exists one, say M_{k_1}, with infinitely many marked descendants. Continuing in this way, we see that there exists a decreasing sequence $\{M_{k_q}\}$ such that every M_{k_q} has infinitely many marked descendants. Since the bounding operation is consistent, we have $\lim_q (\alpha_{k_q} - \beta(M_{k_q})) = 0$. But every M_{k_q} has at least one marked descendant, say $\tilde{M}_q$. Then $\beta(\tilde{M}_q) = \beta_{h_q}$ for some $h_q > k_q$, and since $\tilde{M}_q \subset M_{k_q}$ we must have $\beta(\tilde{M}_q) \ge \beta(M_{k_q})$ (see Step k.4). Thus, $\beta(M_{k_q}) \le \beta_{h_q} \le \alpha_{h_q} \le \alpha_{k_q}$, which implies that $\lim_q (\alpha_{h_q} - \beta_{h_q}) = 0$. Therefore, we have $\alpha = \beta$. ■

Note that for consistent bounding operations, bound improving selection operations are complete. This follows from Theorem IV.3, since the relation $f(x) \geq \beta_{k_q}$ $\forall x \in D$ implies that $\inf f(M \cap D) \geq \beta = \alpha$ for every partition set M.

Convergence of the sequence of current best points x^k now follows by standard arguments.

Corollary IV.2. *Let $f: \mathbb{R}^n \rightarrow \mathbb{R}$ be continuous, D be closed and $C(x^0) := \{x \in D: f(x) \leq f(x^0)\}$ be bounded. In an infinite BB procedure suppose that the bounding operation is consistent and the selection operation is complete. Then every accumulation point of $\{x^k\}$ solves problem (P).*

Proof. The set $C(x^0)$ is bounded and, by continuity of f, $C(x^0)$ is closed and therefore compact. By construction, we have $f(x^{k+1}) \leq f(x^k)$ $\forall k$; thus $\{x^k\} \subset C(x^0)$. Hence $\{x^k\}$ possesses accumulation points. Corollary IV.2. then follows immediately from Theorem IV.2. ■

Several BB procedures proposed earlier for solving special global optimization problems define the iteration sequence in a different way. Instead of choosing the best feasible point x^k obtained so far, a point $\tilde{x}^k$ is considered which is associated with the lower bound β_k. Usually $\beta(M)$ is calculated by minimizing or maximizing a certain subfunction of f over $D \cap M$ or M. Then, in several procedures, $\tilde{x}^k$ is a point where β_k is attained (cf., e.g., Falk and Soland (1969), Horst (1976 and 1980), Benson (1982)). If $\tilde{x}^k$ is feasible, then $\tilde{x}^k$ will enter the set S_M of feasible points known in M. Since, however, x^k is the best feasible point known in Step k, we will then have $f(x^k) \leq f(\tilde{x}^k)$, i.e.,, x^k will be the better choice. Note that for all k we have $f(x^{k+1}) \leq f(x^k)$ whereas $f(\tilde{x}^{k+1}) \leq f(\tilde{x}^k)$ does not necessarily hold.

If in these algorithms the iteration sequence $\{\tilde{x}^k\}$ is replaced by the sequence of best feasible points $\{x^k\}$, then, for continuous f and compact M_0, convergence is preserved in the following sense: whenever every accumulation point of $\{\tilde{x}^k\}$ solves (P), then every accumulation point of $\{x^k\}$ solves (P) as well. To see this, let $\bar{x}$ be an accumulation point of $\{x^k\}$ and $\{x^q\}_{q \in I}$ be a subsequence of $\{x^k\}$ satisfying $x^q \xrightarrow[(q \in I)]{} \bar{x}$. By continuity of f, we have $f(x^q) \xrightarrow[(q \in I)]{} f(\bar{x})$.

Now consider the corresponding subsequence $\{\tilde{x}^q\}_{q \in I}$ of $\{\tilde{x}^q\}$. Since M_0 is compact and $\{\tilde{x}^q\}_{q \in I} \subset M_0$, there is a convergent subsequence $\{\tilde{x}^r\}_{r \in \tilde{I}}$, $\tilde{I} \subset I$, with $f(\tilde{x}^r) \xrightarrow[(r \in \tilde{I})]{} f(x^*)$, where $x^* \in \operatorname{argmin} f(D)$. For the corresponding subsequence $\{x^r\}_{r \in \tilde{I}}$, we then have $f(x^r) \xrightarrow[(r \in \tilde{I})]{} f(\bar{x}) \geq f(x^*)$. But, because $f(\tilde{x}^r) \geq f(x^r)$, we also have $f(\bar{x}) \leq f(x^*)$.

For several classes of problems, however, it is not possible to guarantee that a sequence of best feasible points x^k and the associated upper bounds α_k can be obtained in such a way that consistency holds (e.g., Horst (1988 and 1989), Horst and Dien (1987), Horst and Thoai (1988), Pinter (1988), cf. the comments in Section IV.1.). In this case we propose to consider the above sequence $\{\tilde{x}^k\}$ or sequences closely related to $\{\tilde{x}^k\}$ (cf. Section IV.5).

Example IV.2. We want to minimize the concave function $f(x_1, x_2) = -(x_2 - 10)^2$ on the set

$$D = \{(x_1, x_2)\} \in \mathbb{R}^2 : 0 \leq x_2 \leq 10,\ x_2 \geq 10x_1,\ x_2 \geq -10x_1\}.$$

Obviously, $\min f(D) = -100$ is attained at $\bar{x} = (0,0)$. Suppose that we apply a BB procedure that uses two-dimensional simplices M defined as the convex hull of their known vertex sets V(M), i.e., $M = \operatorname{conv} V(M)$. Moreover, suppose that all operations required by the algorithm are carried out on V(M). For example, let the lower bounds $\beta(M)$ be the minimum of f on V(M), i.e., $\beta(M) = \min f(V(M))$, and let the upper bounds $\alpha(M)$ be the minimum of f taken over the feasible vertices of M, i.e., $S_M = V(M) \cap D$, i.e.,

$$\alpha(M) = \begin{cases} \min f(V(M \cap D)), & \text{if } V(M) \cap D \neq \emptyset , \\ \infty , & \text{if } V(M) \cap D = \emptyset . \end{cases}$$

By concavity of f, we have

$$\beta(M) = \min f(M) \leq \inf f(M \cap D) \leq \alpha(M) \text{ (cf. Section I.2, Theorem 1).}$$

It is easy to define refining procedures and selection rules such that the branch and bound procedure generates a decreasing sequence $\{M_n\}$ of simplices

$$M_n = \mathrm{conv}\{(1/n, 1/n), (-1/n, 1/n), (0, -1/n)\}, \quad n \in \mathbb{N},$$

satisfying

$$V(M_n) \cap D = \emptyset , \forall n \in \mathbb{N} , \quad \lim_{n\to\infty} M_n = \{\bar{x}\} .$$

Obviously, we have

$$\lim_{n\to\infty} \beta(M_n) = f(\bar{x}) , \lim_{n\to\infty} v^n = \bar{x},$$

for any sequence of points $v^n \in M_n$.

The algorithm converges in the sense that every accumulation point of such a sequence $\{v^n\}$ solves the problem. However, because $V(M_n) \cap D = \emptyset$, we have $\alpha(M_n) = \infty$, $\forall n \in \mathbb{N}$, and the iteration sequence of best feasible points as used in the prototype algorithm is empty. Consistency does not hold, and convergence cannot be established by the theory developed above.

A related question is that of deleting infeasible partition sets in an appropriate way. As mentioned in Section IV.1, in many cases it is too difficult to decide exactly whether $M \cap D = \emptyset$ for all partition sets M from the information available. Therefore, we have to invent simple techniqes for checking infeasibility that, though possibly incorrect for a given M, lead to correct and convergent algorithms when incorporated in Step k.3 of the prototype BB procedure.

Example IV.3. In Example IV.2, we could have decided precisely whether $M \cap D = \emptyset$ or $M \cap D \neq \emptyset$. For example, we define the convex function

$$g(x) := \max \{-x_2, x_2 - 10 , 10x_1 - x_2 , -10x_1 - x_2\}$$

and solve the convex optimization problem

$$\begin{array}{l} \text{minimize } g(x) . \\ \text{s.t. } x \in M \end{array}$$

We have $x \in D$ if and only if $g(x) \leq 0$. Hence $M \cap D = \emptyset$ if and only if $\min g(M) > 0$.

Note that for the convex set $D := \{x: g_i(x) \leq 0, i = 1,...,m\}$ defined by the convex functions $g_i: \mathbb{R}^n \longrightarrow \mathbb{R}$ $(i = 1,...,m)$, the function $g(x) = \max \{g_i(x), i=1,...,m\}$ is nonsmooth, and it may be numerically expensive to solve the above minimization problem for each M satisfying $V(M) \cap D = \emptyset$. Moreover, whenever $\min g(M) > 0$ is small, the algorithms available for solving this problem may in practice produce a result that leads to a wrong decision on M.

The situation is, of course, worse for more complicated feasible sets D. For example, let D be defined by a finite number of inequalities $g_i(x) \leq 0$, $i \in I$, where $g_i: \mathbb{R}^n \longrightarrow \mathbb{R}$ is Lipschitzian on a set M_0 containing D $(i \in I)$. Then the problem of deciding whether we have $M \cap D = \emptyset$ or $M \cap D \neq \emptyset$ is clearly almost as difficult as the original problem (1).

Definition IV.7. *A lower bounding operation is called* ***strongly consistent****, if at every step any undeleted partition element can be further refined, and if any infinite decreasing sequence* $\{M_{k_q}\}$ *of successively refined partition elements possesses a subsequence* $\{M_{k_{q'}}\}$ *satisfying*

$$\overline{M} \cap D \neq \emptyset , \quad \beta (M_{k_{q'}}) \xrightarrow[q' \to \infty]{} \min f(\overline{M} \cap D) , \quad \text{where } \overline{M} = \bigcap_q M_{k_q} .$$

Note that the notions of consistency and strong consistency involve not only the calculation of bounds but, obviously, also the subdivision of partition elements. Moreover, in order to ensure strong consistency of the lower bounding operation, the deletion of infeasible sets in Step k.3 of the BB procedure has to guarantee that $M \cap D \neq \emptyset$ holds for the limit M of every nested sequence of partition elements generated by the algorithm.

Definition IV.8. *The "deletion by infeasibility" rule used in Step k.3 throughout a BB procedure is called* **certain in the limit** *if for every infinite decreasing sequence* $\{M_{k_q}\}$ *of successively refined partition elements with limit* $\overline{M}$ *we have* $\overline{M} \cap D \neq \emptyset$.

Corollary IV.3. *In the BB procedure suppose that the lower bounding operation is strongly consistent and the selection is bound improving. Then we have*

$$\beta = \lim_{k \to \infty} \beta_k = \min f(D).$$

Proof. As in the proof of Theorem IV.3, it follows by the bound improving selection that there must be a decreasing sequence $\{M_{k_q}\}$ satisfying $\beta_{k_q} = \beta(M_{k_q})$. Clearly, $\beta = \lim_{k \to \infty} \beta_k \leq \min f(D)$ exists (the limit of a nondecreasing sequence bounded from above). Hence we have

$$\beta = \lim_{q \to \infty} \beta(M_{k_q}) = \min f(\overline{M} \cap D) \leq \min f(D).$$

Since $D \supset \overline{M} \cap D$, this is only possible if $\min f(\overline{M} \cap D) = \min f(D)$. ■

In the following sections, some concrete types of partition sets, refining operations, and bounding and deletion rules are presented that illustrate the wide range of applicability of the BB procedure.

3. TYPICAL PARTITION SETS AND THEIR REFINEMENT

As mentioned above, for the partition sets M it is natural to use most simple polytopes or convex polyhedral sets such as simplices, rectangles and polyhedral cones.

3.1. Simplices

Suppose that $D \subset \mathbb{R}^n$ is robust and convex. Furthermore, let M_0 and all partition elements be n-dimensional simplices (n-simplices). An initial simplex M_0 can usually be determined as described in Section II.4..

Definition IV.9. *Let M be an n–simplex with vertex set* $V(M) = \{v^0, v^1,..,v^n\}$. *Choose a point* $w \in M$, $w \notin V(M)$ *which is uniquely represented by*

$$w = \sum_{i=0}^{n} \lambda_i v^i \,,\ \lambda_i \ge 0\ (i=0,...,n),\ \sum_{i=0}^{n} \lambda_i = 1, \tag{3}$$

and for each i such that $\lambda_i > 0$ *form the simplex* $M(i,w)$ *obtained from M by replacing the vertex i by w, i.e.,* $M(i,w) = \text{conv}\,\{v^0,..,v^{i-1},w,v^{i+1},.., v^n\}$.

This subdivision is called a ***radial subdivision*** *(Fig. IV.4).*

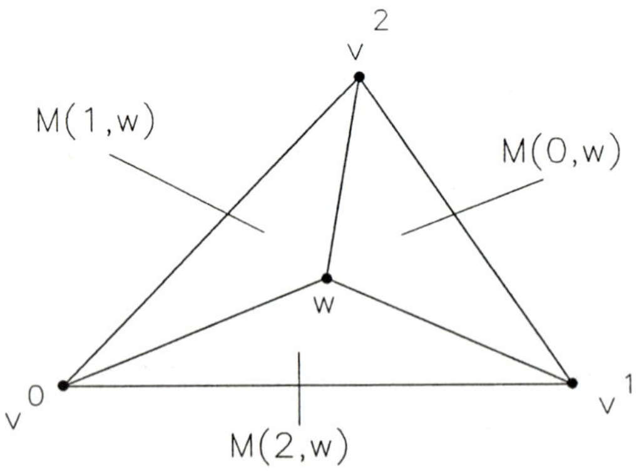

Fig. IV.4. Radial subdivision of a simplex

Radial subdivisions of simplices were introduced in Horst (1976) and subsequently used by many authors.

Proposition IV.1. *The set of subsets M(i,w) that can be constructed from an n–simplex M by an arbitrary radial subdivision forms a partition of M into n–simplices.*

Proof. It is well known from elementary linear algebra that, given an affinely independent set of points $\{v^0,\ldots, v^n\}$ and a point w represented by (3), then $\{v^0,\ldots,v^{i-1},w,v^{i+1},\ldots,v^n\}$ is a set of affinely independent points whenever we have $\lambda_i > 0$ in (3). Thus, all the sets M(i,w) generated by a radial subdivision of an n–simplex M are n–simplices.

Let $x \in M(i,w)$, i.e.,

$$x = \sum_{\substack{j=0\\ j\neq i}}^{n} \mu_j v^j + \mu_i w,\ \mu_j \geq 0 \quad (j=0,\ldots,n), \quad \mu_i + \sum_{\substack{j=0\\ j\neq i}}^{n} \mu_j = 1. \tag{4}$$

Inserting (3) yields

$$x = \sum_{\substack{j=0\\ j\neq i}}^{n} \mu_j v^j + \mu_i \sum_{k=0}^{n} \lambda_k v^k, \quad \mu_j \geq 0\ (j=0,\ldots,n),\ \lambda_k \geq 0\ (k=0,\ldots,n),$$

$$\mu_i + \sum_{\substack{j=0\\ j\neq i}}^{n} \mu_j = 1, \qquad \sum_{k=0}^{n} \lambda_k = 1. \tag{5}$$

This can be expressed as

$$x = \sum_{\substack{j=0\\ j\neq i}}^{n} (\mu_j + \mu_i\lambda_j)v^j + \mu_i\lambda_i v^i, \tag{6}$$

where all of the coefficients in (6) are greater than or equal to zero, and, by (5), their sum equals $1 - \mu_i + \mu_i(1-\lambda_i) + \mu_i\lambda_i = 1$. Hence, x is a convex combination of the vertices $v^0,\ldots,v^n$ of M and we have $x \in M$. Consequently, $M(i,w) \subset M$.

To show that $M \subset \cup M(i,w)$, let $x \in M$, $x \neq w$ and consider the ray $\rho(w,x) = \{\alpha(x-w)+w, \alpha \geq 0\}$ from w through x. Let y be the point where $\rho(w,x)$ intersects the boundary of M. Then we have

$$y = \sum_{i \in I} \mu_i v^i,\ \mu_i \geq 0\ (i \in I),\ \sum_{i \in I} \mu_i = 1,\ I \subset \mathbb{N},\ |I| < n+1 \tag{7}$$

(y is an element of a face of M of dimension smaller than n).

By construction, we also have

$$x = \bar{\alpha} y + (1 - \bar{\alpha}) w,\ 0 < \bar{\alpha} \leq 1.$$

Hence, using (7), we obtain

$$x = \sum_{i \in I} \bar{\alpha} \mu_i v^i + (1 - \bar{\alpha}) w, \tag{8}$$

where $\bar{\alpha}\mu_i \geq 0$ $(i \in I)$, $(1 - \bar{\alpha}) \geq 0$ and $\sum_{i \in I} \bar{\alpha}\mu_i + (1 - \bar{\alpha}) = 1$. It follows that $x \in \cup M(i,w)$.

Finally, for $i \neq k$, let $x \in M(i,w) \cap M(k,w)$, so that

$$x = \sum_{\substack{j=0 \\ j \neq i,k}}^{n} \lambda_j v^j + \lambda_k v^k + \lambda_i w = \sum_{\substack{j=0 \\ j \neq i,k}}^{n} \mu_j v^j + \mu_i v^i + \mu_k w\ , \tag{9}$$

$$\lambda_j \geq 0,\ \mu_j \geq 0 \quad (j=0,\ldots,n),\quad \sum_{j=0}^{n} \lambda_i = \sum_{j=0}^{n} \mu_j = 1.$$

In (9), we cannot have $\lambda_k > 0$ or $\mu_i > 0$. To see this, suppose that $\lambda_k > 0$. Then express in (9) w as a convex combination of $v^0,\ldots,v^n$, and subtract v^i from each term of the second equality. It follows that $v^k - v^i$ can be written as a linear combination of the remaining $v^j - v^i$ $(j \neq i,k)$. This contradicts the affine independence of the vertices $v^0,\ldots,v^n$. Finally, the case $\mu_i > 0$ can be treated in a similar way. The details are left to the reader. ■

In order to establish consistency or strong consistency for a bounding operation in a BB procedure using radially subdivided simplices, we have to make sure that the points w in Definition IV.9 are chosen in a way which guarantees convergence of every decreasing sequence of simplices generated by the algorithm to a simple set $\overline{M}$, where min $f(\overline{M})$ is known. If no specific properties of f are known that could be exploited, then we would like to ensure that $\overline{M}$ is a singleton.

Denote by $\delta(M)$ **the diameter of M** (measured by the Euclidean distance). For the simplex M, $\delta(M)$ is also the **length of a longest edge of M**.

Definition IV.10. *A subdivision is called* ***exhaustive*** *if* $\delta(M_q) \xrightarrow[(q\to\infty)]{} 0$ *for all decreasing subsequences* $\{M_q\}$ *of partition elements generated by the subdivision.*

The notion of exhaustiveness was introduced in Thoai and Tuy (1980) for a similar splitting procedure for cones (see also Horst (1976), and it was further investigated in Tuy, Katchaturov and Utkin (1987). Note that exhaustiveness, though often intuitively clear, is usually not easy to prove for a given radial subdivision procedure. Moreover, some straightforward simplex splitting procedures are **not** exhaustive.

Example IV.4. Let $v^i_{M_q}$ (i=0,..,n) denote the vertices of a simplex M_q. Let $n > 1$, and in Definition IV.9 choose the **barycenter** of M_q, i.e.,

$$w = w_q = \frac{1}{n+1} \sum_{i=0}^{n} v^i_{M_q} . \tag{10}$$

Construct a decreasing subsequence $\{M_q\}$ of simplices using radial subdivision with w given by (10), and suppose that for all q, M_{q+1} is obtained from M_q by replacing $v^n_{M_q}$ by the barycenter w_q of M_q. Then, clearly, every simplex M_q contains the face conv $\{v^0_{M_1},...,v^{n-1}_{M_1}\}$ of the initial simplex M_1 , and thus $\overline{M} = \bigcap_{q=1}^{\infty} M_q$ has positive diameter.

A large class of exhaustive radial subdivisions is discussed in Tuy, Katchaturov and Utkin (1987). We present here only the most frequently used **bisection**, introduced in Horst (1976), where, in Definition IV.10, w is the **midpoint of one of the longest edges of M**, i.e.,

$$w = \frac{1}{2}(v_M^r + v_M^s) \tag{11}$$

where $[v_M^r, v_M^s]$ is a longest edge of M. In this case, M is obviously subdivided into two n–simplices having equal volume. The exhaustiveness of any decreasing sequence of simplices produced by successive bisection follows from the following result.

Proposition IV.2. *Let $\{M_q\}$ be any decreasing sequence of n–simplices generated by the bisection subdivision process. Then we have*

(i) $\delta(M_{n+q}) \leq \frac{\sqrt{3}}{2}\delta(M_q) \quad \forall q,$

(ii) $\delta(M_q) \xrightarrow[(q\to\infty)]{} 0.$

Proof. Consider a sequence $\{M_q\}$ such that M_{q+1} is always obtained from M_q by bisection. Let $\delta(M_q) = \delta_q$. It suffices to prove (i) for q = 1. Color every vertex of M_1 "black", color "white" every vertex of M_r with $r > 1$ which is not black. Let d_r denote the longest edge of M_r that is bisected. Let p be the smallest index such that d_p has at least one white endpoint.
Since a black vertex is replaced by a white one at each bisection before p, we must have $p \leq n+1$.

Let $d_p = [u,v]$ with u white. Then u is the midpoint of some d_k with $k < p$. Let $d_k = [a,b]$. If a or b coincides with v, then obviously $\delta_p = \frac{1}{2}\delta_k \leq \frac{1}{2}\delta_1$ and (i) holds.

Otherwise, consider the triangle conv {a,b,v}. Since $v \in M_p \subset M_k$ and $\delta_k = \|d_k\|$ is the diameter of M_k, we must have $\|v-a\| \leq \delta_k$, $\|v-b\| \leq \delta_k$. Since u is the midpoint of [a,b], we deduce from the "parallelogram rule"

$$2\|b-u\|^2 + 2\|u-v\|^2 = \|v-a\|^2 + \|v-b\|^2$$

and the relation

$$\|b-u\|^2 = \frac{1}{4}\|a-b\|^2$$

that

$$2\|u-v\|^2 = \|v-a\|^2 + \|v-b\|^2 - \frac{1}{2}\|a-b\|^2 \le 2\delta_k^2 - \frac{1}{2}\delta_k^2 = \frac{3}{2}\delta_k^2 ,$$

and therefore $\delta_p \le \frac{\sqrt{3}}{2}\delta_k$. Since $\delta_{n+1} \le \delta_p$ and $k \ge 1$, we then have (i).

(ii) is an immediate consequence of (i). ■

Though very natural and simple, bisection is not necessarily the most efficient way to subdivide simplices, since it does not take into account the structure of a given optimization problem.

More sophisticated subdivision procedures for simplices that are especially useful in linearly constrained concave minimization problems will be presented in Chapter VII of this book.

Note that the notion of a radial subdivision can be defined similarly for any sequence of polytopes, and the definition of exhaustiveness is obviously not restricted to simplices.

3.2. Rectangles and Polyhedral Cones

The further development of BB methods to be presented later shows that in most cases the whole vertex set V(M) of a partition set M is needed to compute bounds and "deletion by infeasibility" rules. Since an n–simplex has the least number of vertices among all n–dimensional polytopes, it is frequently natural to choose n–simplices.

For some classes of problems, however, rectangles $M = \{x: a \le x \le b\}$, $a,b \in \mathbb{R}^n$, $a < b$, are a more natural choice. Note that M is uniquely determined by its "lower left" vertex $a = (a_1,...,a_n)^T$ and its "upper right" vertex $b = (b_1,...,b_n)^T$. Each of the 2^n vertices of the rectangle M is of the form

$$a + c$$

where c is a vector with components 0 or $(b_i - a_i)$ $(i \in \{1,...,n\})$.

Moreover, an initial rectangle $M_0 \supset D$ is often known by given bounds on the variables.

Rectangular partition sets have been used to solve certain Lipschitzian optimization problems (e.g., Strongin (1984), Pinter (1986, 1986 a, 1987), Horst (1987 and 1988), Horst and Tuy (1987), Neferdov (1987) cf. Chapter XI).

Most naturally, rectangular sets are suitable if the functions involved in the problem are **separable**, i.e., the sum of n functions of one variable, since in this case appropriate lower bounds are often readily available (e.g., Falk and Soland (1969), Soland (1971), Falk (1972), Horst (1978), Kalantari and Rosen (1987), Pardalos and Rosen (1987)). We will return to separable problems in several later chapters.

Let M be an n–rectangle and let $w \in M$, $w \notin V(M)$, where V(M) again denotes the vertex set of M. Then a radial subdivision of M using w (defined in the same way as in the case of simplices) does not partition M into n–rectangles but rather into more complicated sets.

Therefore, the subdivision of n–rectangles is usually defined via hyperplanes passing through w parallel to the facets ((n–1) dimensional faces) of M, so that M is partitioned into up to 2^n rectangles.

For most algorithms, the subdivision must be **exhaustive**. An example is the **bisection**, where w is the midpoint of one of the longest edges of M and M is subdivided into two n–rectangles having equal volume and such that w is a vertex of both new n–rectangles. It can be shown in a manner similar to Proposition IV.2, that the bisection of n–rectangles is exhaustive.

Polyhedral cones are frequently used for concave minimization problems with (possibly unbounded) robust convex feasible sets (e.g., Thoai and Tuy (1980), Tuy, Thieu and Thai (1985), Horst, Thoai and Benson (1988)).

Assume that D possesses an interior point y^0. Let S be an n–simplex containing y^0 in its interior. Consider its n+1 facets F_i. Each F_i is an (n–1)–simplex.

For each F_i let C_i be the convex polyhedral cone vertexed at y^0 and having exactly n edges defined by the halflines from y^0 through the n vertices of F_i. Then $\{C_i: i = 1,...,n+1\}$ is a conical partition of $\mathbb{R}^n$.

We construct a partition of such a cone C by means of the corresponding facet F. Any radial subdivision of the simplex F defines a partition of C into cones vertexed at y^0 and having exactly n edges, namely the halflines from y^0 through the vertices of the corresponding partition element of F. Whenever the subdivision of the (n–1)–simplices F is exhaustive, then any nested sequence $\{C_q\}$ of cones associated to the subdivision of the corresponding sequence $\{F_q\}$ of (n–1)–simplices converge to a ray from y^0 through the point $\bar{x}$ satisfying $F_q \xrightarrow[(q\to\infty)]{} \{\bar{x}\}$.

4. LOWER BOUNDS

In this section, we discuss some examples of bounding operations. Given a family of partition sets, a subdivision procedure (e.g., as outlined in the preceding section) and an appropriate "deletion by infeasibility" mechanism, we find a group of consistent (strongly consistent) bounding operations. Since the calculation of lower bounds $\beta(M)$ depends on the data of the optimization problem to be solved, we first discuss some ideas and classes of problems which occur frequently.

Note that not all the bounding operations discussed below are necessarily monotonic as required in Step k.4 of the Prototype BB–Procedure. Let $M \subset M'$ be an element of a partition of Mq and suppose that we have $\beta(M) < \beta(M')$. Then let us agree to use

$$\bar{\beta}(M) = \max\{\beta(M), \beta(M')\}$$

instead of $\beta(M)$ as defined below.

4.1. Lipschitzian Optimization

Natural lower bounds on Lipschitzian functions have already been mentioned in Section I.4.:

Let f be **Lipschitzian on M**, i.e., assume that there is a constant $L > 0$ such that

$$|f(z)-f(x)| \leq L\|z-x\| \quad \forall x,z \in M , \tag{12}$$

where $\|\cdot\|$ denotes the Euclidean norm.

Suppose that an **upper bound A for L** is known.

Finding a good upper bound $A \geq L$ is, of course, difficult in general, but without such a bound branch and bound cannot be applied for Lipschitzian problems. On the other hand, there are many problems, where A can readily be determined. Note that in a branch and bound procedure "local" bounds for L on the partition sets M should be used instead of global bounds on M_0.

Moreover, suppose that the **diameter $\delta(M)$** of M is known.

Recall that for a simplex M, $\delta(M)$ is the length of the longest edge of M. For a rectangle $M = \{x \in \mathbb{R}^n: a \leq x \leq b\}$; $a,b \in \mathbb{R}^n$, $a < b$, $\delta(M)$ is the length of the diagonal [a,b] joining the "lower left" vertex a and the "upper right" vertex b (all inequalities are to be understood in the sense of the componentwise ordering in $\mathbb{R}^n$).

By (12), we then have

$$f(z) \geq f(x) - L\|z-x\| \geq f(x) - A\delta(M) \quad \forall x,z \in M. \tag{13}$$

Let V'(M) be a nonempty subset of the vertex set V(M) of M. Then

$$\beta(M) := \max\{f(x): x \in V'(M)\} - A\delta(M) \tag{14}$$

is a natural lower bound.

In (14), V'(M) can be replaced by any known subset of M.

If $M = \{x \in \mathbb{R}^n: a \leq x \leq b\}$ is a hyperrectangle, then

$$\beta(M) := f(\tfrac{1}{2}(a+b)) - \tfrac{A}{2}\delta(M) \tag{14'}$$

might be a better choice than (14).

4.2. Vertex Minima

Let M be a polytope. Then, for certain classes of objective functions, lower bounds for $\inf f(M \cap D)$ or $\inf f(M)$ can be determined simply by minimizing a certain function related to f over the finite vertex set V(M) of M.

For example, if **f is concave on M**, then

$$\inf f(M \cap D) \geq \min f(M) = \min f(V(M)),$$

and we may choose

$$\beta(M) = \min f(V(M)). \tag{15}$$

Tighter bounds can be obtained by cutting off parts of $M \setminus D$ by means of some steps of an outer approximation (relaxation) method by cutting planes (cf. Chapter II). The resulting polytope P will satisfy $M \cap D \subseteq P \subset M$, hence

$$\inf f(D \cap M) \geq \min f(V(P)) \geq \min f(M), \tag{16}$$

and $\beta(M) = \min f(V(P))$ is, in general, a tighter bound than $\min f(V(M))$.

Another example is the calculation of lower bounds for a d.c.–function $f(x) = f_1(x) + f_2(x)$, where f_1 is concave and f_2 is convex on a polytope M.

Choose any $v^* \in M$ and let $\partial f_2(v^*)$ denote the subdifferential of f_2 at v^*. Let $p^* \in \partial f_2(v^*)$ and determine

$$\bar{v} \in \operatorname{argmin} \{f_1(v) + f_2(v^*) + p^*(v-v^*): v \in V(M)\} . \tag{17}$$

Then, by the definition of a subgradient, we have

$$\ell(x) := f_2(v^*) + p^*(x-v^*) \le f_2(x) \quad \forall x \in \mathbb{R}^n,$$

and hence $f_1(x) + \ell(x) \le f(x) \quad \forall x \in \mathbb{R}^n$.

But $f_1(x) + \ell(x)$ is concave and attains its minimum on M at a vertex of M. Consequently,

$$\beta(M) := f_1(\bar{v}) + f_2(v^*) + p^*(\bar{v}-v^*) \tag{18}$$

is a lower bound for $\min f(M) \le \inf f(M \cap D)$.

4.3. Convex Subfunctionals

A commonly used way of calculating lower bounds $\beta(M)$ for $\min f(D \cap M)$ or $\min f(M)$ is by minimizing a suitable convex subfunctional of f over $D \cap M$ or over M. A **convex subfunctional** of f on M is a convex function that never exceeds f on M. A convex subfunctional φ is said to be the **convex envelope** of f on M, if no other convex subfunctional of f on M exceeds φ at any point $x \in M$ (i.e., φ is the pointwise supremum of all convex subfunctionals of f on M). Convex envelopes play an important role in optimization and have been discussed by many authors. We refer here mainly to Falk (1969), Rockafellar (1970), Horst (1976, 1976a and 1979).

The convex envelope φ is the uniformly best convex approximation of f on M from below. For our purposes, it is sufficient to consider lower semicontinuous functions f: $M \longrightarrow \mathbb{R}$, where $M \subset \mathbb{R}^n$ is compact and convex. We are led thereby to the following definition.

Definition IV.11. *Let $M \subset \mathbb{R}^n$ be convex and compact, and let $f: M \longrightarrow \mathbb{R}$ be lower semicontinuous on M. A function $\varphi: M \longrightarrow \mathbb{R}$ is called the* ***convex envelope*** *of f on M if it satisfies*

(i) $\varphi(x)$ *is convex on M,*

(ii) $\varphi(x) \le f(x) \;\; \forall x \in M,$

(iii) *there is no function $\Psi: M \longrightarrow \mathbb{R}$ satisfying (i), (ii) and $\varphi(\bar{x}) < \Psi(\bar{x})$ for some point $\bar{x} \in M$.*

Thus we have $\varphi(x) \geq \Psi(x)$ for all $x \in M$ and all convex subfunctionals Ψ of f on M.

By (iii), it is easily seen that the convex envelope is uniquely determined, if it exists.

Theorem IV.4. *Let $f: M \longrightarrow \mathbb{R}$ be lower semicontinuous on the convex compact set $M \subset \mathbb{R}^n$. Let φ be the convex envelope of f over M. Then we have*

a) $\min \varphi(M) = \min f(M)$

b) $\operatorname{argmin} \varphi(M) \supset \operatorname{argmin} f(M).$

Proof. Let $\bar{x} \in \operatorname{argmin} f(M)$. Then we must have $\varphi(\bar{x}) \le f(\bar{x})$ (by Definition IV.11 (ii)). But we cannot have $\varphi(\bar{x}) < f(\bar{x})$, since if this were the case, then the constant function $\Psi(x) \equiv f(\bar{x})$ would satisfy (i), (ii) of Definition IV.11; but the relation $\Psi(\bar{x}) > \varphi(\bar{x})$ would contradict (iii).

Moreover, it follows by the same argument that we have

$$\varphi(x) \geq \min f(x) \;\; \forall x \in M,$$

hence a) and b) are both proved. ■

By Theorem IV.4, many nonconvex minimization problems could be replaced by a convex minimization problem if the convex envelope φ of f were available.

However, for arbitrary convex compact sets and arbitrary lower semicontinuous f, computing φ is at least as difficult as solving the original minimization problem. Nevertheless, within the BB method, for sufficiently simple M and certain special forms of f, the convex envelope is readily available and can be used to compute lower bounds $\beta(M)$.

A geometrically more appealing way to introduce convex envelopes is by means of the following characterization (Fig. IV.5).

Recall that the epigraph epi (f):= $\{(x,r) \in M \times \mathbb{R}: r \geq f(x)\}$ of a given function f: $M \longrightarrow \mathbb{R}$ consists of the points in $M \times \mathbb{R}$ on and above the graph of f.

Lemma IV.1. *Let $M \subset \mathbb{R}^n$ be compact and convex, and let $f: M \longrightarrow \mathbb{R}$ be lower semicontinuous on M. Then $\varphi: M \longrightarrow \mathbb{R}$ is the convex envelope of f if and only if*

$$\mathit{epi}\,(\varphi) = \mathit{conv}\,(\mathit{epi}\,(f)) \tag{19}$$

or, equivalently

$$\varphi(x) = \inf\{r: (x, r) \in \mathit{conv}\,(\mathit{epi}\,(f))\}\ . \tag{20}$$

Proof. The proof is an immediate consequence of the definition of the convex hull conv (A) of a set A as the smallest convex set containing A. ■

Note that epi (f) and conv (epi(f)) are closed sets, since f is lower semicontinuous (cf., e.g., Rockafellar (1970), Blum and Oettli (1975)).

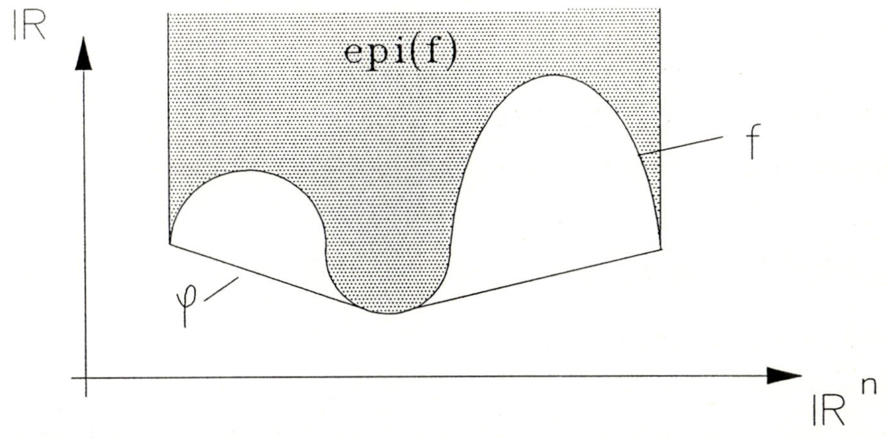

Fig. IV.5. Convex envelope

A useful result can be derived by means of Carathéodory's Theorem which states that every point of a compact convex set M in $\mathbb{R}^n$ is the convex combination of at most n+1 extreme points of M (cf. Rockafellar (1970).

Corollary IV.4. *Let $M \subset \mathbb{R}^n$ be compact and convex, and let $f: M \longrightarrow \mathbb{R}$ be lower semicontinuous on M. Then φ and f coincide at the extreme points of M.*

Proof. Applying Carathéodory's Theorem to (20), we see that $\varphi(x)$ can be expressed as

$$\varphi(x) = \inf \left\{ \sum_{i=1}^{n+1} \lambda_i f(x^i),\ \sum_{i=1}^{n+1} \lambda_i = 1,\ \sum_{i=1}^{n+1} \lambda_i x^i = x; \right.$$

$$\left. \lambda_i \ge 0,\ x^i \in M\ (i=1,\dots,n+1) \right\}.$$

Since, for extreme points x, x = x is the only representation of x as a convex combination of points in M, it follows that $\varphi(x) = f(x)$. ∎

The convex envelope φ can also be characterized as well by means of the

conjugate function as defined by Fenchel (1949 and 1951) (see also, e.g., Rockafellar (1970)).

Definition IV.12. *Let $f\colon M \to \mathbb{R}$ be lower semicontinuous on the convex, compact set $M \subset \mathbb{R}^n$. Then*

$$f^*(t) = \max_{x \in M} \{xt - f(x)\} \tag{21}$$

is called the ***conjugate function*** *of f.*

Note that the maximum in (21) always exists, since f is lower semicontinuous and M is compact. Hence, f^* is defined for all $t \in \mathbb{R}^n$. If we replace the max operator in (21) by the sup operator then the conjugate can be defined for arbitrary functions on M. The domain of f^* is then

$$D(f^*) = \{t\colon \sup_{x \in M} \{xt - f(x)\} < \infty\} . \tag{22}$$

It is easily seen that f^* is convex (the pointwise supremum of a family of affine functions).

The same operation can be performed on f^* to yield a new convex function f^{**}: this is the so–called second conjugate of f. The function f^{**} turns out to be identical to the convex envelope φ (cf., e.g., Falk (1969), Rockafellar (1970)). Our proof of the following theorem follows Falk (1969).

Theorem IV.5. *Let $f\colon M \to \mathbb{R}$ be lower semicontinuous on the compact convex set $M \subset \mathbb{R}^n$. Denote by f^{**} the second conjugate of f and by φ the convex envelope of f on M. Then we have*

$$f^{**}(x) = \varphi(x) \quad \forall x \in M \tag{23}$$

Proof. We first show that f^{**} is defined throughout M, i.e., $D(f^{**}) \supset M$.
Let $x^0 \in M$. Then, by (21), we have

$$x^0t - f(x^0) \le f^*(t) \quad \forall t \in \mathbb{R}^n .$$

Hence,

$$x^0t - f^*(t) \le f(x^0) \quad \forall t \in \mathbb{R}^n ,$$

and

$$f^{**}(x^0) = \sup_{t \in \mathbb{R}^n} \{x^0t - f^*(t)\} \le f(x^0) < \infty , \tag{24}$$

i.e., $x^0 \subset D(f^{**})$.

Note that it can actually be shown that $D(f^{**}) = M$ (cf., e.g., Falk (1969), Rockafellar (1970)).

Since f^{**} is convex on M and, by (24), $f^{**}(x) \le f(x) \;\; \forall x \in M$, it follows from the definition of φ that $f^{**}(x) \le \varphi(x) \;\; \forall x \in M$. Suppose that $f^{**}(x^0) < \varphi(x^0)$ for some $x^0 \in M$. Then we have $(x^0, f^{**}(x^0)) \notin \text{epi}\,(\varphi)$. Note that epi (φ) is a closed convex set. Thus, there is a hyperplane strictly separating the point $(x^0, f^{**}(x^0))$ from epi φ, i.e., there is a vector $(s,\sigma) \in \mathbb{R}^{n+1}$ satisfying

$$\sigma\varphi(x) + xs > \sigma f^{**}(x^0) + x^0s \quad \forall x \in M . \tag{25}$$

If $\sigma = 0$, then $xs > x^0s \;\; \forall x \in M$, which is impossible, since $x^0 \in M$.

Now, since $\sigma \neq 0$, we may divide (25) by $-\sigma$ and set $s \longleftarrow -s/\sigma$. Then from (25) it follows that either

$$-\varphi(x) + xs > -f^{**}(x^0) + x^0s \quad \forall x \in M \tag{26}$$

or

$$-\varphi(x) + xs < -f^{**}(x^0) + x^0s \quad \forall x \in M. \tag{27}$$

If (26) holds, then in particular it must hold for $x = x^0$, so that

$$-\varphi(x^0) > -f^{**}(x^0).$$

But this is equivalent to $f^{**}(x^0) > \varphi(x^0)$, which we have seen above to be false. If (27) holds, then, since $\varphi(x) \le f(x)$, it follows that

$$-f(x) + xs < -f^{**}(x^0) + x^0 s \quad \forall x \in M.$$

But because $f^{**}(x^0) \ge x^0 s - f^*(s)$ this implies that

$$-f(x) + xs < f^*(s) \quad \forall x \in M . \tag{28}$$

This contradicts the definition of f^*, which implies that in (25) equality (rather than inequality) holds for some $x \in M$. ■

Further useful results on convex envelopes have been obtained for special classes of functions f over special polytopes. The following two results are due to Falk and Hoffman (1976) and to Horst (1976 and 1976a).

Theorem IV.6. *Let M be a polytope with vertices $v^1,\dots,v^k$, and let $f\colon M \longrightarrow \mathbb{R}$ be concave on M. Then the convex envelope φ of f on M can be expressed as*

$$\begin{aligned} \varphi(x) &= \min_{\alpha} \sum_{i=1}^{k} \alpha_i f(v^i) \\ \text{s.t. } &\sum_{i=1}^{k} \alpha_i v^i = x \,, \; \sum_{i=1}^{k} \alpha_i = 1, \; \alpha \ge 0 \,, \end{aligned} \tag{29}$$

where $\alpha = (\alpha_1,\dots,\alpha_k)$.

Proof. The function φ defined in (29) is convex. To see this, let $0 \le \lambda \le 1$ and $x^1,x^2 \in M$. Let α^1,α^2 solve (29) for $x = x^1$ and $x = x^2$, respectively. Then

$$\varphi(\lambda x^1 + (1-\lambda)x^2) \le \sum_{i=1}^{k} (\lambda\alpha_i^1 + (1-\lambda)\alpha_i^2) f(v^i)$$

$$= \lambda \sum_{i=1}^{k} \alpha_i^1 f(v^i) + (1-\lambda) \sum_{i=1}^{k} \alpha_i^2 f(v^i) = \lambda\varphi(x^1) + (1-\lambda)\varphi(x^2),$$

where the inequality follows from the feasibility of $(\lambda\alpha^1 + (1-\lambda)\alpha^2)$.

For every α that is feasible for the above problem (29), the concavity of f implies that

$$\varphi(x) \le \sum_{i=1}^{k} \alpha_i f(v^i) \le f(x) \quad \forall\, x \in M .$$

Finally, suppose that Ψ is a convex function on M which underestimates f over M, and suppose that $\varphi(\bar{x}) < \Psi(\bar{x})$ for some $\bar{x} \in M$. Let $\bar{\alpha}$ solve (29) for $x = \bar{x}$. Then we have

$$\varphi(\bar{x}) < \Psi(\bar{x}) = \Psi\,(\sum_{i=1}^{k} \bar{\alpha}_i v^i) \le \sum_{i=1}^{k} \bar{\alpha}_i \Psi(v^i) \le \sum_{i=1}^{k} \bar{\alpha}_i f(v^i) = \varphi(\bar{x})$$

which is a contradiction. ■

Theorem IV.7. *Let* $M = conv\, \{v^0,..,v^n\}$ *be an n–simplex with vertices* $v^0,..,v^n$*, and let* $f\colon M \longrightarrow \mathbb{R}$ *be concave on M. Then the convex envelope of f on M is the affine function*

$$\varphi(x) = ax + \alpha, \quad a \in \mathbb{R}^n, \quad \alpha \in \mathbb{R} \tag{30}$$

which is uniquely determined from the system of linear equations

$$f(v^i) = av^i + \alpha \quad (i=0,1,\dots,n)\,. \tag{30'}$$

Proof. (30') constitutes a system of (n+1) linear equations in the n+1 unknown $a \in \mathbb{R}^n$, $\alpha \in \mathbb{R}$.

Subtracting the first equation from all of the n remaining equations yields

$$a(v^i - v^0) = f(v^i) - f(v^0) \quad (i=1,\dots,n).$$

The coefficient matrix V^T whose columns are the vectors $v^i - v^0$ (i=1,...,n) is nonsingular, since the vectors $v^i - v^0$ (i=1,...,n) are linearly independent. Thus, the α are uniquely determined by (30').

But $\varphi(x) = ax + \alpha$ is affine, and hence convex.

Let $x \in M$, hence $x = \sum_{i=0}^{n} \lambda_i v^i$, $\sum_{i=0}^{n} \lambda_i = 1$, $\lambda_i \geq 0$ $(i=0,\ldots,n)$.

From the concavity of f it follows that

$$\varphi(x) = \sum_{i=0}^{n} \lambda_i \varphi(v^i) = \sum_{i=0}^{n} \lambda_i f(v^i) \leq f(x), \text{ hence}$$

$\varphi(x) \leq f(x)$ $\forall x \in M$, and φ is a convex subfunctional of f on M.

Now suppose that there is another convex subfunctional Ψ of f on M and a point $\bar{x} \in M$ satisfying $\Psi(\bar{x}) > \varphi(\bar{x})$.

Then $\bar{x} = \sum_{i=0}^{n} \mu_i v^i$, $\sum_{i=0}^{n} \mu_i = 1$, $\mu_i \geq 0$ $(i = 0,\ldots,n)$, and

$$\Psi(\bar{x}) = \Psi(\sum_{i=0}^{n} \mu_i v^i) \leq \sum_{i=0}^{n} \mu_i \Psi(v^i) \leq \sum_{i=0}^{n} \mu_i f(v^i) = \sum_{i=0}^{n} \mu_i \varphi(v^i) = \varphi(\bar{x}),$$

which is a contradiction. ■

Note that Theorem IV.7 can also be derived from Theorem IV.6. Each point x of an n–simplex M has a **unique** representation as a convex combination of the n+1 affinely independent vertices $v^0,\ldots,v^n$. To see this, consider

$$x = \sum_{i=1}^{n} \alpha_i v^i + (1 - \sum_{i=1}^{n} \alpha_i) v^0 \,, \quad \alpha_i \geq 0 \quad (i=1,\ldots,n) \,,$$

i.e.,

$$x = \sum_{i=1}^{n} \alpha_i (v^i - v^0) + v^0 = V\alpha + v^0 \,, \quad \alpha \geq 0 \,,$$

where $\alpha = (\alpha_1,\ldots,\alpha_n)^T$ and V is the nonsingular matrix with rows $v^i - v^0$ $(i=1,\ldots,n)$. It follows that $\alpha = V^{-1}(x - v^0)$ is uniquely determined by x. Hence, by Theorem IV.6, we have

$$\varphi(x) = \sum_{i=0}^{n} \alpha_i f(v^i) \,, \tag{30''}$$

where $x = \sum_{i=0}^{n} \alpha_i v^i$, $\sum_{i=0}^{n} \alpha_i = 1$, $\alpha_i \geq 0$ (i=1,...,n) is the unique representation of x in the barycentric coordinates $\alpha_0,\ldots,\alpha_n$ of M. It is very easy to see that this function coincides with (30).

It follows from (30") that, whenever the barycentric coordinates of M are used, the system of linear equations (30') does not need to be solved in order to determine $\varphi(x)$.

By Theorem IV.7, the construction of convex envelopes is especially easy for concave functions of one real variable $f: [a,b] \longrightarrow \mathbb{R}$ over an interval [a,b]. The graph of the convex envelope φ then is simply the line segment passing through the points (a, f(a)) and (b, f(b)).

Another useful result is due to Falk (1969) and Al–Khayyal (1983).

Theorem IV.8. *Let $M = \prod_{i=1}^{r} M_i$ be the product of r compact n_i–dimensional rectangles M_i (i=1,...,r) satisfying $\sum_{i=1}^{r} n_i = n$. Suppose that $f: M \longrightarrow \mathbb{R}$ can be decomposed into the form $f(x) = \sum_{i=1}^{r} f_i(x^i)$, where $f_i: M_i \longrightarrow \mathbb{R}$ is lower semicontinuous on M_i (i=1,...,r). Then the convex envelope φ of f on M is equal to the sum of the convex envelopes φ_i of f_i on M_i, i.e.,*

$$\varphi(x) = \sum_{i=1}^{r} \varphi_i(x^i) \quad .$$

Proof. We use Theorem IV.5. Let $t^i \in \mathbb{R}^{n_i}$ (i=1,...,r). Then we have

$$f^*(t) = \max_{x \in M} \{xt - f(x)\} = \sum_{i=1}^{r} \max_{x^i \in M_i} \{x^i t^i - f_i(x^i)\} = \sum_{i=1}^{r} f_i^*(t^i) .$$

$$\varphi(x) = f^{**}(x) = \sup_{t \in \mathbb{R}^n} \{xt - f^*(t)\} = \sum_{i=1}^{r} \sup_{t^i \in \mathbb{R}^{n_i}} \{x^i t^i - f_i^*(t^i)\}$$

$$= \sum_{i=1}^{r} f_i^{**}(x^i) = \sum_{i=1}^{r} \varphi_i(x^i).$$ ■

Theorem IV.8 is often used for **separable functions f**, where r=n and the $M_i = [a_i,b_i]$ are one-dimensional intervals, i.e., we have $f(x) = \sum_{i=1}^{n} f_i(x_i)$, $f_i: [a_i,b_i] \to \mathbb{R}$ $(i=1,...,n)$.

Note that Theorem IV.8 cannot be generalized to arbitrary sums of functions. For example, the convex envelope of the function $f(x) = x^2 - x^2$ over the interval $[0,1]$ is the zero function, while the sum of the convex envelopes is $x^2 - x$.

The following theorem was given in Al-Khayyal (1983).

Theorem IV.9. *Let $f: M \to \mathbb{R}$ be lower semicontinuous on the convex compact set $M \subset \mathbb{R}^n$ and let $h: \mathbb{R}^n \to \mathbb{R}$ be an affine function. Then*

$$\varphi_{f+h} = \varphi_f + h,$$

where φ_f and φ_{f+h} denote the convex envelopes of f and (f+h) on M, respectively.

Proof. By the definition of a convex envelope, we have

$$f(x) + h(x) \geq \varphi_{f+h}(x) \geq \varphi_f(x) + h(x) \quad \forall x \in M,$$

with the last inequality holding because the right-hand side is a convex subfunctional of f+h. Hence,

$$f(x) \geq \varphi_{f+h}(x) - h(x) \geq \varphi_f(x) \quad \forall x \in M.$$

Since the middle expression is a convex function, equality must hold in the second inequality. ■

More on convex envelopes and attempts to determine φ can be found in McCormick (1976 and 1983). Convex envelopes of bilinear forms over rectangular sets are discussed in Al–Khayyal (1983), Al–Khayyal and Falk (1983). Convex envelopes of negative definite quadratic forms over the parallelepipeds defined by the conjugate directions of the quadratic form are derived in Kalantari and Rosen (1987). We will return to some of these problems in subsequent chapters.

4.4. Duality

Consider the so–called primal optimization problem

$$\text{(P)} \qquad \begin{array}{l} \text{minimize } f(x) \\ \text{s.t. } x \in C,\ g_i(x) \le 0 \ (i=1,\dots,m) \end{array} \tag{31}$$

where f, g_i: $\mathbb{R}^n \to \mathbb{R}$. Assume that g_i is convex (i=1,...,m), f is lower semicontinuous and C is convex and compact.

A **dual** problem to (31) is defined by

$$\text{(D)} \qquad \max_{u \in \mathbb{R}^m_+} \{\inf_{x \in C} [f(x) + ug(x)]\}, \tag{32}$$

where $g(x) := (g_1(x),\dots,g_m(x))$ and $\mathbb{R}^m_+ := \{(u_1,\dots,u_m) \in \mathbb{R}^m : u_i \ge 0,\ i = 1,\dots,m\}$.

Problem (32) has the objective function

$$d(u) := \inf_{x \in C} [f(x) + ug(x)]$$

which is the pointwise infimum of a collection of functions affine in u and hence concave on the feasible region $\{u \in \mathbb{R}^m_+ : d(u) > -\infty\}$ of (32).

Let inf (P) and sup (D) denote the optimal value of (P) and (D), respectively. Then we have the following familiar result (so–called weak duality).

Lemma IV.2. *We always have inf (P) ≥ sup (D).*

Proof. Let $\bar{u} \geq 0$ satisfy $d(\bar{u}) > -\infty$, and let $\bar{x} \in C$ satisfy $g(\bar{x}) \leq 0$. Then it follows that

$$d(\bar{u}) = \inf_{x \in C} [f(x) + \bar{u}g(x)] \leq f(\bar{x}) + \bar{u}g(\bar{x}) \leq f(\bar{x}). \quad \blacksquare$$

Note that Lemma IV.2 holds without the convexity and continuity assumptions made on C, g_i and f, respectively.

Let $M = C \cap \{x: g_i(x) \leq 0 \ (i=1,...,m)\}$. Then, by Lemma IV.2, any feasible point $\bar{u}$ of the dual problem could be a candidate for deriving the lower bounds $\beta(M) = d(\bar{u})$ for min f(M).

It is well known that for convex f we have inf (P) = sup (D) whenever a suitable "constraint qualification" holds. The corresponding duality theory can be found in many textbooks on optimization, cf. also Geoffrion (1971) for a thorough exposition. For nonconvex f, however, a "duality gap" inf (P) − sup (D) > 0 has to be expected and we would like to have an estimate of this duality gap (cf. Bazaraa (1973), Aubin and Ekeland (1976)). A very easy development is as follows (cf. Horst (1980a)).

Let φ be the convex envelope of f on C. Replacing f by φ in the definition of problems (P) and (D), we obtain two new problems, which we denote $(\bar{P})$ and $(\bar{D})$, with $\bar{d}(u)$ being the objective function of $(\bar{D})$. Obviously, since $\varphi(x) \leq f(x) \ \forall x \in C$, one has $\bar{d}(u) \leq d(u) \ \forall u \in \mathbb{R}^m_+$, and we obtain the following lemma as a trivial consequence of the definition of φ.

Lemma IV.3. *inf $(\bar{P})$ ≤ inf (P), sup $(\bar{D})$ ≤ sup (D).*

Convex duality applies to $(\bar{P})$ and $(\bar{D})$ (cf., e.g., Geoffrion (1971)):

Lemma IV.4. *If a so-called "constraint qualification" is fulfilled in $(\overline{P})$, then inf $(\overline{P})$ = max $(\overline{D})$.*

One of the most popular "constraint qualifications" is **Slater's condition:**

(I) **There exists a point $x^0 \in C$ such that $g_i(x^0) < 0$ (i=1,...,m).**

(I) depends only on the constraints; hence its validity for $(\overline{P})$ can be verified on (P).

Another "constraint qualification" applies for linear constraints.

(II) **$g(x) = Ax - b$, $A \in \mathbb{R}^{m\times n}$, $b \in \mathbb{R}^m$, C is a convex polytope and φ can be extended to an open convex set $\overline{C} \supset C$ such that the extension of φ is convex on $\overline{C}$** (cf. Blum and Oettli (1975)).

Combining the preceding lemmas we can easily obtain an estimate for the duality gap inf (P) – sup (D).

Theorem IV.10. *Suppose that $C \subset \mathbb{R}^n$ is nonempty, convex and compact, $f: \mathbb{R}^n \longrightarrow \mathbb{R}$ is lower semicontinuous on C and each $g_i: \mathbb{R}^n \longrightarrow \mathbb{R}$ is convex (i=1,...,m). Moreover, suppose that a* ***"constraint qualification"*** *is fulfilled. Then*

$$0 \leq \inf(P) - \sup(D) \leq \inf(P) - \inf(\overline{P}) \leq \sup_{x\in C}\{f(x) - \varphi(x)\}.$$

Proof. The function φ exists and Lemmas IV.2, IV.3, IV.4 yield

$$\inf(P) \geq \sup(D) \geq \sup(\overline{D}) = \inf(\overline{P}). \tag{33}$$

By Lemma IV.4, $M = C \cap \{x: g_i(x) \leq 0 \ (i=1,...,m)\}$ is non-empty and inf $(\overline{P})$ is finite. Hence, by (33), we have inf (P) $\neq \pm \infty$, and the first two inequalities in the assertion are fulfilled.

By the definition of sup and inf, we have

$$\inf(\mathrm{P}) - \inf(\overline{\mathrm{P}}) \le \sup_{x \in M} \{f(x) - \varphi(x)\},$$

and obviously (since $M \subset C$) we also have

$$\sup_{x \in M} \{f(x) - \varphi(x)\} \le \sup_{x \in C} \{f(x) - \varphi(x)\}.$$ ■

The quantity $\sup_{x \in C} \{f(x) - \varphi(x)\}$ may be considered as a measure of the lack of convexity of f over C.

An interesting result in connection with BB procedures is that if, in addition to the assumptions of Theorem IV.10, the constraint functions $g_i(x)$ are affine, then we have sup (D) = sup ($\overline{\mathrm{D}}$), i.e., instead of minimizing φ on M, we can solve the dual of the original problem min $\{f(x): x \in M\}$ without calculating φ (cf., e.g., Falk (1969)). However, since (D) is usually a difficult problem, until now this approach has been applied only for some relatively simple problems (cf., e.g., Falk and Soland (1969), Horst (1980a)).

Theorem IV.11. *Suppose that the assumptions of Theorem IV.10 are fulfilled and, in addition, $g(x) = Ax - b$, $A \in \mathbb{R}^{m \times n}$, $b \in \mathbb{R}^m$. Then we have*

$$sup\ (D) = sup\ (\bar{D}) = inf\ (\bar{P})\ .$$

Proof. The last equation is Lemma IV.4. To prove the first equation, we first observe that

$$f^*(t) = \max_{x \in C} \{xt - f(x)\}$$

is defined throughout $\mathbb{R}^n$, by the assumptions concerning f and C. Moreover, f^* is a pointwise maximum of affine functions, and hence a convex function. Thus, f^* is continuous everywhere and equals its convex envelope, so that we may apply Theorem IV.5 to obtain

$$f^{***}(t) = f^*(t) = \varphi^*(t) \ \forall t \in \mathbb{R}^n. \tag{34}$$

Consider the objective function d(u) of the dual (D):

$$d(u) = \min_{x \in C} \{f(x) + u(Ax - b)\}$$

$$= \min_{x \in C} \{f(x) + (A^T u)x\} - ub$$

$$= -\max_{x \in C} \{(x(-A^T u) - f(x)\} - ub$$

$$= -f^*(-A^T u) - ub.$$

From (34) it follows that

$$d(u) = -\varphi^*(-A^T u) - ub$$

$$= -\max_{x \in C} \{x(-A^T u) - \varphi(x)\} - ub$$

$$= \min_{x \in C} \{\varphi(x) + u(Ax - b)\}.$$

But this is the objective function of $(\overline{D})$, and we have in fact shown that the objective functions of (D) and $(\overline{D})$ coincide, this clearly implies the assertion. ■

We would like to mention that several authors have proposed using interval arithmetic to solve certain global optimization problems (see, e.g., Hanson, E. (1979 and 1980), Ratschek and Rokne (1984 and 1988), Ratscheck (1985) and references there). In principle, the lower bounds computed there could also be incorporated into BB procedures.

Some other bounding operations that are closely related to specific properties of special problems will be introduced later.

4.5. Consistency

In this section, we show that for many important classes of global optimization problems, the examples discussed in the preceding sections yield consistent or strongly consistent bounding operations whenever the subdivision is exhaustive and the deletion of infeasible partition elements is certain in the limit (cf. Definitions IV.6, IV.7, IV.8, IV.10).

Recall that f denotes the objective function and D denotes the closed feasible set of a given optimization problem. Let $S_M \subset M \cap D$ be the set introduced in Step k.4 of the prototype BB procedure, and recall that we suppose that $\alpha(M) = \min f(S_M)$ is available whenever $S_M \neq \emptyset$.

The following Lemma is obvious.

Lemma IV.5. *Suppose that $f: M_0 \longrightarrow \mathbb{R}$ is continuous and the subdivision procedure is exhaustive. Furthermore, assume that every infinite decreasing sequence $\{M_q\}$ of successively refined partition elements satisfies $\emptyset \neq S_{M_q} \subset M_q \cap D$.*
Then every strongly consistent lower bounding operation yields a consistent bounding operation.

Proof. Exhaustiveness and strong consistency imply that $M_q \xrightarrow[q\to\infty]{} \{\bar{x}\}$, $\bar{x} \in D$ and $\beta(M_q) \xrightarrow[q\to\infty]{} f(\bar{x})$. The sequence of upper bounds $\alpha(M_q)$ associated with M_q is defined by $\alpha(M_q) = \min f(S_{M_q})$ and continuity of f implies that $\alpha(M_q) \xrightarrow[q\to\infty]{} f(\bar{x})$. ■

Suppose that M_0 and all of the partition sets M are polytopes. Recall that the BB procedure requires that

$$-\infty < \beta(M) \leq \min f(M \cap D), \text{ if M is known to be feasible,}$$

$$-\infty < \beta(M) \leq \min f(M) \quad , \text{ if M is uncertain.}$$

Of course, if the set M is known to be infeasible, then it will be deleted.

Looking at the examples discussed in the preceding sections, we see that $\beta(M)$ is always determined by a certain optimization procedure. **A suitably defined function $\phi: M \longrightarrow \mathbb{R}$ is minimized or maximized over a known subset T of M:**

$$\beta(M) = \min \{\phi(x): x \in T\} \tag{35}$$

or

$$\beta(M) = \max \{\phi(x): x \in T\} . \tag{36}$$

Examples.

1. Lipschitzian Optimization (cf. Section IV.4.1):

Let f be Lipschitzian on M_0, let A be an upper bound for the Lipschitz constant of f on M, and let $\delta(M)$ be the diameter of M.
Then we may set

$$\phi_1(x) = f(x) - A\delta(M),\ T = V'(M),\ \beta(M) = \max\{\phi_1(x): x \in T\}, \tag{37}$$

where V'(M) is a nonempty subset of the vertex set V(M) (see also (14'))..

2. Concave Objective Function – Vertex Minima (cf. Section IV.4.2):

Let f be concave on an open set containing M_0. Then we may set

$$\phi_2(x) = f(x),\ T = V(M),\ \beta(M) = \min \{\phi_2(x): x \in T\} . \tag{38}$$

3. D.C. Programming – Vertex Minima (cf. Section IV.4.2):

Let $f = f_1 + f_2$, with f_1 concave and f_2 convex on an open set containing M_0. Then we have

$$\phi_3(x) = f_1(x) + f_2(v^*) + p^*(x-v^*),\ T = V(M),\ \beta(M) = \min\{\phi_3(x): x \in T\}\ , \qquad (39)$$

where $v^* \in V(M)$ and $p^* \in \partial f_2(v^*)$.

4. Convex Envelopes (cf. Sections IV.4.3, IV.4.4):

Suppose that the convex envelope φ_M of f over M is available, and D is such that $\min \varphi_M(D \cap M)$ can be calculated. Then we may set

$$\left.\begin{array}{l} \phi_4(x) = \varphi_M(x),\ \beta(M) = \min\{\phi_4(x): x \in T\} \text{ with} \\ \quad T = D \cap M, \text{ if M is known to be feasible ,} \\ \quad T = M, \quad \text{ if M is uncertain.} \end{array}\right\} \qquad (40)$$

Now suppose that φ_M exists but is not explicitly available. Let D be a polytope. Then, by Theorem IV.10, $\beta(M)$ in (40) can be obtained by solving the **dual** to $\min\{f(x): x \in T\}$. Since, however, this dual problem is difficult to solve, this approach seems to be applicable only in special cases, e.g., if M is a rectangle and D is defined by a few separable constraints (cf., e.g., Horst (1980a)).

Let $M \subset M'$ be an element of a partition of M'. Then it is easily seen that the bounding operations (37), (39) do not necessarily fulfill the monotonicity requirement $\beta(M) \geq \beta(M')$ in Step k.4 of the prototype BB procedure, whereas (38), (40) yield monotonic bounds. Recall that in the case of nonmonotonic lower bounds we agreed to use $\bar{\beta}(M) := \max\{\beta(M), \beta(M')\}$ instead of $\beta(M)$.

Since $\beta(M) \leq \bar{\beta}(M) \leq f(x)\ \forall x \in M$, obviously $\bar{\beta}$ is strongly consistent whenever β is strongly consistent.

Similarly, it is clear that a strongly consistent lower bounding operation β can be replaced by any lower bounding operation β' satisfying $\beta(M) \leq \beta'(M) \leq f(x)$ $\forall x \in M$. For example, in (37), whenever possible, A should be replaced by tighter local upper bounds $A_M < A$ on the Lipschitz constant of f on M.

Let ϕ_q denote the functions ϕ on M_q defined in (35), (36). Let $\beta(M)$ be determined by (35), (36), respectively, and associate to each bound $\beta(M)$ a point $\tilde{x} \in M$ as follows

$$\tilde{x} \in \operatorname{argmin} \phi(T), \quad \text{if (35) holds,} \tag{41}$$

$$\tilde{x} \in \operatorname{argmax} \phi(T), \quad \text{if (36) holds.} \tag{42}$$

In each step of the BB procedure choose $\tilde{x}^k$ to be the point associated to β_k in this way.

Now consider the bounding methods in the above example, respectively using (37), (38), (39), (40). In the first three cases, f is obviously continuous on M_0. Suppose that **f is also continuous** on M_0 in case 4 (convex envelopes, (40)).

Proposition IV.3. *Suppose that at every step any undeleted partition element can be further refined. Furthermore, suppose that the "deletion by infeasibility" rule is certain in the limit and the subdivision is exhaustive. Then each bounding operation in the above example given by (37), (38), (39), (40), respectively, is strongly consistent.*

Proof. Let $\phi_{i,q}\colon M_q \longrightarrow \mathbb{R}$ (i=1,2,3) denote the functions defined in (37), (38), (39), i.e., we have

$$\phi_{1,q}(x) = f(x) - A_q \delta(M_q); \text{ f Lipschitzian,} \tag{43}$$

$$\phi_{2,q}(x) = f(x); \text{ f concave,} \tag{44}$$

$$\phi_{3,q}(x) = f_1(x) + f_2(v^{*q}) + p^{*q}(x - v^{*q}); \; f_1 \text{ concave, } f_2 \text{ convex.} \tag{45}$$

It is sufficient to demonstrate that, for every decreasing sequence $\{M_q\}$ of successively refined partition elements generated by an exhaustive subdivision such that $M_q \longrightarrow \{\overline{x}\}$, there is a subsequence $\{M_{q'}\}$ satisfying

$$\beta(M_{q'} = \phi_{i,q'}(\tilde{x}^{q'}) \underset{q'\to\infty}{\longrightarrow} f(\overline{x}) \quad (i=1,2,3)\ .$$

Since $\overline{\beta}(M_q)$ is a nondecreasing sequence satisfying $\beta(M_q) \le \overline{\beta}(M_q) \le f(\overline{x})$, it follows that $\bar{\beta}(M_q) \xrightarrow[q\to\infty]{} f(\bar{x})$. (Note that in the case of monotonic $\beta(M_q)$, we have $\beta(M_q) = \bar{\beta}(M_q)$.)

The assumption concerning the deletion rule then implies that $\bar{x} \in D$, and hence we have strong consistency.

In the case i=1, we can assume that $A_q \le A_0$ $\forall q$. Then we have

$$\phi_{1,q}(\bar{x}^q) \ge f(\bar{x}^q) - A_0\delta(M_q) ,$$

which, by the continuity of f and the exhaustiveness, implies that $\phi_{1,q}(\bar{x}^q) \xrightarrow[q\to\infty]{} f(\bar{x})$.

The case i=2 is trivial since $\phi_{2,q} = f$, and f is continuous.

In the case i=3, note that $v^{*q} \xrightarrow[q\to\infty]{} \bar{x} \in D$ since $M_q \xrightarrow[q\to\infty]{} \{\bar{x}\}$. Moreover, since $\{\partial f_2(x)\colon x \in M_0\}$ is compact (cf., e.g., Rockafellar (1970)), there exists a subsequence $p^{*q'} \xrightarrow[q'\to\infty]{} \bar{p}^*$. Hence, we have

$$\phi_{3,q'}(\bar{x}^{q'}) = f_1(\bar{x}^{q'}) + f_2(v^{*q'}) + p^{*q'}(\bar{x}^{q'}-v^{*q'}) \xrightarrow[q'\to\infty]{} f_1(\bar{x}) + f_2(\bar{x}) + \bar{p}^*(\bar{x}-\bar{x}) = f(\bar{x}).$$

Now consider the case (40) (convex envelopes). By Theorem IV.4, we have $\min f(M_q) = \min \varphi_{M_q}(M_q)$, and hence

$$\left.\begin{array}{ll} \min f(M_q) \le \beta(M_q) \le \min f(D \cap M_q) & \text{if } M_q \text{ is known to be feasible,} \\ \min f(M_q) = \beta(M_q) & \text{if } M \text{ is uncertain.} \end{array}\right\} \quad (46)$$

From the assumptions we know that $M_q \xrightarrow[q\to\infty]{} \{\bar{x}\}$, $\bar{x} \in D$. If $D \cap M_{q'} \neq \emptyset$ is known for infinitely many q', then $D \cap M_{q'} \xrightarrow[q'\to\infty]{} \{\bar{x}\}$. If $M_{q'}$ is uncertain for all but finitely many q', then $\beta(M_{q'}) = \min f(M_{q'})$. Finally, the continuity of f and (46) imply that $\beta(M_{q'}) \xrightarrow[q'\to\infty]{} f(\bar{x})$, and hence we have strong consistency. ■

The following Corollary IV.5 is an immediate consequence of Corollary IV.3.

Corollary IV.5. *In the BB procedure suppose that the lower bounding operation is strongly consistent, and the selection is bound improving. Suppose that f is continuous on M_0. Then every accumulation point of $\{\tilde{x}^k\}$ solves problem (P).*

Note that it is natural to require that all partition sets where β_k is attained are refined in each step, since otherwise $\beta_{k+1} = \beta_k$ holds and the lower bound is not improved.

5. DELETION BY INFEASIBILITY

In this section, following Horst (1988) certain rules are proposed for deleting infeasible partition sets M. These rules, properly incorporated into the branch and bound concept, will lead to convergent algorithms (cf. Section IV.4.5). Since the infeasibility of partition sets depends upon the specific type of the feasible set D, three cases which frequently arise will be distinguished:

Convex feasible sets D, intersections of a convex set with finitely many complements of convex sets (reverse convex programming), feasible sets defined by Lipschitzian inequalities.

Again suppose that the partition sets M are convex polytopes defined by their vertex sets V(M).

Deletion by Certainty.

Clearly, whenever we have a procedure that, for each partition set M, can definitely decide after a reasonable computational effort whether we have $M \cap D = \emptyset$ or $M \cap D \neq \emptyset$, then this procedure should be applied (deletion by certainty, cf. Example IV.3).

Example IV.5. Let $D := \{x \in \mathbb{R}^n: h(x) \geq 0\}$, where $h: \mathbb{R}^n \longrightarrow \mathbb{R}$ is convex. Then a partition set M is deleted whenever its vertex set V(M) satisfies

$$V(M) \subset \{x \in \mathbb{R}^n: h(x) < 0\}. \tag{47}$$

Because of the convexity of the polytope M and the convexity of the set $\{x \in \mathbb{R}^n: h(x)<0\}$, we obviously have $M \cap D = \emptyset$ if and only if (47) holds.

Convex Feasible Sets.

Let

$$D := \{x \in \mathbb{R}^n: g(x) \leq 0\}, \tag{48}$$

where $g: \mathbb{R}^n \longrightarrow \mathbb{R}$ is convex, e.g., $g(x) = \sup \{g_i(x): i \in I\}$ with $g_i: \mathbb{R}^n \longrightarrow \mathbb{R}$ convex, $i \in I \subset \mathbb{N}$. Suppose that a point y^0 satisfying $g(y^0) < 0$ is known (the Slater condition) and that D is compact.

Let M be a partition set defined by its vertex set V(M). If there is a vertex $v \in V(M)$ satisfying $v \in D$, then trivially $D \cap M \neq \emptyset$. However, $V(M) \cap D = \emptyset$ does not imply $M \cap D = \emptyset$ (cf., e.g., Example IV.2).

Suppose that $V(M) \cap D = \emptyset$. Let p be an arbitrary point of $M \setminus D$.

Compute the point z where the line segment $[y^0,p]$ intersects the boundary of D. By convexity of D, we have

$$z = \lambda y^0 + (1-\lambda)p, \tag{49}$$

where λ is the unique solution of the univariate convex programming problem

$$\min \{\mu \in [0,1]: \mu y^0 + (1-\mu)p \in D\}. \tag{50}$$

Equivalently, λ is the unique solution of the equation

$$g(\mu y^0 + (1-\mu)p) = 0, \ \mu \geq 0. \tag{51}$$

Let $s(z) \in \partial g(z)$ be a subgradient of g at z. If M is strictly separated from D by the hyperplane $(s(z), x-z) = 0$ supporting D at z, then M is deleted, i.e., we have the first deletion rule

(DR 1) Delete a partition element M if its vertex set V(M) satisfies

$$V(M) \subset \{x: s(z)(x-z) > 0\} . \tag{52}$$

where s(z) and z are defined above.

Clearly, a partition element that is deleted according to (DR 1) must be infeasible. However, it is easy to see that infeasible sets M may exist that do not satisfy (52).

Intersections of a Convex Set with Finitely Many Complements of a Convex Set.

Let

$$D = D_1 \cap D_2, \tag{53}$$

where

$$D_1 = \{x \in \mathbb{R}^n: g(x) \leq 0\} \text{ is compact,} \tag{54}$$

$$D_2 = \{x \in \mathbb{R}^n: h_j(x) \geq 0 \ (j=1,\dots,r)\} , \tag{55}$$

and $g, h_j: \mathbb{R}^n \to \mathbb{R}$ are convex $(j=1,\dots,r)$.

Assume that a point y^0 satisfying $g(y^0) < 0$ is known.

Recall that a typical feasible set arising from **reverse convex programming** can be described by (53), (54), and (55).

Let

$$C_j := \{x \in \mathbb{R}^n: h_j(x) < 0\} \ (j=1,\dots,r) . \tag{56}$$

(DR 2) **Delete a partition set M if its vertex set V(M) satisfies either (DR 1) applied to D_1 (in place of D) or if there is a $j \in (1,...,r)$ such that we have**

$$V(M) \subset C_j.$$

Again, it is easy to see that by (DR 2) only infeasible sets will be deleted, but possibly not all infeasible sets.

Lipschitz Constraints.

Let

$$D = \{x \in \mathbb{R}^n : g_j(x) \le 0 \ \ (j=1,...,m)\}, \tag{57}$$

where $g_j: \mathbb{R}^n \to \mathbb{R}$ are Lipschitzian on the partition set M with Lipschitz constants L_j (j=1,...,m). Let $\delta(M)$ be the diameter of M, and let A_j be the upper bounds for L_j (j=1,...,m). Furthermore, let V'(M) be an arbitrary nonempty subset of the vertex set V(M). Then we propose

(DR 3) **Delete a partition element M whenever there is a $j \in (1,...,m)$ satisfying**

$$\max \{g_j(x): x \in V'(M)\} - A_j \delta(M) > 0. \tag{58}$$

Again, if $M = \{x \in \mathbb{R}^n : a \le x \le b\}$ one might use

$$g_j(\tfrac{1}{2}(a + b)) - \frac{A_j}{2} \delta(M) > 0 \tag{58'}$$

rather than (58).

Proposition IV.4. *Suppose that the subdivision is exhaustive. Then the "deletion by infeasibility" rules (DR 1) – (DR 3) are certain in the limit.*

Proof. a) Let $D = \{x: g(x) \le 0\}$, where $g: \mathbb{R}^n \to \mathbb{R}$ is convex. Suppose that we have y^0 satisfying $g(y^0) < 0$. Apply deletion rule (DR 1). Let $\{M_q\}$ be a decreasing sequence of partition sets and let $p^q \in M_q \setminus D$. Consider the line segment $[y^0, p^q]$. By exhaustiveness, there is an $\bar{x}$ satisfying $p^q \xrightarrow[q\to\infty]{} \bar{x}$.

Suppose that we have $\bar{x} \notin D$. Then by the convexity of g on $\mathbb{R}^n$ we have continuity of g on $\mathbb{R}^n$ (e.g., Rockafellar (1970)), and $\mathbb{R}^n \setminus D$ is an open set. It follows that there exists a ball $B(\bar{x},\varepsilon) := \{x: \|x-\bar{x}\| \leq \varepsilon\}$, $\varepsilon > 0$, satisfying $B(\bar{x}, \varepsilon) \subset \mathbb{R}^n \setminus D$, such that $M_q \subset B(\bar{x},\varepsilon)$ $\forall q > q_0$, where $q_0 \in \mathbb{N}$ is sufficiently large.

Consider the sequence of points $z^q \in \partial D$, where $[y^0,p^q]$ intersects the boundary ∂D of D, $p^q \in M_q$:

$$z^q = \lambda_q y^0 + (1 - \lambda_q)p^q, \ \lambda_q \in [0,1] \tag{59}$$

Since ∂D is compact and $\lambda_q \in [0,1]$, there is a subsequence (that we denote by q'), satisfying

$$z^{q'} \xrightarrow[q'\to\infty]{} \bar{z} \in \partial D, \ \lambda_{q'} \longrightarrow \bar{\lambda} \in [0,1], \ \bar{z} = \bar{\lambda} y^0 + (1-\bar{\lambda})\bar{x}. \tag{60}$$

Consider the associated sequence of subgradients $s(z^{q'}) \in \partial g(z^{q'})$. Since $\{\partial g(x): x \in \partial D\}$ is compact (cf. Rockafellar (1970)), we may assume (possibly passing to another subsequence which we again denote by q') that $s(z^{q'}) \xrightarrow[q'\to\infty]{} \bar{s}$. But we know that the set-valued mapping $x \longrightarrow \partial g(x)$ is closed on ∂D (cf. Rockafellar (1970)), and hence we have $\bar{s} \in \partial g(\bar{z})$. By the definition of a subgradient, it follows that

$$0 > g(y^0) \geq \bar{s}(y^0 - \bar{z}) + g(\bar{z}) = \bar{s}(y^0 - \bar{z}) \ . \tag{61}$$

But (60) implies that we have

$$(1 - \bar{\lambda}) \ (\bar{x} - \bar{z}) = -\bar{\lambda}(y^0 - \bar{z}) \ . \tag{62}$$

Note that $\bar{\lambda} > 0$ since otherwise (60) would imply that $\bar{x} = \bar{z} \in \partial D \subset D$ which contradicts the assumption that $\bar{x} \notin D$. Likewise, it follows that we have $\bar{\lambda} < 1$, since $\bar{z} = y^0$ is not possible.

Obviously, by (62), we have $\bar{s}(\bar{x} - \bar{z}) = \frac{-\bar{\lambda}}{1-\bar{\lambda}} \bar{s}(y^0 - \bar{z})$, $0 < \bar{\lambda} < 1$.

Hence, from (61) it follows that

$$(\bar{s}, \bar{x} - \bar{z}) > 0. \tag{63}$$

However, (63) implies that for sufficiently large q_1 we have

$$s(z^{q'})(p^{q'} - z^{q'}) > 0 \quad \forall q' > q_1. \tag{64}$$

Since $p^{q'}$ is an arbitrary point of $M_{q'}$ and $M_{q'} \longrightarrow \{\bar{x}\}$, we see that (64) also holds for all vertices of $M_{q'}$, q' sufficiently large. Hence, according to deletion rule (DR 1), $M_{q'}$ was deleted, and we contradict the assumptions. Thus, we must have $\bar{x} \in D$.

b) Let $D = D_1 \cap D_2$, where

$$D_1 = \{x: g(x) \le 0\},$$

$$D_2 = \{x: h_j(x) \ge 0, j = 1,\dots,r\}.$$

and g, h_j: $\mathbb{R}^n \longrightarrow \mathbb{R}$ are convex (j=1,...,r).

Apply deletion rule (DR 2).

By part a) above, we have $\bar{x} \in D_1$.

Let $C_j := \{x \in \mathbb{R}^n: h_j(x) < 0\}$, j=1,...,r. Suppose that we have $\bar{x} \notin D_2$. Then there is a $j \in \{1,\dots,r\}$ such that $\bar{x} \in C_j$, and in a manner similar to the first part of a) above, we conclude that $M_q \subset C_j$ for sufficiently large q, since C_j is an open set. This contradicts the deletion rule (DR 2).

c) Finally, let $D = \{x: g_j(x) \le 0 \ (j=1,\dots,m)\}$, where all g_j: $\mathbb{R}^n \longrightarrow \mathbb{R}$ are Lipschitzian on M_0. Suppose that the local overestimators $A_j(M_q)$ of the Lipschitz constants $L_j(M_q)$ are known (j=1,...,m). Since the overestimator $A_j(M_q)$ is an overestimator for $L_j(M_{q'})$ whenever $M_{q'} \subset M_q$, we may assume that there is a bound A satisfying

$$A \ge A_j(M_q) \quad \forall q, j=1,\dots,m. \tag{65}$$

Apply deletion rule (DR 3), and suppose that we have $\bar{x} \notin D$. Since $M_q \longrightarrow \{\bar{x}\}$, by the continuity of g_j $(j=1,...,m)$ and by (65) it follows that for every sequence of nonempty sets $V'(M_q) \subset V(M_q)$, we have

$$\max \{g_j(x)\colon x \in V'(M_q)\} - A_j(M_q)\, \delta(M_q) \underset{q\to\infty}{\longrightarrow} g_j(\bar{x}) \ (j=1,...,m)\ . \qquad (66)$$

Since $\bar{x} \notin D$, there is at least one $j \in \{1,...,m\}$ satisfying $g_j(\bar{x}) > 0$. Taking into account the boundedness of $\{A_j(M_q)\}$, the limit $\delta(M_q) \longrightarrow 0$ and the continuity of g_j, we then see from (66) that there is a $q_0 \in \mathbb{N}$ such that

$$\max \{g_j(x)\colon x \in V'(M_q)\} - A_j(M_q)\, \delta(M_q) > 0 \ \ \forall q \geq q_0.$$

This contradicts deletion rule (DR 3). ■

Combining the bounding procedures discussed in Section III.4.5 and the above deletion rules yields a family of BB procedures for many important classes of global optimization problems (cf. Part B and C of this book).

6. RESTART BRANCH AND BOUND ALGORITHM

In this section the combination of outer approximation and branch and bound will be discussed in the following sense.

Suppose that we apply an outer approximation algorithm to our global optimization problem

(P) minimize $f(x)$
subject to $x \in D$

and we would like to solve each relaxed problem

(Q_ν) minimize $f(x)$
subject to $x \in D_\nu$

by a BB–Procedure.

Clearly, since in the basic outer approximation method, each of these problems differs from the previous one by only a single additional constraint, when solving $(Q_{\nu+1})$ we would like to be able to take advantage of the solution of (Q_ν). In other words, in order to handle efficiently the new constraints to be added to the current feasible set, the algorithm selected for solving the relaxed problems in an outer approximation scheme should have the capability of being restarted at its previous stage after each iteration of the relaxation scheme.

An algorithm which proceeds strictly according to the prototype BB procedure would clearly not satisfy this requirement.

Indeed, suppose we are at Step k of the BB–Procedure for solving (Q_ν). Let $\alpha_{\nu,k-1}$ denote the value of the objective at the best feasible point in D_ν obtained so far. Then certain partition sets M may be deleted in Step k.1, since $\beta(M) \geq \alpha_{\nu,k-1}$. These sets, however, may not qualify for deletion when the BB procedure is applied to solve a subproblem (Q_μ), $\mu > \nu$, since $D_\mu \subset D_\nu$ and possibly $\alpha_{\mu,k-1} > \alpha_{\nu,k-1}$ for all steps k of the BB procedure used to solve (Q_μ).

This defect can often be avoided: it suffices, when solving any relaxed problem (Q_ν) to make sure that $S_M \subset M \cap D$ and not just $S_M \subset M \cap D_\nu$ (cf. Step k.4 of the BB procedure), so that x^k is always the current best feasible solution to (P) and not just to (Q_ν).

Since often outer approximation by convex polyhedral sets is used and in many applications (Q_ν) can be solved in finitely many steps, we assume that the BB procedure for solving (Q_ν) is finite.

Restart Branch and Bound – Outer Approximation Procedure (RBB–R)

Let $\mathcal{F}$ be the family of sets D_ν admitted in the outer approximation procedure (cf. Chapter II).

Choose $D_1 \in \mathcal{F}$ such that $D_1 \supset D$. Set $\nu = 1$.

Apply the finite BB procedure of Section IV.1 to problem (Q_ν) with the conditions in Step k.4 replaced by

$$S_M \subset M \cap D, \ \beta(M) \leq \inf (M \cap D_\nu) \tag{67}$$

and with step k.5 modified as follows.

a) If $\alpha_k = \beta_k$, then stop: x^k solves (P).

b) If $\alpha_k > \beta_k$ and $\beta_k = f(z^\nu)$ for some $z^\nu \in D_\nu \setminus D$, then construct a constraint $l_\nu(x) \leq 0$ satisfying $\{x \in D_\nu: l_\nu(x) \leq 0\} \in \mathcal{F}$, $l_\nu(z^\nu) > 0$, $l_\nu(x) \leq 0 \ \ \forall x \in D$ (cf. Chapter II), and let

$$D_{\nu+1} = \{x \in D_\nu: l_\nu(x) \leq 0\} . \tag{68}$$

Set $v \leftarrow v+1$ and go to Step k+1 of the BB procedure (applied from now on to problem $(Q_{\nu+1})$).

c) If neither a) nor b) occurs, then go to Step k+1 of the BB–Procedure (with ν unchanged).

Theorem IV.12. *If the family $\mathcal{F}$ is finite, then the (RBB–R)–algorithm terminates after finitely many steps at an optimal solution. If conditions (i) and (ii) of Theorem III.1 (the convergence theorem for the basic outer approximation procedure) are satisfied and the (RBB–R)–Algorithm generates an infinite sequence $\{z^\nu\}$, then every cluster point of $\{z^\nu\}$ solves (P).*

Proof. Since $D_\nu \supset D$, in view of (67) we have

$$\alpha_k = f(x^k) \geq \min f(D) \geq \min f(D_\nu) \geq \beta_k .$$

Therefore, if it happens that $\alpha_k = \beta_k$ (which has to be the case for some k, if the family $\mathcal{F}$ is finite), then $f(x^k) = \min f(D)$, i.e., x^k solves (P).

Now suppose that the algorithm generates an infinite sequence. Then we have

$\alpha_k > \beta_k$ at every Step k of the BB procedure within the (RBB–R)–algorithm. As long as ν is unchanged, since $D \subset D_\nu$ and we have (67), we are in fact applying the BB procedure to problem (Q_ν). Since we assume that this algorithm is finite, after finitely many steps we must have $\beta_k = \min f(D_\nu)$, i.e., $\beta_k = f(z^\nu)$ for some $z^\nu \in D_\nu$. Note that $z^\nu \notin D$, since otherwise we would have $\alpha_k = \beta_k$. So the case b) must occur after finitely many steps. We then replace D_ν by $D_{\nu+1}$ according to (68) and go to the next Step k+1 with ν changed to $\nu+1$. That is, we apply the BB–Procedure to problem (P) starting from the most recent partition and bounds obtained in solving (Q_ν).

Consequently, if the (RBB–R)–algorithm generates an infinite sequence $\{z^\nu\}$, then every z^ν solves (Q_ν). Theorem IV.12 now follows from Theorem III.1. ■

Note that the (RBB–R)–algorithm may be regarded as a single BB procedure as well, and hence Theorem IV.12 can also be verified by proving consistency of the bounding operation and completeness of the selection.

Applications of the (RBB–R) algorithm will be presented in Part B.

PART B

CONCAVE MINIMIZATION

Many applications lead to minimizing a concave function over a convex set (cf. Chapter I). Moreover, it turns out that concave minimization techniques also play an important role in other fields of global optimization.

Part B is devoted to a thorough study of methods for solving concave minimization problems and some related problems having reverse convex constraints.

The methods for concave minimization fall into three main categories: cutting methods, successive approximation methods, and successive partition methods. Although most methods combine several different techniques, cutting planes play a dominant role in cutting methods, relaxation and restriction are the main aspects of successive approximation, and branch and bound concepts usually serve as the framework for successive partition.

Aside from general purpose methods, we also discuss decomposition approaches to large scale problems and specialized methods adapted to problems with a particular structure, such as quadratic problems, separable problems, bilinear programming, complementarity problems and concave network problems.

CHAPTER V

CUTTING METHODS

In this chapter we discuss some basic cutting plane methods for concave minimization. These include concavity cuts and related cuts, facial cuts, cut and split procedures and a discussion of how to generate deep cuts. The important special case of concave quadratic objective functions is treated in some detail.

1. A PURE CUTTING ALGORITHM

The basic concave programming (BCP) problem to be studied here is

$$\text{minimize } f(x) \tag{1}$$

$$\text{(BCP)} \qquad \text{s.t. } Ax \le b\,, \tag{2}$$

$$x \ge 0\,, \tag{3}$$

where A is an m×n matrix, x is an n–vector, b is an m–vector, and $f: \mathbb{R}^n \longrightarrow \mathbb{R}$ is a concave function. For ease of exposition, in this chapter we shall assume that the feasible domain $D = \{x\colon\ Ax \le b,\ x \ge 0\}$ is bounded (i.e., is a polytope) with $\operatorname{int} D \neq \emptyset$, and that for any real number α the level set $\{x\colon \mathbb{R}^n\colon f(x) \ge \alpha\}$ is bounded.

Note that the assumption that f(x) is defined and finite throughout $\mathbb{R}^n$ is not fulfilled in various applications. However, we can prove the following result on the extension of concave functions.

Proposition V.1. *Let $f_0: D \longrightarrow \mathbb{R}$ be any concave function which is continuous on D. If int $D \neq \emptyset$ and $\|\nabla f_0(x)\|$ is bounded on the set of all $x \in D$ where $f_0(x)$ is differentiable ($\nabla f_0(x)$ denotes the gradient of $f_0(x)$ at the point x), then f_0 can be extended to a concave function $f: \mathbb{R}^n \longrightarrow \mathbb{R}$.*

Proof. It is well known that the set C of points $x \in D$ where $f_0(x)$ is differentiable is dense in int D (cf. Rockafellar (1970), Theorem 25.5). For each point $y \in C$, consider the affine function $h_y(x) = f_0(y) + \nabla f_0(y)(x-y)$. Then the function $f(x) = \inf \{h_y(x): y \in C\}$ is concave on $\mathbb{R}^n$ (as the pointwise infimum of a family of affine functions). Since by assumption $\|\nabla f_0(y)\|$ is bounded on C, it is easily seen that $-\infty < f(x) < +\infty$ for all $x \in \mathbb{R}^n$. Moreover, for any $x,y \in C$ we have $h_y(x) \geq f_0(x)$, while $h_x(x) = f_0(x)$. Since $f(x) = f_0(x)$ for all $x \in C$ and C is dense in int D, continuity implies that $f(x) = f_0(x)$ for all $x \in D$. ■

Note that if $\Phi(x)$ is any other concave extension of f_0, then for any $x \in \mathbb{R}^n$ and any $y \in C$, we have $\Phi(x) \leq h_y(x)$, hence $\Phi(x) \leq f(x)$. Thus, f(x) is the maximal extension of $f_0(x)$. Also, observe that the condition on boundedness of $\|\nabla f_0(x)\|$ is fulfilled if, for example, $f_0(x)$ is defined and finite on some open set containing D (cf. Rockafellar (1970), Theorem 24.7).

1.1. Valid Cuts and a Sufficient Condition for Global Optimality

Let x^0 be the feasible solution with the least objective function value found so far by some method. A fundamental question in solving our problem is how to check whether x^0 is a global solution.

Clearly, by the very nature of global optimization problems, any criterion for global optimality must be based on global information about the behaviour of the objective function on the whole feasible set. Standard nonlinear programming methods use only local information, and hence cannot be expected to provide global optimality criteria.

However, when a given problem has some particular structure, by exploiting this structure, it is often possible to obtain useful sufficient conditions for global optimality. For the problem (BCP) formulated above, the structural property to be exploited can be expressed in of the following forms:

I. *The global minimum of f(x) over any polytope is always attained at some vertex (extreme point) of the polytope* (see Theorem I.1).

Therefore, the problem is equivalent to minimizing f(x) over the vertex set of D.

II. *For any polytope P with vertices* $u^1,u^2,\ldots,u^s$ *the number* $\min\{f(u^1),f(u^2),\ldots,f(u^s)\}$ *is a lower bound for* $f(D \cap P)$.

Here the points u^i might not belong to D. Thus, the values of f(x) outside D can be used to obtain information on the values inside.

These observations underlie the main idea of the cutting method we are going to present.

First of all, in view of Property I, we can assume that the point x^0 under consideration is a vertex of D.

Definition V.1. *Let* $\gamma = f(x^0)$. *For any* $x \in \mathbb{R}^n$ *satisfying* $f(x) \geq \gamma$ *the point* $x^0 + \theta(x-x^0)$ *such that*

$$\theta = \sup\{t\colon t \geq 0,\ f(x^0 + t(x-x^0)) \geq \gamma\} \tag{4}$$

is called the ***γ-extension*** *of x (with respect to* x^0*).*

From the concavity of f(x) and the boundedness of its upper level sets it is immediate that $1 \leq \theta < +\infty$.

Let $y^1,y^2,...,y^s$ denote the vertices of D adjacent to x^0 ($s \geq n$). We may assume that

$$\min \{f(y^1),...,f(y^s)\} \geq \gamma \tag{5}$$

(otherwise we know that x^0 is not optimal). For each $i=1,2,...,s$ let $z^i = x^0 + \theta_i(y^i-x^0)$ be the γ-extension of y^i. Normally, these points lie outside D. By Proposition III.1, we know that any solution π of the system of linear inequalities

$$\theta_i \pi(y^i - x^0) \geq 1 \quad (i=1,2,...,s) \tag{6}$$

provides a γ-valid cut for (f,D). In other words, we have the following information on the values of f(x) inside the polytope D.

Theorem V.1. *(Sufficient condition for global optimality). Let π be a solution of the system (6).*
Then

$$\pi(x-x^0) > 1 \quad \text{for all } x \in D \text{ such that } f(x) < \gamma \,. \tag{7}$$

Hence, if

$$\max \{\pi(x - x^0): x \in D\} \leq 1 \,, \tag{8}$$

then x^0 is a global optimal solution of (BCP).

Proof. Theorem V.1. is an obvious consequence of Proposition III.1. ■

Thus, to check the global optimality of x^0 we can solve the linear program

$$\mathrm{LP}(x^0,\pi,D) \qquad \max \{\pi(x-x^0): x \in D\} \,, \tag{9}$$

where $\pi(x-x^0) \geq 1$ is a valid cut for (f,D). If the optimal value of this program does not exceed 1, then x^0 is a global optimal solution. Otherwise, we know that any feasible point that is better than x^0 must be sought in the region $D \cap \{x: \pi(x-x^0) \ge 1\}$ left over by the cut.

Clearly, we are interested in using the deepest possible cut, or at least a cut which is not dominated by any other one of the same kind. Such cuts correspond to the basic solutions of (6), and can be obtained, for example, by solving the linear program

$$\min \sum_{i=1}^{s} \theta_i \pi(y^i - x^0) \quad \text{s.t.}\ \theta_i \pi(y^i - x^0) \ge 1 \quad (i=1,2,...,s) . \tag{10}$$

When x^0 is a nondegenerate vertex of D, i.e., $s = n$, the system (6) has a unique basic solution π, satisfying

$$\theta_i \pi(y^i - x^0) = 1 \quad (i=1,...,n) . \tag{11}$$

This yields

$$\pi = eQ^{-1} , \quad Q = (z^1-x^0,...,z^2-x^0,...,z^n-x^0) , \tag{12}$$

where $e = (1,1,...,1)$ and z^i is the γ–extension of the i–th vertex y^i of D adjacent to x^0. The corresponding cut $\pi(x - x^0) \ge 1$ is then the γ–valid concavity cut as defined in Definition III.3.

In the general case, where degeneracy may occur, solving the linear program (10) may be time consuming. Therefore, as pointed out in Section III.2, the most convenient approach is to transform the problem to the space of nonbasic variables relative to the basic solution x^0. More specifically, in the system (2), (3) let us introduce the slack variables $s = b - Ax \in \mathbb{R}^m$ and denote $t = (s,x)$. Let $t_B = (t_i, i \in B)$ and $t_N = (t_i, i \in N)$ be the basic and nonbasic variables, respectively, relative to the basic solution $t^0 = (s^0,x^0)$, $s^0 = b - Ax^0$, that corresponds to the vertex x^0 of D. Then, expressing the basic variables in terms of the nonbasic ones, we obtain from (2), (3) a system of the form

$$t_B = t_B^0 - Wt_N \,,\; t_B \geq 0,\; t_N \geq 0 \,, \tag{13}$$

where $t_B^0 \geq 0$ (since $t_N = 0$ corresponds to the basic solution x^0) and W is an m×n matrix. The objective function becomes a certain concave function of the nonbasic variables t_N.

Writing (13) as: $Wt_N \leq t_B^0$, $t_N \geq 0$, and changing the notation ($x \leftarrow t_N$, $A \leftarrow W$, $b \leftarrow t_B^0$), we can thus assume that the original constraints (2), (3) have been given such that $x^0 = 0$.

In the sequel, when the constraints have the form (2) (3) with the origin 0 at a vertex x^0 of D, we shall say that the BCP problem is in ***standard form with respect to x^0***. Under these conditions, if $\gamma < f(x^0)$ (e.g., $\gamma = f(x^0)-\varepsilon$, $\varepsilon > 0$ being the tolerance), and

$$\theta_i = \max \{t: f(te^i) \geq \gamma\} > 0 \quad (i=1,2,...,n) \,, \tag{14}$$

then a γ-valid cut is given by

$$\sum_{i=1}^{n} x_i/\theta_i \geq 1 \,. \tag{15}$$

More generally, if a cone $K_0 \supset D$ is available that is generated by n linearly independent vectors $u^1,u^2,...,u^n$, and if $\gamma < f(0)$, and

$$\theta_i = \max \{t: f(tu^i) \geq \gamma\} > 0 \quad (i=1,2,...,n) \,, \tag{16}$$

then a γ-valid cut is given by

$$eQ^{-1}x \geq 1 \,, \tag{17}$$

where $Q = (\theta_1 u^1, \theta_2 u^2, ..., \theta_n u^n)$ (cf. Theorem III.1 and Corollary III.1).

1.2. Outline of the Method

The above sufficient condition for global optimality suggests the following cutting method for solving the problem (BCP).

Since the search for the global minimum can be restricted to the vertex set of D, we first compute a vertex x^0 which is a local minimizer of f(x) over D. Such a vertex can be found, e.g., as follows: starting from an arbitrary vertex v^0, pivot from v^0 to a better vertex v^1 adjacent to v^0, then pivot from v^1 to a better vertex v^2 adjacent to v^1, and so on, until a vertex $v^n = x^0$ is obtained which is not worse than any vertex adjacent to it. From the concavity of f(x) it immediately follows that $f(x^0) \le f(x)$ for any x in the convex hull of x^0 and the adjacent vertices; hence x^0 is actually a local minimizer.

Let $\gamma = f(x^0)$. In order to test x^0 for global optimality, we construct a γ–valid cut for x^0:

$$\pi^0(x-x^0) \ge 1 , \tag{18}$$

and solve the linear program $LP(x^0,\pi^0,D)$. If the optimal value of this linear program does not exceed 1, then by Theorem V.1, x^0 is a global minimizer and we stop. Otherwise, let ω^0 be a basic optimal solution of $L(x^0,\pi^0,D)$, and consider the residual polytope left over by the cut (18), i.e., the polytope

$$D_1 = D \cap \{x: \pi^0(x-x^0) \ge 1\} . \tag{19}$$

By Theorem V.1, any feasible solution better than x^0 must be sought only in D_1. Therefore, starting from ω^0, we find a vertex x^1 of D_1 which is a local minimizer of f(x) over D_1 (then $f(x^1) \le f(\omega^0)$). It may happen that $f(x^1) < \gamma$. Then, by Theorem V.1, x^1 must satisfy (18) as a strict inequality; hence, since it is a vertex of D_1, it must also be a vertex of D. In that case the same procedure as before can be repeated, but with x^1 and $\gamma_1 = f(x^1)$ replacing x^0 and $\gamma = f(x^0)$.

More often, however, we have $f(x^1) \ge \gamma$. In this case, the procedure is repeated with $x^0 \leftarrow x^1$, $D \leftarrow D_1$, while γ is unchanged.

In this manner, after one iteration we either obtain a better vertex of D, or at least reduce the polytope that remains to be explored. Since the number of vertices of D is finite, the first situation can occur only finitely often. Therefore, if we can also ensure the finiteness of the number of occurrences of the second situation, the method will terminate successfully after finitely many iterations (Fig. V.1).

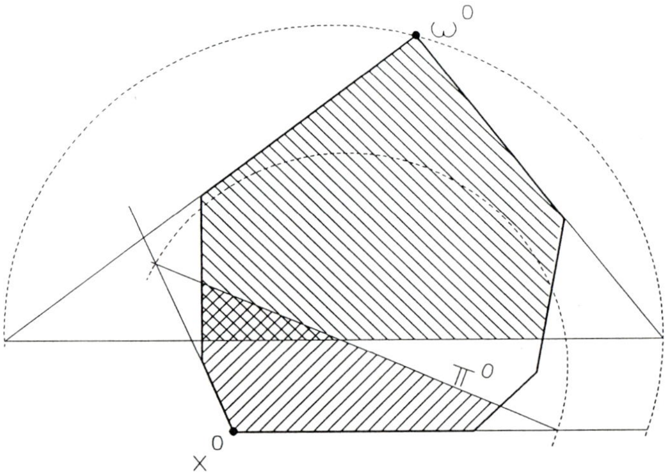

Fig. V.1

Algorithm V.1.

Initialization:

Search for a vertex x^0 which is a local minimizer. Set $\gamma = f(x^0)$, $D_0 = D$.

Iteration k = 0,1,...:

1) At x^k construct a γ–valid cut π^k for (f,D_k).

2) Solve the linear program

$LP(x^k,\pi^k,D_k)$ $\quad \max \pi^k(x-x^k)$ s.t. $x \in D_k$.

Let ω^k be a basic optimal solution of this linear program. If $\pi^k(\omega^k - x^k) \le 1$, then stop: x^0 is a global minimizer. Otherwise, go to 3).

3) Let $D_{k+1} = D_k \cap \{x: \pi^k(x-x^k) \ge 1\}$. Starting from ω^k find a vertex x^{k+1} of D_{k+1} which is a local minimizer of f(x) over D_{k+1}. If $f(x^{k+1}) \ge \gamma$, then go to iteration k+1. Otherwise, go to 4).

4) Set $\gamma \leftarrow f(x^{k+1})$, $x^0 \leftarrow x^{k+1}$, $D_0 \leftarrow D_{k+1}$, and go to iteration 0.

Theorem V.2. *If the sequence $\{\pi^k\}$ is bounded, then the above cutting algorithm is finite.*

Proof. The algorithm consists of a number of cycles of iterations. During each cycle the incumbent x^0 is unchanged, and each occurrence of Step 4) marks the end of one cycle and the beginning of a new one. As noticed above, in view of the inequality $f(x^{k+1}) < \gamma$ in Step 4), x^{k+1} satisfies all the previous cuts as strict inequalities; hence, since x^{k+1} is a vertex of D_{k+1}, it must be a vertex of D, distinct from all the vertices of D previously encountered. Since the vertex set of D is finite, it follows that the number of occurrences of Step 4), i.e., the number of cycles, is finite.

Now during each cycle a sequence of cuts $\ell_k(x) := \pi^k(x-x^k)-1 \ge 0$ is generated such that

$$\ell_k(x^k) < 0 \,,\ \ell_h(x^k) \ge 0 \qquad (h = 0,1,\ldots,k-1).$$

Since the sequence $\{\pi^k\}$ is bounded, we conclude from Corollary III.2 that each cycle is finite. Hence the algorithm itself is finite. ■

Thus, to ensure finiteness of Algorithm V.1 we should select the cuts π^k so as to have $\|\pi^k\| \le C$ for some constant C. Note that $1/\|\pi^k\|$ is the distance from x^k to the hyperplane $\pi^k(x-x^k) = 1$, so these distances (which measure the depth of the cuts) must be bounded away from 0. Though there is some freedom in the choice of π^k

(condition (6)), it is generally very difficult to enforce the boundedness of this sequence. In the sections that follow we shall discuss various methods for overcoming this difficulty in cutting plane algorithms.

2. FACIAL CUT ALGORITHM

An advantage of concavity cuts as developed above is that they are easy to construct. Unfortunately, in practical computations, it has been observed that these cuts, when used alone in a pure cutting algorithm, often tend to become shallower and shallower as the algorithm proceeds, thus making the convergence very difficult to achieve. Therefore, it is of interest to study other kinds of cuts which may be more expensive but have the advantage that they guarantee finiteness and can be suitably combined with concavity cuts to produce reasonably practical finite algorithms.

2.1. The Basic Idea

A problem closely related to the concave minimization problem (BCP) is the following:

Vertex problem. *Given two polyhedra D and M, find a vertex of D lying in M, or else establish that no such vertex exists.*

If we know some efficient procedure for solving this problem, then the concave programming problem (BCP) can be solved as follows.

Start from a vertex x^0 of D which is a local minimizer. At step $k = 0,1,\ldots$,let γ_k be the best feasible value of the objective function known thus far, i.e.,

$\gamma_k = \min \{f(x^0),....,f(x^k)\}$. At x^k construct a γ_k–valid cut $\pi^k(x - x^k) \geq 1$ for (f,D) and let

$$M_k = \{x \in \mathbb{R}^n: \pi^i(x-x^i) \geq 1 \quad (i=0,1,...,k)\} ,$$

where π^i defines a γ_i–valid cut for (f,D) at x^i.

Solve the vertex problem for D and M_k. If M_k contains no vertex of D, then stop: $\gamma_k = \min f(D)$. Otherwise, let x^{k+1} be a vertex of D lying in M_k. Go to Step k+1.

Since each cut eliminates at least one vertex of D, the above procedure is obviously finite.

Despite its attractiveness, this procedure cannot be implemented. In fact, the vertex problem is a very difficult one, and up to now there has been no reasonably efficient method developed to solve it. Therefore, following Majthay and Whinston (1974), we replace the vertex problem by an easier one.

Definition V.2. *A face F of a polyhedron D is called an **extreme face** of D relative to a polyhedron M if*

$$\emptyset \neq F \cap M \subset ri\, F . \tag{20}$$

For example, in Fig. V.2, page 186, F_1, F_2 and the vertex x are extreme faces of the polytope D relative to the polyhedral cone M. Since the relative interior of a point coincides with the point itself, any vertex of D lying in M is an extreme 0–dimensional face of D relative to M.

The following problem should be easier than the vertex problem:

Extreme face problem. *Given two polyhedra D and M, find an extreme face of D relative to M, or else prove that no such face exists.*

Actually, as it will be seen shortly, this problem can be treated by linear programming methods.

A cutting scheme that uses extreme faces can be realized in the following way: at each step k find an extreme face of D relative to M_k and construct a cut eliminating this extreme face without eliminating any possible candidate for an optimal solution. Since the number of faces of D is finite, this procedure will terminate after finitely many steps.

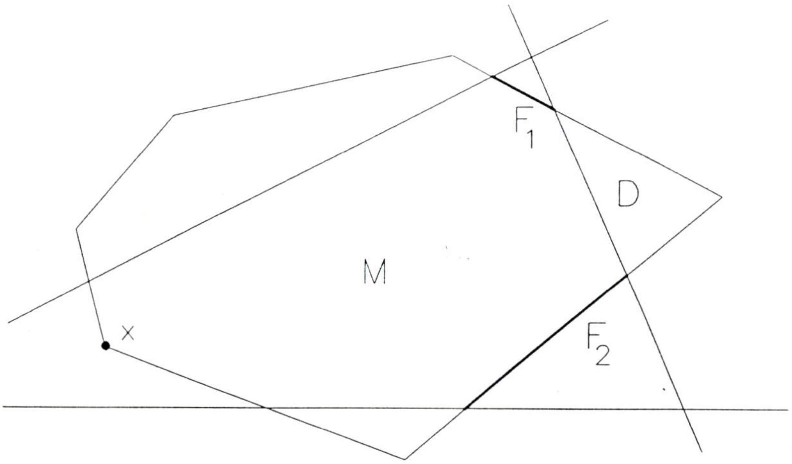

Fig. V.2

2.2. Finding an Extreme Face of D Relative to M

Assume now that the constraints defining the polytope D have been given in the canonical form

$$x_i = p_{io} - \sum_{j \in J} p_{ij} x_j \quad (i \in B) \tag{21}$$

$$x_k \geq 0 \quad (k=1,\dots,n) , \tag{22}$$

where B is the index set of basic variables ($|B| = m$) and J is the index set of nonbasic variables ($|J| = n-m$).

The following consideration makes use of the fact that a face of D is described in (21), (22) by setting some of the variables x_k equal to zero.

For any $x \in \mathbb{R}^n$ let $Z(x) = \{j: x_j = 0\}$. Then we have the following characteristic property of an extreme face of D relative to M.

Proposition V.2. *Let* $x^0 \in D \cap M$, $F_0 = \{x \in D: x_j = 0 \quad \forall j \in Z(x^0)\}$. *Then* F_0 *is an extreme face of D relative to M if and only if for any* $i \in \{1,2,...,n\} \setminus Z(x^0)$:

$$0 < \min \{x_i: x \in D \cap M, x_j = 0 \quad \forall j \in Z(x^0)\} .$$

Proof. Obviously, F_0 is a face of D containing x^0, so $F_0 \cap M \neq \emptyset$.

If F_0 is an extreme face of D relative to M, then for any $x \in F_0 \cap M$ we must have $x \in \text{ri } F_0$, hence $x_i > 0$ for any $i \in \{1,2,...,n\} \setminus Z(x^0)$ (since the linear function $x \mapsto x_i$ which is nonnegative on F_0 can vanish at a relative interior point of F_0 only if it vanishes at every point of F_0). In view of the compactness of $F_0 \cap M$, this implies that $0 < \min \{x_i: x \in F_0 \cap M\}$ for any $i \in \{1,2,...,n\} \setminus Z(x^0)$.

Conversely, if the latter condition is satisfied, then for any $x \in F_0 \cap M$ we must have $x_i > 0$ for all $i \notin Z(x^0)$, hence $x \in \text{ri } F_0$, i.e., $F_0 \cap M \subset \text{ri } F_0$, and F_0 is an extreme face. ■

Note that the above proposition is equivalent to saying that a face F_0 of D that meets M is an extreme face of D relative to M if and only if $Z(x) = Z(x')$ for any $x,x' \in F_0 \cap M$.

On the basis of this proposition, we can solve the extreme face problem by the following procedure of Majthay and Whinston (1974).

Let M be given by the system of linear inequalities

$$\sum_{j=1}^{n} c_{ij}x_j \geq d_i \quad (i=n+1,\ldots,\hat{n}) .$$

Introducing the slack variables x_i $(i=n+1,\ldots,\hat{n})$ and using (21), (22) we can describe the polytope $D \cap M$ by a canonical system of the form

$$x_i = \hat{p}_{io} - \sum_{j\in\hat{J}} \hat{p}_{ij}x_j \quad (i \in \hat{B}) , \tag{23}$$

$$x_i \geq 0 \quad (i=1,\ldots,\hat{n}), \tag{24}$$

where $\hat{B}$ is the index set of basic variables $(|\hat{B}| = m + (\hat{n}-n))$, $\hat{J}$ is the index set of nonbasic variables $(|\hat{J}| = n-m)$.

In the sequel, the variables x_i $(i=1,\ldots,n)$ will be referred to as "original", the others $(x_i, i > n)$ as "nonoriginal".

Let $x^0 = (x_1^0,\ldots,x_n^0) \in D \cap M$ be the point in the space $\mathbb{R}^n$ of original variables obtained by setting $x_j = 0$ for all $j \in \hat{J}$ in (23) (24), i.e., $x_i^0 = \hat{p}_{io}$ $(i \in \hat{B}, i \leq n)$, $x_j^0 = 0$ $(j \in \hat{J}, j \leq n)$.

If $\hat{J} \subset Z(x^0) := \{j \in \{1,2,\ldots,n\}: x_j^0 = 0\}$, i.e., if all nonbasic variables are original, then x^0 is a vertex of D, since in this case the system (23) (24) restricted to $i \leq n$ gives a canonical representation of D.

In the general case, let

$$\{1,2,\ldots,n\} \setminus Z(x^0) = \{i_1,i_2,\ldots,i_s\} .$$

Then we can derive the following procedure.

Procedure I.

Starting from k=1, solve

$$(P_k) \qquad \min x_{i_k} \quad \text{s.t. (23), (24) and } x_j = 0 \quad \forall\, j \in Z(x^{k-1}) \tag{25}$$

Let ξ_k be the optimal value and x^k be a basic optimal solution of (P_k). Set $k \leftarrow k+1$ and repeat the procedure until $k = s$.

Clearly, $Z(x^{k-1}) \subset Z(x^k)$ (k=1,2,...,s) and $Z(x^s) \supset Z(x^0) \cup \{i_k: \xi_k = 0\}$.

Proposition V.3. *If $Z(x^s) \neq \emptyset$, then $F = \{x \in D: x_j = 0 \quad \forall j \in Z(x^s)\}$ is an extreme face of D relative to M. Otherwise, there is no extreme face of D relative to M other than D itself.*

Proof. If $i \in \{1,2,...,n\} \setminus Z(x^s)$, then $i = i_k$ for some $i_k \notin Z(x^s)$, hence $\xi_k > 0$, i.e., $0 < min\,\{x_i: x \in D \cap M,\ x_j = 0 \quad \forall j \in Z(x^s)\}$. Therefore, F is an extreme face by Proposition V.2 provided $Z(x^s) \neq \emptyset$. Moreover, if $Z(x^s) = \emptyset$, then $0 < \min\,\{x_i: x \in D \cap M\}$, from which it easily follows that the only extreme face is D itself. ■

Remarks V.1. (i) A convenient way to solve (P_k) for k=1 is as follows.

Recall that x^0 is given by (23), (24). If $Z(x^0) = J \cap \{1,2,...,n\}$ (i.e., all the variables x_j, $j \in Z(x^0)$, are nonbasic), then to solve (P_1) we apply the simplex procedure to the linear program

$$(P_1^*) \qquad \min x_{i_1} \quad \text{subject to (23), (24)*}, \tag{25*}$$

where the asterisk means that all nonbasic original variables in (23), (24) should be omitted.

However, it may occur that $Z(x^0) \cap \hat{B} \neq \emptyset$ (i.e., $x_i^0 = 0$ for certain $i \in \hat{B} \cap \{1,2,...,n\}$). In this case, we must first remove the variables x_i, $i \in Z(x^0) \cap \hat{B}$, from the basis whenever possible. To do this, we observe that, since $x_i^0 = 0$, we must have

$$0 = \min\,\{x_i: (23)\ (24) \text{ and } x_j = 0 \quad \forall j \in \hat{J} \cap \{1,2,...,n\}\}\ ,$$

hence $\hat{p}_{i,j} \geq 0$ for all $j \in \hat{J} \setminus \{1,2,...,n\}$. If $\hat{p}_{ij} > 0$ for at least one $j \in \hat{J} \setminus \{1,2,...,n\}$, then by pivoting on this element (i,j) we will force x_i out of the basis. On the other hand, if $\hat{p}_{ij} = 0$ for all $j \in \hat{J} \setminus \{1,2,...,n\}$, this means that x_i depends only on the variables x_j, $j \in \hat{J} \cap \{1,2,...,n\}$. Repeating the same operation for each $i \in Z(x^0) \cap \hat{B}$, we will transform the tableau (23) into one where the only variables x_i, $i \in Z(x^0)$ which remain basic are those which depend only on the nonbasic original variables. Then and only then we start the simplex procedure for minimizing x_{i_1} with this tableau, where we omit, along with the nonbasic original variables (i.e., all the columns $j \leq n$), also all the basic original variables x_i with $i \in Z(x^0)$ (i.e., all the rows $i \in Z(x^0)$).

In an analogous manner each problem (P_k) is replaced by a problem (P_k^*), starting with a tableau where a variable x_i , $i \in Z(x^{k-1})$, is basic only if it depends upon the nonbasic original variables alone.

(ii) The set $Z(x^k)$ is equal to the index set Z_k of nonbasic original variables in the optimal tableau of (P_k^*) plus the indices of all of the basic original variables which are at level zero in this tableau. In particular, this implies that $F = \{x \in D: x_j = 0 \ \forall j \in Z_s\}$, and therefore that F is a vertex of D if and only if $|Z_s| = n-m$. We can thus stop the extreme face finding process when $k = s$ or $|Z_k| = n-m$.

2.3. Facial Valid Cuts

Let $F = \{x \in D: x_j = 0 \ \forall j \in Z\}$ be an extreme face of D relative to M that has been obtained by the above method. If $F = D$ (i.e., $Z = \emptyset$), then, by the definition of an extreme face, $D \cap M \subset \text{ri}\, D$, so that M does not contain any vertex of D. If F is a vertex of D, then we already know how to construct a cut which eliminates it without eliminating any better feasible point.

Now consider the case where F is a proper face but not a vertex of D $(0 < |Z| < n-m)$.

Definition V.3. *Let F be a proper face but not a vertex of D. A linear inequality $\ell(x) \geq 0$ is a* ***facial cut*** *if it eliminates F without eliminating any vertex of D lying in M.*

A facial cut can be derived as follows.

Let α_j $(j \in Z)$ be prechosen positive numbers and for each $h \in \{1,2,...,n\} \setminus Z$ consider the parametric linear program

$$(P_h(q)) \qquad \min \{x_h: x \in D \cap M, \sum_{j\in Z} \alpha_j x_j \le q\} , \tag{26}$$

where q is a nonnegative parameter. From Proposition V.2 it follows that

$$0 < \min \{x_h: x \in D \cap M, x_j = 0 \quad \forall j \in Z\} , \tag{27}$$

i.e., the optimal value in (26) for q = 0 is positive. Therefore, if $q_h = \sup \{q: 0 <$ optimal value of $(P_h(q))\}$, then $q_h > 0$, and hence

$$p := \min \{q_h: h \in \{1,2,...,n\} \setminus Z\} > 0 . \tag{28}$$

Proposition V.4. *Let $0 < |Z| < n-m$. If $p < +\infty$, then the inequality*

$$\sum_{j\in Z} x_j \ge p \tag{29}$$

defines a facial valid cut.

Proof. It is obvious that the cut eliminates F. If $x \in D \cap M$ violates (29), then, since $p \le q_h$, it follows that $x_h > 0$ for all $h \in \{1,2,...,n\} \setminus Z$. Hence, $Z \supset \{i: x_i = 0\}$; and, since $|Z| < n-m$, x cannot be a vertex of D. ■

If $p = +\infty$ we say that the facial cut is ***infinite***. Obviously, in that case there is no vertex of D in M.

Remark V.2. For each h the value $q_h = \sup \{q: 0 <$ optimal value in $(P_h(q))\}$ can be computed by parametric linear programming methods.

Of course, the construction of a facial cut is computationally rather expensive, even though it only involves solving linear programs. However, such a cut eliminates an entire face of D, i.e., all the vertices of D in this face, and this is sometimes worth the cost. Moreover, if all we need is a valid (even shallow) cut, it is enough for each h to choose any $q = q_h > 0$ for which the optimal value in $(P_h(q))$ is positive.

2.4. A Finite Cutting Algorithm

Since a facial cut eliminates a face of D, the maximal number of facial valid cuts cannot exceed the total number of distinct faces of D. Therefore, a cutting procedure in which facial valid cuts are used each time after finitely many steps must be finite.

The following modification of Algorithm V.1 is based upon this observation.

Algorithm V.2.

Initialization:

Search for a vertex x^0 which is a local minimizer. Set $\gamma = f(x^0)$, $D_0 = D$.

Select two numbers $\delta_o \geq 0$, $N > 1$. Set $d_0 = +\infty$.

Iteration k = 0,1,... :

0) If $k = 0$ or $d_{k-1} \geq \delta_{k-1}$, go to 1a); otherwise, go to 1b).

1a) Construct at x^k a γ-valid cut $\ell_k(x) := \pi^k(x-x^k) - 1 \geq 0$ for (f,D_k).
Set $d_k = 1/\|\pi^k\|$, $\delta_k = \delta_{k-1}$ and go to 2).

1b) Starting from x^k, identify an extreme face F_k of D relative to D_k (the inter section of D with all previously generated cuts).
If $F_k = D$, then stop: x^0 is an optimal solution of (BCP).

If F_k is a vertex of D (i.e., $F_k = x^k$), then construct at x^k a γ-valid cut $\ell_k(x) := \pi^k(x-x^k) - 1 \geq 0$ for (f,D). Set $d_k = 1/\|\pi^k\|$, $\delta_k = \frac{1}{N}\delta_{k-1}$ and go to 2).

If F_k is a proper face but not a vertex of D, construct a facial valid cut $\ell_k(x) \geq 0$.

If this cut is infinite, then stop: x^0 is a global optimal solution; otherwise, set $d_k = +\infty$, $\delta_k = \frac{1}{N}\delta_{k-1}$ and go to 3).

2) Solve the linear program

$$\max \ell_k(x) \quad \text{subject to } x \in D_k$$

to obtain a basic optimal solution ω^k of this problem. If $\ell_k(\omega^k) \le 0$, then stop: x^0 is a global optimal solution. Otherwise, go to 3).

3) Let $D_{k+1} = D_k \cap \{x: \ell_k(x) \ge 0\}$. Find a vertex x^{k+1} of D_{k+1} which is a local minimizer of f(x) over D_{k+1}. If $f(x^{k+1}) \ge \gamma$, go to iteration k+1. Otherwise, go to 4).

4) Set $\gamma \leftarrow f(x^{k+1})$, $x^0 \leftarrow x^{k+1}$, $D_0 \leftarrow D_{k+1}$ and return to iteration 0.

Theorem V.3. *The above algorithm is finite.*

Proof. Like Algorithm V.1, the above procedure consists of a number of cycles of iterations, each of which results in a vertex of D (the point x^{k+1} in Step 4) which is better than the incumbent one. Therefore, it suffices to show that each cycle is finite. But within a given cycle, since the number of facial cuts is finite, there must exist a k_0 such that Step 1a) occurs in all iterations $k \ge k_0$. Then $d_k \ge \delta_k = \delta_{k_0}$ for all $k > k_0$, i.e., $\|\pi^k\| \le 1/\delta_{k_0}$. So the sequence $\|\pi^k\|$ is bounded, which, by Theorem V.2, implies the finiteness of the algorithm. ■

Remark V.3. The above algorithm differs from Algorithm V.1 only in Step 1b), which occurs in iteration k when $d_{k-1} \le \delta_{k-1}$. Since $d_{k-1} = 1/\|\pi^{k-1}\|$ is the distance from x^{k-1} to the cutting hyperplane (when a γ–valid cut has been applied in iteration k–1), this means that a facial cut is introduced if the previous cut was a γ–valid cut that was too shallow; moreover, in that case δ_k is decreased to $\frac{1}{N}\delta_{k-1}$, so that a facial cut will have less chance of being used again in subsequent iterations.

Roughly speaking, the (δ,N) device allows facial cuts to intervene from time to time in order to prevent the cutting process from jamming, while keeping the frequency of these expensive cuts to a low level.

Of course, the choice of the parameters δ_0 and N is up to the user, and N may vary with k. If δ_0 is close to zero, then the procedure will degenerate into Algorithm V.1; if δ_0 is large, then the procedure will emphasize on facial cuts.

In practice, it may be simpler to introduce facial cuts periodically, e.g., after every N cuts, where N is some natural number. However, this method might end up using more facial cuts than are actually needed.

Example V.1. We consider the problem

$$\text{minimize } -(x_1-1.2)^2-(x_2-0.6)^2$$

$$\begin{aligned}\text{subject to } \quad -2x_1+x_2 &\le 1 ,\\ x_2 &\le 2 ,\\ x_1+x_2 &\le 4 ,\\ x_1 \qquad &\le 3 ,\\ 0.5x_1-x_2 &\le 1 ,\\ x_1 \ge 0,\ x_2 &\ge 0 .\end{aligned}$$

Suppose that in each cycle of iterations we decide to introduce a facial cut after every two concavity cuts. Then, starting from the vertex $x^0 = (0,0)$ (a local minimizer with $f(x^0) = -1.8$), after two cycles of 1 and 3 iterations, respectively, we find the global minimizer $x^* = (3,1)$ with $f(x^*) = -3.4$.

Cycle 1.

Iteration 0: $x^0 = (0,0)$, $\gamma = -1.8$; concavity cut: $0.417x_1 + 0.833x_2 \ge 1$.

Cycle 2.

Iteration 0: $x^0 = (3,1)$, $\gamma = -3.4$;

concavity cut $1.964x_1 + 1.250x_2 \le 6.143$;

Iteration 1: $x^1 = (0.5,2)$; concavity cut: $-5.261x_1 + 0.751x_2 \leq 0.238$.

Iteration 2: $x^2 = (2.856,0.43)$.

Extreme face containing x^2: $0.5x_1 - x_2 = 1$.

Facial cut: $0.5x_1 - x_2 \leq 1-p$, where $p > 0$ can be arbitrarily large.

Since the facial cut is infinite, the incumbent $x^* = (3,1)$ is the global optimizer (Fig. V.3).

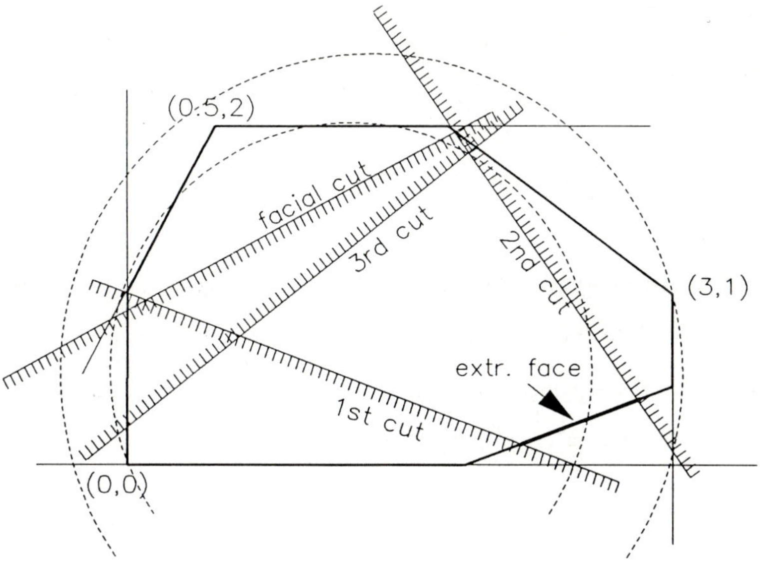

Fig. V.3

3. CUT AND SPLIT ALGORITHM

A pure cutting plane algorithm for the BCP problem can be made convergent in two different ways: either by introducing special (usually expensive) cuts from time to time, for example, facial cuts, that will eliminate a face of the polytope D; or by the use of deep cuts that at one stoke eliminates a sufficiently "thick" portion of D.

A drawback of concavity cuts is that, when used repeatedly, these cuts tend to degrade and become shallow. To overcome this phenomenon, we could attempt to strengthen these cuts: later, in Section V.4, we shall examine how this can be done in certain circumstances.

In the general case, a procedure which has proven to be rather efficient for making deep cuts is in an appropriate manner to combine cutting with splitting (partitioning) the feasible domain.

3.1. Partition of a Cone

Let us first describe a construction which will frequently be used in this and subsequent chapters.

To simplify the language, in the sequel, unless otherwise stated, a *cone* always means a convex polyhedral cone vertexed at the origin 0 and generated by n linearly independent vectors. A cone K generated by n vectors $z^1,z^2,\ldots,z^n$ forming a nonsingular matrix $Q = (z^1,z^2,\ldots,z^n)$ will be denoted: $K = \text{con}(Q)$.

Since a point x belongs to K if and only if $x = \Sigma\lambda_i z^i = Q\lambda^T$ with $\lambda = (\lambda_1,\lambda_2,\ldots,\lambda_n) \geq 0$, we have

$$K = \{x: Q^{-1}x \geq 0\} . \tag{30}$$

Now let u be any point of K, so that $Q^{-1}u = (\lambda_1,\lambda_2,\ldots,\lambda_n)^T \geq 0$. Let $I = \{i: \lambda_i > 0\}$. For each $i \in I$, z^i can be expressed as a linear combination of z^j $(j \neq i)$ and u, namely: $z^i = (u - \sum_{j \neq i} \lambda_j z^j) / \lambda_i$; hence the matrix $Q_i = (z^1,\ldots,z^{i-1},u,z^{i+1},\ldots,z^n)$ obtained from Q by substituting u for z^i is still nonsingular. Denoting $K_i = \text{con}(Q_i)$, we then have the following fact whose proof is analogous to that of Proposition IV.1:

$$(\text{int } K_i) \cap (\text{int } K_j) = \emptyset \quad (j \neq i) ;$$

$$K = \cup \{K_i : i \in I\} .$$

In other words, the cones K_i ($i \in I$) constitute a partition of K. In the sequel, this partition will be referred to as the ***partition (splitting) of the cone K*** with respect to u.

In actual computations, we work with the matrices rather than with the cones. Therefore, it will often be more convenient to speak of matrix partitioning instead of cone partitioning. Thus, we shall say that the matrices Q_i ($i \in I$), with $Q_i = (z^1,\ldots,z^{i-1},u,z^{i+1},\ldots,z^n)$, form the ***partition of*** $Q = (z^1,\ldots,z^n)$ with respect to u. Note that the partitions of a matrix Q with respect to λu for different $\lambda > 0$, lead to different matrices, although the corresponding partitions of the cone $K = \mathrm{con}(Q)$ are the same.

3.2. Outline of the Method

We consider problem (BCP) in the form (1), (2), (3).

Let $\varepsilon \geq 0$ be a prescribed tolerance. Our aim is to find a ***global ε–optimum***, i.e., a feasible solution x^* such that

$$f(x^*) - \varepsilon \leq \min f(D).$$

Let us start with a vertex x^0 of D which is a local minimizer of f(x) over D. Set $\gamma = f(x^0)$, $\alpha = \gamma - \varepsilon$. Without loss of generality we may always assume that the problem is in standard form with respect to x^0 (cf. Section V.1), so that $x^0 = 0$ and condition (14) holds.

Let G_α denote the set of all points $z \in \mathbb{R}^n$ such that z is the α–extension of some $y \neq 0$. Clearly, since f(x) is assumed to have bounded level sets, G_α is a compact set and

$$G_\alpha = \{z: f(z) = \alpha,\ f(\lambda z) < \alpha \quad \forall \lambda > 1\}\ .$$

For each i=1,2,...,n let z^i be the point where G_α meets the positive x_i–axis. Then, as already seen from (17), the vector

$$\pi = eQ^{-1} \text{ with } e = (1,1,...,1)\ ,\ Q = (z^1,z^2...,z^n)\ ,$$

defines an α–valid cut for (f,D) at x^0.

Consider the linear program

$$LP(Q,D) \qquad \max eQ^{-1}x \quad \text{s.t. } x \in D\ ,$$

and let $\omega = \omega(Q)$ be a basic optimal solution and $\mu = \mu(Q)$ be the optimal value of this linear program (i.e., $\mu(Q) = eQ^{-1}\omega$).

If $\mu \le 1$, then x^0 is a global ε–optimizer. On the other hand, if $f(\omega) < \alpha$, then we can find a vertex x^1 of D such that $f(x^1) \le f(\omega) < \alpha$. Replacing the polytope D by $D \cap \{x: eQ^{-1}x \ge 1\}$, we can restart from x^1 instead of x^0. So the case that remains to be examined is when $\mu > 1$, but $f(\omega) \ge \alpha$ (Fig. V.4).

As we already know, a feasible point x with $f(x) < \gamma-\varepsilon$ should be sought only in the residual polytope $D \cap \{x: eQ^{-1}x \ge 1\}$ left over by the cut. In order to cut off more of the unwanted portion of D, instead of repeating for this residual polytope what was done for D (as in Algorithm V.1) we now construct the α–extension $\hat{\omega}$ of ω and split the cone K = con(Q) with respect to $\hat{\omega}$ (Fig. V.4).

Let Q_i , $i \in I$, be the corresponding partition of the matrix Q:

$$Q_i = (z^1,...,z^{i-1},\hat{\omega},z^{i+1},...,z^n).$$

Now note that for each subcone $K_i = con(Q_i)$ the cut through $z^1,...,z^{i-1},\hat{\omega}$, $z^{i+1},...,z^n$ does not eliminate any point x of K_i with $f(x) < \gamma-\varepsilon$. (This can be seen in the same way that one sees that the cut through $z^1,z^2,...,z^n$ does not eliminate any point $x \in K$ with $f(x) < \gamma-\varepsilon$.) Therefore, to check whether there is a feasible point x with $f(x) < \gamma-\varepsilon$ in any subcone K_i, for each $i \in I$ we solve the linear program

$LP(Q_i,D) \quad \max eQ_i^{-1}x \quad s.t. \quad x \in D, Q_i^{-1}x \geq 0 .$

Note that the constraint $Q_i^{-1}x \geq 0$ simply expresses the condition that $x \in K_i = con(Q_i)$ (see (30)). If all of these linear programs have optimal values ≤ 1, this indicates that no point $x \in D$ in any cone K_i has $f(x) < \gamma-\varepsilon$, and hence, that x^0 is a global ε-optimal solution. Otherwise, each subcone K_i for which the linear program $LP(Q_i,D)$ has an optimal value > 1 can be further explored by the same splitting method that was used for the cone K.

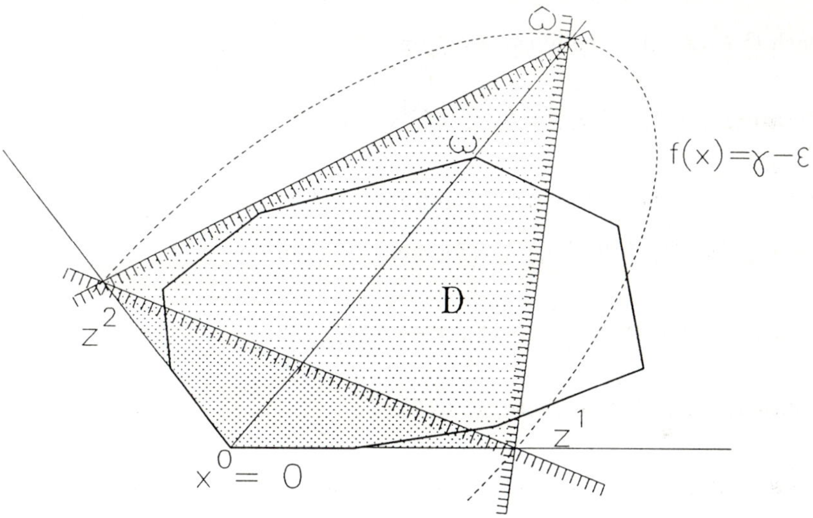

Fig. V.4

In a formal way we can give the following algorithm.

Algorithm V.3.

Select $\varepsilon \geq 0$.

Initialization:

Compute a point $z \in D$. Set $M = D$.

Phase I.

Starting from z search for a vertex x^0 of M which is a local minimizer of f(x) over M.

Phase II.

0) Let $\gamma = f(x^0)$, $\alpha = \gamma - \varepsilon$. Rewrite the problem in standard form with respect to x^0. Construct $Q_0 = (z^{01}, z^{02}, \ldots, z^{0n})$, where z^{0i} is the intersection of G_α with the i-th edge of K_0. Set $\mathcal{M} = \{Q_0\}$.

1) For each $Q \in \mathcal{M}$ solve the linear program

$$LP(Q,M) \quad \max eQ^{-1}x \quad \text{s.t. } x \in M \ , \ Q^{-1}x \geq 0$$

to obtain a basic optimal solution $\omega(Q)$ and the optimal value $\mu(Q) = eQ^{-1}\omega(Q)$. If $f(\omega(Q)) < \alpha$ for some Q, then set

$$M \leftarrow M \cap \{x: eQ_0^{-1}x \geq 1\} \ , \ z \leftarrow \omega(Q) \ ,$$

and return to Phase I. Otherwise, go to 2).

2) Let $\mathcal{R} = \{Q \in \mathcal{M}: \mu(Q) > 1\}$. If $\mathcal{R} = \emptyset$, then stop: x^0 is a global ε-optimal solution. Otherwise, go to 3).

3) For each $Q \in \mathcal{R}$ construct the α-extension $\hat{\omega}(Q)$ of $\omega(Q)$ and split Q with respect to $\hat{\omega}(Q)$. Replace Q by the resulting partition and let $\mathcal{M}'$ be the resulting collection of matrices. Set $\mathcal{M} \leftarrow \mathcal{M}'$ and return to 1).

3.3. Remarks V.4

(i) The algorithm involves a sequence of cycles of iterations. Every "return to Phase I" indicates the passage to a new cycle. Within a cycle the polytope M that remains to be explored and the incumbent x^0 do not change, but from one cycle to the next the polytope M is reduced by a cut

$$eQ_0^{-1}x \geq 1 ,$$

while x^0 changes to a better vertex of D. (Note that $Q_0 = (z^{01},z^{02}...,z^{0n})$ is the matrix formed in Step 0 of Phase II, and hence determines an α–valid cut for (f,M) at x^0; the subsequent cuts $eQ_0^{-1}x \geq 1$ in this cycle cannot be used to reduce M, because they are not α–valid for (f,M) at x^0.) Under these conditions, it is readily seen that at each stage the current x^0 satisfies every previous cut as a strict inequality. Therefore, since x^0 is a vertex of M, it must also be a vertex of D. Moreover, since the vertex set of D is finite, the number of cycles of iterations must also be finite.

(ii) The value of $\varepsilon \geq 0$ is selected by the user. If ε is large, then few iterations will be needed but the accuracy of the solution will be poor; on the other hand, if ε is small, the accuracy will be high but many iterations will be required. Also, since the minimum of f(x) is achieved at least at one vertex, if ε is smaller than the difference between the values of f at the best and the second best vertex, then a vertex x^0 which is globally ε–optimal will actually be an exact global optimal solution. Therefore, for ε small enough, the solution given by Algorithm V.3 is an exact global optimizer.

(iii) The linear programs LP(Q,M) can be given in a more convenient form which does not require computing the inverse matrices Q^{-1}. Indeed, since the problem is in standard form with respect to x^0, the initial cone K_0 in Phase II is the nonnegative orthant. Then any cone $K = \text{con}(z^1,z^2,...,z^n)$ generated in Phase II is a subcone of K_0 , and the constraints $Q^{-1}x \geq 0$ (i.e., $x \in K$) imply that $x \geq 0$. Therefore, if $M = D \cap \{x: Cx \leq d\}$, where $Cx \leq d$ is the system formed by the previous cuts, then the constraints of the linear program LP(Q,M) are

$$Ax \leq b, \quad Cx \leq d , \quad Q^{-1}x \geq 0 .$$

Thus, in terms of the variables $(\lambda_1,\lambda_2,...,\lambda_n) = Q^{-1}x$, this linear program can be written as

$$\text{LP(Q,M)} \qquad \max \sum_{j=1}^{n} \lambda_j$$

$$\text{s.t.} \sum_{j=1}^{n} \lambda_j(Az^j) \le b\,, \quad \sum_{j=1}^{n} \lambda_j(Cz^j) \le d, \quad \lambda_1,\lambda_2...,\lambda_n \ge 0\,.$$

In this form, all the linear programs LP(Q,M) corresponding to different cones $K = \text{con}(Q)$ have the same objective function. If $u \in \text{con}(Q)$ and Q' is obtained from Q by replacing a certain column z^i of Q by u, then LP(Q',M) is derived from LP(Q,M) simply by replacing Az^i,Cz^i by Au, Cu respectively.

Moreover, if $z^{0j} = \theta_j e^j$, where e^j denotes the j–th unit vector, then $Q_0^{-1} = \text{diag}(\theta_1,\theta_2...,\theta_n)$, and $eQ_0^{-1}x = \Sigma\, x_j/\theta_j$. Consequently, the linear programs LP(Q,M) can easily be derived, and we do not need to compute the corresponding inverse matrices Q^{-1}. Furthermore, if the dual simplex method is used to solve these linear programs, then the optimal solution of an LP(Q,M) is dual feasible to any LP(Q',M) corresponding to an immediate successor Q' of Q, and therefore can be used to start the solution process for LP(Q',M).

(iv) For ease of exposition, we assumed that the function f(x) has bounded level sets. If this is not the case, we cannot represent a cone by a matrix $Q = (z^1,...,z^n)$, where each z^i is the intersection of the i–th edge with G_α (because this intersection may not exist). Therefore, to each cone we associate a matrix $Q = (z^1,...,z^n)$, where each z^i is the intersection point of the i–th edge with G_α if this intersection exists, or the direction of the i–th edge otherwise. Then, in the linear subproblem LP(Q,M) the vector e should be replaced by a column vector with its i–th component equal to 1 if z^i is a point or equal to 0 if z^i is a direction. If $I = \{i\colon z^i \text{ is a point}\}$, this subproblem can similarly be written as

$$\max \sum_{i \in I} \lambda_i$$

$$\text{s.t. } \sum_{i=1}^{n} \lambda_i (Az^i) \le b \ , \ \sum_{i=1}^{n} \lambda_i (Cz^i) \le d \ ,$$

$$\lambda_i \ge 0 \quad (i = 1, 2, \ldots, n).$$

With these modifications, Algorithm V.3 still works for an arbitrary concave function f: $\mathbb{R}^n \longrightarrow \mathbb{R}$.

(v) An algorithm which is very similar to Algorithm V.3, but with $\varepsilon = 0$ and the condition $Q_i^{-1}x \ge 0$ omitted in $LP(Q_i,M)$, was first given in Tuy (1964). This algorithm was later shown to involve cycling (cf. Zwart (1973)). Zwart then developed a modified algorithm by explicitly incorporating the condition $Q_i^{-1}x \ge 0$ into the linear program $LP(Q_i,M)$ and by introducing a tolerance parameter $\varepsilon \ge 0$. When $\varepsilon = 0$ Zwart's algorithm and Algorithm V.3 coincide, but when $\varepsilon > 0$ the two algorithms differ, but only by the use of this parameter.

While Zwart's algorithm for $\varepsilon > 0$ is computationally finite, it may give an incorrect solution, which is not even an ε–approximate solution in the sense of Zwart (cf. Tuy (1987b)).

On the other hand, as it stands, Algorithm V.3 is not guaranteed, theoretically, to terminate in finitely many steps. However, the computational experiments reported by Zwart (1974) as well as more recent ones seem to suggest that the algorithm will be finite in most cases encountered in practice. Moreover, it turns out that, theoretically, the algorithm can be made finite through the use of an appropriate anti–jamming device. The resulting algorithm, called the Normal Conical Algorithm, will be discussed in Chapter VII.

Let us also mention two other modifications of Tuy's original algorithm: one by Bali (1973), which is only slightly different from Zwart's algorithm (for $\varepsilon = 0$), the other by Gallo and Ülkücü (1977), in the application to the bilinear programming problem. The algorithm of Gallo and Ülkücü has been proved to cycle in an example of Vaish and Shetty (cf. Vaish and Shetty (1976 and 1977)). The convergence of

Bali's algorithm, like that of the $\varepsilon = 0$ version of Zwart's algorithm or Algorithm V.3, is problematic, although no counter–example has been found yet.

The only result in this regard that has been established is stated in the following proposition (cf. Jacobsen (1981)):

Proposition V.5. *Let $Q_i = (z^{i1}, z^{i2}, \ldots, z^{in})$, $i=1,2,\ldots$, be the matrices generated in a Phase II of Algorithm V.3, where the index system is such that $i < j$ if Q_i is generated before Q_j. Suppose that $\varepsilon = 0$. If $eQ_j^{-1}z^{ik} \le 1$ $(k=1,2,\ldots,n)$ for any $i < j$, then Phase II is finite, unless x^0 is already a global optimizer.*

Proof. Suppose that a Phase II is infinite, while there is a feasible solution x^* better than x^0. Denote $\hat{\omega}^j = \hat{\omega}(Q_j)$, $L_j = \{x: eQ_j^{-1}x \le 1\}$. By construction, we have $\hat{\omega}^j \notin L_j$, and by hypothesis $\hat{\omega}^i \in L_j$ for any $j > i$. Hence,

$$\hat{\omega}^i \in \cap \{L_j: j > i\} .$$

By Lemma III.2, this implies that $d(\hat{\omega}^i, L^i) \longrightarrow 0$ as $i \longrightarrow \infty$. Now for any i, x^* must belong to some cone $K_j = \mathrm{con}(Q_j)$ with $j > i$, and from the definition of $\omega^j = \omega(Q_j)$, we have $d(x^*, L_j) \le d(\hat{\omega}^j, L_j)$. Therefore, $d(x^*, L_j) \longrightarrow 0$ as $j \longrightarrow \infty$.

On the other hand, by hypothesis, the halfspace L_j entirely contains the polytope spanned by $0, z^{01}, z^{02}, \ldots, z^{0n}$ and $\hat{\omega}^0 = \hat{\omega}(Q_0)$. Hence, the distance from $x^0 = 0$ to the hyperplane $H_j = \{x: eQ_j^{-1}x = 1\}$, i.e., $1/\|Q_j^{-1}\|$, must be bounded below by some positive constant δ.

Let y^j denote the intersection of H_j with the halfline from 0 through x^*, and note that $\|x^* - y^j\|/d(x^*, L_j) = \|y^j\|/d(0, H_j)$, which implies that $\|x^* - y^j\| \le \|y^j\|\ \delta\, d(x^*, L_j) \longrightarrow 0$. But since y^j belongs to the simplex $[z^{j1}, z^{j2}, \ldots, z^{jn}]$ with $f(z^{jk}) = \gamma$ $(j=1,2,\ldots,n)$, we must have $f(y^j) \ge \gamma$. This contradicts the assumption that $f(x^*) < \gamma$, since we have just established that $x^* - y^j \longrightarrow 0$ as $j \longrightarrow \infty$. ■

For every cone $K = \mathrm{con}(Q)$ generated by Algorithm V.3, let $\Delta(K) = K \cap \{x: eQ^{-1}x \le 1\}$. Then the convergence condition stated in the above proposition is equivalent to requiring that at the completion of any iteration in Phase II, the union of all $\Delta(K)$ corresponding to all of the cones K that have been generated so far (in the current Phase II) is a polytope. In the next chapter, we shall present an algorithm which realizes this condition (the Polyhedral Annexation Algorithm).

4. GENERATING DEEP CUTS: THE CASE OF CONCAVE QUADRATIC FUNCTIONALS

The basic construction in a cutting method for solving (BCP) is for a given feasible value γ (the current best value) of the objective function f(x) to define a cutting plane which will delete as large as possible a subset of $D \cap \{x: f(x) \geq \gamma\}$. Therefore, although shallow cuts which can delete some basic constituent of the feasible set (such as a face) may sometimes be useful, we are more often interested in generating a deep cut.

4.1. A Hierarchy of Valid Cuts

Consider a vertex x^0 of D which is a local minimizer of f(x) over D, and let $\alpha \leq f(x^0)$. As usual, we may assume that the problem is in standard form with respect to x^0 and that (14) holds. So $x^0 = 0$, $D \subset \mathbb{R}^n_+$ and

$$\theta_i = \max \{\tau : f(\tau e^i) \geq \alpha\} > 0 \quad (i=1,2,...,n),$$

where e^i is the i–th unit vector. We already know that an α–valid cut for x^0 is furnished by the concavity cut

$$\Sigma x^i/\theta_i \geq 1 . \tag{31}$$

Our aim is to develop α–valid cuts at x^0 which may be stronger than the concavity cut.

Suppose that we know a continuous function $F(u,v)$: $\mathbb{R}^n \times \mathbb{R}^n \longrightarrow \mathbb{R}$ such that:

(i) $$F(x,x) = f(x) \quad \forall x \in D; \tag{32}$$

(ii) $$F(u,v) \geq \min \{f(u), f(v)\} \quad \forall u,v; \tag{33}$$

(iii) $F(u,v)$ is concave in u for every fixed v and affine in v for every fixed u. (34)

For every $t = (t_1,t_2,...,t_n) > 0$, let

$$\Delta(t) = \{x \in D: \Sigma x_i/t_i \leq 1\} ,$$

$$M(t) = \{x \in D: \Sigma x_i/t_i \geq 1\} ,$$

and define the function

$$\Phi_t(x) = \min \{F(x,v): v \in M(t)\} \tag{35}$$

(where we set $\Phi_t(x) = -\infty$ if $F(x,v)$ is not bounded from below on $M(t)$).

Proposition V.6. *The function $\Phi_t(x)$ is concave and satisfies*

$$\min \{\Phi_t(x): x \in M(t)\} = \min \{f(x): x \in M(t)\} . \tag{36}$$

Proof. The concavity of the function in (35) follows from assumption (iii). Since, by assumption (ii), $F(x,v) \geq \min \{f(x), f(v)\} \geq \min f(M(t))$ for any $x,v \in M(t)$, we have:

$$\Phi_t(x) \geq \min f(M(t)) \quad \forall x \in M(t). \tag{37}$$

But, $\min \{F(x,v): v \in M(t)\} \leq F(x,x)$; hence, in view of (i):

$$\Phi_t(x) \leq f(x) \quad \forall x \in M(t).$$

This, together with (37), yields (36). ■

Theorem V.4. *Let $t > 0$ define an α-valid cut:*

$$\Sigma\, x_i / t_i \geq 1\,, \tag{38}$$

and let $t_i^* = max\,\{\tau : \Phi_t(\tau e^i) \geq \alpha\}$ $(i=1,2,...,n)$.

If

$$\Phi_t(t_i e^i) \geq \alpha \quad (i=1,2,...,n) \tag{39}$$

then $t^* \geq t$ *and the inequality*

$$\Sigma x_i / t_i^* \geq 1 \tag{40}$$

is an α-valid cut for (f,D).

Proof. The inequality $t^* \geq t$ follows from the definition of t^*. Then $\Delta(t^*) \cap M(t)$ is contained in a polytope with vertices $u^i = t_i e^i$, $v^i = t_i^* e^i$ $(i=1,2,...,n)$. Since we have $\Phi_t(u^i) \geq \alpha$, $\Phi_t(v^i) \geq \alpha$ $(i=1,2,...,n)$,by (39) and the definition of t^*, it follows from the concavity of $\Phi_t(x)$ that $\Phi_t(x) \geq \alpha$ $\forall x \in \Delta(t^*) \cap M(t)$.

Therefore,

$$f(x) = F(x,x) \geq \Phi_t(x) \geq \alpha \quad \forall x \in \Delta(t^*) \cap M(t).$$

Moreover, the α–validity of the cut (38) implies

$$f(x) \geq \alpha \ \ \forall x \in \Delta(t)\,.$$

Hence, $f(x) \geq \alpha$ $\forall x \in \Delta(t^*)$, which proves the theorem. ■

Referring to the concavity cut (31), we obtain the following version of Theorem V.4.

Corollary V.1. *Let* $\theta = (\theta_1,\theta_2,\ldots,\theta_n)$ *define the concavity cut (31). For each* $i = 1,2,\ldots,n$ *let*

$$t_i = max\,\{\tau\colon \Phi_\theta(\tau e^i) \geq \alpha\}\,, \tag{41}$$

$$y^i \in argmin\,\{F(\theta_i e^i,v)\colon v \in M(\theta)\}\,. \tag{42}$$

If $f(y^i) \geq \alpha$ $(i=1,2,\ldots,n)$, *then* $t_i \geq \theta_i$ $(i=1,2,\ldots,n)$ *and the inequality*

$$\Sigma x_i / t_i \geq 1 \tag{43}$$

is an α*-valid cut.*

Proof. $\Phi_\theta(\theta_i e^i) = F(\theta_i e^i, y^i) \geq \min\,\{f(\theta_i e^i),\ f(y^i)\} \geq \alpha$. This shows that the conditions of Theorem V.4 are satisfied. ■

Remark V.5. If property (ii) holds with strict inequality in (33) whenever $u \neq v$, i.e., if we have

$$F(u,v) < \min\,\{f(u),f(v)\} \quad \forall u \neq v\,, \tag{44}$$

then under the conditions in the above corollary we will have $t_i > \theta_i$, provided that $\theta_i e^i \notin D$, because then $y^i \neq \theta_i e^i$ and consequently $\Phi_\theta(\theta_i e^i) > \min\,\{f(\theta_i e^i),f(y^i)\} \geq \alpha$. This means that the cut (43) will generally be strictly deeper than the concavity cut (Fig. V.5).

Moreover, note that for the purpose of solving the problem (BCP), a cut construction is only a means to move towards a better feasible solution than the incumbent one whenever possible. Therefore, in the situation of the corollary, if $f(y^i) < \alpha$ for some i, then the goal of the cut is achieved, because the incumbent has improved.

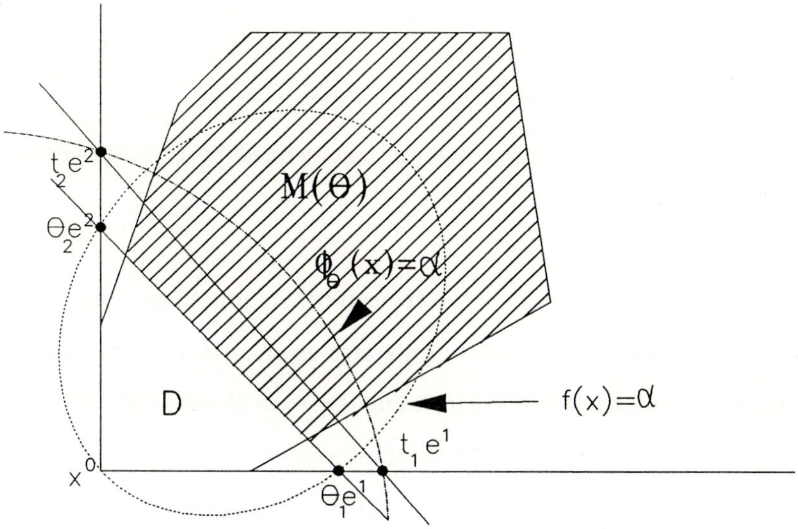

Fig. V.5

These results lead to the following procedure.

Iterative Cut Improvement Procedure:

Start with the concavity cut (31).

Iteration 1:

Compute the points y^i $(i=1,2,...,n)$ according to (42) (this requires solving a linear program for each i, since $F(.,v)$ is affine in v).

If $f(y^i) < \alpha$ for some i, then the incumbent has improved.

Otherwise, compute the values t_i $(i=1,2,...,n)$ according to (41). Then by Corollary V.1, $t \geq \theta$, so that the cut defined by t is deeper than the concavity cut provided $t \neq \theta$ (which is the case if (44) holds).

Iteration k > 1:

Let (38) be the current cut. Compute $t_i^* = \max \{\tau : \Phi_t(\tau e^i) \geq \alpha\}$ (i=1,2,...,n). Since $t \geq \theta$, we have $M(t) \subset M(\theta)$, hence $\Phi_t(t_i e^i) \geq \Phi_\theta(t_i e^i) \geq \alpha$ (i=1,2,...,n). Therefore, by Theorem V.4, $t^* \geq t$ and the cut defined by t^* is deeper than the current one provided $t^* \neq t$.

Thus, if condition (44) holds, then the above iterative procedure either leads to an improvement of the incumbent, or else generates an increasing sequence of vectors $\theta \leq t \leq t^* \leq t^{**} \leq \ldots$, which define an increasing sequence of valid cuts (the procedure stops when successive cuts do not differ substantially).

Of course, this scheme requires the availability of a function F(u,v) with properties (i), (ii), (iii).

In the next section we shall see that this is the case if f(x) is quadratic. For the moment, let us mention a result which sometimes may be useful.

Proposition V.7. *If a function F(u,v) satisfies (ii) and (iii), then the function $f(x) = F(x,x)$ is quasiconcave.*

Proof. For any $x,y \in \mathbb{R}^n$ and $\lambda,\mu \geq 0$ such that $\lambda+\mu = 1$, we have $f(\lambda x+\mu y) = F(\lambda x+\mu y,\lambda x+\mu y) = \lambda F(\lambda x+\mu y,x) + \mu F(\lambda x+\mu y,y) \geq \lambda[\lambda F(x,x) + \mu F(y,x)] + \mu[\lambda F(x,y) + \mu F(y,y)]$. Hence, if $\gamma := \min \{f(x),f(y)\}$, then $f(\lambda x+\mu y) \geq \lambda[\lambda\gamma+\mu\gamma] + \mu[\lambda\gamma+\mu\gamma] = \lambda\gamma + \mu\gamma = \gamma$. This proves the quasiconcavity of f(x). ■

Concerning the applicability of this proposition, note that most of the material discussed in this chapter can easily be extended to quasiconcave minimization problems.

Another important issue when implementing the above scheme is how to compute the values t_i according to (41) (or the values t_i^* in Theorem V.4). For this we consider the linear program

$$\min \{F(\tau e^i,v): \ Av \le b, \ v \ge 0, \ \Sigma v_i/\theta_i \ge 1\} \tag{45i}$$

whose objective function is a concave function of the parameter τ by assumption (iii). Since the optimal value $\Phi_\theta(\tau e^i)$ of this linear program is concave in τ, by Proposition V.6, and since $\Phi_\theta(\theta_i e^i) \ge \alpha$, it follows that t_i is the value such that $\Phi_\theta(\tau e^i) \ge \alpha$ for all $\tau \in [\theta_i,t_i]$, while $\Phi_\theta(\tau e^i) < \alpha$ for all $\tau > t_i$. Therefore, this value can be computed by parametric linear programming techniques.

More specifically, let v^i be a basic optimal solution of (45i) for $\tau = \theta_i$. Then the reduced costs (in linear programming terminology) for this optimal solution must be ≥ 0. Clearly, these reduced costs are concave functions of the parameter τ, and one can determine the maximal value τ_i such that these reduced costs are still ≥ 0 for $\theta_i \le \tau \le \tau_i$. If $\Phi_\theta(\tau_i e^i) = F(\tau_i e^i,v^i) < \alpha$, then t_i is equal to the unique value $\tau \in [\theta_i,\tau_i]$ satisfying $F(\tau e^i,v^i) = \alpha$. Otherwise, $F(\tau_i e^i,v^i) \ge \alpha$, in which case at least one of these reduced costs will become negative for $\tau > \tau_i$, and by pivoting we can pass to a basic solution v^2 which is optimal for $\tau > \tau_i$ sufficiently close to τ_i.

The procedure is then repeated, with τ_i and v^2 instead of θ and v^i. (For the sake of simplicity we assumed τ_i to be finite, but the case $\tau_i = +\infty$ can be considered similarly.)

4.2. Konno's Cutting Method for Concave Quadratic Programming

The cut improvement technique presented above was first developed in a more specialized form by Konno (1976a) for solving the concave quadratic programming problem (CQP). Results similar to those of Konno have also been obtained by Balas and Burdet (1973) using the theory of generalized outer polars (the corresponding cuts are sometimes called polar cuts).

The concave quadratic programming (CQP) problem is an important special case of the problem (BCP). It consists in finding the global minimum of a concave quadratic function f(x) over a polyhedron D. Assuming that the polyhedron D is bounded and int $D \neq \emptyset$, let x^0 be a vertex of D which achieves a local minimum of f(x) over D. Writing the problem in standard form with respect to x^0 (see Section V.1.1), we have

$$\text{(CQP)} \qquad \text{minimize } f(x) = 2px - x(Cx) \tag{46}$$

$$\text{s.t. } Ax \le b \, , \tag{47}$$

$$x \ge 0 \, , \tag{48}$$

where p is an n–vector, $C = (c_{ij})$ is a symmetric positive semidefinite n×n matrix, A is an m×n matrix and b is an m–vector. Then the conditions $x^0 = 0$ (so $f(x^0) = 0$), $D \subset \mathbb{R}^n_+$ (hence $b \ge 0$) are satisfied.

Furthermore, $p \ge 0$ because $x^0 = 0$ is a local minimizer of f(x) over D (if $p_i < 0$ for some i we would have $f(\lambda e^i) = 2\lambda p_i - \lambda^2 c_{ii} < 0$ for all sufficiently small $\lambda > 0$).

Let us associate with f(x) the bilinear function

$$F(u,v) = pu + pv - u(Cv) \quad (u,v \in \mathbb{R}^n) \tag{49}$$

Proposition V.8. *The bilinear function (49) satisfies the conditions (32) (33) (34) for the concave quadratic function (46). If the matrix C is positive definite, then*

$$F(u,v) > \min \{f(u),f(v)\} \quad \forall u \neq v \, . \tag{50}$$

Proof. Clearly, f(x) = F(x,x), and F(u,v) is affine in u for fixed v and affine in v for fixed u. For any u, v we can write

$$F(u,v) - F(u,u) = p(v-u) - u(C(v-u)) \, ,$$

$$F(u,v) - F(v,v) = p(u-v) - v(C(u-v)) \, .$$

Hence, $[F(u,v)-F(u,u)] + [F(u,v)-F(v,v)] = (v-u)(C(v-u)) \ge 0$ since C is positive semidefinite. Therefore, at least one of the two differences F(u,v)–F(u,u),

F(u,v)–F(v,v) must be ≥ 0. This proves that

$$F(u,v) \geq \min \{f(u),f(v)\}.$$

If C is positive definite, then $(v-u)(C(v-u)) > 0$ for $u \neq v$ and one of the two differences must be positive, proving (50). ■

It follows from this proposition that the cut improvement scheme can be applied to the problem (CQP). We now show that this scheme can be implemented.

Let $\alpha < f(x^0)$. First, the α–valid concavity cut for $x^0 = 0$

$$\Sigma x_i / \theta_i \geq 1 \tag{51}$$

can easily be determined on the basis of the following result.

Proposition V.9. *The quantity θ_i is the larger root of the quadratic equation:*

$$2p_i\lambda - c_{ii}\lambda^2 = \alpha \quad (i=1,...,n) . \tag{52}$$

Proof. Since $\alpha < f(x^0) = 0$, $p_i \geq 0$ and $c_{ii} > 0$, it is easily seen that the equation (52) always has a positive root which is also the larger one. Obviously, $f(\theta_i e^i) = \alpha$, and the proposition follows. ■

Next, the points y^i in Corollary V.1 are obtained by solving the linear programs:

$$\min F(\theta_i e^i,v) \quad \text{s.t.} \quad \Sigma v_j/\theta_j \geq 1,\ Av \leq b,\ v \geq 0. \tag{53i}$$

If $f(y^i) \geq \alpha$ (i=1,2,...,n), then, according to Corollary V.1, a stronger cut than the concavity cut can be obtained by computing the values

$$t_i = \max \{\tau : \Phi_\theta(\tau e^i) \geq \alpha\} \quad (i=1,2,...,n).$$

Proposition V.10. *The value t_i is the optimal value of the linear program*

(*)

$$\begin{aligned} \text{minimize } & pz - \alpha z_0 \\ \text{s.t. } & -Az + z_0 b \geq 0 , \\ & \sum_{j=1}^{n} z_j/\theta_j - z_0 \geq 0 , \\ & \sum_{j=1}^{n} c_{ij} z_j - p_i z_0 = +1 , \\ & z_j \geq 0 \quad (j = 0, 1, \ldots, n) . \end{aligned}$$

Proof. Since $F(\tau e^i, v) = \tau p_i + pv - \tau \sum_j c_{ij} v_j$, the optimal value of (53i) is equal to $g_i(\tau) + \tau p_i$, where

$$g_i(\tau) = \min \{pv - \tau \sum_{j=1}^{n} c_{ij} v_j : -Av \geq -b, \sum_{j=1}^{n} v_j/\theta_j \geq 1, v \geq 0\} .$$

Since the constraint set of this linear program is bounded, by the duality theorem in linear programming we have

$$g_i(\tau) = \max\{-bs+r: -A^T s + r \operatorname{diag}(1/\theta_1,\ldots,1/\theta_n) \leq p - \tau C_i^T, s \geq 0, r \geq 0\}$$

where C_i is the i–th row of C.

Thus

$$t_i = \max \{\tau : g_i(\tau) + \tau p_i \geq \alpha\} .$$

Hence,

$$t_i = \max\{\tau : -bs+r + \tau p_i \geq \alpha, -A^T s + r \operatorname{diag}(1/\theta_1,\ldots,1/\theta_n) \leq p - \tau C_i^T, s \geq 0 , r \geq 0\} .$$

By passing again to the dual program and noting that the above program is always feasible, we obtain the desired result (with the understanding that $t_i = +\infty$ if the dual program is infeasible). ■

Since $p \geq 0$ and $\alpha < 0$, it follows that $(z, z_0) = (0,0)$ is a dual feasible solution of

($*$) with only one constraint violated, and it usually takes only a few pivots to solve ($*$) starting from this dual feasible solution. Moreover, the objective function of ($*$) increases monotonically while the dual simplex algorithm proceeds, and we can stop when the objective function value reaches a satisfactory level.

After the first iteration, if a further improvement is needed, we may try a second iteration, and so forth. It turns out that this iterative procedure quite often yields a substantially deeper cut than the concavity cut. For example, for the problem

$$\begin{aligned}
\text{minimize} \quad & 2x_1 + 3x_2 - 2x_1^2 + 2x_1x_2 - 2x_2^2 \\
\text{s.t.} \quad & -x_1 + x_2 \le 1, \\
& x_1 - x_2 \le 1, \\
& -x_1 + 2x_2 \le 3, \\
& 2x_1 - x_2 \le 3, \\
& x_1 \ge 0 \ , x_2 \ge 0 \ ,
\end{aligned}$$

(Konno (1976a)), the concavity cut and the cuts produced by this method are shown in Fig. V.6 for $\alpha = -3$.

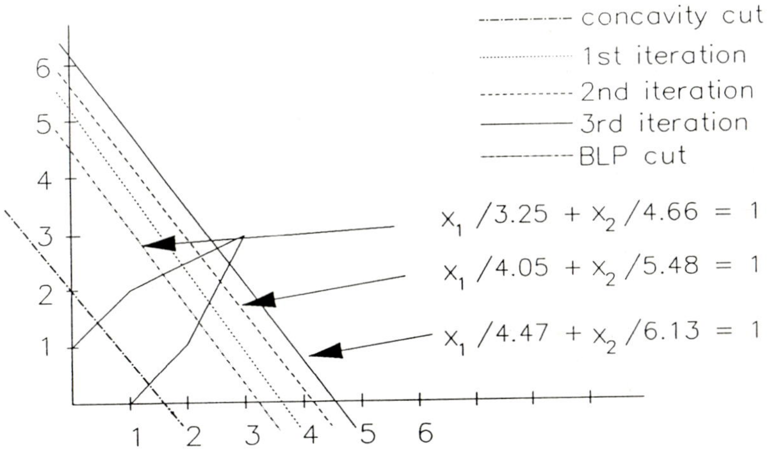

Fig. V.6

The cutting method of Konno (1976a) for the problem (CQP) is similar to Algorithm V.1, except that in Step 1 a deep cut of the type discussed above is generated rather than any γ–valid cut.

The convergence of this method has never been established, though it has seemed to work successfully on test problems. Of course, the method can be made convergent by occasionally introducing facial cuts as in Algorithm V.2, or by combining cuts with splitting as in the Normal Conical Algorithm which will be introduced in Section VII.1.

4.3. Bilinear Programming Cuts

The cut improvement procedure may be started from the concavity cut or the bilinear programming (BLP) cut constructed in the following way.

Define the function

$$\Phi_0(x) = \min \{F(x,v)\colon v \in D\} \tag{54}$$

(so that $\Phi_0(x)$ has the form (35), with $M(0) = D$).

By the same argument as in the proof of Proposition V.6 it follows that the function Φ_0 is concave and satisfies

$$\min \{\Phi_0(x)\colon x \in D\} = \min \{f(x)\colon x \in D\} \tag{55}$$

(i.e., Proposition V.6 extends to $t = 0$).

Since $\Phi_0(0) = \min \{p^T v\colon v \in D\} \geq 0 \geq \alpha$, we can define

$$\theta_i^* = \max \{\tau : \Phi_0(\tau e^i) \geq \alpha\} .$$

Then $\theta_i^* > 0$ $\forall i$ and the cut

$$\Sigma\, x_i / \theta_i^* \geq 1 \tag{56}$$

which is nothing but the α-valid concavity cut for the problem min $\{\Phi_0(x): x \in D\}$, will be also α-valid for the problem (CQP) because $f(x) = F(x,x) \geq \min \{F(x,v): v \in D\} = \Phi_0(x) \quad \forall x \in D$. This cut is sometimes referred to as the ***bilinear programming (BLP) cut.***

If $f(y^{i*}) \geq \alpha$ $(i=1,2,...,n)$, where $y^{i*} \in \operatorname{argmin} \{F(\theta_i e^i,v): v \in D\}$, and if C is positive definite, then the BLP cut is strictly deeper than the concavity cut (for the problem CQP).

Note that $F(\tau e^i,v) = \tau p_i + p^T v - \tau \sum_j c_{ij} v_j$. With the same argument as in the proof of Proposition V.10, one can easily show that θ_i^* is equal to the optimal value of the linear program

$$\text{minimize } pz - \alpha z_0$$

$$\text{s.t.} \quad -Az + z_0 b \geq 0$$

$$\sum_j c_{ij} z_j - z_0 p_i = +1$$

$$z_j \geq 0 \quad (j=1,2,...,n).$$

The BLP cut for the previously given example is shown in Fig. V.6.

Another interpretation is as follows.

Denote

$$D_F^*(\alpha) = \{x: F(x,y) \geq \alpha \ \ \forall y \in D\} .$$

Since $D_F^*(\alpha) = \{x: \Phi_0(x) \geq \alpha\}$, this set is closed and convex and contains the origin as an interior point. Therefore, the BLP cut can also be defined as a valid cut for $\mathbb{R}_+^n \setminus D_F^*(\alpha)$ (i.e., a convexity cut relative to $D_F^*(\alpha)$, see Section III.4).

The set $D_F^*(\alpha)$ is sometimes called the *polaroid* (or *generalized outer polar*) *set* of D with respect to the bilinear function F(u,v) (cf. Balas (1972) and Burdet (1973)).

For a given valid cut $\Sigma x_i / t_i \geq 1$, the improved cut $\Sigma x_i / t_i^* \geq 1$ provided by Theorem V.4 can also be viewed as a convexity cut relative to the polaroid of $M(t) = D \cap \{x: \Sigma x_i / t_i \geq 1\}$.

CHAPTER VI

SUCCESSIVE APPROXIMATION METHODS

In the cutting plane methods discussed in the previous chapter, the feasible domain is reduced at each step by cutting off a feasible portion that is known to contain no better solution than the current best solution.

In the successive approximation methods, to which this chapter is devoted, we construct a sequence of problems which are used to approximate the original one, where each approximating problem can be solved by the methods already available and, as we refine the approximation, the optimal solution of the approximating problem will get closer and closer to a global optimal solution of the original one.

The approximation may proceed in various ways: ***outer approximation (relaxation)*** of the constraint set by a sequence of nested polyhedra, ***inner approximation*** of the constraint set by a sequence of expanding polyhedra (***polyhedral annexation***), or ***successive underestimation*** of the objective function by convex or polyhedral functions.

1. OUTER APPROXIMATION ALGORITHMS

Consider the concave programming (CP) problem

(CP) minimize $f(x)$ subject to $x \in D$,

where D is a closed convex subset of $\mathbb{R}^n$ and f: $\mathbb{R}^n \longrightarrow \mathbb{R}$ is a concave function.

In Chapter II the general concept of outer approximation for nonlinearly constrained problems was discussed.

In this section, we shall deal with the application of outer approximation methods to the CP problem (cf. Hoffman (1981), Thieu, Tam and Ban (1983), Thieu (1984), Tuy (1983), Horst, Thoai and Tuy (1987 and 1989), Horst, Thoai and de Vries (1988)).

1.1. Linearly Constrained Problem

Let us first consider the problem (BCP) as defined in Section V.1., i.e., the problem of minimizing a concave function f: $\mathbb{R}^n \longrightarrow \mathbb{R}$ over a polyhedron D given by a system of linear inequalities of the form

$$A_i x \le b_i \quad (i=1,2,\ldots,m) \,, \tag{1}$$

$$x_j \ge 0 \quad (j=1,2,\ldots,n) \,, \tag{2}$$

where $b_i \in \mathbb{R}$ and A_i is the i–th row of an m×n matrix A.

The outer approximation method for solving this problem, when the feasible set D may be **unbounded**, relies upon the following properties of a concave function f: $\mathbb{R}^n \longrightarrow \mathbb{R}$.

Proposition VI.1. *If the concave function f(x) is bounded from below on some halfline, then it is bounded from below on any parallel halfline.*

Proof. Define the closed convex set (the hypograph of f)

$$G = \{(x, t): x \in \mathbb{R}^n, f(x) \ge t\} \subset \mathbb{R}^{n+1} \,.$$

Suppose that f(x) is bounded from below on a halfline from x^0 in the direction y, i.e.,

$f(x) \geq c$ for all $x = x^0 + \lambda y$ with $\lambda \geq 0$. Then the halfline $\{(x^0 + \lambda y, c): \lambda \geq 0\}$ (in $\mathbb{R}^{n+1}$) lies entirely in G. Hence, by a well–known property of closed convex sets (cf. Rockafellar (1970), Theorem 8.3), (y,0) is a recession direction of G, so that $\{(\bar{x}+\lambda y, f(\bar{x})): \lambda \geq 0\} \subset G$ for any $\bar{x}$, which means that $f(\bar{x} + \lambda y) \geq f(\bar{x})$ for all $\lambda \geq 0$. Thus, f(x) is bounded from below on any halfline parallel to y. ∎

Proposition VI.2. *Let M be any closed convex set in* $\mathbb{R}^n$. *If the concave function f(x) is bounded from below on every extreme ray of M, then it is bounded from below on any halfline contained in M.*

Proof. Denote by K the recession cone of M, and let $x^0 \in M$. Then $K + x^0 \subset M$ and by Proposition VI.1 it follows from the hypothesis that f(x) is bounded from below on every edge of $K + x^0$. Therefore, by a well–known property of concave functions (cf. Rockafellar (1970), Theorem 32.3), f(x) is bounded from below on $K + x^0$. Since any halfline contained in M is parallel to some halfline in $K + x^0$, it follows, again by Proposition VI.1, that f(x) is bounded from below on any halfline contained in M. ∎

Corollary VI.1. *Let* $M \neq \emptyset$ *be a polyhedron in* $\mathbb{R}^n$ *that contains no line. Either f(x) is unbounded from below on an unbounded edge of M, or else the minimum of f(x) over M is attained at some vertex of M.*

Proof. Suppose that f(x) is bounded from below on every unbounded edge of M. Then, by Proposition VI.2, f(x) is bounded from below on any halfline contained in M. By a well–known property of concave functions (cf. Rockafellar (1970), Corollaries 32.3.3 and 32.3.4) it follows that f(x) is bounded from below on M and attains its minimum over M at some vertex of M (see Theorem I.1). ∎

On the basis of the above propositions, it is now easy to describe an outer approximation method for solving the problem (BCP).

If the polyhedron D is ***bounded***, then the KCG–method proceeds as follows (cf. Section II.2).

Start with a simplex D_1 such that $D \subset D_1 \subset \mathbb{R}^n_+$. At iteration $k = 1,2,\ldots$, one has a polytope D_k such that $D \subset D_k \subset \mathbb{R}^n_+$. Solve the relaxed problem

$$(Q_k) \qquad \text{minimize } f(x) \quad \text{s.t. } x \in D_k \,,$$

obtaining an optimal solution x^k. If $x^k \in D$, then stop: x^k is a global optimal solution of (BCP). Otherwise, x^k violates at least one of the constraints (1). Select

$$i_k \in \arg\max\{A_i x^k - b_i : i=1,\ldots,m\} \tag{3}$$

and form

$$D_{k+1} = D_k \cap \{x: A_{i_k} x - b_{i_k} \le 0\} \,. \tag{4}$$

Go to iteration k+1. ■

Since each new constraint (4) cuts off x^k, and hence is different from all of the previous ones, the finiteness of this procedure is an immediate consequence of the finiteness of the number of constraints (1).

The implementation of this method and its extension to the case of an **unbounded** feasible set D require us to examine two questions:

(i) When the polyhedron D defined by (1), (2) is unbounded, we cannot start with a simplex D_1 since all of the polyhedra D_k are unbounded. Then problems (1), (2) may possess an optimal solution, but the relaxed problem (Q_k) may not have an optimal solution. According to the above Corollary VI.1, we may find an unbounded edge of D_k over which $f(x) \longrightarrow \infty$. How should we proceed further in such a case?

Let u^k be the direction of the edge mentioned above. If u^k is a recession direction for D, i.e., if $A_i u^k \le 0$ $(i=1,2,\ldots,m)$, then we stop, since by Proposition VI.1, either $D = \emptyset$ or f(x) is unbounded from below over any halfline emanating from a feasible point in the direction u^k.

Otherwise, u^k violates one of the inequalities $A_i u \leq 0$ $(i=1,...,m)$. Let

$$i_k \in \arg\max\{A_i u^k : i=1,...,m\} . \tag{5}$$

If we define

$$D_{k+1} = D_k \cap \{x: A_{i_k} x \leq b_{i_k}\} \tag{6}$$

then, since $A_{i_k} u^k > 0$, it follows that u^k is no longer a recession direction for D_{k+1}.

(ii) Each relaxed problem (Q_k) is itself a concave minimization problem over a polyhedron. How can it be solved? Taking into account that (Q_k) differs from (Q_{k-1}) by just one additional linear constraint, how can we use the information obtained in solving (Q_{k-1}) in order to solve (Q_k)?

Denote the vertex set of D_k by V_k and the extreme direction set of D_k by U_k: $V_k = \text{vert } D_k$, $U_k = \text{extd } D_k$, respectively. Since the initial polyhedron D_1 is subject to our choice, we may assume that V_1 and U_1 are known. At iteration k–1, knowing the sets V_{k-1} and U_{k-1}, and the constraint to be adjoined to S_{k-1} to form S_k, we can compute V_k and U_k by the procedures presented in Section II.4.2. Thus, we embark upon iteration k with knowledge of the sets V_k and U_k. This is sufficient information for solving (Q_k).

Indeed, by the above Corollary VI.1, a concave function which is bounded from below on a halfline (emanating from the origin) must attain its minimum over this halfline at the origin. Therefore, if there exists $u^k \in U_k$ such that we have $f(\lambda u^k) < f(0)$ for some $\lambda > 0$, then f(x) is unbounded from below on the extreme ray of D_k in the direction u^k. Otherwise, the minimum of f(x) over D_k must be attained at one of the vertices of D_k, namely, at $x^k \in \arg\min\{f(x): x \in V_k\}$.

We can thus state the following procedure.

Algorithm VI.1.

Initialization:

Take a (generalized) n–simplex D_1 containing D. Set $V_1 = \text{vert}\, D_1$, $U_1 = \text{extd}\, D_1$.

(For instance, one can take $D_1 = \mathbb{R}^n_+$. Then $V_1 = \{0\}$, $U_1 = \{e^1,e^2,...,e^n\}$ where e^i is the i–th unit vector of $\mathbb{R}^n$.)

Set $I_1 = \{1,...,m\}$.

Iteration k = 1,2,... :

1) For each $u \in U_k$ check whether there exist $\lambda > 0$ such that $f(\lambda u) < f(0)$. If this occurs for some $u^k \in U_k$, then:

a) If $A_i u^k \leq 0$ $(i \in I_k)$, then stop: either D is empty or the problem has no finite global optimal solution and f(x) is unbounded from below on any halfline parallel to u^k contained in D.

b) Otherwise, compute

$$i_k \in \arg\max\{A_i u^k : i \in I_k\} \tag{7}$$

and go to 3).

2) If no $u \in U_k$ exists such that $f(\lambda u) < f(0)$ for some $\lambda > 0$, then find $x^k \in \arg\min\{f(x): x \in V_k\}$.

a) If $x^k \in D$, i.e., if $A_i x^k \leq b_i$ $(\forall i \in I_k)$, then terminate: x^k is a global optimal solution.

b) Otherwise, compute

$$i_k \in \arg\max\{A_i x^k - b_i : i \in I_k\} . \tag{8}$$

and go to 3).

3) Form

$$D_{k+1} = D_k \cap \{x: A_{i_k} x \le b_{i_k}\} . \tag{9}$$

Compute the vertex set V_{k+1} and the extreme direction set U_{k+1} of S_{k+1} from knowledge of V_k and U_k. Set $I_{k+1} = I_k \setminus \{i_k\}$ and go to iteration k+1.

Theorem VI.1. *Suppose that $D \neq \emptyset$. Then Algorithm VI.1 terminates after at most m iterations, yielding a global optimal solution of (BCP) or a halfline in D on which* $f(x) \longrightarrow \infty$.

Proof. It is easy to see that each i_k is distinct from all $i_1,\ldots,i_{k-1}$. Indeed, if 1b) occurs, then we have $A_i u^k \le 0$ $(i=i_1,\ldots,i_{k-1})$, $A_{i_k} u^k > 0$, while if 2b) occurs, then we have $A_i x^k \le b_i$ $(i=i_1,\ldots,i_{k-1})$, $A_{i_k} x^k > b_{i_k}$. Since $I_k \subset \{1,\ldots,m\}$, it follows that the algorithm must terminate after at most m iterations. ∎

Remarks. VI.1. (i) An alternative procedure for solving (Q_k) which has restart capability and does not necessarily involve a complete inspection of all of the vertices of S_k , will be discussed in Chapter VII (the Modified ES Algorithm).

(ii) A drawback of the above outer approximation algorithm is that all of the intermediate approximate solutions x^k, except the last one, are infeasible. In order to have an estimate of the accuracy provided by an intermediate approximate solution x^k, we can proceed in the following way. Suppose that D is full dimensional. At the beginning we take an interior point z^0 of D; then, at iteration k, when Step 2 occurs, we compute $y^k \in \arg\min\{f(x): x \in \Gamma(x^k)\}$, where $\Gamma(x^k)$ denotes the intersection of D with the halfline from x^k through z^0 (y^k is one endpoint of this line segment). If $\bar{x}^k$ denotes the best among all of the feasible points y^h $(h \le k)$ obtained in this way until Step k, then $f(x^k) \le \min f(D) \le f(\bar{x}^k)$. Therefore, the difference $f(\bar{x}^k) - f(x^k)$ yields an estimate of the accuracy attained; in particular, if

$f(\bar{x}^k) - f(x^k) \le \varepsilon$, then $\bar{x}^k$ is a global ε–optimal solution.

(iii) In the worst case, the algorithm may terminate only when all the constraints of D have been used, i.e., when D_k coincides with D. However, when applying the outer approximation strategy, one generally hopes that a global optimal solution can be found before too many constraints have been used. In the computational experiments on problems of size up to 14 × 16 reported by Thieu et al. (1983), on the average only about half of the constraints of D (not including the nonnegativity constraints) were used.

Example VI.1.

Consider the problem

$$\text{minimize } f(x) = \frac{x_1 x_2}{x_1+x_2} - 0.05(x_1 + x_2)$$

$$\text{subject to } -3x_1 + x_2 - 1 \le 0\,, \quad -3x_1 - 5x_2 + 23 \le 0\,,$$

$$x_1 - 4x_2 - 2 \le 0\,, \quad -x_1 + x_2 - 5 \le 0\,,$$

$$x_1 \ge 0\,, \quad x_2 \ge 0\,.$$

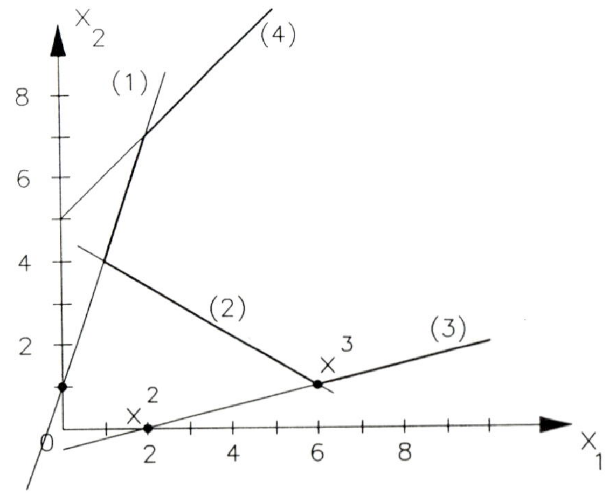

Fig. VI.1

Initialization:

Start with $D_1 = \mathbb{R}^2_+$, $V_1 = \{0,0\}$, $U_1 = \{(1,0),(0,1)\}$. $I_1 = \{1,2,3,4\}$.

Iteration 1:

For $x = (t,0)$, $f(x) = -0.05t \longrightarrow -\infty$ as $t \longrightarrow +\infty$. Hence, $u^1 = (1,0)$.
Values $A_i u^1$ (i=1,2,3,4): –3, –3, 1, –1. Since the largest of these values is 1, we have $i_1 = 3$, so that

$$D_2 = D_1 \cap \{x: x_1 - 4x_2 - 2 \le 0\} ,$$

with $V_2 = \{(0,0), (2,0)\}$, $U_2 = \{(0,1), (4,1)\}$. $I_2 = \{1,2,4\}$

Iteration 2:

For $x = (0,t)$, $f(x) = -0.05t \longrightarrow -\infty$ as $t \longrightarrow +\infty$. Hence, $u^2 = (0,1)$.
Values $A_i u_i$ (i=1,2,4): 1, –5, 1. Since the largest of these values is 1, we select $i_2 = 1$, so that

$$D_3 = D_2 \cap \{x: -3x_1 + x_2 - 1 \le 0\}$$

with $V_3 = \{(0,0), (2,0), (0,1)\}$, $U_3 = \{(4,1), (1,3)\}$, $I_3 = \{2,4\}$.

Iteration 3:

$f(x)$ is bounded from below on each direction $u \in U_3$. Therefore, we compute

$$\min f(V_3) = \min \{0, -0.1, -0.05\} = -0.1. \text{ Thus } x^3 = (2,0).$$

Values $A_i x^3 - b_i$ (i=2,4): 17, –7. Since the largest of these values is 17, we have $i_3 = 2$, so that

$$D_4 = D_3 \cap \{x: -3x_1 - 5x_2 + 23 \le 0\}$$

with $V_4 = \{(6,1), (1,4)\}$, $U_4 = U_3$, $I_4 = \{4\}$.

Iteration 4:

$f(x)$ is bounded from below on each direction $u \in U_4$. Therefore, we compute

$$\min f(V_4) = \min \{0.507143, 0.55\} = 0.507143. \text{ Thus, } x^4 = (6,1).$$

Since $A_4 x^4 = -10 < 0$, this is the optimal solution (Fig. VI.1).

1.2. Problems with Convex Constraints

Now let us consider the concave minimization problem in the general case when the feasible set D is a closed convex set given by an inequality of the form

$$g(x) \le 0 , \tag{10}$$

where $g: \mathbb{R}^n \longrightarrow \mathbb{R}$ is a continuous convex function.

Suppose that D is **compact**. Then, according to the general outer approximation method discussed in Chapter II, in order to solve the problem (CP) we can proceed as follows.

Start with a polytope D_1 containing D. At iteration $k = 1,2,...$, solve the relaxed problem

$$(Q_k) \qquad \text{minimize } f(x) \quad \text{s.t. } x \in D_k ,$$

obtaining an optimal solution x^k. If $x^k \in D$, then stop: x^k is a global optimal solution of (CP). Otherwise we have $g(x^k) > 0$. Construct a hyperplane strictly separating x^k from D, i.e., construct an affine function $\ell_k(x)$ such that

$$\ell_k(x) \le 0 \quad \forall x \in D , \tag{11}$$

$$\ell_k(x^k) > 0 . \tag{12}$$

Form

$$D_{k+1} = D_k \cap \{x: \ell_k(x) \leq 0\} ,$$

and go to iteration k+1.

The convergence of this procedure depends on the choice of the affine function $\ell_k(x)$ in iteration k. Let K be a compact subset of the interior of D (K may be empty, for example, when D has no interior point). If we again choose a point $y^k \in \text{conv}(K \cup \{x^k\}) \setminus \text{int } D$, and let

$$\ell_k(x) = p^k (x - y^k) + g(y^k) , \tag{13}$$

with $p^k \in \partial g(y^k) \setminus \{0\}$, then by Theorem II.2, $\ell_k(x)$ satisfies (11), (12) and the procedure will converge to a global optimal solution, in the sense that whenever the algorithm is infinite, any accumulation point of the generated sequence x^k is such a solution.

For $K = \emptyset$, this is the KCG–method applied to concave programming; for $|K| = 1$ and $y^k \in \partial D$ this is the method of Veinott (1960) which has been applied to concave minimization by Hoffman (1981). Notice that here each relaxed problem (Q_k) is solved by choosing $x^k \in \text{argmin}\{f(x): x \in V_k\}$, where $V_k = \text{vert } D_k$; the set V_k is derived from V_{k-1} by one of the procedures indicated in Section II.4.2 (V_1 is known at the beginning).

Extending the above algorithm to the general case when D may be **unbounded** is not trivial. In fact, the relaxed problem (Q_k) may now have no finite optimal solution, because the polyhedron D_k may be unbounded. We cannot, however, resolve this difficulty in the same way as in the linearly constrained case, since the convex constraint $g(x) \leq 0$ is actually equivalent to an infinite system of linear constraints.

The algorithm below was proposed in Tuy (1983).

Assume that D has an interior point.

Algorithm VI.2.

Initialization:

Select an interior point x^0 of D and a polyhedron D_1 containing D. Set $V_1 = \text{vert } D_1$, $U_1 = \text{extd } D_1$.

Iteration $k = 1,2,\ldots$:

1) Solve the relaxed problem

(Q_k) minimize $f(x)$ s.t. $x \in D_k$

by a search through $V_k = \text{vert } D_k$, $U_k = \text{extd } D_k$. If a direction $u^k \in U_k$ is found on which f(x) is unbounded from below, then:

a) If $g(x^0 + \lambda u^k) \le 0$ for all $\lambda \ge 0$ (i.e., the halfline Γ_k from x^0 in the direction u^k lies entirely in D), then stop: the function f(x) is unbounded from below on $\Gamma_k \subset D$.

b) Otherwise, compute the intersection point y^k of Γ_k with the boundary ∂D of D, find $p^k \in \partial g(y^k)$ and go to 3).

2) If $x^k \in V_k$ is an optimal solution of (Q_k), then:

a) If $g(x^k) \le 0$ (i.e., $x^k \in D$), then terminate: x^k is a global optimal solution of (P).

b) Otherwise, compute the intersection point y^k of the line segment $[x^0,x^k]$ with the boundary ∂D of D, find $p^k \in \partial g(y^k)$, and go to 3).

3) Form

$$\ell_k(x) = p^k(x - y^k)\ ,$$

$$D_{k+1} = D_k \cap \{x: \ell_k(x) \le 0\}\ .$$

Compute $V_{k+1} = \text{vert } D_{k+1}$, $U_{k+1} = \text{extd } D_{k+1}$, and go to iteration k+1.

Theorem VI.2. *Assume that for some $\alpha < f(x^0)$ the set $\{x \in \mathbb{R}^n: f(x) = \alpha\}$ is bounded. Then the above algorithm either terminates after finitely many iterations (with a finite global optimal solution of (CP) or with a halfline in D on which f(x) is unbounded from below), or it is infinite. In the latter case, the algorithm either generates a bounded sequence $\{x^k\}$, every accumulation point $\bar{x}$ of which is a global optimal solution of (CP), or it generates a sequence $\{u^k\}$, every accumulation point $\bar{u}$ of which is a direction of a halfline in D on which f(x) is unbounded from below.*

The proof of Theorem VI.2 uses the following lemmas.

Lemma VI.1. *Under the assumptions of Theorem VI.2, let $u^k \longrightarrow u$ $(k \longrightarrow \infty)$ and let f be unbounded from below on each halfline $\Gamma_k = \{x^0 + \lambda u^k: \lambda \geq 0\}$. Then f is also unbounded from below on the halfline $\Gamma = \{x^0 + \lambda u: \lambda \geq 0\}$.*

Proof. On each Γ_k take a point z^k such that $f(z^k) = \alpha$. By hypothesis, the sequence $\{z^k\}$ is bounded; hence, by passing to a subsequence if necessary, we may assume that the z^k converge to some z. Because of the continuity of f, we then have $f(z) = \alpha < f(x^0)$. Since $z \in \Gamma$, it follows by Corollary VI.1 that f is unbounded from below on $M = \Gamma$. ■

Lemma VI.2. *Under the assumptions of Theorem VI.2, if the algorithm generates an infinite sequence x^k, then this sequence is bounded and any accumulation point of the sequence is a global optimal solution of (CP).*

Proof. Suppose that the sequence $\{x^k\}$ is unbounded, so that it contains a subsequence $\{x^{k_q}\}$ satisfying $|x^{k_q}| > q$ (q=1,2,...). Let $\hat{x}$ be an accumulation point of the sequence $x^{k_q}/|x^{k_q}|$. Since $f(x^{k_q}) < f(x^0)$, Corollary VI.1 implies that f is unbounded from below on the halfline from x^0 through x^{k_q}. Hence, by the previous lemma, f is unbounded from below on the halfline from x^0 through $\hat{x}$, and we can find a point z on this halfline such that $f(z) < f(x^1)$. Let B be a ball around z such that $f(x) < f(x^1)$ for all $x \in B$. Then for all sufficiently large q, the halfline from x^0 through x^{k_q} meets B at some point z^q such that $f(z^q) < f(x^1) \le f(x^{k_q})$. Because of the concavity of f(x), this implies that x^{k_q} lies on the line segment $[x^0, z^q]$. Thus, all of the x^{k_q} with q large enough belong to the convex hull of x^0 and B, contradicting the assumption that $|x^{k_q}| > q$. Therefore, the sequence $\{x^k\}$ is bounded. Since the conditions of Theorem II.2 are satisfied, it then follows that any accumulation point of this sequence solves the problem (P). ■

Lemma VI.3. *Under the assumptions of Theorem VI.2, if the algorithm generates an infinite sequence $\{u^k\}$, then every accumulation point of this sequence yields a recession direction of D on which f(x) is unbounded from below.*

Proof. Let $u = \lim_{q \to \infty} u^{k_q}$ and denote by Γ_k (resp. Γ) the halfline emanating from x^0 in the direction u^k (resp. u). Suppose that Γ is not entirely contained in D and let $\bar{y}$ be the intersection point of Γ with ∂D. It is not hard to see that $y^{k_q} \to \bar{y}$ (y^k is the point defined in Step 1b of the algorithm). Indeed, denoting by φ the gauge of the convex set $D - x^0$, by the continuity of φ we have:

$$y^{k_q} - x^0 = \frac{u^{k_q}}{\varphi(u^{k_q})} \to \frac{u}{\varphi(u)} = \bar{y} - x^0 ,$$

since $\varphi(y^{k_q} - x^0) = \varphi(\bar{y} - x^0) = 1$.

Now, since $p^k \in \partial g(y^k)$, we can write

$$p^k(x^0 - y^k) \le g(x^0) - g(y^k) = g(x^0) < 0 . \tag{13}$$

Let $z^k = 2y^k - x^0$. Clearly $z^k = y^k + (y^k - x^0) \in \Gamma_k$ and from (13) it follows that

$$p^k(z^k - y^k) = p^k(y^k - x^0) \ge -g(x^0) > 0 . \tag{14}$$

But the sequence $\{y^{k_q}\}$ is convergent, and hence bounded. It follows that the sequence $p^{k_q} \in \partial g(y^{k_q})$ is also bounded (cf. Rockafellar (1970), Theorem 24.7). Therefore, by passing to subsequences if necessary, we may assume that $p^{k_q} \longrightarrow p \in \partial g(\bar{y})$. Obviously, $z^{k_q} \longrightarrow \bar{z} = 2\bar{y} - x^0$, and, by (14), $p(\bar{z}-\bar{y}) \geq -g(x^0) > 0$. Since $p^{k_q}(z^{k_s}-y^{k_q}) \longrightarrow p(\bar{z}-\bar{y})$ as $s \longrightarrow \infty$, $q \longrightarrow \infty$, it follows that for all sufficiently large q and all $s > q$ we must have

$$p^{k_q}(z^{k_s} - y^{k_q}) > 0 . \tag{15}$$

However, for $s > q$, we have

$$\Gamma_{k_s} \subset D_{k_s} \subset D_{k_q+1} \subset \{x: p^{k_q}(x - y^{k_q}) + g(y^{k_q}) \leq 0\} ,$$

which implies that $p^{k_q}(x - y^{k_q}) \leq 0$ for all $x \in \Gamma_{k_s}$. This contradicts (15), since $z^{k_s} \in \Gamma_{k_s}$.

Therefore, Γ must lie entirely in D. Finally, since f(x) is unbounded from below on each halfline Γ_k, it follows from Lemma VI.1 that it must be unbounded on Γ. This completes the proof of Lemma VI.3, and with it the proof of Theorem VI.2. ■

1.3. Reducing the Sizes of the Relaxed Problems

A convex constraint can actually be viewed as an infinite system of linear constraints. Outer approximation by polyhedral sets can simply be regarded as a method for generating these constraints one by one, as they are needed. For problems with many variables, usually a large number of constraints has to be generated before a satisfactory approximate solution can be obtained. Accordingly, the sets V_k, U_k increase rapidly in size, making the computation of these sets more and more difficult as the algorithm proceeds. In practice, for problems with about 15 variables $|V_k|$ may exceed several thousand after 8–10 iterations.

To alleviate this difficulty, a common idea is from time to time to drop certain constraints that are no longer indispensable. Some constraint dropping strategies were discussed in Section II.3. Here is another constraint dropping strategy which is more suitable for the present problem:

Let K denote the index set of all iterations k in which the relaxed problem (Q_k) has a finite optimal solution x^k. Then for $k \in K$ we have

$$\ell_k(x^k) > 0 \, , \, \ell_k(x^j) \le 0 \quad \forall j > k \, . \tag{16}$$

For $k \notin K$, a direction u^k is generated. Let us define

$$x^k = 2y^k - x^0$$

(to avoid confusion this point was denoted by z^k in the proof of Lemma VI.3). Recall from (14) that

$$\ell_k(x^k) = p^k(x^k - y^k) > 0 \, ,$$

whereas $x^j \in \Gamma_j \subset D_k$ for all $j > k$ (Γ_j is the halfline from x^0 parallel to u^j). Therefore, (16) holds for all $k = 1,2,\ldots$.

Now, choose a natural number N. At each iteration k let ν_k denote the number of points x^j with $j < k$ such that $\ell_k(x^j) > 0$ (i.e., the number of previously generated points that violate the current constraint). Let N_0 be a fixed natural number greater than N. Then we may modify the rule for forming the relaxed problems (Q_k) as follows:

(*) At every iteration $k \ge N_0$, if $\nu_{k-1} \ge N$, then form (Q_{k+1}) by adjoining the newly generated constraint $\ell_k(x) \le 0$ to (Q_k); otherwise, form (Q_{k+1}) by adjoining the newly generated constraint to (Q_{k-1}).

It is easily seen that in this way any constraint $\ell_k(x) \le 0$ with $\nu_k < N$, $k > N_0$, is used just once (in the (k+1)–th relaxed problem) and will be dropped in all subsequent iterations. Intuitively, only those constraints are retained that are sufficiently efficient in the sense of having discarded at least N previously generated points.

Since (Q_{k+1}) is constructed by adding just one new constraint to (Q_k) or (Q_{k-1}), the sets V_{k+1}, U_{k+1} can be computed from V_k , U_k or V_{k-1} , U_{k-1} , respectively (of course, this requires that at each iteration one stores the sets V_k, U_k, V_{k-1}, U_{k-1}, as well as the previously obtained points x^i, $i < k$).

Proposition VI.3. *With the above modification, Algorithm VI.1 still converges under the same conditions as in Theorem VI.2.*

Proof. Observe that instead of (16) we now have

$$\ell_k(x^k) > 0\ ;\ \ell_k(x^j) \le 0 \quad \forall j > k \text{ such that } \nu_k \ge N\ . \tag{17}$$

Let us first show that any accumulation point $\bar{x}$ of the sequence $\{x^k\}$ belongs to D. Assume the contrary, i.e., that there is a subsequence $x^{k_q} \to \bar{x} \notin D$. Without difficulty one can prove, by passing to subsequences if necessary, that $y^{k_q} \to \bar{y} \in \partial D$, $p^{k_q} \to \bar{p} \in \partial g(\bar{y})$ (cf. the proof of Lemma VI. 3).

Let $\ell(x) = \bar{p}(x-\bar{y})$. Since $g(x^0) < 0$, we have $\ell(x^0) = \bar{p}(x^0-\bar{y}) \le g(x^0)-g(\bar{y}) = g(x^0) < 0$, and consequently, $\ell(\bar{x}) > 0$ (because $\ell(\bar{y}) = 0$). Noting that $\ell_{k_q}(x^j) - \ell_{k_q}(\bar{x}) = p^{k_q}(x^j-\bar{x})$, for any q, j we can write

$$\ell_{k_q}(x^j) \ge \ell_{k_q}(\bar{x}) - C\|x^j-\bar{x}\|\ ,$$

where C is a constant such that $\|p^{k_q}\| \le C$. Since $x^j \to \bar{x}$ $(j \to +\infty)$, $\ell_{k_q}(\bar{x}) \to \ell(\bar{x})$ $(q \to +\infty)$, there exist j_0 and q_0 such that for $j \ge j_0$ and $q \ge q_0$:

$$C\|x^j-\bar{x}\| < \tfrac{1}{2}\ell(\bar{x})\ ,\quad \ell_{k_q}(\bar{x}) > \tfrac{1}{2}\ell(\bar{x})\ .$$

Therefore,

$$\ell_{k_q}(x^j) > 0 \quad \forall\, j \geq j_0 , \quad \forall\, q \geq q_0 .$$

In particular, for any q such that $q \geq q_0$ and $k_q > j_0+N$ we have

$$\ell_{k_q}(x^j) > 0 \quad (j = j_0+1,\ldots,j_0+N) .$$

This implies that $\nu_{k_q} \geq N$; hence the constraint $\ell_{k_q}(x) \leq 0$ will be retained, and so $\ell_{k_q}(x^j) \leq 0$ for all $j > k_q$, a contradiction.

Therefore, any accumulation point $\bar{x}$ of $\{x^k\}$ belongs to D. Now suppose that the algorithm generates an infinite sequence $\{x^k, k \in K\}$. Then, by Lemma VI.2, this sequence is bounded, and, by the above, any accumulation point $\bar{x}$ of the sequence must belong to D, and hence must solve (CP).

On the other hand, if the algorithm generates an infinite sequence $\{u^k, k \notin K\}$, then for any accumulation point u of this sequence, with $u = \lim u^{k_q}$, we must have $\Gamma = \{x\colon x^0 + \lambda u\colon \lambda \geq 0\} \subset D$. Indeed, otherwise, using the same notation and the same argument as in the proof of Lemma VI.3, we would have $y^{k_q} \to \bar{y} \in \partial D$, $x^{k_q} \to \bar{x} = 2\bar{y}-x^0$, and hence $2\bar{y}-x^0 \in D$, which is impossible, since $\bar{y} \in \partial D$, $x^0 \in \operatorname{int} D$. This completes the proof of the proposition. ■

The parameters N_0 and N in the above constraint dropping rule can be chosen arbitrarily considering only computational efficiency. While a large value of N allows one to reduce the number of constraints of the relaxed problem more significantly at each iteration this advantage can be offset by a greater number of required iterations.

Though constraint dropping strategies may help to reduce the size of the sets V_k, U_k, when n is large they are often not efficient enough to keep these sets within manageable size. Therefore, outer approximation methods are practical only for CP problems with a relatively small number of variables. Nevertheless, by their structure, outer approximation methods lend themselves easily to reoptimization when handling new additional constraints. Because of this, they are useful, especially in combination with other methods, in decomposing large scale problems or in finding a rough approximate solution of highly nonlinear problems which otherwise would be almost impossible to handle. In Section VII.1.10 we shall present a more efficient method for solving (CP), which combines outer approximation ideas with cone splitting techniques and branch and bound methods.

2. INNER APPROXIMATION

In the outer approximation method for finding min F(D), we approximate the feasible set D from the outside by a sequence of nested polytopes $D_1 \supset D_2 \supset \ldots \supset D$ such that min $f(D_k) \uparrow$ min f(D). Dually, in the inner approximation method, we approximate the feasible set D from the inside by a sequence of expanding polytopes $D_1 \subset D_2 \subset \ldots \subset D$ such that min $f(D_k) \downarrow$ min f(D).

However, since the concavity of the function $f: \mathbb{R}^n \longrightarrow \mathbb{R}$ enables us to use its values outside D to infer estimates for its values inside, it is more advantageous to take $D_k = P_k \cap D$, where $P_1 \subset P_2 \subset \ldots P_h$ is a finite sequence of expanding polytopes satisfying $P_h \supset D$.

The inner approximation approach originated from the early work of Tuy (1964). Later it was developed by Glover (1975) for integer programming and by Vaish and Shetty (1976) for bilinear programming. These authors used the term "polyhedral annexation", referring to the process of enlarging the polytopes P_k by "annexing" more and more portions of the space. Other developments of the inner approximation approach can be found in Istomin (1977) and Mukhamediev (1982).

Below we essentially follow Tuy (1986a and 1988b).

2.1. The (DG)– Problem

Polyhedral annexation is a technique originally devised to solve a special problem which can be formulated as follows:

(DG) *Given a polytope D contained in a cone $K_0 \subset \mathbb{R}^n$ and a compact convex set G with $0 \in \operatorname{int} G$, find a point $y \in D \setminus G$, or else establish that $D \subset G$.*

The (DG)–Problem turns out to be of importance for a large class of global optimization problems, including problems in concave minimization and reverse convex programming.

For example, consider the problem (BCP), i.e., the problem (CP) where the feasible domain is a polytope D determined by the linear constraints (1) (2). Assume that the concave function f(x) has bounded level sets.

As seen in Chapter V, when solving the BCP problem a crucial question is the following one: given a real number $\gamma \in f(D)$ (e.g., γ is the best feasible value of f(x) obtained so far), and given a tolerance $\varepsilon > 0$, find a feasible solution y with $f(y) < \gamma - \varepsilon$ or else establish that no such point exists (i.e. $\min f(D) \geq \gamma - \varepsilon$).

To reduce this question to a (DG)–problem, we compute a vertex x^0 of D such that $f(x^0) \geq \gamma$, and, after translating the origin to x^0, we construct a polyhedral cone $K_0 \supset D$ (this can be done, e.g., by rewriting the problem in standard form with respect to x^0, so that $K_0 = \mathbb{R}^n_+$). Then the above question is just a (DG)–problem, with $G = \{x: f(x) \geq \gamma - \varepsilon\}$ (this set is obviously compact and convex, and it contains $x^0 = 0$ in its interior because $f(x^0) > \gamma - \varepsilon$).

If we know a method for solving (DG), the BCP problem can be solved according to the following two phase scheme:

Start from a feasible solution $z \in D$.

Phase I.

Search for a local minimizer x^0, which is a vertex of D such that $f(x^0) \le f(z)$.

Phase II.

Let $\alpha = f(x^0) - \varepsilon$. Translate the origin to x^0 and construct a cone $K_0 \supset D$. Solve the (DG)–problem for $G = \{x: f(x) \ge f(x^0) - \varepsilon\}$. If $D \subset G$, then terminate: x^0 is a global ε–optimal solution. Otherwise, let $y \in D \setminus G$. Then $f(y) < f(x^0) - \varepsilon$. Set $z \leftarrow y$ and return to Phase I.

Since a new vertex of D is found at each return to Phase I that is better than all of the previous ones, the scheme is necessarily finite.

Thus, the problem (BCP) can always be decomposed into a finite sequence of (DG)–problems.

2.2. The Concept of Polyhedral Annexation

In this and the next sections we shall present the polyhedral annexation method for solving (DG), and hence the problem (BCP).

The idea is rather simple. Since $D \subset K_0$, we can replace G by $G \cap K_0$. Now we start with the n–simplex P_1 spanned by the origin 0 and the n points where the boundary ∂G of G meets the n edges of K_0. We solve the problem:

(DP_1) Find a point $y^1 \in D \setminus P_1$.

(This is easy, since P_1 is an n–simplex.) If no such point exists (which implies that $D \subset G$), or if $y^1 \notin G$ (which means that $y^1 \in D \setminus G$), then we are done.
Otherwise, let z^1 be the point where the halfline from 0 through y^1 meets ∂G (such a point z^1 will be called the *G–extension* of y^1).

Enlarge P_1 to

$$P_2 = \text{conv}\,(P_1 \cup \{z^1\})\ ,$$

and repeat the procedure with P_2 replacing P_1.

In this manner, we generate a sequence of expanding polytopes that approximate $G \cap K_0$ from the interior. We have $P_1 \subset P_2 \subset \dots \subset G \cap K_0$ satisfying

$$\text{con}\,(P_k) = K_0\ ,\quad 0 \in P_k\ ,\quad \text{vert}(P_k) \setminus \{0\} \subset \partial G\ , \tag{18}$$

$$y^k \in D \setminus P_k\ ,\quad P_{k+1} = \text{conv}(P_k \cup \{z^k\})\ , \tag{19}$$

where y^k is obtained by solving (DP_k), while z^k is the G–extension of y^k.

Clearly, when $D \setminus P_k \neq \emptyset$, then $D \setminus P_k$ contains at least one vertex of D, since otherwise, vert $D \subset P_k$ would imply $D \subset P_k$. Therefore, in the (DP_k)–problem we can require that y^k be a vertex of D. Under these conditions, each P_{k+1} contains at least one vertex of D which does not belong to any $P_1, P_2, \dots, P_k$. Since the vertex set of D is finite, the above polyhedral annexation procedure must terminate with a polytope $P_h \supset D$ (proving that $D \subset G$) or with a point $y^h \in D \setminus G$.

Let us examine how to solve the problems

(DP_k) Find a point $y^k \in D \setminus P_k$ which is a vertex of D.

Recall that a ***facet*** of a polytope P is an (n–1)–dimensional subpolytope of P which is the intersection of P with a supporting hyperplane.

Definition VI.1. *A facet of a polytope is said to be **transversal** if its corresponding supporting hyperplane does not contain the origin 0.*

A transversal facet is determined by an equation of the form $vx = 1$, where v is a normal vector and vx denotes the inner product of v and x. To simplify the language, we shall identify a facet with its normal vector and instead of saying "the facet whose hyperplane is $vx = 1$", we shall simply say "the facet v".

In view of the property con $(P_k) = K_0$ (see (18)) we see that, if V_k is the collection of all transversal facets of P_k, then P_k is determined by the inequalities:

$$vx \le 1 \quad (v \in V_k),$$

$$x \in K_0.$$

Now for each $v \in V_k$ let $\mu(v)$ denote the optimal value in the linear program:

LP(v;D) maximize vx
s.t. $x \in D$.

Proposition VI.4. *If $\mu(v) \le 1$ for all $v \in V_k$, then $D \subset P_k$. If $\mu(v) > 1$ for some $v \in V_k$, then any basic optimal solution y^k of LP(v;D) satisfies $y^k \in D \setminus P_k$.*

Proof. Suppose that $\mu(v) \le 1$ for all $v \in V_k$. Then $x \in D$ implies that $vx \le 1$ for all $v \in V_k$. Hence, $x \in D$ implies $x \in P_k$, that is, $D \subset P_k$. On the other hand, if we have $\mu(v) > 1$ for some $v \in V_k$, then a basic optimal solution y^k of the linear program LP(v;D) must satisfy $y^k \in D$ and $vy^k > 1$, hence $y^k \in D \setminus P_k$. This completes the proof of the proposition. ∎

Thus, to solve (DP_k) we have to solve a linear program LP(v;D) for each transversal facet v of P_k. The question that remains is how to find the set V_k of these facets.

2.3. Computing the Facets of a Polytope

The set V_1 is very simple: it consists of a single element, namely the facet whose hyperplane passes through the n intersections of ∂G with the edges of K_0. Since $P_{k+1} = \text{conv}(P_k \cup \{z^k\})$, it will suffice to know how to derive V_{k+1} from V_k. We are thus led to consider the following auxiliary problem:

New Facet Finding Problem.

Given a polytope P of full dimension in $\mathbb{R}^n$ *which contains 0 and whose set of transversal facets is known, and given a point* $z \notin P$*, compute the set of transversal facets of the polytope* $P' = conv\,(P \cup \{z\})$.

However, rather than solving this problem directly, we shall associate it with another problem which is easier to visualize and has already been studied in Section II.4.2.

New Vertex Finding Problem.

Given a polyhedron S of full dimension in $\mathbb{R}^n$ *which contains 0 in its interior and whose vertex set is known, and given a hyperplane* $zx = 1$ *where z is a normal vector, compute the vertex set of the polyhedron* $S' = S \cap \{x: zx \le 1\}$.

It is well known from polyhedral geometry that a duality relationship exists between the above two problems which allows one to reduce the first problem to the second and vice–versa. This relationship is based on the following result.

Proposition VI.5. *Let P be a polytope of full dimension which contains 0, and let* $S = \{x: zx \le 1 \;\; \forall z \in P\}$ *be the polar of P. Then* $0 \in int\, S$ *and each transversal facet* $vx = 1$ *of P corresponds to a vertex v of S and vice versa; each nontransversal facet* $vx = 0$ *of P corresponds to an extreme direction v of S and vice versa.*

Proof. The inclusion $0 \in \text{int S}$ follows from a well–known property of polars of convex sets (cf. Rockafellar (1970), Corollary 14.5.1). Denote by Z the vertex set of P. Since for any x the linear function $z \longrightarrow zx$ attains its maximum over P at some vertex of P, it is easily seen that $S = \{x: zx \le 1 \;\; \forall z \in Z\}$. Now let $vx = 1$ be a transversal facet of P. Since $\dim P = n$, and since the hyperplane of this facet does not pass through 0, the facet must contain at least n linearly independent vertices of P. In other words, the equation $vz = 1$ must be satisfied by at least n linearly independent elements of Z.

Furthermore, since $0 \in P$, the fact that $vx = 1$ is a facet of P implies that $vx \le 1$ for all $x \in P$. Hence, v belongs to S and satisfies n linearly independent constraints of S as equalities. This means that v is a vertex of S.

Conversely, let v be any vertex of S ($v \neq 0$ because $0 \in \text{int } S$). Then v satisfies all of the constraints of S, with equality in n linearly independent constraints. That is, $vz \le 1$ for all $z \in P$ and $vz = 1$ for n linearly independent elements of Z. Therefore, $vx = 1$ is a supporting hyperplane to P that contains n linearly independent vertices of P. Hence, $vx = 1$ is a transversal facet of P.

The assertion about nontransversal facets can be proved similarly (if $vx = 0$ is such a facet then $v \in S$ and $vz \le 0$ for n–1 linearly independent elements of Z). ■

Corollary VI.2. *Let P be a polytope of full dimension which contains 0, and let $z \notin P$. If S denotes the polar of P then each transversal facet $vx = 1$ of the polytope $P' = \text{conv}(P \cup \{z\})$ corresponds to a vertex v of the polyhedron $S' = S \cap \{x: zx \le 1\}$ and vice versa; each nontransversal facet of P' corresponds to an extreme direction of S' and vice versa.*

Proof. Indeed, the polar of P' is precisely S'. ■

To summarize, the computation of the transversal facets of $P_{k+1} = \text{conv}(P_k \cup \{z^k\})$ reduces to the computation of the vertices of the polyhedron $S_{k+1} = S_k \cap \{x: z^k x \le 1\}$, where S_k is the polar of P_k (Fig. VI.2).

Since $P_1 = [0,u^1,...,u^n]$, where u^i is the intersection of ∂G with the i–th edge of K_0, its polar S_1 is the polyhedral cone $\{x: u^i x \le 1, i=1,2,...,n\}$ with a unique vertex $v^1 = eQ_1^{-1}$, where $Q_1 = (u^1,u^2,...,u^n)$. Starting with $V_1 = \{v^1\}$, one can compute the vertex set V_2 of $S_2 = S_1 \cap \{x: z^1 x \le 1\}$, then the vertex set V_3 of $S_3 = S_2 \cap \{x: z_2 x \le 1\}$, i.e., the set of transversal facets of P_3, and so on. To derive V_{k+1} from V_k the best method known up to now seems to be that of Horst, Thoai and de Vries (1988) (cf. Section II.4.2).

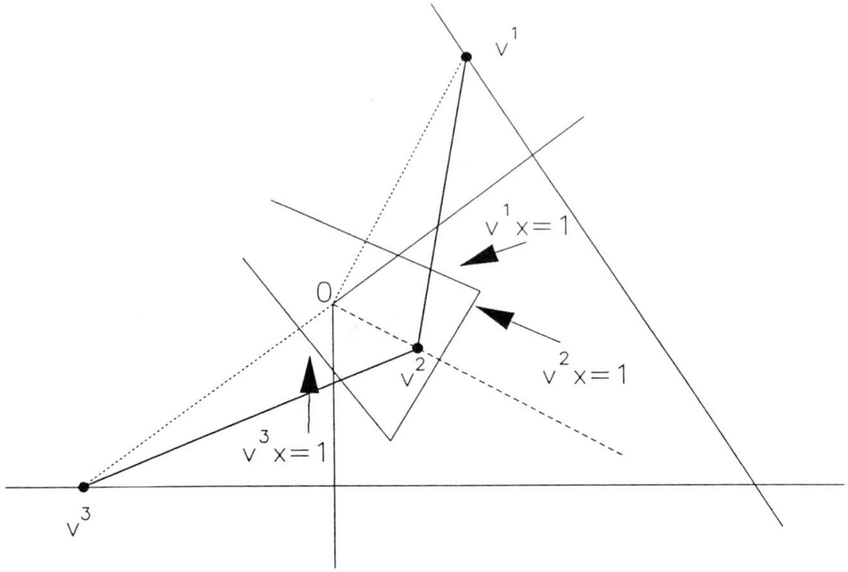

Fig. VI.2

2.4. A Polyhedral Annexation Algorithm

In the preceding section we have developed a polyhedral annexation procedure for solving the (DG)–problem. It now suffices to incorporate this procedure in the two phase scheme outlined in Section VI.2.1 in order to obtain a polyhedral annexation algorithm for the problem (BCP). In the version presented below we describe the procedure through the sequence $S_1 \subset S_2 \subset \ldots$ and add a concavity cut before each return to Phase I.

Algorithm VI.3 (PA Algorithm for BCP).

Select $\varepsilon \geq 0$.

Initialization:

Compute a point $z \in D$. Set $M = D$.

Phase I.

Starting with z search for a vertex x^0 of M which is a local minimizer of f(x) over M.

Phase II.

0) Let $\alpha = f(x^0) - \varepsilon$. Translate the origin to x^0 and construct a cone K_0 containing M (e.g., by rewriting the problem in standard form with respect to x^0; then $K_0 = \mathbb{R}^n_+$).

For each $i=1,2,\ldots,n$ construct the intersection u^i of the i–th edge of K_0 with the surface $f(x) = \alpha$. Let

$$S_1 = \{x: u^i x \leq 1, i=1,2,\ldots,n\} .$$

Compute $v^1 = eQ_1^{-1}$, where $Q_1 = (u^1,u^2,\ldots,u^n)$. Let $V_1 = \{v^1\}$, $V_1^* = V_1$. Set $k = 1$ (For $k > 1$, V_k^* is the set of new vertices of S_k).

1) For each $v \in V_k^*$ solve the linear program

$$\text{LP}(v;M) \quad \begin{array}{l} \text{maximize } vx \\ \text{s.t.} \quad x \in M \end{array}$$

to obtain the optimal value $\mu(v)$ and a basic optimal solution $\omega(v)$.

If we have $f(\omega(v)) < \alpha$ for some $v \in V_k^*$, then set

$$M \leftarrow M \cap \{x: v^1 x \geq 1\} ,$$

where v^1 was defined in Step 0), and return to Phase I. Otherwise, go to 2).

2) Compute $v^k \in \arg\max \{\mu(v): v \in V_k\}$. If $\mu(v^k) \leq 1$, then stop: x^0 is a global ε–optimal solution of (BCP). Otherwise, go to 3).

3) Let z^k be the α–extension of $\omega(v^k)$. Form

$$S_{k+1} = S_k \cap \{x: z^k x \le 1\}.$$

Compute the vertex set V_{k+1} of S_{k+1} and let $V^*_{k+1} = V_{k+1} \setminus V_k$. Set $k \leftarrow k+1$ and return to 1).

Theorem VI.3. *The PA Algorithm terminates after a finite number of steps at a global ε-optimal solution.*

Proof. Phase II is finite because it is the polyhedral annexation procedure for solving a (DG)–problem, with $D = M$ and $G = \{x: f(x) \ge \alpha\}$. Each time the algorithm returns to Phase I, the current feasible set is reduced by a concavity cut $v^1 x \ge 1$. Since the vertex x^0 of M satisfies all the previous cuts as strict inequalities it will actually be a vertex of D. The finiteness of the algorithm follows from the finiteness of the vertex set of D. ■

Remarks VI.2. (i) During a Phase II, all of the linear programs LP(v;M) have the same constraint set M. This makes their solution a relatively easy task.

(ii) It is not necessary to take a local minimizer as x^0. Actually, for the algorithm to work, it suffices that x^0 be a vertex of D and $\alpha < f(x^0)$ (in order that $u^i \neq 0$ $(i=1,2,...,n)$ can be constructed). For example, Phase I can be modified as follows:

Compute a vertex $\bar{x}$ of D such that $f(\bar{x}) \le f(z)$. Let $x^0 \in \operatorname{argmin} \{f(x): x = \bar{x}$ or x is a vertex of D adjacent to $x^0\}$. Then $\alpha = f(x^0) - \varepsilon$ in Phase II, and when $\mu(v^k) \le 1$ (Step 2), $\bar{x}$ is a global ε–optimal solution.

(iii) Each return to Phase I is essentially a ***restart***. Since the cardinality of V_k increases rapidly as Phase II proceeds, an important effect of the restarts is to prevent an excessive growth of V_k. Moreover, since a new vertex x^0 is used as the starting point for the search process at each restart (and the region that remains to be explored is reduced by a cut), another effect of the restarts is to increase the chance for improving the current best feasible solution.

Therefore, the possibility of restarting is a notable advantage to be exploited in practical implementation.

The algorithm prescribes a restart when a point $\omega(v)$ has been found such that $f(\omega(v)) < \alpha$. However, independently of $f(\omega(v))$, one can restart also whenever a point $\omega(v)$ has been found which is a vertex of D (note that each $\omega(v)$ is a vertex of M, but not necessarily a vertex of D): if $f(\omega(v)) \geq \alpha$, simply return to Step 0, with $x^0 \leftarrow \omega(v)$; otherwise, return to Phase I, with $z \leftarrow \omega(v)$. It is easy to see that with this modification, the algorithm will still be finite.

In other cases, when the set V_k approaches a critical size, a restart is advisable even if no $\omega(v)$ is available with the above conditions. But then the original feasible domain D should be replaced by the last set M obtained. Provided that such restarts are applied in limited number, the convergence of the algorithm will not be adversely affected. On the other hand, since the polyhedral annexation method is sensitive to the choice of starting vertex x^0, a restart can often help to correct a bad choice.

(iv) The assumption $\varepsilon > 0$ is to ensure that $u^i \neq 0$ $(i=1,2,...,n)$. However, if the polytope D is nondegenerate (i.e., if any vertex of D is adjacent to exactly n other vertices), then the algorithm works even for $\varepsilon = 0$, because then $\mathbb{R}^n_+$ coincides with the cone generated by the n edges of D emanating from x^0.

Example VI.2.

Consider the problem:

minimize $f(x) = -(x_1-4.2)^2 - (x_2-1.9)^2$

$$\begin{aligned} \text{subject to } -x_1 + x_2 &\leq 3 , \\ x_1 + x_2 &\leq 11 , \\ 2x_1 - x_2 &\leq 16 , \\ -x_1 - x_2 &\leq -1 , \\ x_2 &\leq 5 , \\ x \geq 0 , \ x &\geq 0. \end{aligned}$$

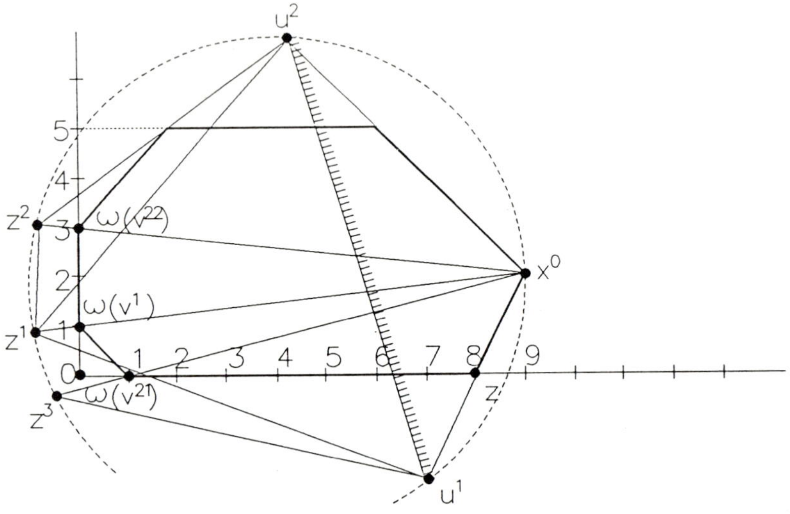

Fig. VI.3

The algorithm is initialized from z = (1.0,0.5).

Phase I: $x^0 = (9.0,2.0)$

Phase II: 0) $\alpha = f(x^0) = -23.05$;

The unique element v^1 of V_1 corresponds to the hyperplane through $u^1 = (7.0,-2.0)$ and $u^2 = (4.3,6.7)$.

Iteration 1:

1) Solution of $LP(v^1,D)$: $\omega(v^1) = (0.0,1.0)$ with $f(\omega(v^1)) = -18.45 > \alpha$.

2) $\mu(v^1) > 1$.

3) α-extension of $\omega(v^1)$: $z^1 = (-0.505,0.944)$.

$V_2 = \{v^{21},v^{22}\}$ corresponds to hyperplanes through u^1,z^1 and through u^2,z^1 respectively.

Iteration 2:

1) LP(v^{21},D): $\omega(v^{21}) = (1.0,0.0)$; LP($v^{22}$,D): $\omega(v^{22}) = (0.0,3.0)$.

2) $v^2 = \text{argmax}\ \{\mu(v): v \in V_2\} = v^{22}$.

3) $V_3^* = \{v^{31},v^{32}\}$ corresponds to hyperplanes through z^1, z^2 and through u^2, z^2 respectively, where z^2 = α–extension of $\omega(v^{22}) = (-0.461,3.051)$.

Iteration 3:

1) LP(v^{31},D): $\mu(v^{31}) < 1$; LP(v^{32},D): $\mu(v^{32}) < 1$.

2) $v^3 = v^{21}$.

3) $z^3 = (-0.082,-0.271)$

$V_4^* = \{v^{41},v^{42}\}$ corresponds to hyperplanes through u^1, z^3 and through z^1, z^3 respectively.

Iteration 4:

1) LP(v^{41},D) and LP(v^{42},D) have $\mu(v) < 1$.

2) $\mu(v^4) < 1$: $x^0 = (9.0,2.0)$ is a global optimal solution (Fig. VI.3).

Thus, the algorithm has actually verified the global optimality of the solution already obtained in the beginning.

Example VI.2.

We consider the problem

minimize f(x) subject to $Ax \le b$, $x \ge 0$,

where $x \in \mathbb{R}^4$,

$$f(x) = -[|x_1 + \tfrac{1}{2}x_2 + \tfrac{2}{3}x_3 + \tfrac{3}{4}x_4|^{\frac{3}{2}} + 0.1(x_1 - 0.5x_2 + 0.3x_3 + x_4 - 4.2)^2] ,$$

$$A = \begin{bmatrix} 1.2 & 1.4 & 0.4 & 0.8 \\ -0.7 & 0.8 & 0.8 & 0.0 \\ 0.0 & 1.2 & 0.0 & 0.4 \\ 2.8 & -2.1 & 0.5 & 0.0 \\ 0.4 & 2.1 & -1.5 & -0.2 \\ -0.6 & -1.3 & 2.4 & 0.5 \end{bmatrix}, \qquad b = \begin{bmatrix} 6.8 \\ 0.8 \\ 2.1 \\ 1.2 \\ 1.4 \\ 0.8 \end{bmatrix}$$

Tolerance $\varepsilon = 10^{-6}$.

First cycle:

$D = \{x: Ax \le b , x \ge 0\}$.

Phase I:

$x^0 = (0, 0, 0, 0)$ (nondegenerate vertex of D).

Adjacent vertices:

$y^{01} = (0.428571, 0.000000, 0.000000, 0,000000)$,

$y^{02} = (0.000000, 0.666667, 0.000000, 0.000000)$,

$y^{03} = (0.000000, 0.000000, 0.333333, 0.000000)$,

$y^{04} = (0.000000, 0.000000, 0.000000, 1.600000)$

Current best point $\bar{x} = (0.000000, 0.666667, 0.000000, 0.000000,)$ (see Remark VI.2(ii)).

Phase II

0) $\alpha = f(\bar{x}) - \varepsilon = -2.055111$. The problem is in standard form with respect to x^0: $K_0 = \mathbb{R}^4_+$.

$u^1 = (1.035485, 0.000000, 0.000000, 0.000000)$,

$u^2 = (0.000000, 0.666669, 0.000000, 0.000000)$,

$u^3 = (0.000000, 0.000000, 29.111113, 0.000000)$,

$u^4 = (0.000000, 0.000000, 0.000000,8.733334)$

The vertex set V_1 of $S_1 = \{x: u^i x \le 1, i=1,...,4\}$ is $V_1 = \{v^1\}$ with

$v^1 = (0.413885, 0.999997, 0.011450, 0.183206)$

Iteration 1:

1) Solution of $LP(v^1,D)$: $(v^1) = (0.413885, 0.999997, 0.011450, 0.183206)$ with

$f(\omega(v^1)) = -1.440295 > \alpha$.

2) $\mu(v^1) = 3.341739 > 1$.

3) α-extension of v^1: $z^1 = (1.616242, 1.654782, 1.088219, 3.144536)$;

$V_2 = \{v^{21}, v^{22}, v^{23}, v^{24}\}$, with

$v^{21} = (-1.162949, 1.499995, 0.034351, 0.114504)$,

$v^{22} = (0.965731, -0.579109, 0.034351, 0.114504)$,

$v^{23} = (0.965731, 1.499995, -3.127204, 0.114504)$,

$v^{24} = (0.965731, 1.499995, 0.034351, -0.979605)$,

Iteration 2:

1) Solution of $LP(v^{24},D)$: $\omega(v^{24}) = (1.083760, 1.080259, 0.868031, 0.000000)$ with $f(\omega(v^{24})) = -2.281489 < \alpha$.

$D \leftarrow D \cap \{x: v^1x \geq 1\}$, with $v^1 = (0.413885, 0.999997, 0.011450, 0.183206)$.

Second cycle (restart):

$M = D \cap \{x: v^1x \geq 1\}$

Phase I:

$x^0 = (1.169415, 1.028223, 0.169811, 4.861582)$ (nondegenerate vertex of M) with adjacent vertices:

$y^{01} = (1.104202, 0.900840, 0.000000, 5.267227)$,

$y^{02} = (1.216328, 1.245331, 0.818956, 2.366468)$,

$y^{03} = (1.134454, 0.941176, 0.000000, 5.151261)$,

$y^{04} = (0.957983, 0.991597, 0.000000, 5.327731)$

Current best point $\bar{x} = (1.083760, 1.080259, 0.868031, 0.000000)$ with $f(\bar{x}) = -2.281489$.

Phase II:

After 12 iterations the algorithm finds the global optimal solution

(1.083760, 1.080259, 0.868031, 0.000000)

with objective function value –2.281489.

Thus, the global optimal solution (v^{24}) is encountered after two iterations of the first cycle, but is identified as such only after a second cycle involving twelve iterations.

A report on some computational experiments with the polyhedral annexation algorithm can be found in Horst and Thoai (1989), where the following types of objective functions are considered:

(1) $x^TCx + 2p^Tx$ (C negative semidefinite n×n matrix),

(2) $-|x_1 + \frac{1}{2}x_2 + \frac{2}{3}x_3 + \dots + \frac{n-1}{n}x_n|^{3/2}$,

(3) $-\sqrt{1 + (x_1 + 2x_2 + \dots + nx_n)^2}$,

(4) $-|x_1 + \frac{1}{2}x_2 + \dots + \frac{x_n}{n}| \ln (1 + |x_1 + \frac{1}{2}x_2 + \dots + \frac{1}{n}x_n|)$,

(5) $-(x_1^2 + x_2^2 + \dots + x_n^2) \ln (1 + x_1^2 + \dots + x_n^2)$,

(6) $-\left[e^{|x_1 + \frac{1}{2}x_2 + \dots + \frac{1}{n}x_n|} + \sqrt{1 + (x_1 + 2x_2 + \dots + nx_n)^2}\right]$,

(7) $\min \{n - x_1^2 + \frac{1}{2}x_2^2 - \dots - \frac{1}{n}x_n^2 ; \arctan(n - x_1^2 - \frac{1}{2}x_2^2 - \frac{1}{n}x_n^2)\}$,

(8) $-\max \{\|x - d^i\|: i=1,\dots,p\}$ with chosen $d^1,\dots,d^p \in \mathbb{R}^n$.

The computational results summarized in the table below, which includes a number of problems with n ranging from 5 to 50, gives an idea of the behaviour of the algorithm. The column "f(x)" indicates the form of the objective function; the column "Res" gives the number of restarts; the column "Lin" gives the number of linear programs solved.

The algorithm was coded in FORTRAN 77, and the computer used was an IBM–PSII, Model 80 (DOS 3.3). The time in seconds includes CPU time and time for printing the intermediate results.

n	m	$f(x)$	Res	Lin	$Time$ $(seconds)$
5	15	(1)	2	19	5.39
5	21	(4)	1	3	2.25
8	21	(1)	3	86	32.48
9	27	(5)	5	417	200.76
9	13	(3)	2	25	15.00
10	12	(2)	1	4	6.38
10	20	(4)	6	275	132.20
10	30	(8)	3	138	84.60
11	17	(3)	2	61	35.00
12	11	(6)	4	129	73.35
12	18	(1)	9	350	220.20
20	18	(1)	6	265	442.52
20	20	(4)	2	128	179.15
20	13	(7)	18	188	950.41
25	18	(2)	2	27	160.63
30	14	(3)	13	893	1338.28
30	22	(6)	7	341	1350.85
32	18	(2)	4	155	1178.16
40	20	(6)	6	275	2001.71
50	21	(6)	8	582	6632.37
50	20	(2)	6	612	4210.24

Table VI.1.

2.5. Relations to Other Methods

There are several interpretations of the PA algorithm which show the relationship between this approach and other known methods.

(i) Polyhedral annexation and outer approximation.

Consider the (DG)–problem as formulated in Section 2.1, and let $G^{\#}$ denote the polar of G, i.e., $G^{\#} = \{v: vx \le 1 \quad \forall x \in G\}$. Since the PA algorithm for (DG) generates a nested sequence of polyhedra $S_1 \supset S_2 \supset \ldots \supset G^{\#}$, each of which is obtained from the previous one by adding just one new constraint, the computational scheme is much like an outer approximation procedure performed on $G^{\#}$. One can even view the PA algorithm for (DG) as an outer approximation procedure for solving the following convex maximization problem

$$\text{maximize } \mu(v) \quad \text{subject to } v \in G^{\#} ,$$

where $\mu(v):= \max v(D)$ (this is a convex function, since it is the pointwise maximum of a family of linear functions $v \longrightarrow v(x)$).

Indeed, starting with the polyhedron $S_1 \supset G^{\#}$, one finds the maximum of $\mu(v)$ over S_1. Since this maximum is achieved at a vertex v^1 of S_1, if $\mu(v^1) > 1$, then $v^1 \notin G^{\#}$ (because $\max v^1(D) = \mu(v^1) > 1$), so one can separate v^1 from $G^{\#}$ by the hyperplane $z^1x = 1$.

Next one finds the maximum of $\mu(v)$ over $S_2 = S_1 \cap \{x: z^1x \leq 1\}$. If this maximum is achieved at v^2, and if $\mu(v^2) > 1$, then $v^2 \notin G^{\#}$, so one can separate v^2 from $G^{\#}$ by the hyperplane $z^2x = 1$, and so on. Note, however, that one stops when a v^k is obtained such that $\mu(v^k) \leq 1$, since this already solves our (DG)–problem.

Thus, the PA algorithm is a kind of dual outer approximation procedure. It would seem, then, that this algorithm should share most of the advantages and disadvantages of the outer approximation methods. However, unlike the usual outer approximation method when applied to the problem (BCP), the polyhedral annexation method at each stage provides a current best feasible solution that monotonically improves as the algorithm proceeds. Furthermore, since restarts are possible, the number of constraints on S_k will have a better chance to be kept within manageable limits than in the usual outer approximation algorithms.

(ii) Polyhedral annexation as a procedure for constructing the convex hull.

In the (DG)–problem we assumed that $0 \in \text{int } G$. However, it can easily be verified that all of the previous results still hold if we only assume that $0 \in \text{int}_{K_0} G$ (the interior of G relative to K_0).

Now suppose that $G = D$, and that 0 is a nondegenerate vertex of D, while K_0 is the cone generated by the n edges of D emanating from 0. Then the PA procedure for (DG) will stop only when $P_k = D$. That is, it will generate a sequence of expanding polytopes, which are the convex hulls of increasing subsets of the vertex set of D; the last of these polytopes is precisely D.

Note that the same procedure can also be performed using the vertex set rather than the constraints of D. In fact, if we know a finite set E such that D = conv E, then instead of defining $\mu(v) = \max\{vx: x \in D\}$ as above (the optimal value of LP(v;D)), we can simply define $\mu(v) = \max\{vx: x \in E\}$. Therefore, the PA algorithm can be used to solve the following problem:

Given a finite set E in $\mathbb{R}^n$, find linear inequalities that determine the convex hull of E.

This problem is encountered in certain applications (cf. Schachtmann (1974)). For instance, if we have to solve a sequence of problems of the form $\min\{c^k x: x \in E\}$, where $c^k \in \mathbb{R}^n$, k=1,2,..., then it may be more convenient first to find the constraints $p^i x \le q^i$, i=1,2,...,r, of the convex hull of E and then solve the linear programs $\min\{c^k x: p^i x \le q^i, i=1,2,...,r\}$.

Assuming that E contains at least n+1 affinely independent points, to solve the above problem we start with the n–simplex P_1 spanned by these n+1 points, and we translate the origin to an interior point of P_1. Then we use the polyhedral annexation procedure to construct a sequence of expanding polytopes $P_1 \subset P_2 \subset \ldots$ terminating with P_h = conv E.

(iii) Polyhedral annexation as a finite cut and split procedure.

For each polytope P_k let $\mathcal{M}_k$ denote the collection of cones generated by the transversal facets of P_k. Clearly, $\mathcal{M}_k$ forms a conical subdivision of the initial cone K_0, and so the PA algorithm may also be regarded as a modification of the cut and split algorithm V.3.

Specifically, each transversal facet vx = 1 of P_k is a concavity cut which in the corresponding cone determines a pyramid M(v) containing only feasible points x with $f(x) \ge \alpha$. If $\max v(D) \le 1$ for all these v (i.e., if the collection of all of the pyramids covers all of D), then the algorithm stops. Otherwise, using $z^k \in \text{argmax} \{v^k x: x \in D\}$ with $v^k \in \text{argmax } v(D)$, we generate a new conical subdivision $\mathcal{M}_{k+1}$, etc. ...

Thus, compared with the cut and split algorithm, the fundamental difference is that the bases of the pyramids M(v) are required to be the transversal facets of a ***convex*** polytope. Because of this requirement, a cone may have more than n edges and $\mathcal{M}_{k+1}$ not necessarily be a refinement of $\mathcal{M}_k$ (i.e., not every cone in $\mathcal{M}_{k+1}$ is a subcone of some cone in $\mathcal{M}_k$). On the other hand, this requirement allows the cones to be examined through linear programs with the same feasible domain throughout a cycle. This property in turn ensures that each z^k is a vertex of D, which is a crucial condition for the finiteness of this kind of procedure.

In the cut and split algorithm, P_{k+1} is obtained by merely adding to P_k the n–simplex spanned by z^k and the vertices of the facet v^k that generated z^k; thus the annexation procedure is very simple. But, on the other hand, the method might not be finite, and we might need some anti–jamming device in order for it to converge. We shall return to this question in Section VII.1.

2.6. Extensions

So far the PA algorithm has been developed under the assumption that the feasible domain D is a polytope and the objective function f(x) is finite throughout $\mathbb{R}^n$ and has bounded level sets.

However, this algorithm can be extended in a straightforward manner to more general situations. Specifically, let us consider the following cases:

(i) ***D is a polyhedron, possibly unbounded but line free, while f(x) has bounded level sets.***

In this case the linear program LP(v;D) might have no finite optimal solution. If $\max v(D) = +\infty$ for some v, i.e., if a halfline in D in the direction $y = \omega(v)$ is found for which $vy > 0$, then, since the set $\{x: f(x) \geq t\}$ is bounded for any real number t, we must have $f(x) \to -\infty$ over this halfline (therefore, the algorithm stops). Otherwise, we proceed exactly as in the case when D is bounded.

(ii) ***D is a polyhedron, possibly unbounded but line free, while inf $f(D) > -\infty$ (but $f(x)$ may have unbounded level sets).***

Under these conditions certain edges of K_0 might not meet the surface $f(x) = \alpha$, so we define

$$S_1 = \{x: u^i x \leq 1 \ (i \in I), \quad u^i x \leq 0 \ (i \notin I)\} ,$$

where I is the index set of the edges of K_0 which meet the surface $f(x) = \alpha$, and if $i \notin I$, then u^i denotes the direction of the i–th edge. Moreover, even if the linear program LP(v;D) has a finite optimal solution $\omega(v)$, this point might have no finite α – extension. That is, at certain steps, the polyhedral annexation process might involve taking the convex hull of the union of the current polyhedron P_k and a point at infinity (i.e. a direction) z^k.

To see how one should proceed in this case observe that Proposition VI.5 still holds when P is an unbounded polyhedron, except that 0 is then a boundary rather than an interior point of S (the polar of P). In particular, an extreme direction of P corresponds to a nontransversal facet of S. Therefore, the polyhedron S_{k+1} in Step 3 of Algorithm VI.3 should be defined as follows:

If the α – extension of $\omega(v^k)$ is a (finite) point z^k, then let

$$S_{k+1} = S_k \cap \{x: z^k x \leq 1\} .$$

Otherwise, either $\omega(v^k)$ or its α – extension is at infinity. In this case let z^k be the direction of $\omega(v^k)$ and

$$S_{k+1} = S_k \cap \{x: z^k x \leq 0\} .$$

(iii) *D is a polytope, while f(x) is finite only on D (f(x) = −∞ outside D).*

Most of the existing concave minimization methods require that the objective function $f: D \longrightarrow \mathbb{R}$ can be extended to a finite concave function on a suitable set A, $A \supset D$.

However, certain problems of practical interest involve an objective function f(x) which is defined only on D and cannot be finitely extended to $\mathbb{R}^n$. To solve these problems, the outer approximation methods, for example, cannot be applied.

However, the PA algorithm might still be useful in these cases. Assume that a nondegenerate vertex of D, say $x^0 = 0$, is available which is a local minimizer. Clearly, for any $\alpha \in f(D)$ the set $G = \{x \in D: f(x) \geq \alpha\}$ is a subset of D; while if $\alpha = \min f(D)$, then it is identical to D. Therefore, the best way to apply Algorithm VI.3 is to keep x^0 fixed (so that M = D) and execute always Step 2 after solving the linear programs LP(v;D), $v \in V_k^*$, regardless of the value $f(\omega(v))$. Thus the algorithm will involve just one cycle and will generate a finite sequence of expanding polytopes, each obtained from the previous one by annexing just one new vertex.

Such a procedure, if carried out completely, will certainly produce all of the vertices of D. However, there are several respects in which this procedure differs from a simple full enumeration.

First, this is a kind of branch and bound procedure, in each step of which all of the vertices that remain to be examined are divided into a number of subsets (corresponding to the facets v of the current polytope P_k), and the subset that has maximal $\mu(v) = \max v(D)$ is chosen for further partitioning. Although $\mu(v)$ is not a lower bound for f(x) on the set $D \cap \{x: vx \leq 1\}$, it does provide reasonable heuristics to guide the branching process.

Second, the best solution obtained up to iteration k is the best vertex of the polytope P_k ; it approximates D monotonically as k increases.

Third, the accuracy of approximation can be estimated with the help of the following

Proposition VI.6. *Let P_k be the approximating polytope at iteration k, and let V_k be the collection of its transversal facets (i.e., the collection of vertices of its polar S_k). If $d(x,P_k)$ denotes the distance from x to P_k, then*

$$\max \{d(x,P_k): x \in D\} \le \rho_k \max\{\mu(v)-1: v \in V_k\},$$

where ρ_k is the maximal distance from x^0 to the vertices of P_k.

Proof. Let $x \in D \setminus P_k$. Then there is a $v \in V_k$ such that x belongs to the cone generated by the facet v. Denoting by y the point where the line segment $[x^0,x]$ meets the hyperplane through the facet v, we see that $y \in P_k$ and

$$\|x - y\| \le [\mu(v) - 1] \, \|x^0 - y\| \le \rho_k[\mu(v) - 1] \ ,$$

and hence

$$d(x,P_k) \le \rho_k \max \{\mu(v)-1: v \in V_k\} \ .$$

Since this holds for arbitrary $x \in D \setminus P_k$, the proposition follows. ∎

3. CONVEX UNDERESTIMATION

It is sometimes more convenient to apply an outer approximation (or relaxation) method not to the original problem itself but rather to a transformed problem. An example of such an approach is the successive underestimation method, which was first introduced by Falk and Soland (1969) in the context of separable, nonconvex programming, and was later used by Falk and Hoffman (1976) to solve the concave programming problems, when the feasible set D is a polytope. Similar methods have also been developed by Emelichev and Kovalev (1970), and by Bulatov (1977) and

Bulatov and Kansinkaya (1982). A method for solving the nonlinearly constrained concave minimization problem by means of convex underestimation was first presented in Horst (1976).

3.1. Relaxation and Successive Underestimation

Consider the general problem

$$\min f(D) , \tag{22}$$

where D is an arbitrary subset of $\mathbb{R}^n$ and f(x) is an arbitrary function defined on some set $S \supset D$. Setting

$$\tilde{G} = \{(x,t): x \in D, \quad f(x) \leq t\} \subset \mathbb{R}^{n+1} ,$$

we see that this problem is equivalent to the problem

$$\min \{t: (x,t) \in \tilde{G}\} . \tag{23}$$

Applying an outer approximation method to the latter problem leads us to construct a sequence of nested sets

$$\tilde{G}_1 \supset \tilde{G}_2 \supset \ldots \supset \tilde{G}_k \supset \ldots \supset \tilde{G}$$

such that each problem

$$\min \{t: (x,t) \in \tilde{G}_k\} \tag{24}$$

can be solved by available algorithms, and its optimal solution approaches an optimal solution of (22) as k increases.

Suppose that $\tilde{G}_k = G_k \cap (D_k \times \mathbb{R})$, where $G_k \supset G_{k+1} \supset G$, $D_k \supset D_{k+1} \supset D$ and G_k is the epigraph of some function φ_k defined on some set $S_k \supset D_k$.

It follows that we must have

$$\varphi_k(x) \le \varphi_{k+1}(x) \le f(x) \quad \forall x \in D \ , \tag{25}$$

i.e., the function $\varphi_k(x)$ defines a nondecreasing sequence of ***underestimators*** (subfunctionals) of f(x).

Problem (24) is equivalent to

$$\min \varphi_k(D_k) \ , \tag{26}$$

and is called a ***relaxation*** of the problem (22). The above variant of outer approximation is called the ***successive relaxation method.***

Usually D_k is a polyhedron, and $\varphi_k(x)$ is an affine, or piecewise affine, or convex underestimator of f(x), so that the relaxed problem (26) can be solved by standard linear or convex programming techniques.

When $D_k = D$ for every k, the method is also called the ***successive underestimation method.*** Thus, the successive underestimation method for solving (22) consists in constructing a sequence of underestimators $\varphi_k(x)$ satisfying (25) and such that the sequence

$$x^k \in \arg\min \{\varphi_k(x): x \in D\}$$

converges to an optimal solution of (22). These underestimators are constructed adaptively, i.e., φ_{k+1} is determined on the basis of the information given by x^k. Specifically, since $\varphi_k(x^k) = \min \varphi_k(D)$ and φ_k underestimates f, it is clear that:

1) if $\varphi_k(x^k) = f(x^k)$, then $f(x^k) = \min f(D)$, i.e. x^k solves (22), and we stop;

2) otherwise, φ_{k+1} must be constructed such that $\varphi_k(x) \le \varphi_{k+1}(x)$ for all $x \in D$ and $\varphi_k \not\equiv \varphi_{k+1}$.

Of course, the convergence of the method crucially depends upon the choice of φ_k, k=1,2,....

Remark VI.3. Another special case of the successive relaxation method is when

$\varphi_k(x) \equiv f(x)$ for every k. Then the sequence D_k is constructed adaptively, based on the results of solving the relaxed problem min $f(D_k)$ at each step: this is just the usual outer approximation method that we discussed in Chapter II and Section VI.1.

3.2. The Falk and Hoffman Algorithm

With the above background, let us now return to the concave programming problem (BCP), i.e., the problem (22) in which D is a polytope of the form

$$D = \{x \in \mathbb{R}^n: Ax \leq b\} ,$$

with $b \in \mathbb{R}^m$, A an m×n matrix, and f(x) is a concave function defined throughout $\mathbb{R}^n$.

Let us apply the successive underestimation method to this problem, using as underestimators of f(x) the ***convex envelopes*** of f(x) taken over polytopes $S \supset D$. Recall from Section IV.4.3 that the convex envelope of f(x) taken over S, is the largest convex underestimator of f over S.

Proposition VI.7. *Let S_k be a polytope with vertices $v^{k,0},v^{k,1},\ldots,v^{k,N(k)}$, and let $\varphi_k(x)$ be the convex envelope of f(x) taken over S_k. Then the relaxed problem min $\{\varphi_k(x): x \in D\}$ is a linear program which can be written as*

$$(Q_k) \qquad \text{minimize} \quad \sum_{j=0}^{N(k)} \lambda_j f(v^{k,j})$$

$$\text{s.t.} \quad \sum_{j=0}^{N(k)} \lambda_j A v^{k,j} \leq b$$

$$\sum_{j=0}^{N(k)} \lambda_j = 1$$

$$\lambda_j \geq 0 \quad (j = 0,\ldots,N(k)) .$$

Proof. Proposition VI.7 follows in a straightforward manner from Theorem IV.6. ■

Now let $\lambda^k = (\lambda_0^k,...,\lambda_{N(k)}^k)$ be an optimal solution of (Q_k), and let $x^k = \sum_{j\in J(k)} \lambda_j^k v^{k,j}$, where $J(k) = \{j: \lambda_j^k > 0\}$. Then

$$x^k \in \operatorname{argmin} \{\varphi_k(x): x \in D\}. \tag{28}$$

If $v^{k,j} \in D$ for all $j \in J(k)$, then $f(v^{k,j}) \geq \min f(D)$ for all $j \in J(k)$, and hence,

$$\varphi_k(x^k) = \sum_{j\in J(k)} \lambda_j^k f(v^{k,j}) \geq \min f(D) \geq \min \varphi_k(D) . \tag{29}$$

In view of (28), this implies that

$$\varphi_k(x^k) = f(v^{k,j}) = \min f(D) \quad \forall j\in J(k). \tag{30}$$

Therefore, any $v^{k,j}$ $(j \in J(k))$ is a global optimal solution of (BCP).

In general, however, $v^{k,j} \notin D$ at least for some $j \in J(k)$, for example, $v^{k,j_k} \notin D$. Then, taking any $i_k \in \{1,...,m\}$ such that

$$A_{i_k} v^{k,j_k} > b_{i_k}$$

(i.e., the i_k–th constraint of D is violated), and defining

$$S_{k+1} = S_k \cap \{x: A_{i_k} x \leq b_{i_k}\} ,$$

we obtain a polytope smaller than S_k but still containing D. Hence, for the convex envelope φ_{k+1} of f over S_{k+1} we have $\varphi_{k+1}(x) \geq \varphi_k(x)$, $\varphi_{k+1} \not\equiv \varphi_k$.

We are thus led to the following successive convex underestimation (SCU) algorithm of Falk and Hoffman (1976):

Algorithm VI.4 (SCU Algorithm).

Initialization:

Select an n–simplex S_0 containing D. Identify the set $V_0 = \text{vert } S_0$. Let $V_0 = \{v^{0,0},...,v^{0,n}\}$. Set $N(0) = n$.

Iteration k = 0,1,... :

1.) Solve problem (Q_k), obtaining an optimal solution λ^k. Let $J(k) = \{j: \lambda_j^k > 0\}$.

2.) If $v^{k,j} \in D$ for all $j \in J(k)$, then stop: any $v^{k,j}$ with $j \in J(k)$ is a global optimal solution of (BCP).

3.) Otherwise, there is a $v^{k,j_k} \notin D$ with $j_k \in J(k)$. Select any $i_k \in \{1,..., m\}$ such that $A_{i_k} v^{k,j_k} > b_{i_k}$ and define

$$S_{k+1} = S_k \cap \{x: A_{i_k} x \le b_{i_k}\} .$$

4.) Compute the vertex set V_{k+1} of S_{k+1} (from knowledge of V_k, using, e.g., one of the procedures in Section II.4). Let $V_{k+1} = \{v^{k+1,0},...,v^{k+1,N(k+1)}\}$. Go to iteration k+1.

Theorem VI.4. *Algorithm VI.4 terminates after at most m iterations, yielding a global optimal solution.*

Proof. As in the proof of Theorem VI.1, it is readily seen that each i_k is distinct from all the previous indices $i_0,...,i_{k-1}$. Hence, after at most m iterations, we have $S_m = D$, and then $v^{m,j} \in D$ for all $j \in J(m)$, i.e., each of these points is a global optimal solution of (BCP). ■

Remarks VI.4. (i) Step 2 can be improved by computing

$$v^{k,j_k} \in \arg\min \{f(v^{k,j}) \colon j \in J(k)\} .$$

If $v^{k,j_k} \in D$, then $f(v^{k,j_k}) \geq \min f(D)$, and hence (29) and (30) hold, i.e., v^{k,j_k} is a global optimal solution. Otherwise, one goes to Step 3.

In Step 3, a reasonable heuristic for choosing i_k is to choose that constraint (not already used in S_k) which is most severely violated by v^{k,j_k}.

(ii) When compared to Algorithm VI.1, a relative advantage of the successive convex underestimation method is that it yields a sequence of ***feasible*** solutions x^k such that $\varphi_k(x^k)$ monotonically approaches the optimal value of the problem (see (25)). However, the price paid for this advantage is that the computational effort required to solve (Q_k) is greater here than that needed to determin $x^k \in \arg\min\{f(x)\colon x \in V_k\}$ in Algorithm VI.1. It is not clear whether the advantage is worth the price.

3.3. Rosen's Algorithm

A method closely related to convex underestimation ideas is that of Rosen (1983), which is primarily concerned with the following concave quadratic programming problem (CQP)

(CQP) $\qquad$ minimize $f(x) := p^T x - \frac{1}{2} x(Cx)$ subject to $x \in D$,

where D is a polytope in $\mathbb{R}^n$, $p \in \mathbb{R}^n$ and C is a symmetric positive definite $n \times n$ matrix.

An important property of a quadratic function is that it can be put into ***separable*** form by a linear transformation. Namely, if $U = [u^1, u^2, \ldots, u^n]$ is a matrix of normalized eigen–vectors of C, so that $U^T CU = \operatorname{diag}(\lambda_1, \lambda_2, \ldots, \lambda_n) > 0$, then by setting $x = Uy$, $F(y) = f(Uy)$ we have:

$$F(y) = qy - \tfrac{1}{2}y[(U^TCU)y] = \sum_{j=1}^{n} F_j(y_j) ,$$

where

$$F_j(y_j) = q_jy_j - \tfrac{1}{2}\lambda_jy_j^2 , \; q = U^Tp . \tag{31}$$

Thus, problem (CQP) becomes

$$\text{minimize} \sum_{j=1}^{n} F_j(y_j) \text{ subject to } y \in \Omega ,$$

where $\Omega = \{y\colon Uy \in D\}$.

Lemma VI.4. *Let* $F(y) = \sum_{j=1}^{n} F_j(y_j)$ *be a separable function and let S be any rectangular domain of the form*

$$S = \{z \in \mathbb{R}^n\colon \beta_{j+n} \le y_j \le \beta_j \quad (j=1,2,\dots,n)\} . \tag{32}$$

Then for every vertex w of S we have

$$F(w) = \sum_{j=1}^{n} F_j(w_j) \text{ with } w_j = \beta_j \text{ or } w_j = \beta_{j+n} . \tag{33}$$

Proof. Lemma VI.4 follows because any vertex w of S must satisfy $w_j = \beta_j$ or $w_j = \beta_{j+n}$. ∎

To simplify the language, in the sequel we shall always use the term rectangle to mean a domain defined by 2n inequalities of the form (32), the faces of which are parallel to the hyperplanes $y_j = 0$.

Now the basic idea of the method can be described as follows.

Compute a starting vertex $\hat{v}$ of Ω which is reasonably good (provided, of course, that we can find it with a relatively small amount of calculation). Then with the help of the above lemma construct two rectangles S, T such that: S is the smallest rectangle containing Ω, while T is the largest rectangle contained in the ellipsoid

$E = \{y \in \mathbb{R}^n: F(y) \geq F(\hat{v})\}$. Denote $\varphi^* = \min \{F(y): y \in \Omega\}$. Then, since $S \supset \Omega$, the number $\underline{\varphi} = \min\{F(y): y \in S\}$ (which is easy to compute using the above lemma) yields a lower bound for φ^* while $F(\hat{v})$ is an upper bound. Furthermore, since $T \subset E$, we will have $F(y) \geq F(\hat{v})$ for all $y \in T$. Hence, if $\Omega \setminus T = \emptyset$, then no feasible point better than $\hat{v}$ can exist, i.e. $\hat{v}$ is a global optimal solution. Otherwise, the interior of T can be excluded from further consideration and it remains to investigate only the set $\Omega \setminus \text{int } T$. Though this set is not convex, it is the union of $r \leq 2n$ convex pieces, and it should be possible to handle each of these pieces efficiently, e.g., by the Falk–Hoffman method, or, alternatively, in the same manner as Ω was treated.

Specifically, let us first compute for each $j=1,...,n$:

$$v^j \in \arg\max \{y_j: y \in \Omega\} ,$$

$$v^{j+n} \in \arg\min \{y_j: y \in \Omega\} .$$

Clearly, since each v^j $(j=1,...,2n)$ is the optimal solution of a linear program, v^j can be assumed to be a vertex of Ω. Let

$$\hat{v} \in \arg\min \{F(v^j): j=1,2,...,2n\} .$$

Then $F(\hat{v})$ is an upper bound for the number $\varphi^* = \min_{y \in \Omega} F(y)$. On the other hand, with $\beta_j = v_j^j$, $\beta_{j+n} = v_j^{j+n}$, the set

$$S = \{y \in \mathbb{R}^n: \beta_{j+n} \leq y_j \leq \beta_j \quad (j=1,2,...,n)\}$$

is the smallest rectangle containing Ω. Therefore, the number

$$\underline{\varphi} = \min \{F(y): y \in S\}$$

furnishes a reasonably tight lower bound for φ^*. If it happens that

$$F(\hat{v}) - \underline{\varphi} \leq \varepsilon ,$$

where ε is the prescribed tolerance, then $\hat{v}$ is accepted as an approximate global optimal solution of (CQP).

Proposition VI.8. *We have*

$$\underline{\varphi} = \sum_{j=1}^{n} \min(\alpha_j, \alpha_{j+n}), \tag{34}$$

where

$$\alpha_j = q_j\beta_j - \frac{1}{2}\lambda_j\beta_j^2, \quad \alpha_{j+n} = q_j\beta_{j+n} - \frac{1}{2}\lambda_j\beta_{j+n}^2 \quad (j=1,2,\ldots,n).$$

Proof. Since the minimum of F(y) over S is attained at some vertex, $\underline{\varphi}$ is equal to the smallest of all of the numbers F(w) where w is a vertex of S. The proposition then follows from the formulas (31) and (33). ■

Now let us construct the second rectangle T. Let $\bar{y}$ be the point of Ω where F(y) attains its maximum over Ω (this point is computed by solving the convex program $\max_{v\in\Omega} F(y)$). Obviously, $\Delta F := F(\bar{y}) - F(\hat{v}) > 0$.

Proposition VI.9. *For each j let γ_j, γ_{j+n} ($\gamma_j > \gamma_{j+n}$) be the roots of the equation*

$$-\frac{1}{2}\lambda_j\gamma^2 + q_j\gamma = F_j(\bar{y}_j) - \frac{\Delta F}{n}. \tag{35}$$

Then the rectangle

$$T = \{y \in \mathbb{R}^n : \gamma_{j+n} \le y_j \le \gamma_j \quad (j=1,\ldots,n)\}$$

has all its vertices lying on the surface $F(y) = F(\hat{v})$.

Proof. Since $F_j(\bar{y}_j) - \frac{\Delta F}{n} < \max_{t\in\mathbb{R}} F_j(t)$ (where $F_j(t) = q_jt - \frac{1}{2}\lambda_jt^2$), the quadratic equation (35) has two distinct roots. By Lemma VI.4 for each vertex w of T we have: $F(w) = \sum_{j=1}^{n}(q_j\bar{\gamma}_j - \frac{1}{2}\lambda_j\bar{\gamma}_j^2)$ with $\bar{\gamma}_j = \gamma_j$ or γ_{j+n}, and hence $F(w) = \sum_{j=1}^{n}(F_j(\bar{y}_j) - \frac{\Delta F}{n}) = F(\bar{y}) - \Delta F = F(\hat{v})$. ■

Note that it can easily be shown that the domain T which we just constructed is the rectangle with maximal volume among all rectangles which can be inscribed in the level ellipsoid $E = \{y: F(y) \geq F(\hat{v})\}$ of $F(x)$.

Since $\underline{\varphi} \leq \varphi^* \leq F(\hat{v}) \leq F(y)$ for all $y \in T$, the optimal solution of (CQP) must be sought among the feasible points lying on the boundary or outside of T, i.e., in the set $\Omega \setminus \text{int } T$ (see Fig. VI.4, papge 270). Since the latter set is nonconvex, it must be treated in a special way. The most convenient method is to use the hyperplanes that determine T to partition $\Omega \setminus \text{int } T$ into 2n convex pieces (some of which may be empty). That is, if

$$H_j = \{y: y_j \geq \gamma_j\},\ H_{j+n} = \{y: y_j \leq \gamma_{j+n}\} \quad (j=1,...,n)\ ,$$

then we must consider the polytopes

$$\Omega_j = H_j \cap \Omega \quad (j=1,...,2n)\ .$$

For each nonempty polytope Ω_j, we already know one vertex, namely v^j. Usually an n–simplex with a vertex at v^j can be constructed that contains Ω_j, so that each problem

$$\min \{F(y): y \in \Omega_j\} \quad (j=1,...,2n) \tag{36}$$

can be treated, for example, by the Falk–Hoffman algorithm. If φ_j^* is the optimal value in (36) then $\varphi^* = \min \{\varphi_1^*,...,\varphi_{2n}^*,F(\hat{v})\}$ with the convention that $\varphi_j^* = +\infty$ if $\Omega_j = \emptyset$. The following example illustrates the above procedure.

Example VI.3. Consider the two dimensional problem whose main features are presented in Fig. VI.4.

The feasible set Ω and a level curve of the objective function $F(y)$ are shown.

First, $v^1 \in \text{argmax } \{y_1: y \in \Omega\}$, $v^3 \in \text{argmin } \{y_1: y \in \Omega\}$, and (similarly) v^2, v^4 are computed, and the rectangle S is constructed. The minimum of $F(v^i)$, i=1,...,4, is attained at $\hat{v} = v^3$ (this is actually the global minimum over Ω but it is not yet recognized as such since $F(\hat{v}) > \underline{\varphi} = \min \{F(y): y \in S\}$).

Next we determine $\bar{y} \in \text{argmax} \{F(y): y \in \Omega\}$ and construct the rectangle T inscribed in the level ellipsoid $F(y) = F(\hat{v})$. The residual domains that remain after the deletion of the interior of T are Ω_3 and Ω_4. Since Ω_4 is a simplex with all vertices inside the level ellipsoid $F(y) = F(\hat{v})$, it can be eliminated from further consideration. By constructing a simplex which has one vertex at v^3 and contains Ω_3, we also see that all of the vertices of this simplex lie inside our ellipsoid. Therefore, the global minimum of F over Ω is identified as $v^* = \hat{v} = v^3$.

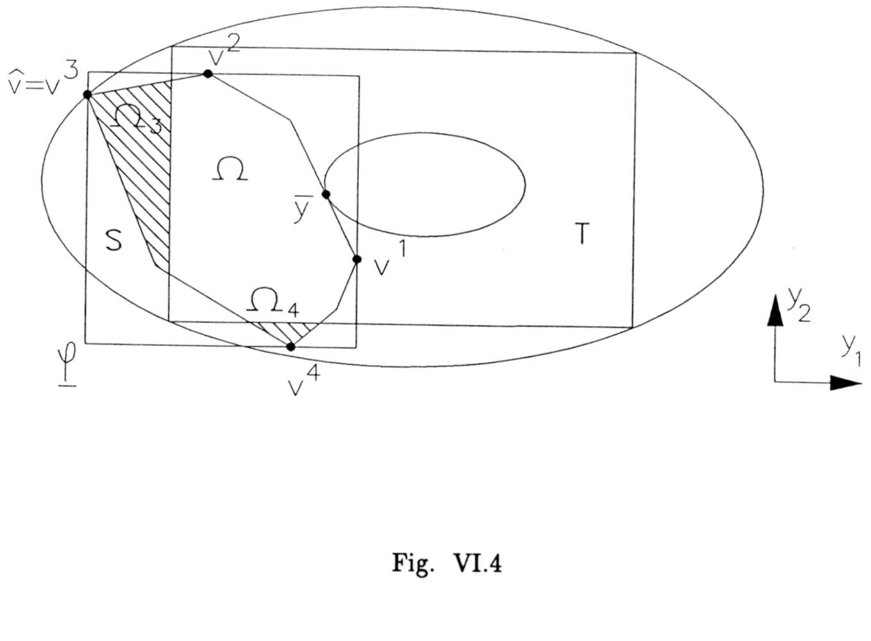

Fig. VI.4

4. CONCAVE POLYHEDRAL UNDERESTIMATION

In this section concave piecewise underestimators are used to solve the concave minimization problem.

4.1. Outline of the Method

We again consider the problem

(BCP) minimize $f(x)$ subject to $x \in D$,

where D is a polytope in $\mathbb{R}^n$ and $f: \mathbb{R}^n \longrightarrow \mathbb{R}$ is a concave function.

Let $M_1 = [v^1,...,v^{n+1}]$ be an n-simplex containing D. For the method to be developed below, it will suffice to suppose that f(x) is concave and finite on M_1. Let

$$G = \{(x,t) \in M_1 \times \mathbb{R}: f(x) \geq t\}$$

denote the ***hypograph*** of f(x). Clearly, G is a convex set. Now consider a finite set $X \subset M_1$ such that conv $X = M_1$. Let

$$Z = \{z = (x,f(x)): x \in X\} \subset M_1 \times \mathbb{R} ,$$

and let P denote the set of points $(x,t) \in M_1 \times \mathbb{R}$ on or below the convex hull of Z. Then we can write

$$P = \text{conv } Z - \Gamma ,$$

where $\Gamma = \{(x,t) \in \mathbb{R}^n \times \mathbb{R}: t > 0\}$ is the positive vertical halfline. We shall call P a ***trunk with base*** *X*. Clearly, P is the hypograph of a certain concave function $\varphi(x)$ on M_1.

Proposition VI.10. *The polyhedral function $\varphi(x)$ with hypograph P is the lowest concave underestimator of $f(x)$ that agrees with $f(x)$ at each point $x \in X$.*

Proof. The function $\varphi(x)$ is polyhedral, since P is a polyhedron. From the construction it follows that $P \subset G$, hence $\varphi(x) \le f(x) \quad \forall x \in D$.

If ψ is any concave function that agrees with f(x) on X, then its hypograph

$$\{z = (x,t) \in M_1 \times \mathbb{R}: \psi(x) \ge t\}$$

must be convex and must contain Z; hence it must contain P. This implies that $\psi(x) \ge \varphi(x)$. ■

We now outline the concave polyhedral underestimation method and discuss the main computational issues involved.

We start with the set

$$X_1 = \{v^1,...,v^{n+1}\} .$$

At the beginning of iteration k=1,2,... we already have a finite set X_k such that $X_1 \subset X_k \subset M_1 = \text{conv } X_k$, along with the best feasible point $\tilde{x}^{k-1}$ known so far. Let P_k be the trunk with base X_k and let $\varphi_k(x)$ be the concave function with hypograph P_k. We solve the relaxed problem

(SP_k) minimize $\varphi_k(x)$ subject to $x \in D$

obtaining a basic optimal solution x^k. Let $\tilde{x}^k$ be the point with the least function value among $\tilde{x}^{k-1}$ and all of the new vertices of D that are encountered while solving (SP_k). Since $\varphi_k(x)$ is an underestimator of f(x), $\varphi_k(x^k)$ yields a lower bound for min f(D). Therefore, if

$$\varphi_k(x^k) = f(\tilde{x}^k) ,$$

then $\tilde{x}^k$ is a global optimal solution of (BCP) and we terminate. Otherwise, we must have $x^k \notin X_k$ (since $x^k \in X_k$ would imply that $\varphi_k(x^k) = f(x^k) \ge f(\tilde{x}^k)$, because the function $\varphi^k(x)$ agrees with f(x) on X_k). Setting $X_{k+1} = X_k \cup \{x^k\}$, we then pass to iteration k+1.

Theorem VI.5. *The procedure just described terminates after finitely many iterations, yielding a global optimal solution of (BCP).*

Proof. We have $\varphi_k(x^k) \le \min f(D) \le f(\tilde{x}^k)$. Therefore, if the k–th iteration is not the last one, then $\varphi_k(x^k) < f(\tilde{x}^k) \le f(x^k) = \varphi_{k+1}(x^k)$. This shows that $\varphi_k \neq \varphi_h$, and hence $X_k \neq X_h$ for all $h < k$. Since each x^k is a vertex of D, the number of iterations is bounded from above by the number of vertices of D. ∎

Thus, finite termination is ensured for this procedure, which can also be identified with a polyhedral annexation (inner approximation) method, where the target set is G and the expanding polyhedra are $P_0, P_1, \dots$ (note that $P_{k+1} = \text{conv}(P_k \cup \{z^k\})$, with $z^k = (x^k, f(x^k)) \notin P_k$).

For a successful implementation of this procedure, the main issue is, of course, how to compute the functions $\varphi_k(x)$ and solve the relaxed problems (SP_k). We proceed to discuss this issue in the sections that follow.

4.2. Computation of the Concave Underestimators $\varphi_k(x)$

Consider any nonvertical facet σ of P_k. Since $\dim \sigma = n$, σ lies in a uniquely determined hyperplane H_σ in $\mathbb{R}^{n+1}$, which is a supporting hyperplane of P_k at every $(x,t) \in \sigma$. This hyperplane is not vertical, so its equation has the form

$$qx + t = q_0, \text{ with } q = q(\sigma) \in \mathbb{R}^n,\ q_0 = q_0(\sigma) \in \mathbb{R}.$$

Now let $\mathcal{F}_k$ denote the collection of all nonvertical facets of P_k.

Proposition VI.11. *We have* $\varphi_k(x) = \min\{\varphi_{k,\sigma}(x) : \sigma \in \mathcal{F}_k\}$, *where*

$$\varphi_{k,\sigma}(x) = q_0(\sigma) - q(\sigma)x .$$

Proof. The result follows because the trunk P_k is defined by the inequalities:

$$x \in M_1,\ t \le q_0(\sigma) - q(\sigma)x \quad \forall \sigma \in \mathcal{F}_k .$$ ■

The above formulas show that the function $\varphi_k(x)$ can be computed, once the equations of the hyperplanes through the nonvertical facets are known. Since

$$\min_{x \in D} \varphi_k(x) = \min_{x \in D} \min_{\sigma \in \mathcal{F}_k} \varphi_{k,\sigma}(x) = \min_{\sigma \in \mathcal{F}_k} \min_{x \in D} \varphi_{k,\sigma}(x) ,$$

the relaxed problem (SP_k) can be solved by separately solving each linear program

LP(σ ;D) minimize $[q_0(\sigma) - q(\sigma)x]$ subject to $x \in D$,

and taking

$$x^k \in \arg\min_{\sigma \in \mathcal{F}_k} \varphi_{k,\sigma}(x^{k,\sigma}) \text{ with} \tag{37}$$

$$x^{k,\sigma} \in \arg\min_{x \in D} [q_0(\sigma) - q(\sigma)x] .$$

Thus, computing the functions φ_k as well as solving (SP_k) are reduced to the computation of the nonvertical facets of P_k, or rather, the associated affine functions $\varphi_{k,\sigma}(x)$.

4.3. Computation of the Nonvertical Facets of P_k.

Observe that the initial trunk P_1 has just one nonvertical facet which is the n–simplex spanned by $(v^i, f(v^i))$, i=1,...,n+1. Therefore, it will suffice to consider the following auxiliary problem:

($\mathcal{F}$) *Let $X_{k+1} = X_k \cup \{x^k\}$, where $x^k \in M_1 \setminus X_k$. Given the collection $\mathcal{F}_k$ of nonvertical facets of the trunk P_k with base X_k, find the collection $\mathcal{F}_{k+1}$ of nonvertical facets of the trunk P_{k+1} with base X_{k+1}.*

By translating if necessary, we may assume that $0 \in \text{int } M_1$ and $f(x) > 0$ for all $x \in M_1$, so that any trunk P with base X contains $0 \in \mathbb{R}^{n+1}$ in its interior. Under these conditions, we shall convert problem ($\mathcal{F}$) into an easier one by using the following result, which is analogous to Proposition VI.4, from which, in fact, it could be derived.

Proposition VI.12. *Let P be a trunk with base X, and let*

$$S = \{(q,q_0) \in \mathbb{R}^n \times \mathbb{R}: q_0 - qx \geq t \quad \forall (x,t) \in \text{vert } P\} .$$

Then each nonvertical facet σ of P whose hyperplane is $qx + t = q_0$ corresponds to a vertex (q,q_0) of S and conversely.

Proof. Consider any nonvertical facet σ of P. Since $0 \in \text{int } P$, the hyperplane through σ does not contain 0, and so σ must contain n+1 linearly independent vertices of P. That is, the equation of the hyperplane through σ : $t = q_0 - qx$ must be satisfied by n+1 linearly independent elements of vert P. Since this hyperplane is a supporting hyperplane for P, we have $t \leq q_0 - qx$ for all $(x,t) \in \text{vert } P$. Hence, $(q,q_0) \in S$, and it satisfies at least n+1 linearly independent constraints on S as strict equalities. This implies that (q,q_0) is a vertex of S.

Conversely, let (q,q_0) be a vertex of S. Then $t \leq q_0 - qx$ for all $(x,t) \in \text{vert } P$, and hence for all $(x,t) \in P$; moreover, of these at least n+1 linearly independent constraints are satisfied as equalities. Hence, the hyperplane $t = q_0 - qx$ is a supporting hyperplane of P, and it passes through at least n+1 linearly independent vertices of P. This implies that the intersection of P with this hyperplane is a facet of P; this facet cannot be vertical, because the coefficient of t in the equation of its hyperplane is 1. ∎

Corollary VI.3. *Let*

$$S_k = \{(q,q_0) \in \mathbb{R}^n \times \mathbb{R}: q_0 - qx \geq t \quad \forall (x,t) \in vert\ P_k\}. \tag{38}$$

Then every nonvertical facet of P_{k+1} *whose hyperplane is* $t = q_0 - qx$ *corresponds to a vertex of*

$$S_{k+1} = S_k \cap \{(q,q_0) \in \mathbb{R}^n \times \mathbb{R}: q_0 - qx^k \geq f(x^k)\} \tag{39}$$

and conversely.

Proof. This follows because vert $P_{k+1} \subset$ vert $P_k \cup \{(x^k, f(x^k))\}$. ■

The above results show that the auxiliary problem ($\mathcal{P}$) is equivalent to the following problem:

($\mathcal{M}$) *Suppose* that *the vertex set* $\mathcal{M}_k$ *of* S_k *(defined in (38)) is known. Compute the vertex set* $\mathcal{M}_{k+1}$ *of* S_{k+1}.

Since S_{k+1} is also obtained by adding just one new linear constraint to S_k (formula (39)), problem ($\mathcal{M}$) can be solved by the available methods (cf. III.4.2). Once the vertices of S_{k+1} have been computed, the equations of the nonvertical facets of P_{k+1}, and hence the function φ_{k+1}, are known.

In more detail, the computation of the nonvertical facets of P_k can be carried out in the following way.

First compute the unique nonvertical facet of P_1, which is given by the unique vertex of

$$S_1 = \{(q,q_0): q_0 - qv^i \geq f(v^i),\ i=1,\dots,n+1\}. \tag{40}$$

Clearly this vertex is a solution of the system

$$(q,q_0) \begin{bmatrix} -v^1 & \dots & -v^{n+1} \\ 1 & & 1 \end{bmatrix} = (f(v^1),\dots,f(v^{n+1})),$$

and hence

$$(q,q_0) = (f(v^1),...,f(v^{n+1}))Q_1^{-1} , \tag{41}$$

where

$$Q_1 = \begin{bmatrix} -v^1 & \dots & -v^{n+1} \\ 1 & \dots & 1 \end{bmatrix} . \tag{42}$$

At iteration k, the vertex set $\mathcal{M}_k$ of S_k is already known. Form S_{k+1} by adding to S_k the new constraint

$$q_0 - qx^k \geq f(x^k) ,$$

where x^k is the point to be added to X_k. Compute the vertex set $\mathcal{M}_{k+1}$ of S_{k+1} (by any available subroutine, e.g., by any of the methods discussed in III.4.2. Then every $(q,q_0) \in \mathcal{M}_{k+1}$ yields a nonvertical facet of P_{k+1} defined by the hyperplane

$$t = q_0 - qx .$$

4.4. Polyhedral Underestimation Algorithm

The above development leads to the following algorithm.

Algorithm VI.5 (PU Algorithm).

Compute a vertex $\bar{x}^0$ of D.

0) Choose an n–simplex $[v^1,v^2,...,v^{n+1}]$ containing D, where $v^1 = \bar{x}^0$. Set $X_1 = \{v^1,v^2,...,v^{n+1}\}$, $S_1 = \{(q,q_0): q_0 - qv^1 \geq f(v^1), i=1,2,...,n+1\}$. Let $\mathcal{M}_1$ be the singleton $\{(f(v^1),...,f(v^{n+1}))Q_1^{-1}\}$, where Q_1 is the matrix (42). Set $\mathcal{N}_1 = \mathcal{M}_1$, k = 1.

1) For each $(q,q_0)) \in \mathcal{N}_k$ solve the linear program

$$\text{minimize } (q_0 - qx) \quad \text{subject to } x \in D ,$$

obtaining a basic optimal solution $\omega(q,q_0)$ and the optimal value $\beta(q,q_0)$.

2) Compute

$$(q^k,q_0^k) \in \arg\min \{\beta(q,q_0): (q,q_0)) \in \mathcal{N}_k\}$$

and let $x^k = \omega(q^k,q_0^k)$, $\beta^k = \beta(q^k,q_0^k)$.

3) Update the current best solution by taking

$$\tilde{x}^k \in \arg\min \{f(\tilde{x}^{k-1}), f(\omega(q,q_0)), (q,q_0) \in \mathcal{N}_k\}.$$

4a) If $f(\tilde{x}^k) = \beta^k$, then terminate: $\tilde{x}^k$ is a global optimal solution of (BCP).

4b) If $f(\tilde{x}^k) < f(\tilde{x}^{k-1})$, set $\tilde{x}^0 \leftarrow \tilde{x}^k$, and return to Step 0.

5) Otherwise, let

$$S_{k+1} = S_k \cap \{(q,q_0): q_0 - qx^k \geq f(x^k)\} ,$$

Compute the vertex set $\mathcal{M}_{k+1}$ of S_{k+1}. Set $\mathcal{N}_{k+1} = \mathcal{M}_{k+1} \setminus \mathcal{N}_k$. Let $k \leftarrow k+1$ and return to 1).

Remarks VI.5. (i) Finiteness of the above algorithm follows from Theorem VI.5. Indeed, since at each return to Step 0 (restart) the new vertex $\tilde{x}^0$ is better than the previous one, Step 4b can occur only finitely many times. That is, from a certain moment on, Step 4b never occurs and the algorithm coincides exactly with the procedure described in Section VI.4.1. Therefore, by Theorem VI.5 it must terminate at a Step 4a, establishing that the last $\tilde{x}^0$ is a global optimal solution.

(ii) A potential difficulty of the algorithm is that the set $\mathcal{M}_k$, i.e., the collection of nonvertical facets of P_k might become very numerous. However, according to Step 4b, when the current best feasible solution is improved, the algorithm returns to

Step 0 with the new best feasible solution as x^0. Such restarts can often accelerate the convergence and prevent an excessive growth of $|\mathcal{M}_k|$.

(iii) Sometimes it may also happen that, while the current best feasible solution remains unchanged, the set $\mathcal{M}_k$ is becoming too large. To overcome the difficulty in that case, it is advisable then to make a restart, after replacing the polyhedron D by $D \cap \{x: \ell(x-\tilde{x}^0) \le 1\}$, where $\ell(x-\tilde{x}^0) \le 1$ is an α–valid cut for (f^0,D) at $\tilde{x}^0$ with $\alpha = f(\tilde{x}^0)$. The finiteness of the algorithm cannot be adversely affected by such a step.

4.5. Alternative Interpretation

The above polyhedral underestimation algorithm was derived by means of an inner approximation of the hypograph of the function f(x). An alternative interpretation is based on the representation of the function f(x) as the pointwise infimum of a collection of affine functions.

Observe that, since f(x) is concave, we have

$$f(x) = \inf \{h(x): h \text{ affine}, h(y) \ge f(y) \quad \forall y \in \mathbb{R}^n\} .$$

Now consider a finite set X_k in $\mathbb{R}^n$ such that $D \subset \operatorname{conv} X_k$ and let

$$\varphi_k(x) = \inf \{h(x): h \text{ affine}, h(v) \ge f(v) \quad \forall v \in X_k\}. \tag{43}$$

Since X_k is finite, φ_k is a polyhedral function and since $D \subset \operatorname{conv} X_k$, we have

$$\varphi_k(x) \le f(x) \quad \forall x \in D ,$$

i.e., φ_k is a polyhedral concave function that underestimates f over D.

Solving the relaxed problem

$$\min \{\varphi_k(x): x \in D\} ,$$

we obtain as optimal solution a vertex x^k of D such that

$$\varphi_k(x^k) \le \min f(D) .$$

Therefore, if $\varphi_k(x^k) = f(\tilde{x}^k)$ for the best feasible point $\tilde{x}^k$ so far encountered, then $\tilde{x}^k$ solves the problem (BCP). Otherwise, since $\varphi_k(x) = f(x)$ for any $x \in X_k$ and $\varphi_k(x^k) < \min f(D)$, we must have $x^k \notin X_k$. Setting $X_{k+1} = X_k \cup \{x^k\}$, we can define

$$\varphi_{k+1}(x) = \inf \{h(x): h \text{ affine}, h(v) \ge f(v) \quad \forall v \in X_{k+1}\} ,$$

and repeat the procedure just described with φ_{k+1} in place of φ_k.
This is exactly the PU algorithm, if we start with $X_1 = \{v^1,...,v^{n+1}\}$, where $M_1 = [v^1,...,v^{n+1}]$ is an n–simplex containing D. To see this, it suffices to observe the following

Proposition VI.13. *The function φ_k defined in (43) is identical to the concave function whose hypograph is the trunk P_k with base X_k.*

Proof. For any $x' \in M_1 = \text{conv } X_k$, let (x',t') be the point where the vertical line through x' meets the upper boundary of P_k. Then (x',t') belongs to some nonvertical facet σ of P_k, so that $t' = q_0 - qx'$, where $t = q_0 - qx$ is the equation of the hyperplane through σ. Since the latter is a supporting hyperplane of P_k at (x',t'), we must have $q_0 - qv \ge f(v)$ for all $v \in X_k$. Moreover, any hyperplane $h(x) = t$ such that $h(v) \ge f(v)$ $\forall v \in X_k$ must meet the vertical through x' at a point (x',t^*) such that $t^* \ge t'$. Therefore, $t' = \min \{t: h(x) = t, h \text{ affine}, h(v) \ge f(v) \ \forall v \in X_k\}$; that is, $\varphi_k(.)$ coincides with the function whose hypograph is just P_k. ■

Remark VI.6. Set $v^{n+i+1} := x^i$, so that $X_k = \{v^i, i \in I_k\}$ with $I_k = \{1,...,n+k+1\}$. By (43), for each $x \in M_1$, $\varphi_k(x)$ is given by the optimal value in the linear program

$L(x;X_k)$ $\qquad$ minimize $q_0 - qx$ $\quad$ s.t. $q_0 - qv^i \ge f(v^i)$ $\ (i \in I_k)$.

Thus, if as before $\mathscr{M}_k$ denotes the vertex set of

$$S_k = \{(q,q_0): q_0 - qv^i \geq f(v^i) \quad (i \in I_k)\}$$

(the feasible set of $L(x;X_k)$), then the relaxed problem min $\varphi_k(D)$ is the same as

$$\min_{(q,q_0)\in \mathscr{M}_k} \min_{x\in D} (q_0 - qx) .$$

From this it is obvious that the crucial step consists in determining the set $\mathscr{M}_k$. On the other hand, instead of determining $\mathscr{M}_k$ one could also determine directly the set $\mathscr{F}_k$ of all nonvertical facets of P_k, i.e., the linearity pieces of φ_k. For this, observe that $\mathscr{F}_1$ is readily available. Once $\mathscr{F}_k$ has been computed, $\mathscr{F}_{k+1} \cap \mathscr{F}_k$ consists of all elements $\sigma \in \mathscr{F}_k$ whose hyperplanes $q_0 - qx = t$ are such that $q_0 - qx^k \geq f(x^k)$, while $\mathscr{F}_{k+1} \setminus \mathscr{F}_k$ is given by the collection of all (q,q_0) that are basic optimal solutions of the linear program $L(x^k; X_{k+1})$.

4.6. Separable Problems

Separable problems form an important class of problems for which the polyhedral underestimation method seems to be particularly suitable. These are problems in which the feasible domain is a polytope D contained in a rectangle $M = \{x \in \mathbb{R}^n: r_j \leq x_j \leq s_j, \ j=1,...,n\}$, while the objective function f(x) has the form

$$f(x) = \sum_{j=1}^{n} f_j(x_j) ,$$

where each $f_j(.)$ is a function of one variable which is concave and finite on the line segment $\Delta_j = [r_j,s_j]$ (but possibly **discontinuous** at the endpoints of the segment, see Fig. VI.5). This situation occurs in practice, e.g., when D represents the set of feasible production programs, and $f_j(t)$ is the cost of producing t units of the j–th product. If a fixed cost and economies of scale are present, then seeking the cheapest production program amounts to solving a problem of the type under consideration.

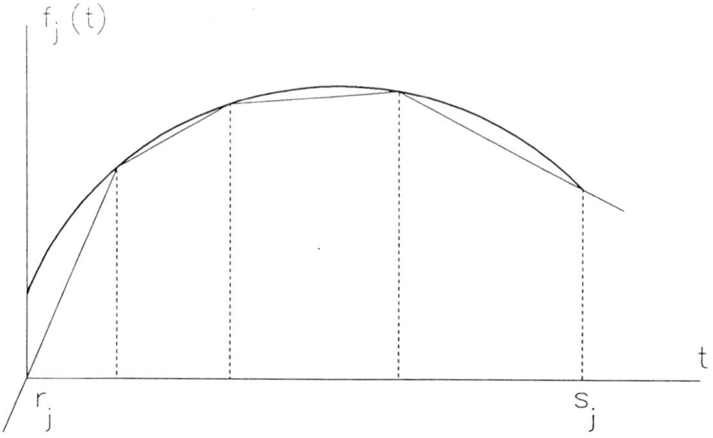

Fig. VI.5

For this problem, many of the methods discussed previously either do not apply or lose their efficiency, because they assume that the function f(x) is extendable to a concave finite function over $\mathbb{R}^n$, which is not the case here.

To apply the PU method, for each j let us choose a finite grid X_{kj} of points in the segment Δ_j for which r_j , $s_j \in X_{kj}$. Consider the piecewise affine function $\varphi_{kj}(t)$, $t \in \mathbb{R}$, that agrees with $f_j(t)$ at each point of X_{kj}. Since we are dealing with functions of one variable, the construction of φ_{kj} presents no difficulty. Clearly, $\varphi_{kj}(t)$ is a concave underestimator of $f_j(t)$ on Δ_j. Therefore, the function

$$\varphi_k(x) = \sum_{j=1}^{n} \varphi_{kj}(x_j)$$

is a concave underestimator of f(x) on M.

Solving the relaxed problem

(SP_k) minimize $\varphi_k(x)$ subject to $x \in D$,

we obtain an optimal solution which is a vertex x^k of D.

Let $\tilde{x}^k$ be the best feasible solution so far available (i.e., the best among $x^1,\ldots,x^k$ and the other vertices of D that may have been encountered during the process of computation). If $\varphi_k(x^k) = f(\tilde{x}^k)$, then $f(\tilde{x}^k) \le \varphi_k(x) \le f(x)\ \ \forall x \in D$, hence $\tilde{x}^k$ solves our problem (BCP) and we stop. Otherwise,

$$\varphi_k(x^k) = \sum_{j=1}^{n} \varphi_{kj}(x_j^k) < f(\tilde{x}^k) \le f(x^k) = \sum_{j=1}^{n} f_j(x_j^k)\ , \tag{44}$$

therefore, $x_{j^*}^k \notin X_{kj^*}$ for at least one j*. Setting $X_{k+1,j^*} = X_{kj^*} \cup \{x_{j^*}^k\}$ (j=1,...,n), $X_{k+1,j} = X_{k,j}$ $(j \neq j^*)$, we can then repeat the procedure, with $k \leftarrow k+1$.

Thus, if we start at $k = 1$ with $X_{1j} = \Delta_j$ (j=1,...,n) and perform the above procedure, we generate a sequence of underestimators $\varphi_1,\ldots,\varphi_k,\ldots$ along with a sequence of vertices of D: $x^k \in \arg\min \{\varphi_k(x): x \in D\}$ (k=1,2,...).

Proposition VI.14. *The above procedure converges to a global optimal solution of (BCP) in a finite number of iterations.*

Proof. It is clear that $X_h \neq X_k$ for $h \neq k$. But X_{k+1} is obtained from X_k by adding to X_{kj^*} a point $x_{j^*}^k \notin X_{kj^*}$. Since x^k is a vertex of D, the set of all possible x_j^k (j=1,...,n; k=1,...} is finite. Hence, the sequence $\{X_1,X_2,\ldots\}$ is finite. ■

An important issue in the implementation of the above procedure is how to solve the relaxed problems (SP_k). Of course, since the objective function $\varphi_k(x)$ in each problem (SP_k) is concave, piecewise affine, and **finite** throughout $\mathbb{R}^n$, these problems can be solved by any of the methods discussed previously. However, in view of the strong connection between (SP_{k+1}) and (SP_k), an efficient method for solving these problems should take advantage of this structure in order to save computational effort. Falk and Soland (1969) suggested the following approach.

Let $X_k = \prod_{j=1}^{n} X_{k,j}$, and denote by $\bar{\mathscr{P}}_k$ the partition of M determined by X_k , i.e., the partition obtained by constructing, for each j=1,2,...,n, all the hyperplanes parallel to the facets of M and passing through the points of $X_{k,j}$ (let us agree to call these hyperplanes partitioning hyperplanes of $\bar{\mathscr{P}}_k$).

For k=1, $\bar{\mathscr{P}}_1 = \{M\}$, so that $\varphi_1(x)$ is affine and solving (SP_1), i.e., finding

$$x^1 \in \operatorname{argmin} \{\varphi_1(x): x \in D \cap M\} ,$$

presents no difficulty. Set $\mathscr{P}_1 = \bar{\mathscr{P}}_1$. At the end of iteration $k = 1,2,\ldots$ we already have:

a) a collection $\mathscr{P}_k$ of rectangles forming a partition of M. These rectangles are of the form $P = \{\ell \leq x \leq L\}$, where ℓ, L belong to X_k (so that either $P \in \bar{\mathscr{P}}_k$, or else P can be subdivided into a finite number of members of $\bar{\mathscr{P}}_k$ by means of partitioning hyperplanes of $\bar{\mathscr{P}}_k$).

b) for each $P \in \mathscr{P}_k$, a point x(P) and a number $\mu(P)$ are determined which are, respectively, a basic optimal solution and the optimal value of the linear program

$$\min \psi_P(x) \quad \text{s.t. } x \in D \cap P , \tag{45}$$

where $P = \{x: \ell \leq x \leq L\}$ and $\psi_P(x) = \sum_{j=1}^{n} \psi_{P,j}(x_j)$, in which $\psi_{P,j}(x_j)$ is the affine function that agrees with $f_j(x_j)$ at the points ℓ_j and L_j .

c) a rectangle $P_k \in \operatorname{argmin} \{\mu(P): P \in \mathscr{P}_k\}$ such that $P_k \in \bar{\mathscr{P}}_k$. Observe that $\psi_P(x)$ always underestimates $\varphi_k(x)$ on P and $\psi_P(x) = \varphi_k(x) \;\; \forall x \in P$ if $P \in \bar{\mathscr{P}}_k$. Therefore, $\mu(P)$ serves as a lower bound for $\min \{\varphi_k(x): x \in D \cap P\}$, and $\mu(P)$ is exactly equal to this minimum if $P \in \bar{\mathscr{P}}_k$.

Now, to pass from iteration k to iteration k+1, we choose an index j* such that $x^k_{j^*} \notin X_{k,j^*}$ and set $X_{k+1,j^*} = X_{k,j^*} \cup \{x^k_{j^*}\}$, $X_{k+1,j} = X_{k,j}$ $(j \neq j^*)$. To solve (SP_{k+1}) we proceed according to the following branch and bound scheme:

0) Set $\mathscr{P}_{k0} = \mathscr{P}_k$, $P_{k0} = P_k$.

1) Split P_{k0} into two subrectangles by means of a partitioning hyperplane of $\bar{\mathscr{P}}_{k+1}$.

Compute x(P), $\mu(P)$ for each of these subrectangles P. Let $\mathscr{P}_{k1}$ be the new partition of M obtained from $\mathscr{P}_{k0}$ by replacing P_{k0} with its subrectangles.

2) Find $P_{k1} \in \operatorname{argmin}\{\mu(P): P \in \mathscr{P}_{k1}\}$. If $P_{k1} \in \bar{\mathscr{P}}_{k+1}$, then stop: $x^{k+1} = x(P_{k1})$, $\mathscr{P}_{k+1} = \mathscr{P}_{k1}$. Otherwise, set $\mathscr{P}_{k0} \leftarrow \mathscr{P}_{k1}$, $P_{k0} \leftarrow P_{k1}$ and return to 1).

It is easily seen that the above process must be finite. Note that in this way it is generally not necessary to investigate all of the members of the partition $\bar{\mathscr{P}}_{k+1}$; nor is it necessary to find $\varphi_{k+1}(x)$ explicitly.

CHAPTER VII

SUCCESSIVE PARTITION METHODS

This chapter is devoted to a class of methods for concave minimization which investigate the feasible domain by dividing it into smaller pieces and refining the partition as needed (successive partition methods, branch and bound).

We shall discuss algorithms that proceed through conical subdivisions (conical algorithms), simplicial subdivisions (simplicial algorithms) or rectangular subdivisions (rectangular algorithms).

1. CONICAL ALGORITHMS

The technique of conical subdivision for concave minimization was introduced by Tuy (1964). Since a conical subdivision induces a partition of the boundary of the feasible set, this technique seems to be appropriate for nonconvex problems where the optimum is achieved at certain boundary points.

Based on rectangular and simplicial branch and bound methods proposed by Falk and Soland (1969) and by Horst (1976), in subsequent conical algorithms, the process of conical subdivision was coupled with a lower bounding or some equivalent operation, following the basic steps of the branch and bound scheme.

A first convergent algorithm of this type was developed by Thoai and Tuy (1980). The algorithm to be presented below is an extended and improved version of both the original algorithm in Tuy (1964) and that of Thoai and Tuy (1980).

1.1. The Normal Conical Subdivision Process

Let us begin with the following (DG) problem which was considered in Section VI.2.

(DG) *Given a polytope D contained in a cone* $K_0 \subset \mathbb{R}^n$ *and a compact convex set G with* $0 \in \text{int } G$, *find a point* $y \in D \setminus G$ *or else establish that* $D \subset G$.

Clearly, if p(x) is a gauge of G, i.e., a convex, positively homogeneous function such that $G = \{x: p(x) \le 1\}$, then $D \subset G$ is equivalent to $\max p(D) \le 1$; in this case, solving the (DG)–problem is reduced to solving the convex maximization problem:

$$\text{maximize } p(x) \quad \text{subject to } x \in D \,. \tag{1}$$

To construct a conical procedure for solving (1), by the branch and bound scheme, we must determine three basic operations: branching, bounding and candidate selection (cf. Section IV.2).

1) **Branching (conical subdivision).** Obviously, any cone can be assumed to be of the form $K = \text{con}(Q)$ with $Q = (z^1,z^2,\dots,z^n)$, $z^i \in \partial G$ (the boundary of G), $i=1,2,\dots,n$. Given such a cone, a subdivision of K is determined by a point $u \in K$ such that $u \neq \lambda z^i$ $\forall \lambda \ge 0$, $i=1,2,\dots,n$. As we saw in Section V.3.1, if $u = \sum_{i \in I} \lambda_i z^i$ ($\lambda_i > 0$) and $\hat{u}$ is the point where the halfline from 0 through u meets ∂G, then the partition (splitting) of K with respect to u consists of the subcones $K_i = \text{con}(Q_i)$, $i \in I$, with

$$Q_i = (z^1,\dots,z^{i-1},\hat{u},z^{i+1},\dots,z^n).$$

Thus, to determine the branching operation, a rule has to be specified that assigns to each cone $K = con(Q)$, $Q = (z^1, z^2, ..., z^n)$, a point $u(Q) \in K$ which does not lie on any edge of K.

2) **Bounding.** For any cone $K = con(Q)$, $Q = (z^1, z^2, ..., z^n)$, the hyperplane $eQ^{-1}x = 1$ passes through $z^1, z^2, ..., z^n$, i.e., the linear function $h(x) = eQ^{-1}x$ agrees with $p(x)$ at $z^1, z^2, ..., z^n$. Hence, $h(x) \geq p(x)$ for all $x \in K$, and the value

$$\mu(Q) = \max \{eQ^{-1}x: x \in K \cap D\}$$

will satisfy $\mu(Q) \geq \max p(K \cap D)$. In other words, $\mu(Q)$ is an upper bound for $p(K \cap D)$ (note that (1) is a maximization problem).

3) **Selection.** The simplest rule is to select (for further splitting) the cone $K = con(Q)$ with largest $\mu(Q)$ among all cones currently of interest.

Once the three basic operations have been defined, a corresponding branch and bound procedure can be described that will converge under appropriate conditions. Since the selection here is bound improving, we know from the general theory of branch and bound algorithms that a sufficient convergence condition is **consistency** of the bounding operation (cf. Section IV.3).

Let us proceed as follows. For each cone $K = con(Q)$ denote by $\omega(Q)$ a basic optimal solution of the linear program

$$\text{LP}(Q;D) \qquad \text{maximize } eQ^{-1}x \quad \text{subject to } x \in D, \ Q^{-1}x \geq 0 . \qquad (2)$$

Note that $\omega(Q)$ is a vertex of $D \cap K$ satisfying $eQ^{-1}\omega(Q) = \mu(Q)$.

Now consider any infinite nested sequence of cones $K_s = con(Q_s)$, $Q_s = (z^{s1}, z^{s2} ..., z^{sn})$, s=1,2,..., generated by the cone splitting process described above. For each s let $\omega^s = \omega(Q_s)$, $u^s = u(Q_s)$ (in general, $u^s \neq \omega^s$ and even $u^s \neq \lambda\omega^s$ for every λ), and denote by q^s and $\hat{\omega}^s$ the points where the halfline from 0 through ω^s meets the simplex $[z^{s1}, z^{s2}, ..., z^{sn}]$ and the boundary ∂G of G, respectively.

Definition VII.1. *A sequence* $K_s = con(Q_s)$, *s=1,2,..., is said to be* ***normal*** *for given D, G if*

$$\lim_{s \to \infty} \|q^s - \hat{\omega}^s\| = 0 . \tag{3}$$

A cone splitting process is said to be ***normal*** *(an NCS process) if any infinite nested sequence of cones that it generates is normal.*

It turns out that normality of a cone splitting process is a sufficient condition for consistency of the bounding operation defined above and thereby for convergence of the resulting conical procedure.

1.2. The Main Subroutine

With the operations of branching, bounding and selection defined above, we now state the procedure for solving (DG), which we shall refer to as the (DG)–procedure and which will be used as a main subroutine in the algorithm to be developed for concave minimization.

(DG)–Procedure:

Select a rule u: $Q \longrightarrow u(Q)$ for the cone splitting operation so as to generate an NCS process (we shall later examine how to select such a rule).

1) Compute the intersections $z^{01}, z^{02}, \ldots, z^{0n}$ of the edges of K_0 with ∂G. Set $Q_0 = (z^{01}, z^{02}, \ldots, z^{0n})$, $\mathcal{M} = \{Q_0\}$, $\mathcal{P} = \mathcal{M}$.

2) For each matrix $Q \in \mathcal{P}$ solve the linear program LP(Q,D) to obtain the optimal value $\mu(Q)$ and a basic optimal solution $\omega(Q)$ of this program. If $\omega(Q) \notin G$ for some Q, then terminate: $y = \omega(Q)$. Otherwise, $\omega(Q) \in G$ for all $Q \in \mathcal{P}$; then go to 3).

3) In $\mathcal{M}$ delete all $Q \in \mathcal{P}$ such that $\mu(Q) \le 1$. Let $\mathcal{R}$ be the collection of remaining matrices. If $\mathcal{R} = \emptyset$, terminate: $D \subset G$. Otherwise, $\mathcal{R} \neq \emptyset$; then go to 4).

4) Choose $Q^* \in \text{argmax}\,\{\mu(Q): Q \in \mathcal{R}\}$, and split $K^* = \text{con}(Q^*)$ with respect to $u^* = u(Q^*)$. Let $\mathcal{P}^*$ be the collection of matrices corresponding to this partition of Q^*.

5) Replace Q^* by $\mathcal{P}^*$ in $\mathcal{R}$ and denote by $\mathcal{M}^*$ the resulting collection of matrices. Set $\mathcal{P} \leftarrow \mathcal{P}^*$, $\mathcal{M} \leftarrow \mathcal{M}^*$ and return to 2).

Proposition VII.1. *If the (DG)–procedure is infinite, then $D \subset G$ and there exists a point $y \in D \cap \partial G$. Consequently, if $D \setminus G$ is nonempty, the (DG)–procedure must terminate after finitely many steps at a point $y \in D \setminus G$.*

Proof. Consider any infinite nested sequence of cones $K_s = \text{con}(Q_s)$, s=1,2,..., generated by the procedure. As previously, let $\omega^s = \omega(Q_s)$, $u^s = u(Q_s)$ and denote by q^s and $\hat{\omega}^s$ the points where the halfline from 0 through ω^s meets the hyperplane $eQ_s^{-1}x = 1$ and ∂G, respectively. Then, by normality, we may assume, by passing to subsequences if necessary, that $\|q^s - \hat{\omega}^s\| \to 0$. Hence, $\|\omega^s - q^s\| \le \|\hat{\omega}^s - q^s\| \to 0$ and since $\|q^s\|$ is bounded, we have

$$\mu(Q_s) = \frac{\|\omega^s\|}{\|q^s\|} = 1 + \frac{\|\omega^s - q^s\|}{\|q^s\|} \to 1 \quad (s \to \infty) .$$

This means that the bounding is consistent (cf. Definition IV.4). Since the selection is bound improving, it follows from Theorem IV.2 that

$$\max p(D) = 1,$$

and hence $D \subset G$. Furthermore, by Corollary IV.1, if the procedure is infinite, then there exists at least one infinite nested sequence $K_s = \text{con}(Q_s)$ of the type just considered. Because of the normality and of the boundedness of the sequence ω^s, we may assume that $q^s - \hat{\omega}^s \to 0$, while the ω^s approach some $y \in D$. Then

$\omega^s - \hat{\omega}^s \longrightarrow 0$, implying that $\hat{\omega}^s \longrightarrow y$. Since $\hat{\omega}^s \in \partial G$, it follows that $y \in \partial G$, which proves the proposition. ■

Proposition VII.2. *Let G' be a compact convex set contained in the interior of G. Then after finitely many steps the (DG) procedure either establishes that $D \subset G$, or else finds a point $y \in D \setminus G'$.*

Proof. Let us apply the (DG)–procedure with the following stopping rule: stop if $\omega(Q) \notin G'$ for some Q in Step 2, or if $\mathcal{R} = \emptyset$ in Step 3 (i.e., $D \subset G$). Suppose that the procedure is infinite. Then, as we saw in the previous proof, there is a sequence $\omega^s \longrightarrow y$, where $y \in D \cap \partial G$. In view of the compactness of both G and G' and the fact that $G' \subset \text{int } G$, we must have $d(y,G') > 0$. Hence, $\omega^s \notin G'$ for some sufficiently large s, and the procedure would have stopped. ■

1.3. Construction of Normal Subdivision Processes

Now we discuss the question of how to construct a normal conical subdivision process.

Definition VII.2. *An infinite nested sequence of cones $K_s = con(Q_s)$, $s=0,1,\ldots$, is said to be* ***exhaustive*** *if the intersection $\bigcap_s K_s$ is a ray (a halfline emanating from 0); it is said to be* ***nondegenerate*** *if $\underline{\lim}_{s\to\infty} \|eQ_s^{-1}\| < \infty$, i.e., if there exists an infinite subsequence $\Delta \subset \{0,1,\ldots\}$ and a constant η such that $\|eQ_s^{-1}\| \le \eta \quad \forall s \in \Delta$. A conical subdivision process is said to be* ***exhaustive (nondegenerate)*** *if all of the infinite nested sequences of cones that it generates are* ***exhaustive (nondegenerate)***.

Based on a related subdivision of simplices (Horst (1976)) the concept of exhaustiveness was introduced by Thoai and Tuy (1980) (see Section IV.3.1). The concept of nondegeneracy is related to but somewhat different from an analogous concept of Hamami and Jacobsen (1987).

Proposition VII.3. *Let $K_s = con(Q_s)$ be an infinite nested sequence of cones with $Q_s = (z^{s1}, z^{s2}, \ldots, z^{sn})$, $z^{si} \in \partial G$ $(i=1,2,\ldots,n)$. For each s let $u^s \in [z^{s+1,1}, \ldots, z^{s+1,n}]$ and let q^s be the point where the halfline from 0 through u^s meets the simplex $[z^{s1}, \ldots, z^{sn}]$. If the sequence K_s is exhaustive or nondegenerate, then*

$$\lim_{s \to \infty} (q^s - u^s) = 0 .$$

Proof. Suppose that the sequence K_s shrinks to a ray Γ. Then each point z^{si} $(i=1,2,\ldots,n)$ approaches a unique point x^* of $\Gamma \cap \partial G$. Hence, both q^s and u^s tend to x^*, i.e., $q^s - u^s \to 0$.

Now suppose that the sequence K_s is nondegenerate and denote by H_{s+1} the hyperplane through $z^{s+1,1}, z^{s+1,2}, \ldots, z^{s+1,n}$ and by L_s the halfspace not containing 0 with bounding hyperplane H_{s+1}.

Then

$$q^s \notin L_{s+1}\text{, while } q^s \in \bigcap_{r \le s} L_r .$$

Since the sequence q^s is bounded, it follows from Lemma III.2 on the convergence of cutting procedures that $d(q^s, L_{s+1}) \to 0$, and hence $d(q^s, H_{s+1}) \to 0$. But the equation of H_{s+1} is $eQ_{s+1}^{-1} x = 1$, and hence we have $d(0, H_{s+1}) = 1/\|eQ_{s+1}^{-1}\|$. Therefore,

$$\|q^s - u^s\| = d(q^s, H_{s+1}) \|eQ_{s+1}^{-1}\| \; \|u_s\| .$$

By the nondegeneracy property, there is a subsequence $\Delta \subset \{0,1,\ldots\}$ such that $\|eQ_{s+1}^{-1}\|$ $(s \in \Delta)$ is bounded. The previous relation then implies that $q^s - u^s \to 0$ $(s \in \Delta, s \to \infty)$. ■

Letting $u^s = \hat{\omega}^s$ we can state the following consequence of the above proposition:

Corollary VII.1. *(Sufficient condition for normality). A conical subdivision process is normal if any infinite nested sequence* $K_s = con\ (Q_s)$ *that it generates satisfies either of the following conditions:*

1) the sequence is exhaustive;

2) the sequence is nondegenerate, and for all but finitely many s the subdivision of K_s *is performed with respect to a basic optimal solution* $\omega^s = \omega(Q_s)$ *of the associated linear program* $LP(Q_s, D)$.

In particular, an exhaustive subdivision process is normal.

Corollary VII.2. *The (DG)–Procedure, where an exhaustive subdivision process is used, can be infinite only if* $D \subset G$ *and* $D \cap \partial G \neq \emptyset$.

This result was essentially established directly by Thoai and Tuy (1980), who pointed out a typical exhaustive subdivision process called ***bisection***.

For simplices, this specific method of subdivision had been introduced earlier by Horst (1976) (cf. Section IV.3.1). If for any cone $K = con(Q) \subset K_0$ we denote by $Z(Q)$ the simplex which is the section of K formed by the hyperplane H_0 through $z^{01}, z^{02}, \ldots, z^{0n}$, then the bisection of K is simply the subdivision of K that is induced in the obvious way by the bisection of the simplex Z(Q). In other words, the bisection of a cone $K = con(Q)$ (or of a matrix Q) is the subdivision of K (or Q) with respect to the midpoint of a longest edge of Z(Q).

A subdivision process consisting exclusively of bisections is called a ***bisection process***. The exhaustiveness of such a subdivision process is shown in Proposition IV.2. In order to construct more efficient subdivision processes, we now extend this result to subdivision processes in which bisection is used infinitely often, but not exclusively.

Given a simplex $Z = [v^1, v^2, \ldots, v^n]$ and a point $w \in Z$, we define

$$\delta(Z) = \max_{i<j} \|v^i - v^j\| \ , \ \delta(w,Z) = \max_{i=1,\ldots,n} \|w - v^i\| \ .$$

Note that $\delta(Z)$ is the diameter of Z, while $\delta(w,Z)$ is the radius of the smallest ball with center w containing Z.

The following lemma is closely related to Proposition IV.2.

Lemma VII.1. *If w is the midpoint of a longest edge of Z, then*

$$\delta(w,Z) \le \frac{\sqrt{3}}{2}\delta(Z) .$$

Proof. Let w be the midpoint of $[v^1,v^2]$, with $\|v^2-v^1\| = \delta(Z)$. Of course,

$$\|w-v^i\| = \frac{1}{2}\delta(Z) \le \frac{\sqrt{3}}{2}\delta(Z) \quad (i=1,2) .$$

For $i>2$, since the line segment $[w,v^i]$ is a median of the triangle $[v^1,v^2,v^i]$ we have

$$2\|w-v^i\|^2 = \|v^1-v^i\|^2 + \|v^2-v^i\|^2 - \frac{1}{2}\|v^1-v^2\|^2 \le \delta^2(Z) + \delta^2(Z) - \frac{1}{2}\delta^2(Z) = \frac{3}{2}\delta^2(Z),$$

whence the desired inequality. ■

Lemma VII.2. *If* $w = \sum_{i=1}^{n} \zeta_i v^i$, $\zeta_i \ge 0$, $\sum_{i=1}^{n} \zeta_i = 1$, *and*

$$max\,\{\|w-v^i\|: \zeta_i = 0\} < \rho\delta(Z) ,$$

$$min\,\{\zeta_i: \zeta_i > 0\} > 1-\rho ,$$

then $\delta(w;Z) < \rho\delta(Z)$.

Proof. For any i such that $\zeta_i > 0$, denote by r^i the point where the halfline from v^i through w meets the facet of Z opposite v^i. Then $w = v^i + \theta(r^i-v^i)$ for some $\theta \in (0,1)$, and $r^i = \sum_{j\neq i} \xi_j v^j$, $\xi_j \ge 0$, $\sum \xi_j = 1$. Hence

$$\sum_{j=1}^{n} \zeta_j v^j = (1-\theta)v^i + \theta(\sum_{j\neq i} \xi_j v^j)$$

and, equating the terms in v^i, we obtain

$$\zeta_i v^i = (1-\theta)v^i ,$$

which shows that $\theta = 1-\zeta_i$. Thus $w = v^i + (1-\zeta_i)(r^i-v^i)$, and consequently, $\|w-v^i\| \le (1-\zeta_i)\|r^i-v^i\| < \rho\delta(Z)$ whenever $\zeta_i > 0$. The lemma follows immediately. ■

Proposition VII.4. *Let* $Z_s = [v^{s1},\dots,v^{sn}]$, $s=0,1,\dots$, *be a nested sequence of* $(n-1)$*-simplices such that for each s,* Z_{s+1} *is obtained from* Z_s *by replacing some* v^{si} *with a point* $w^s \in Z_s$ *satisfying*

$$\delta(w^s,Z_s) \le \rho\delta(Z_s)\,, \tag{4}$$

where $\rho \in (\frac{\sqrt{3}}{2}, 1)$ *is some constant. If for an infinite subsequence* $\Delta \subset \{0,1,\dots\}$ *each* w^s, $s \in \Delta$, *is the midpoint of a longest edge of* Z_s *(i.e.,* Z_{s+1} *is obtained from* Z_s *by a bisection), then the intersection* $\bigcap_{s=0}^{\infty} Z_s$ *is a singleton.*

Proof. Denote the diameter of Z_s by δ_s. Since $\delta_{s+1} \le \delta_s$, δ_s tends to some limit δ as $s \to \infty$. Assume that $\delta > 0$. Then we can choose t such that $\rho\delta_s < \delta$ for all $s \ge t$. From (4) it follows that for all $s > t$

$$\max_{i=1,\dots,n} \|w^s-v^{si}\| \le \rho\delta_s < \delta \le \delta_s\,. \tag{5}$$

Let us colour every vertex of Z_t "black" and color "white" every vertex of any Z_s with $s > t$ which is not black. Then (5) implies that for $s \ge t$ a longest edge of Z_s must have two black endpoints. Consequently, if $s \ge t$ and $s \in \Delta$, then w^s must be the midpoint of an edge of Z_s joining two black vertices, so that Z_{s+1} will have at least one black vertex less than Z_s. On the other hand, Z_{s+1} can never have more black vertices than Z_s. Therefore, after at most n (not necessarily consecutive) bisections corresponding to $s_1 < s_2 < \dots < s_n$ ($s_i \in \Delta$, $s_1 \ge t$), we obtain a simplex Z_s with only white vertices, i.e., according to (5), with only edges of length less than δ. This contradiction shows that we must have $\delta = 0$, as was to be proved. ■

Remarks VII.1. (i) From the proof it is easily seen that the proposition remains valid even if each w^s, $s \in \Delta$, is an arbitrary point (not necessarily the midpoint) of a longest edge of Z_s satisfying (4).

(ii) It is not difficult to construct a nested sequence $\{Z_s\}$ such that Z_{s+1} is obtained from Z_s by a bisection for infinitely many s, and nevertheless the intersection $\bigcap_s Z_s$ is not a singleton (cf. Chapter IV). Thus, condition (4) cannot be omitted in Proposition VII.4.

Returning to conical subdivision processes, let us consider a subcone K = con(Q) of an initial cone $K_0 = con(Q_0)$. Let Z = Z(Q) be the (n–1)–simplex formed by the section of K by the hyperplane $H_0 = \{x: eQ_0^{-1}x = 1\}$. For any point $u \in K$, let w be the point where the halfline from 0 through u meets Z.

In the sequel, we shall refer to the ratio $\delta(w,Z)/\delta(Z)$ as the ***eccentricity*** of Z relative to w, or as the *eccentricity* of K relative to u.

The following corollary then follows from Proposition VII.4:

Corollary VII.3. *Let $K_s = con(Q_s)$, $s=0,1,...$, be an infinite nested sequence of cones, in which K_{s+1} is obtained from K_s either by bisection or by a subdivision with respect to a point $u^s \in K_s$ such that the eccentricity of K_s relative to u^s does not exceed a constant ρ, $\frac{\sqrt{3}}{2} \le \rho < 1$. If the sequence involves infinitely many bisections, then it is exhaustive.*

The importance of this result is that it serves as a basis for "normalizing" any subdivision process satisfying condition (4).

1.4. The Basic NCS Process

To avoid tedious repetitions when considering a conical subdivision process, for any cone K = con(Q) with $Q = (z^1,z^2,...,z^n)$, $z^i \in \partial G$, generated during the process, we shall always use the following notation:

$\omega(Q)$: basic optimal solution of the linear program LP(Q,D) associated with Q;

$u(Q)$: point with respect to which K is split;

$Z(Q)$: (n–1)–simplex which is the section of K by the hyperplane $H_0 = \{x: eQ_0^{-1}x = 1\}$ through $z^{01},z^{02},\ldots,z^{0n}$;

$\sigma(Q)$: eccentricity of K relative to $\omega(Q)$ (i.e., eccentricity of Z(Q) relative to the point where Z(Q) intersects the halfline from 0 through $\omega(Q)$).

When $u(Q) = \omega(Q)$, i.e., K is subdivided with respect to $\omega(Q)$, we shall refer to this subdivision as an *ω–subdivision.*

A subdivision process consisting solely of ω–subdivisions will be called an ***ω–subdivision process.***

This subdivision method obviously depends on the data of the problem and seems to be the most natural one. However, we do not know whether it is normal. By Corollary VII.2, it will be normal whenever it is nondegenerate, i.e., if $\lim_{s\to\infty} \|eQ_s^{-1}\| < \infty$ for any infinite nested sequence of cones $K_s = \text{con}(Q)$ that it generates. Unfortunately, this kind of nondegeneracy is a condition which is very difficult to enforce in practice.

On the other hand, while exhaustive subdivision processes are normal (and hence ensure the convergence of the (DG)–Procedure), so far the most commonly used exhaustive process – the bisection – has not proven to be very efficient computationally. An obvious drawback of this subdivision is that it is defined independently of the problem's data, which partially explains the slow rate of convergence usually observed with the bisection, as compared with the ω–subdivision process in cases when the latter works.

To resolve this conflict between convergence and efficiency, the best strategy suggested by Corollary VII.3 seems to be to generate a "hybrid" process by properly inserting a bisection into an ω–subdivision process from time to time. Roughly speaking, ω–subdivisions should be used most of the time, whereas bisections, because of their convergence ensuring property, can serve as a recourse to get out of possible jams.

It is this strategy which is embodied in the following rule:

Basic NCS Process.

Select $\eta_0 \geq 0$ and $\rho \in (\frac{\sqrt{3}}{2}, 1)$.

Set $\eta(Q_0) = \eta_0$ for the initial cone $K_0 = \text{con}(Q_0)$. At each iteration, compute the eccentricity $\sigma(Q)$ for the cone $K = \text{con}(Q)$ to be subdivided.

a) If $\|eQ^{-1}\| \leq \eta(Q)$ and $\sigma(Q) \leq \rho$, perform an ω-subdivision of $K = \text{con}(Q)$ (i.e., choose $u(Q) = \omega(Q)$) and set $\eta(Q') = \eta(Q)$ for every subcone $K' = \text{con}(Q')$ in the partition;

b) Otherwise, perform a bisection of $K = \text{con}(Q)$ (i.e., choose $u(Q)$ to be the midpoint of a longest edge of $Z(Q)$) and set $\eta(Q') = 2\eta(Q)$ for every subcone $K' = \text{con}(Q')$ in the partition.

Proposition VII.5. *The conical subdivision process just described is normal.*

Proof. Let $K_s = \text{con}(Q_s)$, $s=1,...$, be an infinite nested sequence of cones generated in this process. By completing if necessary, we may assume that K_{s+1} is an immediate successor of K_s for every s. Let $u^s = u(Q_s)$ be the subdivision point of K_s. It is easy to see that this sequence of cones satisfies the conditions of Corollary VII.3. Therefore, if the sequence $\{K_s\}$ is not exhaustive, then it involves at most finitely many bisections. That is, there exists an s_0 such that for all $s \geq s_0$: $u^s = \omega^s$ (ω-subdivision). The NCS rule then implies that $\|eQ_s^{-1}\| \leq \eta(Q_s) = \eta(Q_{s_0})$, and so $\|eQ_s^{-1}\|$ is bounded. Thus, if the sequence is not exhaustive, it is nondegenerate and $u^s = \omega^s$ for all but finitely many s. By Corollary VII.1, the process is normal. ∎

Corollary VII.4. *The (DG)–procedure using the basic NCS process can be infinite only if $D \setminus G \neq \emptyset$. If G' is a compact subset of int G, then after finitely many steps this procedure either establishes that $D \subset G$, or else finds a point $y \in D \setminus G'$.*

Proof. This is a straightforward consequence of Propositions VII.1, VII.2 and VII.5. ■

1.5. The Normal Conical Algorithm

We apply the previous results to the BCP problem:

$$\text{minimize } f(x) \tag{6}$$

$$\text{s.t.} \quad Ax \le b\,, \tag{7}$$

$$x \ge 0\,, \tag{8}$$

where we assume that the constraints (7) (8) define a polytope D, and the concave objective function f: $\mathbb{R} \longrightarrow \mathbb{R}$ has bounded level sets.

In view of Corollary VII.2, we can solve the problem (BCP) by the following two phase scheme which is similar to the one in Section VI.2.2.

Start with a feasible solution $z \in D$.

Phase I. Search for a local minimizer x^0 which is a vertex of D such that $f(x^0) \le f(z)$.

Phase II. Let $\alpha = f(x^0) - \varepsilon$. Translate the origin to x^0 and construct a cone $K_0 \supset D$. Using the basic NCS process, apply the (DG)–procedure for $G = \{x: f(x) \ge f(x^0) - \varepsilon\}$, $G' = \{x: f(x) \ge f(x^0)\}$. If $D \subset G$, then terminate: x^0 is a global ε–optimal solution. Otherwise, a point $y \in D \setminus G'$ is obtained (so that $f(y) < f(x^0)$): set $z \leftarrow y$ and return to Phase I.

As in the PA algorithm (Section VI.2.4), it is not necessary that x^0 be a local minimizer. Also, a concavity cut can be added to the current feasible set before a return to Phase II. Incorporating these observations in the above scheme and replacing cones by their defining matrices, we can state the following procedure.

Algorithm VII.1 (Normal Conical Algorithm for BCP).

Select $\varepsilon \geq 0$.

Initialization:

Compute a point $z \in D$. Set $M = D$.

Phase I.

Starting with z find a vertex x^0 of D such that $f(x^0) \leq f(z)$. Let $\bar{x}$ be the best among x^0 and all the vertices of D adjacent to x^0; let $\gamma = f(\bar{x})$.

Phase II.

Select $\eta_0 \geq 0$ and $\rho \in (\frac{\sqrt{3}}{2}, 1)$.

0) Let $\alpha = \gamma - \varepsilon$. Translate the origin to x^0 and construct a cone $K_0 \supset D$. For each $i=1,2,\ldots,n$ compute the point z^{0i} where the i–th edge of K_0 meets the surface $f(x) = \alpha$. Let $Q_0 = (z^{01}, z^{02}, \ldots, z^{0n})$, $\mathcal{M} = \{Q_0\}$, $\mathcal{P} = \mathcal{M}$, $\eta(Q_0) = \eta_0$.

1) For each $Q = (z^1, \ldots, z^n) \in \mathcal{P}$ solve

$$\text{LP}(Q;M) \qquad \max eQ^{-1}x \qquad \text{s.t. } x \in M,\ Q^{-1}x \geq 0$$

to obtain the optimal value $\mu(Q)$ and a basic optimal solution $\omega(Q)$. If for some Q, $f(\omega(Q)) < \gamma$, set $z \leftarrow \omega(Q)$,

$$M \leftarrow M \cap \{x: eQ_0^{-1}x \geq 1\}$$

and return to Phase I. Otherwise, go to 2).

2) In $\mathcal{M}$ delete all $Q \in \mathcal{P}$ satisfying $\mu(Q) \leq 1$. Let $\mathcal{R}$ be the collection of remaining matrices. If $\mathcal{R} = \emptyset$, then terminate: $\bar{x}$ is a global ε–optimal solution of (BCP).

Otherwise, $\mathscr{R} \neq \emptyset$, go to 3).

3) Choose $Q_* \in \operatorname{argmax} \{\mu(Q): Q \in \mathscr{R}\}$. Compute the eccentricity $\sigma(Q_*)$.

a) If $\|eQ_*^{-1}\| \le \eta(Q_*)$ and $\sigma(Q_*) \le \rho$, split Q_* with respect to $\omega(Q_*)$ (perform an ω-subdivision) and set $\eta(Q) = \eta(Q_*)$ for each member Q of the partition.

b) Otherwise, split Q^* with respect to the midpoint of a longest edge of $Z(Q_*)$ (perform a bisection) and set $\eta(Q) = 2\eta(Q_*)$ for every member Q of the partition.

4) Let $\mathscr{P}_*$ be the partition of Q_*, $\mathscr{M}_*$ the collection obtained from $\mathscr{R}$ by replacing Q_* by $\mathscr{P}_*$. Set $\mathscr{P} \leftarrow \mathscr{P}_*$, $\mathscr{M} \leftarrow \mathscr{M}_*$ and return to 1).

As a consequence of the above discussion we can state the following result.

Theorem VII.1. *For $\varepsilon > 0$ the normal conical algorithm terminates after finitely many steps at a global ε-optimal solution.*

Proof. By Corollary VII.4, where $G = \{x: f(x) \ge \alpha\}$, $G' = \{x: f(x) \ge \gamma\}$, Phase II must terminate after finitely many steps either with $\mathscr{R} = \emptyset$ (the incumbent $\bar{x}$ is a global ε-optimal solution), or else with a point $\omega(Q)$ such that $f(\omega(Q)) < \gamma$. In the latter case, the algorithm returns to Phase I, and the incumbent $\bar{x}$ in the next cycle of iterations will be a vertex of D better than all of the vertices previously encountered. The finiteness of the algorithm follows from the finiteness of the vertex set of D. ■

As previously remarked (cf. Section V.3.3), if ε is sufficiently small, a vertex of D which is a global ε-optimal solution will actually be an exact global optimal solution.

The algorithm will still work for $\varepsilon = 0$, provided that the points $z^{0i} \neq 0$ $(i=1,2,...,n)$ in Step 0 can be constructed. The latter condition holds, for example, if x^0 is a nondegenerate vertex of D (because then the positive i-th coordinate axis will coincide with the i–th edge of D emanating from x^0).

Theorem VII.2. *For $\varepsilon = 0$ the normal conical algorithm either terminates at an exact global optimal solution after finitely many steps, or else it involves an infinite Phase II. The latter case can occur only if the current best solution $\bar{x}$ is actually already globally optimal.*

Proof. Apply the first part of Corollary VII.4. ■

Remark VII.2. Several different algorithms can be derived from the normal conical algorithm by assigning different values to the parameters η_0 and ρ.

If $\eta_0 = 0$, then in Step 3 we always have $\|eQ_*^{-1}\| > \eta(Q_*) = 0$, so that bisections are used throughout and we recover the Thoai – Tuy algorithm (1980).

If $0 < \eta_0 < +\infty$ and ρ is sufficiently close to 1, then $\sigma(Q_*) \le \rho$ is likely to hold most often and the algorithm will generate a hybrid subdivision process, with ω–subdivision in most iterations and bisection occasionally. The larger η_0 is and the closer ρ is to 1, the smaller the frequency of bisections.

If η_0 is very large and ρ is close to 1, then in Step 3 we almost always have $\|eQ_*^{-1}\| \le \eta(Q_*)$, so that the algorithm involves practically only ω–subdivisions, and it is very close to Zwart's algorithm (1974), except for a substantial difference in the use of the tolerance parameter ε.

Thus, the Thoai–Tuy "pure" bisection algorithm and Zwart's "pure" ω–subdivision algorithm appear as two extreme cases in a whole range of algorithms. On the other hand, for η_0 very large and ρ very close to 1, the normal conical algorithm operates much the same as the cut and split algorithm, except that in certain iterations a bisection is used instead of an ω–subdivision. One could say that the (η,ρ) device is a method of forcing the convergence of the cut and split algorithm.

As mentioned earlier, it is an open question whether the algorithm will still be convergent if $\eta_0 = +\infty$ and $\rho = 1$ (the algorithm will then coincide exactly with the cut and split algorithm).

However, the algorithm is convergent since for any finite value of η_0 and any value of $\rho \in (\frac{\sqrt{3}}{2}, 1)$ one can expect that in most cases in practice it will be convergent even if $\eta_0 = +\infty$ and $\rho = 1$. This seems to be in line with the finite termination of Zwart's algorithm for $\varepsilon = 0$ observed in the numerical experiments reported in Zwart (1974).

1.6. Remarks Concerning Implementation

(i) As in the cut and split algorithm (Section V.3.2), when $K_0 = \mathbb{R}^n_+$ (which is the case if the problem is in standard form with respect to x^0), the linear program LP(Q;M) in Step 1) can be solved without having to invert the matrix Q. Actually, if $Q = (z^1, z^2, \ldots, z^n)$ and if the additional constraints (cuts) that define M are $Cx \le d$, then in terms of the variables $(\lambda_1, \lambda_2, \ldots, \lambda_n) = Q^{-1}x$ this program can be written as

$$\text{LP(Q,M)} \qquad \max \sum_{j=1}^{n} \lambda_j \tag{9}$$

$$\text{s.t.} \sum_{j=1}^{n} \lambda_j (Az^j) \le b \,, \; \sum_{j=1}^{n} \lambda_j (Cz^j) \le d \,, \; \lambda_1, \lambda_2, \ldots, \lambda_n \ge 0 \,. \tag{10}$$

If Q' is the successor of Q obtained by replacing some z^i with a vector $u \in \text{con}(Q)$, then LP(Q',M) is derived from LP(Q;) simply by replacing Az^i and Cz^i with Au and Cu, respectively. So to solve LP(Q;M) there is no need to know Q^{-1}.

(ii) As in the PA method (Section VI.2), we return to Phase I (i.e., we ***restart*** a new cycle) whenever an $\omega(Q)$ is found with $f(\omega(Q) < \gamma$. Restarting is also possible when $\omega(Q)$ is a vertex of D, no matter what $f(\omega(Q))$ is. Then, instead of returning to Phase I, one should simply return to Step 0, with $x^0 \leftarrow \omega(Q)$. In this way the new x^0 will be different from all of the previous ones, so that the convergence of the algorithm will still be ensured (recall that a concavity cut should be made before each restart).

Sometimes it may happen that at a given stage, while no $\omega(Q)$ satisfying the above conditions is available, the set $\mathcal{R}$ of cones that must be investigated has become very numerous. In that event, it is also advisable to return to Phase I with $z \leftarrow \omega(Q_0)$, $M \leftarrow M \cap \{x: eQ_0^{-1}x > 1\}$, $D \leftarrow M \cap \{x: eQ_0^{-1}x > 1\}$ (note that not only M, but also the original feasible domain D is reduced). Of course, a finite number of restarts of this kind will not adversely affect the convergence.

Computational experience reported in Horst and Thoai (1989) has shown that a judicious restart strategy can often substantially enhance the efficiency of the algorithm by keeping $\mathcal{R}$ within manageable size and correcting a bad choice of the starting vertex x^0.

(iii) A further question of importance for efficient implementation of the normal conical algorithm is the selection of the subdivision rule.

The above basic NCS rule aims at ensuring the convergence of the algorithm with as few bisections as possible. However, its drawback is that it involves computing eQ_*^{-1} and $\sigma(Q_*)$ at each step, and this may be time consuming.

The following rule ensures convergence without too many bisections:

(*) *Select a natural number N (typically $N \geq 5$) and a number $\rho \in (\frac{\sqrt{3}}{2}, 1)$ such that $1 - \rho$ is small. At the beginning, set $\tau(Q_0) = 0$ for the initial cone $K_0 = con(Q_0)$. At any iteration, if $\tau(Q_*) < N$ and $\sigma(Q_*) \leq \rho$, then perform an ω-subdivision of Q_* and set $\tau(Q) = \tau(Q_*)+1$ for every member Q of the partition. Otherwise, perform a bisection of Q_* and set $\tau(Q) = 0$ for every member Q of the partition.*

Clearly, $\tau(Q_*)$ is the number of ancestors of Q_* (modulo a multiple of N), and so the rule implies that bisection occurs infinitely often in any infinite nested sequence of cones generated by the process. Moreover, $\sigma(Q) \leq \rho$ for all Q in such a sequence. Therefore, by Corollary VII.3, the subdivision process is exhaustive (hence, normal).

If N is sufficiently large, then the condition $\tau(Q_*) < N$ almost always holds (in the computational experiments reported in Zwart (1974), $\tau(Q_*)$ rarely exceeds 5 for problems of up to 15 variables), and the rule $(*)$ just described, in practice amounts to using *ω–subdivisions if $\sigma(Q_*) \le \rho$, and bisections otherwise.*

In practice, based on the essential idea of normalization (i.e., using bisections only to prevent jamming), one could even adopt the simplest rule: operate with ω–subdivisions as long as the algorithm proceeds normally, and use a bisection only when the algorithm slows down and threatens to jam.

If $Q_* = (z^{*1},...,z^{*n})$, $Z(Q_*) = [v^{*1},...,v^{*n}]$, $z^{*i} = \theta_i^* v^{*i}$ (i=1,...,n) and $\omega^* = \omega(Q_*)$, then $\omega^* = \sum_{i=1}^{n} \lambda_i^* z^{*i}$, where $(\lambda_1^*,...,\lambda_n^*)$ is a basic optimal solution of (10) (with $z^i = z^{*i}$). Therefore, the point w^* where the halfline from 0 through ω^* meets $Z(Q_*)$ is $w^* = \frac{\|w^*\|}{\|\omega^*\|} \sum_{i=1}^{n} \lambda_i^* \theta_i^* v_i^*$, i.e., $w^* = \sum_{i=1}^{n} \zeta_i^* v^{*i}$ with $\zeta_i^* = \frac{\|w^*\|}{\|\omega^*\|} \lambda_i^* \theta_i^*$. Since $\omega^* \in D$ and D is bounded, while w^* lies on the fixed hyperplane H_0, which does not pass through the origin 0, there exists a constant ν independent of Q_* such that $\|\omega^*\| \le \nu\|w^*\|$, hence $\zeta_i^* \ge \frac{1}{\nu}\lambda_i^*\theta_i^* \ge \frac{1}{\nu}\lambda_i^*$, and the second condition in Lemma VII.2 is implied by the inequality $\min\{\lambda_i^*: \lambda_i^* > 0\} > (1-\rho)\nu$. This provides some theoretical justification for the following rule, which seems to work quite well in practice (cf. Horst and Thoai (1989)) although it does not necessarily guarantee normality of the subdivision process.

$(**)$ *Choose $c > 0$ sufficiently small. Use an ω–subdivision if $\min\{\lambda_i^*: \lambda_i^* > 0\} \ge c$, and a bisection otherwise.*

The advantage of this rule is that it does not require computing Q_*^{-1} or $\sigma(Q_*)$ at each step.

To sum up, the elaborated (η,ρ) device or even the simplified rule (*) is rarely needed in a practical implementation of the normal conical algorithm.

(iv) It follows from Theorem VII.2 that Algorithm VII.1 with $\varepsilon = 0$ will find an exact global optimal solution after finitely many steps, but it may require infinitely many steps to recognize this global optimal solution as such. A similar situation may occur with $\varepsilon > 0$: though a global ε–optimal solution has already been found at a very early stage, the algorithm might have to go through many more steps to check the global ε–optimality of the solution attained. This is not a peculiar feature, but rather a typical phenomenon in these types of methods.

Finally, the assumption that f(x) has bounded level sets can be removed. In fact, if this assumption is not satisfied, the set $\{x: f(x) \geq \alpha\}$ may be unbounded, and the α–extension of a point may lie at infinity. By analogy with the cut and split algorithm, Algorithm VII.1 can be modified to solve the BCP problem as follows.

To each cone K we associate a matrix $Q = (z^1,z^2,\ldots,z^n)$ where z^i is the intersection of the i–th edge of K with the surface $f(x) = \alpha$, if this intersection exists, or the direction of the i–th edge otherwise. Then, in the problem LP(Q,M), the vector e should be understood as a vector whose i–th component is 1 if z^i is a point, or 0 if z^i is a direction. Also, if $I = \{i: z^i \text{ is a point}\}$, then this linear program can be written as

$$\max \sum_{j\in I} \lambda_j$$

$$\text{s.t.} \quad \sum_{j=1}^{n} \lambda_j (Az^j) \leq b \,,\; \lambda_j \geq 0 \quad (j=1,\ldots,n) \,.$$

Aside from these modifications, the algorithm proceeds exactly as before.

1.7. Example VII.1.

We consider the problem in Example VI.2.4:

$$\text{minimize } f(x) \qquad \text{subject to } Ax \leq b,\; x \geq 0 \,,$$

where $x \in \mathbb{R}^4$,

$$f(x) = -\{|x_1 + \tfrac{1}{2}x_2 + \tfrac{2}{3}x_3 + \tfrac{3}{4}x_4|^{3/2} + 0.1(x_1-0.5x_2+0.3x_3+x_4-4.2)^2\}$$

$$A = \begin{bmatrix} 1.2 & 1.4 & 0.4 & 0.8 \\ -0.7 & 0.8 & 0.8 & 0.0 \\ 0.0 & 1.2 & 0.0 & 0.4 \\ 2.8 & -2.1 & 0.5 & 0.0 \\ 0.4 & 2.1 & -1.5 & -0.2 \\ -0.6 & -1.3 & 2.4 & 0.5 \end{bmatrix}, \quad b = \begin{bmatrix} 6.8 \\ 0.8 \\ 2.1 \\ 1.2 \\ 1.4 \\ 0.8 \end{bmatrix}.$$

Tolerance $\varepsilon = 10^{-6}$.

With the practical rule of subdivision (**) indicated in the previous sections the results are as follows.

First cycle

Initialization $\hat{z} = (0,0,0,0)$ $D = \{x: Ax \le b, x \ge 0\}$.

Phase I. $x^0 = (0,0,0,0)$ (nondegenerate vertex of D)

$y^{01} = (0.428571, 0.000000, 0.000000, 0.000000)$

$y^{02} = (0.000000, 0.666667, 0.000000, 0.000000)$

$y^{03} = (0.000000, 0.000000, 0.333333, 0.000000)$

$y^{04} = (0.000000, 0.000000, 0.000000, 1.600000)$

Current best point: $\bar{x} = (0.000000, 0.666667, 0.000000, 0.000000)$ with $f(\bar{x}) = -2.055110$.

Phase II.

0) $\alpha = -2.055111$. The problem is in standard form with respect to x^0; $K_0 = \mathbb{R}^4_+$.

$z^{01} = (1.035485, 0.000000, 0.000000, 0.000000)$

$z^{02} = (0.000000, 0.666669, 0.000000, 0.000000)$

$z^{03} = (0.000000, 0.000000, 29.111113, 0.000000)$

$z^{04} = (0.000000, 0.000000, 0.000000, 8.733334)$

$Q_0 = (z^{01}, z^{02}, z^{03}, z^{04})$. $\mathcal{M}_0 = \mathcal{P}_0 = \{Q_0\}$.

Iteration 1

1) Solution of LP(Q_0,D): $\omega^0 = (1.216328, 1.245331, 0.818556, 2.366468)$

$$f(\omega^0) = -1.440295$$

$$\mu(Q_0) = 3.341739$$

2) $\mathcal{R}_1 = \mathcal{M}_0$

3) $Q_* = Q_0$. Split $K_0 = \text{con}(Q_0)$ with respect to ω^0.

4) $\mathcal{P}_* = \{Q_{1,1}, Q_{1,2}, \ldots, Q_{1,4}\}$. $\mathcal{M}_1 = \mathcal{P}_*$ $\mathcal{P}_1 = \mathcal{P}_*$.

Iteration 2

1) Solution of LP(Q_{1i}, D) (i=1,2,...,4):

$\omega(Q_{11}) = (1.083760, 1.080259, 0.868031, 0.000000)$

with objective function value $-2.281489 < \alpha$.

Cut: $eQ_0^{-1}x \geq 1$, with $eQ_0^{-1} = (0.413885, 0.999997, 0.011450, 0.183206)$

Second cycle (Restart): $M = D \cap \{x: eQ_0^{-1}x \geq 1\}$

Phase I:

$x^0 = (1.169415, 1.028223, 0.169811, 4.861582)$ (nondegenerate vertex of M),

$y^{01} = (1.104202, 0.900840, 0.000000, 5.267227)$

$y^{02} = (1.216328, 1.245331, 0.818956, 2.366468)$

$y^{03} = (1.134454, 0.941176, 0.000000, 5.151261)$

$y^{04} = (0.957983, 0.991597, 0.000000, 5.327731)$

Current best point $\bar{x} = (1.083760, 1.080259, 0.868031, 0.00000)$ with $f(\bar{x}) = -2.281489$.

Phase II: After 20 iterations (generating 51 subcones in all) the algorithm finds the global optimal solution (1.083760, 1.080259, 0.868031, 0.000000) with objective function value –2.281489.

Thus, the global optimal solution is encountered at the end of the first cycle (with two iterations), but checking its optimality requires a second cycle with twenty more iterations.

For an account of the first computational experiments with the normal conical algorithm we refer the reader to Horst and Thoai (1989). Numerical results for a number of problems with n ranging from 5 to 50 and with different objective functions of the forms mentioned in Section VI.2.4, are summarized in the table below (the column f(x): form of the objective function; Res.: number of cycles; Con: number of cones generated; Bi: number of bisections performed). The time includes CPU time and time for printing intermediate results (the algorithm was coded in FORTRAN 77 and run on an IBM–PS II, Model 80, Dos 3.3).

n	m	$f(x)$	Res	Con	Bi	Time (sec.)
5	15	(1)	3	25	2	6.42
5	21	(4)	1	2	0	1.86
8	21	(1)	3	31	3	22.30
9	27	(5)	5	105	10	90.80
9	13	(3)	2	15	0	14.58
10	12	(2)	1	2	0	7.39
10	20	(4)	12	208	15	180.98
10	30	(8)	3	46	0	68.63
11	17	(3)	4	50	0	54.03
12	11	(6)	4	49	2	74.38
12	18	(1)	5	71	5	110.21
20	18	(1)	8	123	7	436.59
20	20	(4)	2	33	0	150.21
20	13	(7)	20	72	12	1020.72
25	18	(2)	2	14	0	170.02
30	14	(3)	15	317	10	1069.72
30	22	(6)	9	133	0	1800.23
32	18	(2)	5	55	10	856.71
40	20	(6)	6	70	12	2012.38
50	21	(6)	6	172	0	8029.24
50	20	(2)	5	105	0	3838.75

Table VII.1

1.8. Alternative Variants

In this section we discuss two alternative variants of the normal conical algorithm.

I. Branch and Bound Variant

Note that the above Algorithm VII.1 operates in the same manner as a branch and bound algorithm, although the number $\mu(Q)$ associated with each cone $K = \text{con}(Q)$ is not actually a lower bound for $f(M \cap K)$, as required by the conventional branch and bound concept. This is so because Phase II solves a (DG)–problem with $G = \{x: f(x) \geq \alpha\}$ and, as seen in Section VII.1.1, $\mu(Q)$ is an upper bound for $p(M \cap K)$, where $p(x)$ is the gauge of the convex set G.

Because of this observation, we see that an alternative variant of the normal conical algorithm can be developed where a lower bound for $f(M \cap K)$ is used in place of $\mu(Q)$ to determe the least promising cone, to be deleted (Step 2), and the selection of the most promising one, to be subdivided further (Step 3).

Let $Q = (z^1,z^2,...,z^n)$ and $\tilde{z}^i = \mu(Q)z^i$, i.e., $\tilde{z}^i$ is the intersection of the hyperplane through $\omega(Q)$ with the ray through z^i. Since the simplex $[0,\tilde{z}^1,\tilde{z}^2,...,\tilde{z}^n]$ entirely contains $M \cap K$, and $f(\tilde{z}^i) < f(0)$ $(i=1,2,...,n)$, the concavity of $f(x)$ implies that $f(M \cap K) \geq \min \{f(\tilde{z}^i), i=1,2,...,n\}$. Hence, if we define $\beta(Q)$ inductively, starting with (Q_0), by the formula

$$\beta(Q) = \begin{cases} \alpha & \text{if } \mu(Q) \leq 1; \\ \max \{\beta(Q_{anc}), \min [f(\tilde{z}^1), f\tilde{z}^2), ..., f(\tilde{z}^n)]\} & \text{if } \mu(Q) > 1 , \end{cases}$$

where Q_{anc} denotes the immediate ancestor of Q, then clearly $\beta(Q)$ yields a lower bound for $\min f(M \cap K)$ such that $\beta(Q) \leq \beta(Q')$ whenever $\text{con}(Q')$ is a subcone of $\text{con}(Q)$.

Based on this lower bound, we should delete any $Q \in \mathcal{P}$ such that $\beta(Q) \geq \alpha$, and choose for further splitting $Q_* \in \text{argmin} \{\beta(Q): Q \in \mathcal{R}\}$. Note, however, that $\beta(Q) > \alpha$ if and only if $\mu(Q) \leq 1$, so in practice a change will occur only in the selection of Q_*.

Theorem VII.3. *The conclusions of Theorems VII.1 and VII.2 still hold for the variant of the normal conical algorithm where $\beta(Q)$ is used in place of $\mu(Q)$ in Steps 2 and 3.*

Proof. It suffices to show that Phase II can be infinite only if $\varepsilon = 0$ and $\gamma = \min f(D)$. To this end consider any infinite nested sequence of cones $K_s = \text{con}(Q_s)$, s=1,2,... generated in Phase II (by Corollary IV.1, such a sequence exists if Phase II is infinite). Let $\omega^s = \omega(Q_s)$ and, as before, denote by $\hat{\omega}^s$ and q^s, respectively, the α-extension of ω^s and the intersection of the hyperplane $eQ_s^{-1}x = 1$ with the ray through ω^s. Then $f(\omega^s) \geq \gamma = \alpha+\varepsilon$ (see Step 1) and $\omega^s \in [q^s,\hat{\omega}^s]$. Since $q^s - \hat{\omega}^s \longrightarrow 0$ $(s \longrightarrow \infty)$ by the normality condition, we deduce that $\omega^s - \hat{\omega}^s \longrightarrow 0$, and hence, by the uniform continuity of f(x) on the compact set $G = \{x\colon f(x) \geq \alpha\}$, we have $f(\omega^s) - f(\hat{\omega}^s) \longrightarrow 0$, i.e., $f(\omega^s) \longrightarrow \alpha$. We thus arrive at a contradiction, unless $\varepsilon = 0$.

On the other hand, by the same argument as in the proof of Proposition VII.1, we can show, by passing to subsequences if necessary, that $\mu(Q_s) \longrightarrow 1$ $(s \longrightarrow \infty)$. Let $Q_s = (z^{s1},...,z^{sn})$, $\tilde{z}^{si} = \mu(Q_s)z^{si}$. Then $\tilde{z}^{si} - z^{si} \longrightarrow 0$, and hence $f(\tilde{z}^{si}) - f(z^{si}) \longrightarrow 0$, i.e. $f(\tilde{z}^{si}) \longrightarrow \alpha$. Since from the definition of $\beta(Q)$ we have

$$\gamma = \alpha > \beta(Q_s) \geq \min \{f(\tilde{z}^{si})\colon i=1,2,...,n\}$$

it follows that $\beta(Q_s) \longrightarrow \gamma$ $(s \longrightarrow \infty)$. This proves the consistency of the bounding operation.

Since the selection operation is bound improving, we finally conclude from Theorem IV.2 that

$$\min f(D) = \min f(M) = \gamma.$$

■

II. One Phase Algorithm

In Algorithm VII.1 the starting vertex x^0 is changed at each return to Phase I. The advantage of such restarts was discussed in Section VII.1.6. However, sometimes this advantage can be offset by the computational cost of the transformations necessary to start a new cycle. Especially if the old starting vertex x^0 has proven to be a good choice, one may prefer to keep it for the new cycle. For this reason and also in view of the possibility of extending conical procedures to more general situations (see Section VII.1.9 and Chapter X), it is of interest to consider variants of conical algorithms with fixed starting point $x^0 = 0$. Below we present a one phase algorithm which is close to the algorithm of Thoai and Tuy (1980).

For each cone $K = \mathrm{con}(Q)$, $Q = (z^1,z^2,...,z^n)$, with $f(z^i) = \gamma < f(0)$ $(i=1,2,...,n)$, we define

$$\nu(Q,\gamma) = \min \{f(\tilde{z}^i),\ i=1,2,...,n\} ,$$

where $\tilde{z}^i = \mu(Q)z^i$, and $\mu(Q)$ is the optimal value of the linear program

$$\mathrm{LP}(Q,D) \qquad \max eQ^{-1}x \qquad \text{s.t. } x \in D,\ Q^{-1}x \geq 0.$$

Algorithm VII.1*.

Select $\varepsilon \geq 0$ and the two parameters $\eta_0 \geq 0$, $\rho \in (\frac{\sqrt{3}}{2},1)$ of an NCS rule.

0) Translate the origin 0 to a vertex of D. Let $x^0 = \mathrm{argmin}\ \{f(x)$: $x = 0$ or x is a vertex of D adjacent to $0\}$, $\gamma_0 = f(x^0)$ (assume that $f(0) > \gamma_0$).
Construct a cone $K_0 \supset D$. For each $i=1,2,...,n$ compute the point $z^{0i} \neq 0$ where the i–th edge of K_0 meets the surface $f(x) = \gamma_0$. Set $Q_0 = (z^{01},z^{02},...,z^{0n})$, $\mathcal{M}_0 = \mathcal{P}_0 = \{Q_0\}$. Set $k = 0$.

1) For each $Q \in \mathcal{P}_k$ solve LP(Q,D) to obtain the optimal value $\mu(Q)$ and a basic optimal solution $\omega(Q)$ of this program. Compute $\nu(Q,\gamma_k)$, and let

$$\beta(Q) = \max \{\beta(Q_{k-1}),\ \nu(Q,\gamma_k)\} \quad (k \geq 1), \quad \beta(Q_0) = \nu(Q_0,\gamma_0) .$$

2) Let $\mathcal{R}_k = \{Q \in \mathcal{M}_k\colon \beta(Q) < \gamma_k - \varepsilon\}$. If $\mathcal{R}_k = \emptyset$, terminate: x^k is a global ε–optimal solution. Otherwise,

3) Select $Q_k \in \mathrm{argmin}\ \{\beta(Q)\colon Q \in \mathcal{R}_k\}$ and split it by proceeding as in Step 3 of Algorithm VII.1.

4) Let x^{k+1} be the best point among x^k, $\omega(Q)$ ($Q \in \mathcal{P}_k$), and the point (if it exists) where the splitting ray for con(Q_k) meets the boundary of D. Let $\gamma_{k+1} = f(x^{k+1})$.

Let $\mathcal{P}_{k+1}$ be the partition of Q_k obtained in Step 4). For each $Q \in \mathcal{P}_{k+1}$ reset $Q = (z^1,z^2,...,z^n)$ with z^i a point on the i–th edge of $K = \text{con}(Q)$ such that $f(z^i) = \gamma_{k+1}$ (i=1,2,...,n). Let $\mathcal{M}_{k+1} = (\mathcal{R}_k \setminus \{Q_k\}) \cup \mathcal{P}_{k+1}$. Set $k \leftarrow k+1$ and return to 1).

Theorem VII.4. *Algorithm VII.1* can be infinite only if* $\varepsilon = 0$. *In this case* $\gamma_k \downarrow \gamma = \min f(D)$, *and any accumulation point of* $\{x^k\}$ *is a global optimal solution.*

Proof. First observe that Proposition VII.3 remains valid if the condition $z^{si} \in \partial G$ is replaced by the following weaker one: $z^{si} \in \partial G_s$, where G_s is a convex subset of a convex compact set G, and $0 \in \text{int } G_s \subset \text{int } G_{s+1}$. Now, if the procedure is infinite, it generates at least one infinite nested sequence of cones $K_s = \text{con}(Q_s)$, $s \in \Delta \subset \{0,1,2,...\}$, with $Q_s = (z^{s1},z^{s2},...,z^{sn})$ such that $f(z^{si}) = \gamma_s$ (i=1,2,...,n) and $\gamma_s \downarrow \gamma$ ($s \rightarrow \infty$). As in the proof of Proposition VII.5, one can show that this sequence $\{K_s\}$ is either exhaustive or nondegenerate with $\omega^s \in [0,u^s]$ for all but finitely many s, where $\omega^s = \omega(Q_s)$, $u^s = u(Q_s)$. Therefore, by the extended version of Proposition VII.3, where $G_s = \{x: f(x) \geq \gamma_s\}$, we have (taking a subsequence if necessary) $q^s - \omega^s \rightarrow 0$. Hence, as in the proof of Proposition VII.1, $\mu(Q_s) \rightarrow 1$. This implies that $\underset{s}{\underline{\lim}}\ \nu(Q_s,\gamma_s) \geq \gamma$. But since

$$\gamma_s - \varepsilon > \beta(Q_s) \geq \nu(Q_s,\gamma_s) ,$$

this is possible only if $\varepsilon = 0$. Then $\beta(Q_s) \rightarrow \gamma$, i.e., the lower bounding is consistent. Hence, by Theorem IV.2, $\min f(D) = \gamma$. ■

Remarks VII.3. (i) If the problem is in standard form with respect to x^0, so that $x^0 = 0$ is a vertex of the feasible polytope:

$$D = \{x: Ax \le b, x \ge 0\} ,$$

then one can take $K_0 = \mathbb{R}^n_+$ and the linear program LP(Q,D), where $Q = (z^1,...,z^n)$, can be written as

$$\max \sum_j \lambda_j \quad \text{s.t.} \sum_j \lambda_j Az^j \le b, \lambda_j \ge 0 \quad (j=1,2,...,n) .$$

(ii) As with Algorithm VII.1, the efficiency of the procedure critically depends upon the choice of rule for cone subdivision. Although theoretically an NCS rule is needed to guarantee convergence, in practice a simple rule like (**) in Section VII.1.6 usually works.

1.9. Concave Minimization Problem with Convex Constraints

An advantage of the one phase Algorithm VII.1* is that it can easily be extended to the general concave programming problem:

(CP) minimize f(x) subject to $x \in D$,

where $f: \mathbb{R}^n \to \mathbb{R}$ is a concave function and D is a closed convex set defined by the inequality

$$g(x) \le 0 ,$$

with $g: \mathbb{R}^n \to \mathbb{R}$ a convex function.

Assume that the constraint set D is compact and that int $D \neq \emptyset$.

When extending Algorithm VII.1* to problem (CP), the key point is to develop a method for estimating a bound $\beta(Q) \le \min f(D \cap K)$ for each given cone $K = \text{con}(Q)$ such that the lower bounding process is consistent.

Tuy, Thieu and Thai (1985) proposed cutting the cone K by a supporting hyperplane of the convex set D, thus generating a simplex containing $D \cap K$. Then the minimum of f(x) over this simplex provides a lower bound for $\min f(D \cap K)$. This method is simple and can be carried out easily (it does not even involve solving a linear program); furthermore, it applies even if D is unbounded. However, it is practical only for relatively small problems.

A more efficient method was developed by Horst, Thoai and Benson (1988) (see also Benson and Horst (1988)). Their basic idea was to combine cone splitting with outer approximation in a scheme which can roughly be described as a conical procedure of the same type as Algorithm VII.1*, in which lower bounds are computed using an adaptively constructed sequence of outer approximating polytopes $D_0 \supset D_1 \supset ... \supset D$.

The algorithm we are going to present is a modified version of the original method of Horst, Thoai and Benson. The modification consists mainly in using an NCS process instead of a pure bisection process for cone subdivision.

Algorithm VII.2 (Normal Conical Algorithm for CP).

Assume that $f(0) > \min f(D)$ and that a cone K_0 is available such that for any $x \in K_0 \setminus \{0\}$ the ray $\{\tau x: \tau > 0\}$ meets D. Denote by $\bar{x}$ the point θx where $\theta = \sup \{\tau: \tau x \in D\}$. Select $\varepsilon \geq 0$ and two parameters $\eta_0 \geq 0$, $\rho \in (\frac{\sqrt{3}}{2}, 1)$.

0) For each i=1,2,...,n take a point $y^i \neq 0$ on the i–th edge of K_0 and compute the corresponding point $\bar{y}^i = \theta_i y_0$. Let $x^0 \in \text{argmin} \{f(\bar{y}^i), i=1,2,...,n\}$, $\gamma_0 = f(x^0)$, and let z^{0i} be the γ_0–extension of y^i (i=1,...,n) (cf. Definition V.1). Set $Q_0 = (z^{01},...,z^{0n})$ and $\mathscr{M}_0 = \mathscr{P}_0$. Construct a polytope $D_0 \supset D$. Set $k = 0$.

1) For each $Q \in \mathscr{P}_k$, $Q = (z^1,z^2,...,z^n)$, solve the linear program

$$LP(Q,D_k) \quad \max eQ^{-1}x \quad \text{s.t.} \quad x \in D_k, \; Q^{-1}x \geq 0$$

to obtain the optimal value $\mu(Q)$ and a basic optimal solution $\omega(Q)$ of this program; compute $\nu(Q,\gamma_k) = \min\ (f(\tilde{z}^i), i=1,2,...,n\}$, where $\tilde{z}^i = \mu(Q)z^i$, and let

$$\beta(Q) = \max\ \{\beta(Q_{k-1}), \nu(Q,\gamma_k)\}\ (k \geq 1),\quad \beta(Q_0) = \nu(Q_0,\gamma_0)\ .$$

2) Let $\mathcal{R}_k = \{Q \in \mathcal{M}_k: \beta(Q) < \gamma_k - \varepsilon\}$. If $\mathcal{R}_k = \emptyset$, then terminate: x^k is a global ε-optimal solution. Otherwise, go to 3).

3) Select $Q_k \in \operatorname{argmin}\ \{\beta(Q): Q \in \mathcal{R}_k\}$ and split it as in step 3 of Algorithm VII.1.

4) If $\omega^k \in D$, set $D_{k+1} = D_k$. Otherwise, take a vector $p^k \in \partial g(\bar{\omega}^k)$ (recall that $\bar{x} = \theta x$, with $\theta = \sup\{\tau : \tau x \in D\}$), and form

$$D_{k+1} = D_k \cap \{x: p^k(x - \bar{\omega}^k) \leq 0\}\ .$$

5) Let x^{k+1} be the best point among x^k, $\bar{u}(Q_k)$ and all $\bar{\omega}(Q)$ for $Q \in \mathcal{P}_k$; let $\gamma_{k+1} = f(x^{k+1})$.

Denote the partition of Q_k obtained in 3) by $\mathcal{P}_{k+1}$. For each $Q \in \mathcal{P}_{k+1}$ reset $Q = (z^1,z^2,...,z^n)$ with $z^i \neq 0$ a point where the i-th edge of $K = \operatorname{con}(Q)$ meets the surface $f(x) = \gamma_{k+1}$.

Let $\mathcal{M}_{k+1} = (\mathcal{R}_k \setminus \{Q_k\}) \cup \mathcal{P}_{k+1}$. Set $k \leftarrow k+1$ and return to 1).

Theorem VII.5. *Algorithm VII.2 can be infinite only if $\varepsilon = 0$. In this case $\gamma_k \downarrow \gamma$, and every accumulation point of the sequence $\{x^k\}$ is a global optimal solution of (CP).*

Proof. Consider any infinite nested sequence of cones $K_s = \operatorname{con}(Q_s)$, $s \in \Delta \subset \{0,1,2,...\}$, with $Q_s = (z^{s1},z^{s2},...,z^{sn})$, such that $f(z^{si}) = \gamma_s$ (i=1,2,...,n). Let $\gamma = \lim_{k\to\infty} \gamma_k = \lim \gamma_s$ $(s \in \Delta,\ s \to \infty)$.

For each s denote by q^s, $\hat{\omega}^s$ the points where the halfline from 0 through ω^s meets the simplex $[z^{s1},...,z^{sn}]$ and the surface $f(x) = \gamma_s$, respectively; also recall that $\bar{\omega}^s$ is the point where this halfline meets the boundary ∂D of D. Since $f(\hat{\omega}) = f(z^{si}) = \gamma_s$ and $\gamma_s \downarrow \gamma$, as in the proof of Theorem VII.4 it can be seen that the sequence

$K_s = con(Q_s)$ is normal, i.e., $\underline{\lim}\,(q^s - \hat{\omega}^s) = 0$. Hence, in view of the boundedness of D_0, we may assume, by taking subsequences if necessary, that q^s, $\hat{\omega}^s$ tend to a common limit q^∞, while ω^s, $\overline{\omega}^s$ tend to ω^∞, $\overline{\omega}^\infty$, respectively. We claim that $\omega^\infty = \overline{\omega}^\infty$.

Indeed, this is obvious if $\omega^s \in D$ for infinitely many s. Otherwise, for all sufficiently large s we must have $\omega^s \notin D$; and, according to Step 4, each ω^s is strictly separated from D by a hyperplane $\ell_s(x) := p^s(x - \overline{\omega}^s) = 0$ such that

$$\ell_s(\omega^s) > 0\,, \quad \ell_s(x) \le 0 \quad \forall x \in D\,.$$

Therefore, by Theorem II.2, $\lim\limits_{s\to\infty} \omega^s \in D$, i.e., $\omega^\infty = \overline{\omega}^\infty$ in any case. But clearly, $\gamma_{k+1} \le f(\overline{\omega}^k)$ for any k. By passing to the limit, this yields $\gamma \le f(\overline{\omega}^\infty)$, i.e. $\omega^\infty \in [0,q^\infty]$. On the other hand, since $q^s \in [0,\omega^s]$, we have $q^\infty \in [0,\omega^\infty]$. Consequently, $q^\infty = \omega^\infty$, i.e., $q^s - \omega^s \longrightarrow 0$. As illustrated in Fig. VII.1, we then have

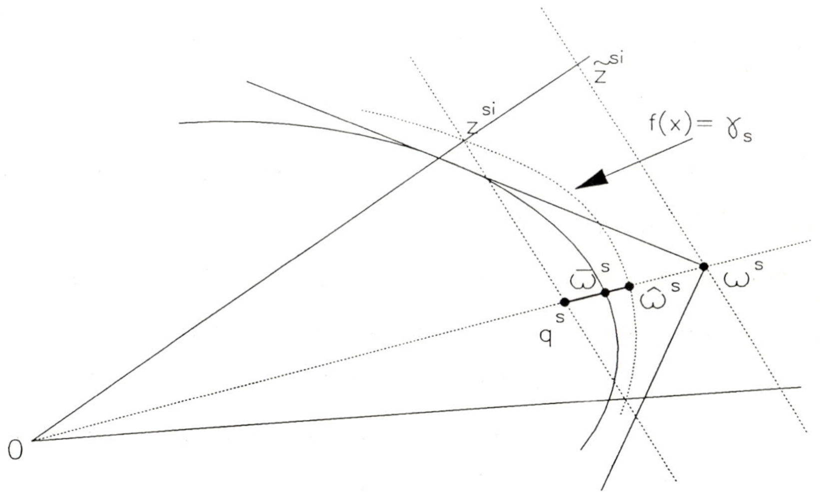

Fig. VII.1

$$\mu(Q_s) = 1 + \frac{\|q^s - \omega^s\|}{\|q^s\|} \longrightarrow 1 \quad (s \longrightarrow \infty).$$

This implies that $\tilde{z}^{si} - z^{si} \longrightarrow 0$, where $\tilde{z}^{si} = \mu(Q_s)z^{si}$. Since $f(z^{si}) = \gamma_s$, it then follows from the definition of $\nu(Q_s,\gamma_s)$ that $\underline{\lim}\ \nu(Q_s,\gamma_s) \geq \gamma$. Hence, noting that $\gamma_s - \varepsilon > \beta(Q_s) \geq \nu(Q_s,\gamma_s)$, we arrive at a contradiction as $s \longrightarrow \infty$, if $\varepsilon > 0$. Therefore, the algorithm can be infinite only if $\varepsilon = 0$. But then $\beta(Q_s) \uparrow \gamma$, i.e., the lower bounding is consistent. Since the candidate selection here is bound improving, the conclusion follows from Theorem IV.2. ∎

Remarks VII.4. (i) The above algorithm differs from Algorithm VII.1* only in the presence of Step 4. If D is a polytope and we take $D_0 = D$, then $D_k = D\ \forall k$, and the above algorithm reduces exactly to Algorithm VII.1*.

(ii) If $D_0 = \{x: Ax \leq b\}$ and $D_k = \{x: Ax \leq b,\ Cx \leq d\}$, then for $Q = (z^1,z^2,\ldots,z^n)$ the linear program $LP(Q;D_k)$ is

$$\max \Sigma\, \lambda_j \quad \text{s.t. } \Sigma\, \lambda_j(Az^j) \leq b,\ \ \Sigma\, \lambda_j(Cz^j) \leq d,\ \ \lambda_j \geq 0\ \ \forall j\,.$$

To avoid inverting matrices like Q, one can use a simplified rule like (**) of Section VII.1.6 for the cone splitting procedure.

(iii) In the general case when an initial cone K_0 as defined above is not available, Step 0) should be modified as follows.

By translating if necessary, assume that $0 \in \text{int } D$. Take an n–simplex $[y^1,\ldots,y^{n+1}]$ containing 0 in its interior. Let $x^0 \in \text{argmin } \{f(\bar{y}^i),\ i=1,\ldots,n+1\}$, $\gamma = f(x^0)$ and let z^{0i} be the γ_0–extension of y^i $(i=1,\ldots,n+1)$. For each $i=1,\ldots,n+1$ let Q_{0i} be the matrix with columns z^{0j}, $j \in \{1,\ldots,n+1\}\setminus\{i\}$. Set $\mathcal{M}_0 = \mathcal{P}_0 = \{Q_{0i},\ i=1,\ldots,n+1\}$. Construct a polytope $D_0 \supset D$. Set k=0.

1.10. Unbounded Feasible Domain

The previous algorithms can be extended to problems with unbounded feasible domain.

Consider, for example, Algorithm VII.2. If D is unbounded, then D_k is unbounded, and in Step 2 the linear program $LP(Q,D_k)$ may have no finite optimal solution. That is, there is the possibility that $\mu(Q) = +\infty$, and $\omega^k = \omega(Q_k)$ is not a finite point but rather the direction of a halfline contained in D_k. If we assume, as before, that the function f(x) has bounded level sets (so that, in particular, it is unbounded from below over any halfline), Algorithm VII.2 will still work, provided we make the following modifications:

When $\omega^k = \omega(Q_k)$ is a direction, the γ_k-extension of ω^k should be understood as a point $\hat{\omega}^k = \lambda\omega^k$ such that $f(\lambda\omega^k) = \gamma_k$. Moreover, in Step 4, if ω^k is a recession direction of D, then the algorithm terminates, because $\inf f(D) = -\infty$ and f(x) is unbounded from below on the halfline in the direction ω^k. Otherwise, the halfline parallel to ω^k intersects the boundary of D at a unique point $\bar{\omega}^k \neq 0$. In this case let $p^k \in \partial g(\bar{\omega}^k)$, and

$$D_{k+1} = D_k \cap \{x: p^k(x-\bar{\omega}^k) \leq 0\} .$$

Theorem VII.6. *Assume that the function f(x) has bounded level sets. For $\varepsilon = 0$ consider Algorithm VII.2 with the above modifications in Steps 2 and 4. Then*

$$f(x^k) \downarrow \inf f(D) .$$

Proof. It suffices to consider the case where $\gamma_k \downarrow \gamma > -\infty$. Let $K_s = con(Q_s)$ be any infinite nested sequence of cones, with $Q_s = (z^{s1},z^{s2},\ldots,z^{sn})$, $f(z^{si}) = \gamma_s$ $(i=1,2,\ldots,n)$. We claim that $\mu(Q_s) = +\infty$ for at most finitely many s. Indeed, suppose the contrary, so that, by taking a subsequence if necessary, we have $\mu(Q_s) = +\infty$ $\forall s$, i.e., each $\omega^s = \omega(Q_s)$ is a direction. Again by passing to subsequences if necessary, we may assume that $\omega^s/\|\omega^s\| \longrightarrow u$. Then, arguing as in the proof of Lemma VI.3, it can be seen that the halfline Γ in the direction u lies entirely in D.

Let W be the intersection of K_0 with a ball around 0, which is small enough so that $W \subset D$, and hence $W+\Gamma \subset D$. Since f(x) is unbounded from below on Γ, we can take a ball B around some point of Γ such that $B \subset W+\Gamma \subset D$ and $f(x) < \gamma \ \forall x \in B$. Then for all sufficiently large s, the ray in the direction ω^s will intersect B at some point $x \in D$ with $f(x) < \gamma$. This implies that $f(\overline{\omega}^s) < \gamma \le \gamma_{s+1}$, contradicting the definition of γ_{s+1}. Therefore, we have $\mu(Q_s) < +\infty$ for all but finitely many s. Arguing as in the first part of the proof of Theorem VII.5, we then can show that $q^s - \hat{\omega}^s \longrightarrow 0$. Since $f(\overline{\omega}^s) \ge \gamma$, the sequence $\{\overline{\omega}^s\}$ is bounded.
If $\|\omega^s - \overline{\omega}^s\| \ge \eta > 0$ for infinitely many s, then, by taking the point $y^s \in [\overline{\omega}^s, \omega^s]$ such that $\|y^s - \overline{\omega}^s\| = \eta$, we have $\ell_s(y^s) > 0$, $\ell_s(x) < 0 \ \forall x \in D$. Hence, by Theorem III.2, $\lim y \in D$, i.e., $y^s - \overline{\omega}^s \longrightarrow 0$, a contradiction. Therefore, $\omega^s - \overline{\omega}^s \longrightarrow 0$, and, as in the last part of the proof of Theorem VII.5, we conclude that $\beta(Q_s) \uparrow \gamma$.

We have thus proved that the lower bounding is consistent. The conclusion of the theorem then follows from the general theory of branch and bound methods (Theorem IV.2). ■

1.11. A Class of Exhaustive Subdivision Processes

In the general theory of branch and bound discussed in Chapter IV, we have seen that exhaustive subdivisions play an important role.

The simplest example of an exhaustive subdivision process is bisection. Since the convergence of conical (and also simplicial) algorithms using bisection is often slow, it is of interest to develop alternative exhaustive subdivision processes.

From Corollary VII.3 we already know an exhaustive subdivision process which involves infinitely many bisections. Now we present another class of exhaustive subdivisions in which a cone is most often split into more than two subcones. This class was introduced by Utkin (see Tuy, Khachaturov and Utkin (1987) and also

Utkin, Khachaturov and Tuy (1988)).

Given an (n–1)–simplex Z_0 which is the section of the initial cone by a fixed hyperplane, we wish to define a subdivision process on Z_0 such that any infinite nested sequence of simplices $\{Z_s\}$ generated by such a process shrinks to a single point. Clearly, the conical subdivision induced by this simplicial subdivision will then be exhaustive.

Recall from Section IV.3 that a radial subdivision process is specified by giving a function w(Z) which assigns to each (n–1)–simplex $Z = [v^1,v^2,...,v^n]$ a point $w(Z) \in Z \setminus V(Z)$, where $V(Z) = \{v^1,v^2,...,v^n\}$. If

$$w(Z) = \sum_{i=1}^{n} \lambda_i v^i \quad , \lambda_i \geq 0 \quad , \sum_{i=1}^{n} \lambda_i = 1 \; ,$$

$$I(Z) = \{i: \lambda_i > 0\} \; ,$$

then Z is divided into subsimplices Z(i,w) that for each $i \in I(Z)$ are formed by substituting w(Z) for v^i in the vertex set of Z. In order to have an exhaustive radial subdivision process, certain conditions must be imposed on the function w(Z). Let us denote

$$\delta(Z) = \max_{i<j} \|v^i - v^j\| \; ,$$

$$\delta(i,Z) = \max_{j} \|v^i - v^j\| \quad (i=1,2,...,n) \; .$$

We see that $\delta(Z)$ is the diameter (the length of the longest edge) of Z, whereas $\delta(i,Z)$ is the length of the longest edge of Z incident to the vertex v^i. The following conditions define an exhaustive class of subdivisions.

There exists a constant ρ, $0 < \rho < 1$, such that for any simplex $Z = [v^1,v^2,...,v^n]$:

$$\|w(Z) - v^i\| \leq \rho\delta(Z) \quad \forall i \in I(Z) \; , \tag{11}$$

while the set I(z) defined above satisfies

$$I(Z) = \{i: \delta(i,Z) > \rho\delta(Z)\} \; . \tag{12}$$

Proposition VII.6. *A radial subdivision process defined by a function $w(Z)$ satisfying the above conditions (11) and (12) is exhaustive.*

The proof of Proposition VII.6 will follow from several lemmas.

Consider any nested sequence of simplices $\{Z_s\}$, each of which is an immediate descendant of the previous one by means of a subdivision defined by $w(.)$. Denote by i_s the index of the vertex of Z_s that is to be replaced by $w(Z_s)$ to generate Z_{s+1}. In other words, if $Z_s = [v^{s1},v^{s2},\ldots,v^{sn}]$, then the vertices of Z_{s+1} are $v^{s+1,i} = v^{si}$ $(i \neq i_s)$, $v^{s+1,i_s} = w(Z_s)$, so that the new vertex in Z_{s+1} receives the index of the vertex of Z_s that it replaces. Obviously, $i_s \in I(Z_s)$.

Lemma VII.3. *For every $s \geq 1$ we have*

$$\delta(i_s, Z_{s+1}) \leq \rho\delta(Z_s).$$

Proof. It suffices to show that $\|w(Z_s)-v^{s+1,i}\| \leq \rho\delta(Z_s)$ for all $i \in \{1,\ldots,n\} \setminus \{i_s\}$. But for $i \in I(Z_s) \setminus \{i_s\}$ this follows from (11), while for $i \in \{1,\ldots,n\} \setminus I(Z_s)$ we have, by (12):

$$\|w(Z_s)-v^{s+1,i}\| = \|\Sigma \lambda_{sj} v^{sj}-v^{si}\| \leq \Sigma \lambda_{sj} \|v^{sj}-v^{si}\|$$

$$\leq \Sigma \lambda_{sj}\rho\delta(Z_s) \leq \rho\delta(Z_s) ,$$

where λ_{sj} are the coefficients expressing $w(Z_s)$ as a convex combination of the vertices of Z_s. ■

Lemma VII.4. *For every s at least one of the following relations holds:*

$$\delta(Z_{s+1}) < \delta(Z_s) , \tag{13}$$

$$|I(Z_{s+1})| < |I(Z_s)| . \tag{14}$$

Proof. Obviously, $\delta(Z_{s+1}) \leq \delta(Z_s)$. Suppose that

$$\delta(Z_{s+1}) = \delta(Z_s) .$$

Since Z_{s+1} is obtained from Z_s by replacing the vertex of index i_s by $w(Z_s)$, for every i we can write:

$$\delta(i,Z_{s+1}) \leq \max \{\delta(i,Z_s), \delta(i_s,Z_{s+1})\} . \tag{15}$$

Hence, if $i \in I(Z_{s+1})$, i.e., $\delta(i,Z_{s+1}) > \rho\delta(Z_{s+1}) = \rho\delta(Z_s)$, then by the previous lemma, $\delta(i,Z_{s+1}) = \delta(i,Z_s)$, and consequently $\rho\delta(i,Z_s) > \delta(Z_s)$, i.e., $i \in I(Z_s)$. Therefore,

$$I(Z_{s+1}) \subset I(Z_s) .$$

But $i_s \in I(Z_s)$, while from (11) we have $\delta(i_s,Z_{s+1}) \leq \rho\delta(Z_s) = \rho\delta(Z_{s+1})$, which implies that $i_s \notin I(Z_{s+1})$. This proves (14). ■

Let us now define

$$I_s = \{i: \delta(i,Z_s) > \rho\delta(Z_1)\} .$$

Clearly, $I_s = \emptyset$ means that $\delta(Z_s) \leq \rho\delta(Z_1)$.

Lemma VII.5. *For $s \geq 1$ suppose that $I_s \neq \emptyset$. Let $h_s = |I(Z_s)| - |I_s| + 1$. Then for all $h \geq h_s$ we have*

$$|I_{s+h}| < |I_s| . \tag{16}$$

Proof. First note that for every $s \geq 1$ one has

$$I_{s+1} \subset I_s \subset I(Z_s) . \tag{17}$$

Indeed, the second inclusion follows from the inequality $\delta(Z_1) \geq \delta(Z_s)$. To check the first, observe that if $i \notin I_s$, i.e., if $\delta(i,Z_s) \leq \rho\delta(Z_1)$, then from (15) and Lemma VII.3 (which implies that $\delta(i_s,Z_{s+1} \leq \rho\delta(Z_1))$ we derive the inequality $\delta(i,Z_{s+1}) \leq \rho\delta(Z_1)$; hence $i \notin I_{s+1}$. From (17) it follows, in particular, that $h_s \geq 1$.

Now suppose that for some $h \geq 1$

$$I_{s+h} = I_s , \tag{18}$$

while

$$\delta(Z_s) = \delta(Z_{s+1}) = \dots = \delta(Z_{s+h}) . \tag{19}$$

From Lemma VII.4 it then follows that $|I(Z_{s+h})| \leq |I(Z_s)| - h$; hence, by (17) and (18),

$$h \leq |I(Z_s)| - |I(Z_{s+h})| \leq |I(Z_s)| - |I_{s+h}| = |I(Z_s)| - |I_s| = h_s - 1 .$$

Therefore, for $h \geq h_s$ either (18) or (19) does not hold. If (18) does not hold, then in view of (17) we must have (16). Otherwise, there exists r such that $s \leq r < s+h$ and

$$\delta(Z_s) = \dots = \delta(Z_r) > \delta(Z_{r+1}) . \tag{20}$$

If $I_r = \emptyset$, then (16) trivially holds, since $I_s \neq \emptyset$. Consequently, we may assume that $I_r \neq \emptyset$. From (20) it follows that $\delta(i_r, Z_r) = \delta(Z_r)$, and hence

$$\delta(i_r, Z_r) \geq \delta(i, Z_r) \quad \forall i .$$

In particular, by taking an $i \in I_r$ we can write

$$\delta(i_r, Z_r) \geq \delta(i, Z_r) > \rho\delta(Z_1) .$$

This implies that $i_r \in I_r$. But, by Lemma VII.3, $\delta(i_r, Z_{r+1}) \leq \rho\delta(Z_r) \leq \rho\delta(Z_1)$, so that $i_r \notin I_{r+1}$. Taking account of (17), we then deduce that $|I_{r+1}| < |I_r|$, and hence (16) holds since $|I_{s+h}| \leq |I_{r+1}|$ and $|I_r| \leq |I_s|$. This completes the proof of Lemma VII.5.

Lemma VII.6. *For $p = n(n-1)/2$ and any $s \geq 1$ we have*

$$\delta(Z_{s+p}) \leq \rho\delta(Z_s) . \tag{21}$$

Proof. Without loss of generality we may assume that $s = 1$. Consider a sequence of integers $s = h_0 < h_1 < \ldots < h_\nu$ such that

$$|I_s| > |I_{h_1}| > \ldots > |I_{h_\nu}| = 0 .$$

According to Lemma VII.5, such a sequence can be obtained inductively by defining

$$h_j = h_{j-1} + |I(Z_{h_{j-1}}| - |I_{h_{j-1}}| + 1 \quad (j=1,2,\ldots,\nu) .$$

Since any nonempty I_t has at least two elements, it follows that $|I_{h_{\nu-1}}| \geq 2$. Hence, $\nu \leq n-1$ and $|I_{h_{j-1}}| \geq 2 + (\nu-j)$. Substituting into the expression for h_j we obtain

$$h_j \leq h_{j-1} + n - (\nu-j) - 1 \quad (j=1,2,\ldots,\nu) .$$

Therefore, $h_\nu \leq h_{\nu-1} + n-1 \leq h_{\nu-2} + (n-2) + (n-1) \leq \ldots \leq h_0 + (n-\nu) + \ldots + (n-2) + (n-1) \leq s + n(n-1)/2$. Thus, $I_{s+p} = \emptyset$, implying the inequality (21). ■

From this lemma it readily follows that $\delta(Z_s) \longrightarrow 0$ as $s \longrightarrow \infty$. Hence, Proposition VII.6 is established.

Example VII.2. Conditions (11) (12) are satisfied, with $\rho = 1-1/n$, if for each (n–1)–simplex Z we let

$I(Z) = \{i\colon \delta(i,Z) > (1-1/n)\delta(Z)\}$, $\quad w(Z) = \Sigma\, \lambda_i v^i$, where $\lambda_i = 0$ for $i \notin I(Z)$, $\lambda_i = 1/|I(Z)|$ for $i \in I(Z)$.

Indeed, condition (12) is obvious, while for $i \in I(Z)$ we have, by setting $\nu = 1/|I(Z)|$:

$$\|w(Z)-v^i\| = \|\Sigma_j \nu v^j - v^i\| \leq \Sigma_j \nu \|v^j - v^i\| \leq \Sigma_j \nu\rho\delta(Z) \leq \rho\delta(Z) ,$$

where Σ_j denotes the sum over all $j \in I(Z)$.

Remark VII.5. The estimate (21) shows that the smaller the constant ρ, the faster the diameters of the subsimplices will go to zero. In the example above, $\rho = 1-1/n \longrightarrow 1$ as $n \longrightarrow \infty$. However, using the Young inequality for the radius of the smallest ball containing a compact set with a given diameter (cf., e.g., Leichtweiß (1980)), one can show the existence of a subdivision process of the type (11) and (12) with $\rho = [(n-1)/2n]^{1/2} \le 2^{-1/2}$.

1.12. Exhaustive Nondegenerate Subdivision Processes

In certain applications, an ε–optimal solution in terms of the objective may not provide an adequate answer to the original problem, so that finite convergence may be a desirable property to achieve in concave minimization over polytopes.

To generate finitely convergent conical algorithms, Hamami and Jacobsen (1988) (cf. also Hamami (1982)) introduced a subclass of exhaustive subdivision processes which satisfy a nondegeneracy condition that is stronger than the one defined in Section VII.1.3.

Given the initial cone $K_0 = \mathbb{R}^n_+$ and the (n–1)–simplex $Z_0 = [e^1,e^2,...,e^n]$, which is the intersection of K_0 with the hyperplane $x_1+x_2+...+x_n = 1$, we associate to each subsimplex $Z_k = [v^{k1},v^{k2},...,v^{kn}]$ of Z_0 the $n \times n(n-1)/2$ matrix

$$B_k = \left[\frac{v^{ki}-v^{kj}}{\|v^{ki}-v^{kj}\|} \ , \ i < j \right] .$$

The columns of B_k represent the unit vectors in the edge directions of the (n–1)–simplex Z_k, and therefore, $\text{rank}(B_k) = n-1$.

Definition VII.3. *An exhaustive subdivision process on Z_0 is said to be* ***nondegenerate*** *(briefly, an END process) if for any infinite nested sequence of simplices $Z_k = [v^{k1},v^{k2},...,v^{kn}]$ obtained by the subdivision method, any convergent subsequence of the associated sequence $\{B_k\}$ converges to a matrix B of rank $n-1$.*

Basically, this property ensures that when $Q_k = (z^{k1}, z^{k2}, \ldots, z^{kn})$, where z^{ki} is the intersection of the ray through v^{ki} with the boundary of a given bounded convex set G whose interior contains 0, then, under mild conditions, the vector eQ_k^{-1} tends to a normal to G at the point $z^* = \lim z^{ki}$ $(k \to \infty)$.

To show this, we need some lemmas.

Lemma VII.7. *Let C be a convex bounded set containing 0 in its interior. Then the mapping π that associates to each point $x \neq 0$ the intersection $\pi(x)$ of the boundary of C with the ray (from 0) through x, is Lipschitzian relative to any compact subset S of $\mathbb{R}^n$ not containing 0.*

Proof. If p(x) denotes the gauge of C, then $\pi(x) = x/p(x)$, so that

$$\|\pi(x') - \pi(x'')\| \leq \frac{\|x'-x''\|}{p(x')} + \|x''\| \frac{\|p(x'')-p(x')\|}{p(x')p(x'')} .$$

From this the lemma follows, since S is compact, the convex function p(x) is Lipschitzian relative to S (cf. Rockafellar (1970), Theorem 10.4) and, moreover, $\|x\|$ is bounded from above, while $\|p(x)\|$ is bounded from below on S. ∎

Now consider any infinite nested sequence of simplices $Z_k = [v^{k1}, v^{k2}, \ldots, v^{kn}]$ generated in an END subdivision process on the initial simplex Z_0. Assume that $\gamma < f(0)$ (f(x) is the objective function of the BCP problem) and let $z^{ki} = \theta_{ki} v^{ki}$ be the γ-extension of v^{ki}. Since the subdivision is exhaustive, the sequences $\{v^{ki}\}$ (i=1,2,...,n) tend to a common limit v^* and if z^* is the γ-extension of v^* then $z^* = \lim z^{ki}$ $(k \to \infty)$, i=1,2,...,n.

As a consequence we have the following lemma.

Lemma VII.8. *There exist two positive numbers η_{ij} and ρ_{ij} such that for $i < j$ and every k:*

$$\eta_{ij} \leq \frac{\|z^{ki} - z^{kj}\|}{\|v^{ki} - v^{kj}\|} \leq \rho_{ij} .$$

Proof. The second inequality is due to the Lipschitz property of the mapping π, where $\pi(x)$ denotes the intersection of the boundary of the convex set $G = \{x: f(x) \geq \gamma\}$ with the ray through x. The first inequality is due to the Lipschitz property of the mapping σ, where $\sigma(x)$ denotes the intersection of Z_0 with the ray through x (Z_0 can be embedded in the boundary of a convex set containing 0 in its interior). ■

Also note the following simple fact:

Lemma VII.9. *There exist θ_1 and θ_2 such that $0 < \theta_1 \leq \theta_{ki} \leq \theta_2$ for $i=1,2,\dots,n$ and every k.*

Proof. We have $\theta_{ki} = 1/p(v^{ki})$, where p(x) is the gauge of the convex set $G = \{x: f(x) \geq \gamma\}$. The lemma then follows from the continuity of p(x) on Z_0, since any continuous function on a compact set must have a maximum and a minimum on this set. ■

Now for each simplex $Z_k = [v^{k1}, v^{k2}, \dots, v^{kn}]$ define the matrix

$$U_k = \left[\frac{z^{ki} - z^{kj}}{\|z^{ki} - z^{kj}\|}, i < j \right].$$

Lemma VII.10. *There exists a subsequence $\Delta \subset \{k\}$ such that U_s, $s \in \Delta$, tends to a matrix U with $rank(U) = n-1$.*

Proof. By the definition of an END process, there exists a subsequence $\Delta \subset \{k\}$ such that $\lim B_s = B$ as $s \to \infty$, $s \in \Delta$, with $rank(B) = n-1$. Let b^{ij}, $(i,j) \in I$, be a set of n–1 linearly independent columns of B. We may assume that $(z^{si} - z^{sj})/\|z^{si} - z^{sj}\|$ converges to u^{ij} for $i < j$. Let us show that the vectors u^{ij}, $(i,j) \in I$, are linearly independent. Suppose that $\sum_{(i,j)\in I} \alpha_{ij} u^{ij} = 0$.

We may write

$$\frac{z^{si}-z^{sj}}{\|z^{si}-z^{sj}\|} = \theta_{si}\,\frac{\|v^{si}-v^{sj}\|}{\|v^{si}-v^{sj}\|}\,\frac{(v^{si}-v^{sj})}{\|z^{si}-z^{sj}\|} + \frac{(\theta_{si}-\theta_{sj})}{\|z^{si}-z^{sj}\|}\,v^{sj}\,. \tag{22}$$

and therefore

$$\sum_{(i,j)\in I} \alpha_{ij}(z^{si} - z^{sj})/\|z^{si} - z^{sj}\|$$

$$= \sum_{(i,j)\in I} \beta^{s}_{ij}(v^{si} - v^{sj})/\|v^{si} - v^{sj}\|$$

$$+ \sum_{(i,j)\in I} \alpha_{ij}((\theta_{si}-\theta_{sj})/\|z^{si} - z^{sj}\|)v^{sj}\,, \tag{23}$$

where

$$\beta^{s}_{ij} = \alpha_{ij}\theta_{si}\|v^{si} - v^{sj}\|/\|z^{si} - z^{sj}\|\,. \tag{24}$$

By Lemma VII.9 and Lemma VII.8, we may assume that this latter sequence converges. Then from equation (22) we may also assume that $(\theta_{si}-\theta_{sj})/\|z^{si} - z^{sj}\|$ converges. Multiplying both sides of (23) by $e = (1,1,\dots,1)$ and letting $s \longrightarrow \infty$, we obtain

$$0 = e\Big(\sum_{(i,j)\in I} \alpha_{ij}u^{ij}\Big) = \lim_{s\to\infty} \sum_{(i,j)\in I} \alpha_{ij}(\theta_{si}-\theta_{sj})/\|z^{si} - z^{sj}\|\,.$$

Now, taking the limit in equation (23) as $s \longrightarrow \infty$, we deduce from the above equality that

$$0 = \sum_{(i,j)\in I} \alpha_{ij}u^{ij} = \sum_{(i,j)\in I} \beta_{ij}b^{ij}\,,\ \text{where } \beta_{ij} = \lim_{s\to\infty} \beta^{s}_{ij}\,.$$

Since the b^{ij}, $(i,j) \in I$, are linearly independent, this implies that $\beta_{ij} = 0$ for $(i,j) \in I$. But from (24) it follows, by Lemma VII.9 and Lemma VII.8, that the only way $\beta_{ij} = 0$ can hold is that $\alpha_{ij} = 0$ for all $(i,j) \in I$. Therefore, the vectors u^{ij}, $(i,j) \in I$, are linearly independent. ■

We are now in a position to prove the main property of END processes. Recall that for any simplex $Z_k = [v^{k1},v^{k2},\dots,v^{kn}]$, Q_k denotes the matrix of columns $z^{k1},z^{k2},\dots,z^{kn}$.

Proposition VII.7. *Let $\{Z_k\}$ be any infinite nested sequence of simplices generated in an END process. If the function $f(x)$ is continuously differentiable (at least in a neighbourhood of z^*), then there exists a subsequence $\Delta \subset \{k\}$ such that*

$$\frac{eQ_s^{-1}}{\|eQ_s^{-1}\|} \longrightarrow -\frac{\nabla f(z*)}{\|\nabla f(z*)\|} \quad (s \longrightarrow \infty,\ s \in \Delta)\ . \tag{25}$$

Proof. Let $q^{si} = \nabla f(z^{si})/\|\nabla f(z^{si})\|$, $q^* = \nabla f(z^*)/\|\nabla f(Z^*)\|$. From the continuous differentiability of f(x) we have

$$f(z^{sj}) = f(z^{si}) + q^{si}(z^{sj}-z^{si}) + 0(\|z^{sj}-z^{si}\|).$$

But $f(z^{sj}) = f(z^{si}) = \gamma$, and therefore, dividing by $\|z^{sj}-z^{si}\|$ and letting $s \longrightarrow \infty$ we obtain

$$q^*u^{ij} = 0 \quad \forall i < j\ ,$$

i.e., $q^*U = 0$, where $\text{rank}(U) = n-1$. On the other hand, one can assume that the subsequence $\Delta = \{s\}$ has been chosen in such a way that the vectors $q^s = eQ_s^{-1}/\|eQ_s^{-1}\|$ converge. Since $eQ_s^{-1}(z^{sj}-z^{si}) = 0$, it follows that if $q = \lim q^s$, then we also have $qu^{ij} = 0\ \forall i < j$, i.e., $qU = 0$. This implies that $q = \pm\, q^*$. But $eQ_s^{-1}z^{si} = 1$ for any s, hence $qz^* = 1$, and so we must have $q = -q^*$, proving the proposition. ■

Because of the above property, END subdivision processes can be used to ensure finite convergence in conical algorithms for concave minimization over polytopes. More specifically, let us select an END process and consider Algorithm 1* for the BCP problem, in which the following subdivision method is used:

a) If the optimal solution $\omega(Q_k)$ of the linear program $LP(Q_k;D)$ is a vertex of D (the feasible polytope), then split $\text{con}(Q_k)$ with respect to this vertex, i.e., select $u(Q_k)$ to be the γ_k–extension of $\omega(Q_k)$.

b) Otherwise, subdivide $\text{con}(Q_k)$ using the END process.

Theorem VII.7. *Assume that the function f(x) has bounded level sets and is continuously differentiable (at least in a neighbourhood of any global optimal solution of (BCP)). Furthermore, assume that any global optimal solution of (BCP) is a strict local minimum. Then for $\varepsilon = 0$ Algorithm VII.1* with the above subdivision method is finite.*

Proof. If the algorithm is infinite, it must generate an infinite nested sequence of cones: $\text{con}(Q_s) = [s^{s1}, z^{s2}, \ldots, z^{sn}]$. Because of the exhaustiveness of the subdivision, the algorithm is convergent (see Theorem VII.4): $\gamma_s \longrightarrow \gamma = \min f(D)$; and since $f(z^{si}) = \gamma_s$, $z^{si} \longrightarrow z^*$, it follows that z^* is a global optimal solution. By Proposition VII.7, we may assume that (25) holds (although γ_k here is not constant, the previous reasoning applies because $\gamma_k \downarrow \gamma$).

But since z^* is a strict local minimizer, it must be a vertex of D; moreover, the tangent hyperplane to the surface $f(x) = \gamma$ at z^* must be a supporting hyperplane of D which contains no other point of D. That is, any linear program $\max \{px: x \in D\}$, where $p / \|p\|$ is sufficiently near to $-\nabla f(z^*) / \|\nabla f(z^*)\|$, will have z^* as its unique optimal solution. In view of (25), then, for large enough s, say for $s \geq s_0$, the linear program $LP(Q_s;D)$, i.e., $\max \{eQ_s^{-1}x: x \in D \cap \text{con}(Q_s)\}$ will have z^* as its unique optimal solution, i.e., $\omega(Q_s) = z^*$. Then $\text{con}(Q_s)$ is subdivided with respect to z^*. But z^* is common to all $\text{con}(Q_s)$; hence for all $s > s_0$ we must have $z^* \in \{z^{s1}, \ldots, z^{sn}\}$ and $eQ_s^{-1}z^* = 1$. Since z^* solves $LP(Q_s;D)$, it follows that $\mu(Q_s) = \max \{eQ_s^{-1}x: x \in D \cap \text{con}(Q_s)\} = 1$, contradicting the fact that $\mu(Q_k) > 1$ for any k. This proves finiteness of the algorithm. ■

Remarks VII.6. (i) An example of an END process is the following: Given a simplex $Z_k = [v^{k1}, v^{k2}, \ldots, v^{kn}]$, we divide it into 2^{n-1} subsimplices by the (n–2)–hyperplanes through the midpoints of the edges (for each i there is an (n–2)–hyperplane through the midpoints of the n–1 edges of Z_k emanating from v^{k1}). Clearly, the edge directions of each subsimplex are the same as those of Z_k. Therefore, the sequence $\{B_k\}$ will be constant.

But of course there are subdivision processes (even exhaustive subdivision processes) which do not satisfy the END condition. An example is the classical barycentric subdivision method used in combinatorial topology. Fig. VII.2 illustrates the case $n = 2$: the edge lengths of Z_{k+1} are less than 2/3 those of Z_k, and therefore the process is exhaustive. It is also easy to verify that the sequence $\{B_k\}$ converges to a matrix of rank 1.

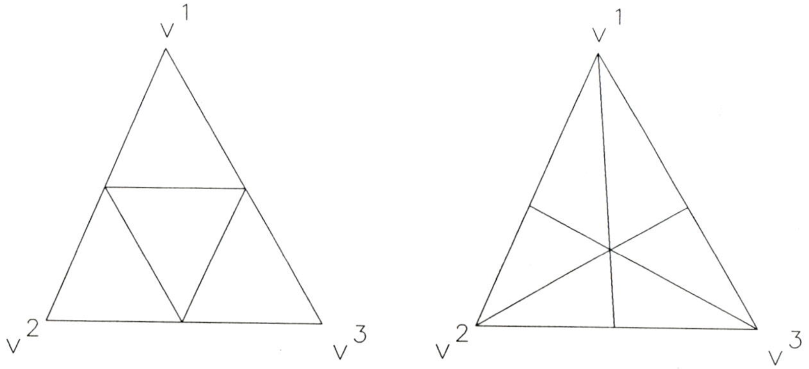

Fig. VII.2

(ii) The difficulty with the above method of Hamami and Jacobsen is that no sufficiently simple END process has yet been found. The authors conjecture that bisection should be an END process, but this conjecture has never been proved or disproved. In any case, the idea of using hybrid subdivision processes, with an emphasis on subdivision with respect to vertices, has inspired the development of further significant improvements of conical algorithms for concave minimization.

(iii) From Proposition VII.7 it follows that if the function f(x) is continuously differentiable, then the END process is nondegenerate in the sense presented in Section VII.1.3. Thus, the condition of an END process is generally much stronger than the nondegeneracy condition discussed in normal conical algorithms.

2. SIMPLICIAL ALGORITHMS

In this section we present branch and bound algorithms for solving problems (BCP) and (CP), in which branching is performed by means of simplicial subdivisions rather than conical subdivisions.

The first approach for minimizing a concave function f: $\mathbb{R}^n \longrightarrow \mathbb{R}$ over a nonpolyhedral compact convex set D was the simplicial branch and bound algorithm by Horst (1976) (see also Horst (1980)). In this algorithm the partition sets M are simplices that are successively subdivided by a certain process. A lower bound $\beta(M)$ is determined by constructing the convex envelope φ_M of f over M and subsequently minimizing φ_M over $D \cap M$. Recall from Theorem IV.7 that φ_M is the affine function that is uniquely determined by the set of linear equations that arises from the requirements that φ_M and f have to coincide at the vertices of M. Note that this linear system does not need to be solved explicitly, since (in barycentric coordinates) we have

$$\varphi_M(x) = \sum_{i=1}^{n+1} \lambda_i f(v^i) \, , \; \sum_{i=1}^{n+1} \lambda_i = 1 \, , \; \lambda_i \geq 0 \quad (i=1,...,n) \, ,$$

where v^i (i=1,...,n) denotes the vertices of M and $x = \sum_{i=1}^{n+1} \lambda_i v^i$. The lower bound $\beta(M)$ is then determined by the convex program

$$\text{minimize } \varphi_M(x) \qquad \text{s.t. } x \in D \cap M \, ,$$

which is a linear program when D is a polytope.

Whenever the selection operation is bound improving, the convergence of this algorithm follows from the discussion in Chapter IV. An example that illustrates this approach is Example IV.1.

We discuss certain generalizations and improvements of this algorithm and begin by introducing "normal" simplicial algorithms.

Since the conical subdivision discussed in the preceding sections are defined by means of subdivisions of simplices the normal simplicial subdivision processes turn out to be very similar to the conical ones.

We first treat the case (BCP) where D is a polytope.

2.1. Normal Simplicial Subdivision Processes

Let M_0 be an n–simplex containing D such that a vertex x^0 of M_0 is an extreme point of D.

For any n–subsimplex $M = [v^1,...,v^{n+1}]$ of M_0 denote by $\varphi_M(x)$ the convex envelope of f over M; it is the affine function that agrees with f(x) at each vertex of M (cf. Theorem IV.7). Associate to M the linear program

$$\text{LP(M,D)} \quad \min \varphi_M(x) \quad \text{s.t.} \quad x \in D \cap M \ .$$

Let $\omega(M)$ be a basic optimal solution of this linear program, and let $\beta(M)$ be its optimal value.

Then we know from Section IV.4.3 that

$$\beta(M) \le \min f(D \cap M) \ ,$$

i.e., $\beta(M)$ is a lower bound for f(x) on $D \cap M$. If M' is a subsimplex of M then the value of $\varphi_{M'}$ at any vertex v of M' is $f(v) \ge \varphi_M(v)$ (by the concavity of f); hence $\varphi_{M'}(x) \ge \varphi_M(x)$ $\forall x \in M'$. Therefore, this lower bounding satisfies the monotonicity condition:

$$\beta(M') \ge \beta(M) \text{ whenever } M' \subset M \ .$$

Now consider a simplicial subdivision process in which:

1) The initial simplex is M_0.
2) The subdivision of any n–simplex M is a radial subdivision with respect to a point $w = w(M) \in M$ distinct from any vertex of M (cf. Section IV.3.1).

Let M_s, s=0,1,..., be any infinite nested sequence of simplices generated by the process, i.e., such that M_{s+1} is obtained by a subdivision of M_s. For each s let $\omega^s = \omega(M_s)$, $w^s = w(M_s)$ (note that in general $w^s \neq \omega^s$).

Definition VII.4. *A nested sequence M_s, $s=0,1,...$, is said to be* ***normal for a given*** *(f,D) if*

$$\underline{lim}_{s\to\infty} |f(\omega^s) - \beta(M_s)| = 0 . \tag{26}$$

A simplicial subdivision process is said to be ***normal*** *if any infinite nested sequence of simplices that it generates is normal.*

We shall later examine how to construct a normal simplicial subdivision (NSS) process.

A rule for simplicial subdivision that generates an NSS process is called an NSS rule.

2.2. Normal Simplicial Algorithm

Incorporating a normal simplicial subdivision (NSS) process and the above lower bounding into a branch and bound scheme yields the following generalization of the algorithm in Horst (1976).

Algorithm VII.3 (Normal Simplicial Algorithm).

Select a tolerance $\varepsilon \geq 0$ and an NSS rule for simplicial subdivision.

Initialization:

Choose an n–simplex $M_0 \supset D$ such that a vertex of M_0 is also a vertex of D. Let x^0 be the best feasible solution available. Set $\mathcal{M}_0 = \mathcal{N}_0 = \{M_0\}$.

Iteration k = 0,1,...:

1) For each $M \in \mathcal{N}_k$ form the affine function $\varphi_M(x)$ that agrees with f(x) at the vertices of M, and solve the linear program

LP(M,D) $\min \varphi_M(x)$ s.t. $x \in D \cap M$

to obtain a basic optimal solution $\omega(M)$ and the optimal value $\beta(M)$.

2) Delete all $M \in \mathcal{M}_k$ such that $\beta(M) \geq f(x^k) - \varepsilon$. Let $\mathcal{R}_k$ be the remaining collection of simplices.

3) If $\mathcal{R}_k = \emptyset$, terminate: x^k is an ε–optimal solution of (BCP). Otherwise,

4) Select $M_k \in \operatorname{argmin} \{\beta(M): M \in \mathcal{R}_k\}$ and subdivide it according to the chosen NSS process.

5) Update the incumbent, setting x^{k+1} equal to the best of all feasible solutions known so far.

Let $\mathcal{N}_{k+1}$ be the collection of subsimplices of M_k provided by the subdivision in Step 4, and let $\mathcal{M}_{k+1} = (\mathcal{R}_k \setminus \{M_k\}) \cup \mathcal{N}_{k+1}$. Set $k \leftarrow k+1$ and return to Step 1.

Theorem VII.8. *The normal simplicial algorithm can be infinite only if $\varepsilon = 0$, and in this case any accumulation point of the generated sequence $\{x^k\}$ is a global optimal solution of (BCP).*

Proof. Consider an infinite nested sequence M_s, $s \in \Delta \subset \{0,1,...\}$, generated by the algorithm. Since the sequence $f(x^k)$ is nonincreasing, while the sequence $\beta(M_k)$ is nondecreasing (by virtue of the monotonicity of the lower bounding and the selection criterion), there exist $\gamma = \lim f(x^k) = \lim f(x^{k+1})$ and $\beta = \lim \beta(M_k)$. Further– more, by the normality condition, we may assume (taking subsequences if necessary) that $\lim |f(\omega^s) - \beta(M_s)| = 0$, and hence $\lim f(\omega^s) = \beta$, where

$\omega^s = \omega(M_s)$. But, from the selection of M_k in Step 2 we find that $\beta(M_s) < f(x^s) - \varepsilon$, and from the definition of the incumbent in Step 6 we have $f(x^{k+1}) \le f(\omega^k)$ for any k. Therefore, $\beta \le \gamma - \varepsilon$, while $\gamma \le \lim f(\omega^s) = \beta$. This leads to a contradiction unless $\varepsilon = 0$, and in the latter case we obtain $\lim (\gamma_k - \beta(M_k)) = 0$. The conclusion then follows by Theorem IV.3. ■

Remark VII.7. Given a feasible point z we can find a vertex of D which corresponds to an objective function value no greater than f(z). Therefore, we may suppose that each x^k is a vertex of D. Then for $\varepsilon = 0$, we have $f(x^k) = \min f(D)$ when k is sufficiently large.

2.3. Construction of an NSS Process

Of course, the implementation of the above algorithm requires us to construct an NSS process.

For any n–simplex $M = [v^1, v^2, \ldots, v^n]$ in $\mathbb{R}^n$ let

$$\varphi_M(x) = \pi(M)x + r(M),\ \pi(M) \in \mathbb{R}^n,\ r(M) \in \mathbb{R},$$

where, as before, $\varphi_M(x)$ is the affine function that agrees with f(x) at the vertices of M.

Definition VII.5. *An infinite nested sequence of simplices M is said to be* ***nondegenerate*** *if* $\underline{\lim}_{s \to \infty} \|\pi(M_s)\| < \infty$, *i.e., if there exists a subsequence* $\Delta \subset \{1,2,\ldots\}$ *and a constant* η *such that* $\|\pi(M_s)\| \le \eta\ \ \forall s \in \Delta$. *A simplicial subdivision process is* ***nondegenerate*** *if any infinite nested sequence of simplices that it produces is nondegenerate.*

Proposition VII.8. *Let* $M_s = [v^{s1}, v^{s2}, \ldots, v^{sn}]$, *s=1,2,..., be an infinite nested sequence of simplices such that* M_{s+1} *is obtained from* M_s *by means of a radial subdivision with respect to a point* w^s. *If the sequence is nondegenerate, then*

$$\underline{\lim}_{s\to\infty} |f(w^s) - \varphi_{M_s}(w^s)| = 0 .$$

Proof. Denote by H_s the hyperplane in $\mathbb{R}^{n+1}$ that is the graph of the affine function $\varphi_{M_s}(x)$, and denote by L_s the halfspace that is the epigraph of this function. Let

$$y^s = \begin{bmatrix} w^s \\ \varphi_{M_s}(w^s) \end{bmatrix} , \quad z^s = \begin{bmatrix} w^s \\ f(w^s) \end{bmatrix} .$$

For j < s, from the concavity of f(x) it is easily seen that

$$y^s \notin L_{s+1}, \text{ while } y^s \in \bigcap_{j\leq s} L_j .$$

Since the sequence y^s is bounded, it follows by Lemma III.2 that $d(y^s, L_{s+1}) \longrightarrow 0$, and hence $d(y^s, H_{s+1}) \longrightarrow 0$. Noting that $H_{s+1} = \{(x,t) \in \mathbb{R}^n \times \mathbb{R}: t = \pi(M_{s+1})x + r(M_{s+1})\}$, we can write

$$d(y^s, H_{s+1}) = \frac{|\pi(M_{s+1})w^s + r(M_{s+1}) - \varphi_{M_s}(w^s)|}{(1 + \|\pi(M_{s+1})\|^2)^{1/2}} .$$

But $z^s \in H_{s+1}$, i.e., $f(w^s) = \pi(M_{s+1})w^s + r(M_{s+1})$, and hence

$$d(y^s, H_{s+1}) = \frac{|f(w^s) - \varphi_{M_s}(w^s)|}{(1 + \|\pi(M_{s+1})\|^2)^{1/2}} .$$

Now, from the degeneracy assumption, it follows that there exist an infinite sequence $\Delta \subset \{1,2,\ldots\}$ and a constant η such that $\|\pi(M_{s+1})\| \leq \eta$ $\forall s \in \Delta$. Therefore,

$$|f(w^s) - \varphi_{M_s}(w^s)| \le (1+\eta^2)^{1/2} d(y^s, H_{s+1}) \longrightarrow 0,$$

as $s \longrightarrow \infty$, $s \in \Delta$, proving the proposition. ∎

Corollary VII.5. *A simplicial subdivision process is normal if any infinite nested sequence of simplices M_s that it generates satisfies either of the following conditions:*

1) it is exhaustive;

2) it is nondegenerate and satisfies $w^s = \omega^s$ for all but finitely many s.

Proof. If the sequence $\{M_s\}$ is exhaustive, so that M_s shrinks to a point x^* as $s \longrightarrow \infty$, then $\omega^s \longrightarrow x^*$, and the relation (26) readily follows from the inequalities

$$\min f(M_s) \le \beta(M_s) \le \min f(D \cap M_s) \le f(\omega^s) .$$

If the sequence M_s is nondegenerate and $w^s = \omega^s$ for all but finitely many s, then the relation (26) follows from Proposition VII.8, since $\beta(M_s) = \varphi_{M_s}(\omega^s)$ for sufficiently large s. ∎

2.4. The Basic NSS Process

According to the previous corollary, if bisection is used throughout Algorithm VII.3, then convergence is assured. On the other hand, if the ***ω–subdivision process*** (i.e., subdivision with respect to $w(M_k) = \omega(M_k)$) is used throughout, then convergence will be achieved when the subdivision process is nondegenerate. However, just as with conical algorithms, nondegeneracy is a difficult property to realize, whereas pure bisection is expected to give slow convergence. Therefore, to obtain an algorithm that is convergent and at the same time more efficient than when using pure bisection, we can combine ω–subdivisions with bisections in a suitable manner, using the same (η,ρ) device that was used earlier for conical algorithms.

Specifically, select two numbers $\eta_0 \ge 0$, $\rho \in (\frac{\sqrt{3}}{2}, 1)$ and adopt the following rule for the subdivision process:

Set $\eta(M_0) = \eta_0$ for the initial simplex M_0. At iteration $k = 0,1,2,\ldots$, if M_k is the simplex to be subdivided, then:

a) If $\|\pi(M_k)\| \le \eta(M_k)$ and $\sigma(M_k) \le \rho$, then subdivide M_k with respect to $w^k = \omega^k$ (perform an ω-subdivision of M_k) and set $\eta(M) = \eta(M_k)$ for each subsimplex M of this subdivision;

b) Otherwise, bisect M_k, i.e., subdivide it with respect to w^k = midpoint of a longest edge of M_k, and set $\eta(M) = 2\eta(M_k)$ for each subsimplex M of this subdivision.

Proposition VII.9. *The simplicial subdivision process just defined is an NSS process.*

Proof. Let $\{M_s\}$ be any infinite nested sequence of simplices generated by the subdivision process. By Proposition VII.4, if this sequence is not exhaustive, then it must involve at most finitely many bisections. That is, for all sufficiently large s, say $s \ge s_0$, one must have $\|\pi(M_s)\| \le \eta(M_s) = \eta(M_{s_0})$, and so this sequence is nondegenerate. Moreover, $w^s = \omega^s$, which shows that the sequence is normal, according to Corollary VII.5.

■

We shall refer to the subdivision process constructed above as the **Basic NSS Process**. Of course, the choice of the parameters η_0 , ρ is user specified and must be based upon computational considerations.

When $\eta_0 = 0$ the subdivision process consists exclusively of bisections: the corresponding variant of Algorithm VII.3 is just the algorithm of Horst (1976). When η_0 is very large and ρ is very close to 1, the subdivison process consists essentially of ω-subdivisions.

Remark VII.8. Let D be given by the system (7) (8). Recall that for any simplex $M = [v^1, v^2, \ldots, v^{n+1}]$, the affine function that agrees with f(x) at the vertices of M is $\varphi_M(x) = \Sigma \lambda_i f(v^i)$, with

$$\Sigma \lambda_i v^i = x, \ \Sigma \lambda_i = 1, \ \lambda_i \geq 0 \quad (i=1,2,\ldots,n+1) .$$

Hence, the associated linear program LP(M,D) is

$$\min \Sigma \lambda_i f(v^i)$$

$$\text{s.t. } \Sigma \lambda_i A v^i \leq b \, , \ \Sigma \lambda_i = 1, \ \lambda_i \geq 0 \quad (i=1,2,\ldots,n+1) .$$

If we denote

$$Q = \begin{bmatrix} v^1 & v^2 & \ldots & v^{n+1} \\ 1 & 1 & \ldots & 1 \end{bmatrix} ,$$

then $(\lambda_1, \lambda_2, \ldots, \lambda_{n+1})^T = Q^{-1} \begin{bmatrix} x \\ 1 \end{bmatrix}$, and so

$$\varphi_M(x) = (f(v^1), f(v^2), \ldots, f(v^{n+1})) Q^{-1} \begin{bmatrix} x \\ 1 \end{bmatrix} .$$

From this one can determine the vector $\pi(M)$ used in the (η,ρ) rule for subdivision, namely:

$$(\pi(M), r(M)) = (f(v^1), f(v^2), \ldots, f(v^{n+1})) Q^{-1} .$$

The computation of Q^{-1} and $\sigma(M)$ required by the (η,ρ) rule is time consuming. For this reason it is often preferable to replace the (η,ρ) rule by a simpler one which is not necessarily normal, but which, nevertheless, has worked successfully in practical cases. An example is the rule (**) mentioned in Section VII.1.6.

2.5. Normal Simplicial Algorithm for Problems with Convex Constraints

Like the normal conical algorithm, the normal simplicial algorithm can be extended to problems where the constraint set D is a convex nonpolyhedral set defined by an inequality of the form

$$g(x) \leq 0 .$$

The extension is based upon the same method that was used for conical algorithms. Namely, outer approximation is combined with a branch and bound technique according to the scheme proposed by Horst, Benson and Thoai (1988) and Benson and Horst (1988).

For simplicity, let us assume that D is bounded and that the convex function g(x) is finite throughout $\mathbb{R}^n$ (hence, it is subdifferentiable at every point x).

Algorithm VII.4 (Normal Simplicial Algorithm for CP).

Select a tolerance $\varepsilon \geq 0$ and an NSS rule for simplicial subdivision.

Initialization:

Choose an n–simplex $M_0 \supset D$ and translate the origin to an interior point of D. Let x^0 be the best available feasible solution. Set $D_0 = M_0$, $\mathcal{M}_0 = \mathcal{N}_0 = \{M_0\}$, $\eta(M_0) = \eta_0$.

Iteration k = 0,1,...:

1) For each $M \in \mathcal{N}_k$ form the affine function $\varphi_M(x)$ and solve the linear program

$$LP(M,D_k) \quad \min \varphi_M(x) \quad \text{s.t. } x \in D_k \cap M$$

to obtain a basic optimal solution $\omega(M)$ and the optimal value $\beta(M)$.

2) Delete all $M \in \mathcal{M}_k$ such that $\beta(M) \geq f(x^k)-\varepsilon$. Let $\mathcal{R}_k$ be the remaining collection of simplices.

3) If $\mathcal{R}_k = \emptyset$, terminate: x^k is an ε–optimal solution of (CP). Otherwise,

4) Select $M_k \in \operatorname{argmin} \{\beta(M): M \in \mathcal{R}_k\}$ and subdivide it according to the

chosen NSS rule.

5) Let $\omega^k = \omega(M_k)$. If $\omega^k \in D$, set $D_{k+1} = D_k$. Otherwise, let $p^k \in \partial g(\overline{\omega}^k)$, where $\overline{x}$ denotes the intersection of the boundary of D with the ray through x. Set

$$D_{k+1} = D_k \cap \{x: p^k(x - \overline{\omega}^k) \le 0\} .$$

6) Update the incumbent, setting x^{k+1} equal to the best among x^k, $\overline{u}^k$ and all $\overline{\omega}(M)$ corresponding to $M \in \mathscr{N}_k$.
Let $\mathscr{N}_{k+1}$ be the partition of M_k, $\mathscr{M}_{k+1} = (\mathscr{R}_k \setminus \{M_k\}) \cup \mathscr{N}_{k+1}$. Set $k \leftarrow k+1$ and return to Step 1.

Theorem VII.9. *Algorithm VII.4 can be infinite only if $\varepsilon = 0$ and in this case any accumulation point of the generated sequence $\{x^k\}$ is a global optimal solution.*

Proof. Consider any infinite nested sequence of simplices M_s, $s \in \Delta$, generated by the algorithm. By the normality condition, we may assume (passing to subsequences if necessary) that $\lim |f(\omega^s) - \beta(M_s)| = 0$, where $\omega^s = \omega(M_s)$. Let $\gamma = \lim f(x^k) = \lim f(x^{k+1})$, $\beta = \lim \beta(M_k)$. Then $\beta = \lim f(\omega^s)$. But $f(x^{s+1}) \le f(\overline{\omega}^s)$ (see Step 6) and $f(x^{s+1}) \le f(x^0) \le f(0)$; Hence if $\omega^s \in D$ (i.e., $\omega^s \in [0,\overline{\omega}^s]$) for infinitely many s, then, by the concavity of f(x), we have $f(x^{s+1}) \le f(\omega^s)$, which, if we let $s \to \infty$, gives $\gamma \le \beta$.

On the other hand, if $\omega^s \notin D$ for infinitely many s, then, reasoning as in the proof of Theorem VII.5, we see that $\omega^s - \overline{\omega}^s \to 0$, and since $f(x^{s+1}) \le f(\overline{\omega}^s)$ again it follows that $\gamma \le \beta$. Now, from the selection of M_k in Step 4, we have $\beta(M_s) < f(x^s) - \varepsilon$, and hence $\beta \le \gamma - \varepsilon$. This contradicts the above inequality $\gamma \le \beta$ unless $\varepsilon = 0$. Then $\gamma = \beta$, i.e., $\lim (\beta(M_s) - \gamma_s) = 0$, and the conclusion follows by Theorem IV.2.

■

3. AN EXACT SIMPLICIAL ALGORITHM

The normal simplicial algorithm with $\varepsilon = 0$ is convergent but may be infinite. We now present an exact (finite) algorithm based upon a specific simplicial subdivision of the feasible set. The original idea for this method is due to Ban (1983, 1986) (cf. also Tam and Ban (1985)). Below we follow a modified presentation of Tuy and Horst (1988).

3.1. Simplicial Subdivision of a Polytope

We shall assume that the feasible set is a polytope $D \subset \mathbb{R}^n_+$ defined by the inequalities

$$g_i(x) := \sum_{j=1}^{n} a_{ij}x_j - b_i \geq 0 \;,\; (i=1,...,m) \;,\; x \geq 0 \;. \tag{27}$$

Definition VII.6. *A simplex $M = [u^1,u^2,...,u^r] \subset \mathbb{R}^n_+$ with vertices $u^1,u^2,...,u^r$ $(r \leq n+1)$ is said to be* ***trivial*** *(more precisely, D–trivial, or trivial with respect to the system (27)) if for every i=1,...,m:*

$$g_i(u^p) \cdot g_i(u^q) \geq 0 \quad \forall p,q \;. \tag{28}$$

The motivation for defining this notion stems from the following property.

Proposition VII.10. *If a simplex M is trivial then $M \cap D$ is simply the face of M spanned by those vertices of M that lie in D:*

$$M \cap D = conv\, \{u^j\colon u^j \in D\} \;. \tag{29}$$

Proof. Let $x \in M$, so that $x = \sum_{j\in J} \lambda_j u^j$ with $J \subset \{1,...,r\}$, $\lambda_j > 0$, $\Sigma\lambda_j = 1$. If $x \in D$ then for every i=1,...,m:

$$g_i(x) = \sum_{j \in J} \lambda_j g_i(u^j) \geq 0 \quad .$$

In view of (28), this implies that for all $j \in J$

$$g_i(u^j) \geq 0 \; (i=1,...,m) \quad ,$$

i.e., $u^j \in D$. Therefore, $M \cap D \subset \text{conv} \{u^j: u^j \in D\}$. Since the reverse inclusion is obvious, we must have (29). ■

Now suppose that a simplex $M = [u^1,u^2,...,u^r]$ is **nontrivial**. Then there is an index s (which, for convenience, will be called the **test index** of M) such that (21) holds for all $i=1,...,s-1$, while there are p,q such that

$$g_s(u^p) \cdot g_s(u^q) < 0 \quad . \tag{30}$$

Define

$$\rho(M) = s + \text{the number of indices j such that } g_s(u^j) = 0.$$

From (30) and the linearity of g_s, there is a unique $v = \lambda u^p)+(1-\lambda)u^q$ $(0 < \lambda < 1)$ satisfying

$$g_s(v) = 0 \quad . \tag{31}$$

Let M_1 (resp. M_2) be the simplex whose vertex set is obtained from that of M by replacing u^p (resp. u^q) by v (i.e., M_1, M_2 form a radial subdivision of M with respect to v).

Proposition VII.11. *The set $\{M_1 , M_2\}$ is a partition of M. If each M_ν $(\nu=1,2)$ is nontrivial, then $\rho(M_\nu) > \rho(M)$.*

Proof. The first assertion follows from Proposition IV.1. To prove the second assertion, observe that for each $i < s$ and each j:

$$g_i(v) \cdot g_i(u^j) = \lambda g_i(u^p) \cdot g_i(u^j) + (1-\lambda) g_i(u^q) \cdot g_i(u^j) \geq 0 \; .$$

Therefore, when M_ν is still nontrivial, its test index s_ν is at least equal to s. If $s_\nu = s$, then it follows from (30) and (31) that the number of vertices u satisfying $g_s(u) = 0$ has increased by at least one. Consequently, $\rho(M_\nu) > \rho(M)$. ■

The operation of dividing M into M_1, M_2 that was just described will be called a **D–bisection** (or a bisection with respect to the constraints (27)). Noting that $\rho(M) \le m+r$, we can conclude the following corollary.

Corollary VII.6. *Any sequence of simplices M_k such that M_{k+1} is obtained from M_k by a D–bisection is finite.*

Corollary VII.7. *Any simplex can be partitioned into trivial subsimplices by means of a finite number of successive D–bisections.*

In particular, if we start with a simplex M_0 containing D, then after finitely many D–bisections M_0 will be decomposed into trivial subsimplices. If we then take the intersections of D with these subsimplices, we obtain a partition of D into simplices, each of which is entirely contained in a face of D, by Proposition VII.10. It is easily seen that every vertex of D must be a vertex of some simplex in the decomposition. Therefore, this simplicial subdivision process will eventually produce **all** of the vertices of the polytope D (incidentally, we thus obtain a method for generating all of the vertices of a polytope). The point, however, is that this subdivision process can be incorporated into a branch and bound scheme. Since deletions will be possible by bounding, we can hope to find the optimal vertex well before all the simplices of the decomposition have been generated.

3.2. A Finite Branch and Bound Procedure

To every simplex $M = [u^1,u^2,\dots,u^r] \subset \mathbb{R}^n_+$ let us assign two numbers $\alpha(M)$, $\beta(M)$ defined as follows:

$$\alpha(M) = \min f(D \cap \text{vert } M) , \tag{32}$$

$$\beta(M) = \begin{cases} \min f(D \cap \text{vert } M) & \text{if } M \text{ is D–trivial} \\ \min f(\text{vert } M) & \text{otherwise} \end{cases} . \tag{33}$$

Clearly, by Proposition VII.10,

$$\alpha(M) \geq \min f(M \cap D) \geq \beta(M) \tag{34}$$

and equality must hold everywhere in (34) if M is a trivial simplex. Furthermore, $\beta(M') \geq \beta(M)$ whenever M' is contained in a nontrivial simplex M. Therefore, the bounding (34) can be used along with the above simplicial subdivision process to define a branch and bound algorithm according to the Prototype BB–Procedure (cf. Section IV.1).

Proposition VII.12. *The BB procedure just defined, with the selection rule*

$$\mathcal{P}_k = \{M_k\},\ M_k \in \arg\min \{\beta(M) : M \in \mathcal{R}_k\} \quad , \tag{35}$$

is finite.

Proof. By Theorem IV.1, it suffices to show that the bounding is finitely consistent. For any trivial simplex M, since equality holds everywhere in (34), we obviously have

$$\beta(M) \geq \alpha_k := \min \{\alpha(M): M \in \mathcal{M}_k\} \quad .$$

Hence, any trivial simplex is removed (cf. Section IV.1). This means that every unremoved simplex M is nontrivial, and consequently is capable of further refinement by the above simplicial subdivision. Then by Corollary VII.6 the bounding is finitely consistent. ■

In order to speed up the convergence of this BB procedure (which we shall refer to as the **Exact Simplicial (ES) Algorithm**), we may use the following rules in Step k.3 to reduce or delete certain simplices $M \in \mathcal{P}'_k$.

Rule 1. Let $M = [u^1, u^2, \ldots, u^r]$. If for some index i there is just one p such that $g_i(u^p) > 0$, then reduce M by replacing each u^q for which $g_i(u^q) < 0$ by the point $v^q \in [u^p, u^q]$ that satisfies $g_i(v^q) = 0$.

Rule 2. If there is a p such that $g_i(u^p) < 0$ for some i satisfying (28), then replace M by the proper face of M spanned by those u^j with $g_i(u^j) = 0$. If for some i we have $g_i(u^j) < 0 \ \forall j$, then delete M.

The hypothesis in Rule 1 means that exactly one vertex u^p of the simplex M lies on the positive side of the hyperplane $g_i(x) = 0$. It is then easily seen that, when we replace M by the smaller simplex M' which is the part of M on the nonnegative side of this hyperplane, we do not lose any point of $M \cap D$ (i.e. $M \cap D \subset M'$). This is just the adjustment prescribed by Rule 1.

The hypothesis in Rule 2 means that all vertices of M lie on the nonpositive side of the hyperplane $g_i(x) = 0$ and at least one vertex u^p lies on the negative side. It is then obvious that we do not lose any point of $M \cap D$ when we replace M by its intersection with this hyperplane. This is just the adjustment prescribed by Rule 2. If every vertex of M lies on the negative side of the hyperplane $g_i(x) = 0$, then $M \cap D = \emptyset$, and M can therefore be deleted. ∎

3.3. A Modified ES Algorithm

In the ES algorithm, a simplex M_k must always be subdivided with respect to the test constraint, i.e., the first constraint in (27) that does not satisfy (28). Since this might not be the constraint that is the most violated by the current approximate solution or the most frequently violated by vertices of M_k, more subdivisions than necessary may have to be performed before reaching the trivial simplex of interest. On the other hand, if M_k is subdivided with respect to some constraint other than the test one, the finiteness of the procedure may not be guaranteed.

To overcome this drawback, we modify the algorithm so as to favour subdivision with respect to the constraint that is the most violated by the current approximate solution. For this we use the strategy of Restart Branch and Bound–Outer Approximation that was discussed in Section IV.6.

Consider the following outer approximation procedure for solving (BCP):

Start with a simplex M_0 such that $D \subset M_0 \subset \mathbb{R}^n_+$. Set $D_0 = M_0$. At iteration $\nu = 0,1,...$ use the ES algorithm to solve the relaxed problem

$$(Q_\nu) \qquad\qquad \min f(D_\nu) \ .$$

Let $\tilde{z}^\nu$ be the vertex of D_ν which solves (Q_ν). If $\tilde{z}^\nu \in D$, stop: $\tilde{z}^\nu$ is a global optimal solution of (P).

Otherwise, form $D_{\nu+1}$ by adjoining to D_ν the constraint of D that is the most violated by $\tilde{z}^\nu$. Go to iteration ν+1.

By incorporating the ES Algorithm into the above outer approximation scheme we obtain the following RBB–OP procedure (cf. Section IV.6).

Algorithm VII.5 (Modified ES Algorithm).

Iteration 0:

Choose a simplex M_0 such that $D \subset M_0 \subset \mathbb{R}^n_+$ and $D \cap \text{vert } M_0 \neq \emptyset$. Let

$$\mathscr{M}_0 = \{M_0\} \ ,$$

$$\alpha_0 = \min f(D \cap \text{vert } M_0) = f(x^0),\ x^0 \in D \ ,$$

$$\beta_0 = \min f(\text{vert } M_0) = f(z^0), \quad z^0 \in \text{vert } M_0 \ .$$

If $\alpha_0 = \beta_0$, then stop: x^0 is a global optimal solution. Otherwise, let

$$i_0 \in \arg\min \{g_i(x^0) : i=1,\dots,m\} .$$

Form the polytope D_1 by adjoining to M_0 the i_0–th constraint in the system (27):

$$D_1 = \{x \in M_0 \colon g_{i_0}(x) \geq 0\} .$$

Set $\nu = 1$ and go to iteration 1.

Iteration k = 1,2,... :

k.1. Delete all $M \in \mathcal{M}_{k-1}$ such that

$$\beta(M) \geq \alpha_{k-1} \quad .$$

Let $\mathcal{R}_k$ be the collection of remaining members of $\mathcal{M}_{k-1}$.

k.2. Select $M_k \in \arg\min \{\beta(M)\colon M \in \mathcal{R}_k\}$. Bisect M_k with respect to the polytope D_ν (i.e. with respect to the constraints $i_0,\dots,i_{\nu-1}$).

k.3. Reduce or delete any newly generated simplex that can be reduced or deleted according to Rules 1 and 2 (with respect to the polytope D_ν). Let $\mathcal{M}_k'$ be the resulting collection of newly generated simplices.

k.4. For each $M \in \mathcal{M}_k'$ compute

$$\alpha(M) = \min f(D_\nu \cap \operatorname{vert} M)$$

$$\beta(M) = \begin{cases} \min f(D_\nu \cap \operatorname{vert} M) & \text{if } M \text{ is } D_\nu\text{–trivial} \\ \min f(\operatorname{vert} M) & \text{otherwise} \end{cases} .$$

k.5. Let $\mathcal{M}_k = (\mathcal{R}_k \setminus \{M_k\}) \cup \mathcal{M}_k'$. Compute

$$\alpha_k = \min \{\alpha(M)\colon M \in \mathcal{M}_k\} = f(x^k), \quad x^k \in D \quad ,$$

$$\beta_k = \min \{\beta(M)\colon M \in \mathcal{M}_k\} = f(z^k) \quad .$$

k.6. If $\alpha_k = \beta_k$, then stop: x^k is a global optimal solution of (BCP).

If $\alpha_k > \beta_k$ and $z^k \in D_\nu$, choose

$$i_\nu \in \arg\min \{g_i(z^k): i=1,...,m\} \quad .$$

Let $D_{\nu+1}$ be the polytope which is determined by adjoining to D_ν the i_ν–th constraint of the system (27). Set $\nu \leftarrow \nu+1$ and go to iteration k+1.

Otherwise, go to iteration k+1 with ν unchanged.

Theorem VII.10. *The modified ES algorithm terminates after finitely many steps.*

Proof. Clearly $D \subset D_\nu$ and the number ν is bounded from above by the total number of constraints in (27), i.e., $\nu \leq m$. As long as ν is unchanged, we are applying the ES Algorithm to minimize f(x) over D_ν. Since this algorithm is finite, the finiteness of the Modified ES Algorithm follows from Theorem IV.12. ■

Remarks VII.9. (i) For each simplex $M = [u^1,u^2,...,u^r]$ we must consider the associated matrix

$$(g_{ij}) \; , \quad i=1,...,m; \; j=1,...,r \quad , \tag{36}$$

where $g_{ij} = g_i(u^j)$. M is trivial if no row of this matrix has two entries with opposite sign. The test index s is the index of the first row that has two such entries, i.e., $g_{sp} > 0$, $g_{sq} < 0$. Then M is subdivided with respect to

$$v = \frac{g_{ip}u^q - g_{iq}u^p}{g_{ip} - g_{iq}} \; ,$$

and the matrices associated with the two subsimplices are obtained by replacing column p (resp. q) in the matrix (36) with the column of entries

$$\frac{g_{ip}g_{iq} - g_{iq}g_{ip}}{g_{ip} - g_{iq}} \; , \quad i=1,...,m \quad .$$

(ii) When the constraints of D are

$$Ax = b \quad , \quad x \geq 0 \,,$$

we can use standard transformations to rewrite this system in the form

$$x_i = \alpha_{i0} - \sum_{k=m+1}^{n} \alpha_{ik} x_k \geq 0 \quad (i=1,\ldots,n)$$

$$x_{m+1} \geq 0,\ldots, x_n \geq 0 \,.$$

Viewing the polytope D as a subset of the (n–m)–dimensional space of $x_{m+1},\ldots,x_n$, we can then apply the above method with $g_i(x) = x_i$. In this case the matrix associated with a simplex $M = [u^1,u^2,\ldots,u^r]$ is given simply by the first m rows of the matrix (u_i^j), $i=1,\ldots,n$; $j=1,\ldots,r$.

3.4. Unbounded Feasible Set

The above method can also be interpreted as a conical algorithm. Indeed, introducing an additional variable t, we can rewrite the constraints on D as follows:

$$h_i(y) := \sum_{j=1}^{n} a_{ij} x_j - b_i t \geq 0 \quad (i=1,\ldots,m) \,,$$

$$y = (x,t) \in \mathbb{R}_+^{n+1} \,,$$

$$t = 1 \,.$$

Then D is embedded in a subset of the hyperplane $t = 1$ in $\mathbb{R}^{n+1}$, and each simplex $M = [u^1,u^2,\ldots,u^r]$ in $\mathbb{R}_+^n$ may be regarded as the intersection of the hyperplane $t = 1$ with the cone $K = \mathrm{con}(y^1,y^2,\ldots,y^r)$, where $y^i = (u^i,1)$. A simplicial subdivision of D may be regarded as being induced by a conical subdivision of $\mathbb{R}_+^{n+1}$.

With this interpretation the method can be extended to the case of an unbounded polyhedron D.

Specifically, for any $y = (x,t) \in \mathbb{R}_+^{n+1}$ define $\pi(y) = \frac{x}{t}$ if $t > 0$, $\pi(y) = x$ if $t = 0$.

Proposition VII.13. *For any cone $K = con(y^1,y^2,\dots,y^r) \subset \mathbb{R}^{n+1}_+$ with $y^i = (u^i,t^i)$ and $I = \{i:t^i > 0\}$, the set $\pi(K)$ is a generalized simplex with vertices $\pi(y^i)$, $i \in I$, and extreme directions $\pi(y^i)$, $i \notin I$.*

Proof. Let $z = \pi(y)$ with $y = (x,t) \in K$, i.e. $y = \sum_{i=1}^{r} \lambda_i(u^i,t^i)$, $\lambda_i \geq 0$. If $t = \sum_{i\in I} \lambda_i t^i > 0$, then

$$z = x/t = \sum_{i\in I} (\lambda_i t^i/t)u^i/t^i + \sum_{i\notin I} (\lambda_i/t)u^i$$

$$= \sum_{i\in I} \mu_i \pi(y^i) + \sum_{i\notin I} \mu_i \pi(y^i) ,$$

with $\sum_{i\in I} \mu_i = 1$, $\mu_i \geq 0$ $(i=1,\dots,r)$. If $t = 0$, i.e., $\lambda_i = 0$ $(i\in I)$, then $z = x = \sum_{i\notin I} \lambda_i u^i = \sum_{i\notin I} \lambda_i \pi(y^i)$ with $\lambda_i \geq 0$. Thus, in any case $z \in \pi(K)$ implies that z belongs to the generalized simplex generated by the points $\pi(y^i)$, $i \in I$, and the directions $\pi(y^i)$), $i \notin I$. The converse is obvious. ■

A cone $K = con(y^1,y^2,\dots,y^r) \subset \mathbb{R}^{n+1}_+$ is said to be **trivial** if for every $i=1,\dots,m$ we have

$$h_i(y^p) \cdot h_i(y^q) \geq 0 \quad \forall p,q .$$

The following propositions are analogous to those established in Section VII.3.1.

Proposition VII.14. *If a cone $K = con(y^1,y^2,\dots,y^r)$ is trivial, then $\pi(K) \cap D$ is the face of $\pi(K)$ whose vertices are those points $\pi(y^i)$, $i \in I = \{i: t^i > 0\}$, which are vertices of D and whose extreme directions are those vectors $\pi(y^i)$, $i \notin I$, which are extreme directions of D.*

Now if a cone $K = con(y^1,y^2,\dots,y^r)$ is nontrivial, we define its test index to be the smallest $s \in \{1,\dots,m\}$ such that there exist p, q satisfying

$$h_s(y^p)\cdot h_s(y^q) < 0 ,$$

and we let $\rho(K) = s +$ the number of indices j such that $h_s(y^j) = 0$.

Proposition VII.15. *Let K_1, K_2 be the subcones of K arising in the subdivision of K with respect to the point $z \in [y^p,y^q]$ such that $h_s(z) = 0$. Then $\rho(K_\nu) > \rho(K)$ for $\nu = 1,2$, unless K_ν is trivial.*

This subdivision of K into K_1,K_2 is called a **D–bisection** (or a bisection with respect to the constraints (27)).

As in the bounded case, it follows from this proposition that:

1) Any sequence of cones generated by successive D–bisections is finite.

2) Any cone in $\mathbb{R}_+^{n+1}$ can be partitioned into trivial subcones by means of a finite number of successive D–bisections. It follows that one can generate a conical partition of $\mathbb{R}_+^{n+1}$ by means of a finite sequence of D–bisections of $\mathbb{R}_+^{n+1}$ which induce a partition of D into generalized simplices.

In order to define a branch and bound algorithm based on this partition method, it remains to specify the bounds.

Let $K = \text{con}(y^1,y^2,\dots,y^r)$, with $y^i = (u^i,t^i) \in \mathbb{R}_+^{n+1}$, $I = \{i\colon t^i > 0\}$. Take an arbitrary $i^* \in I$. If there is a $j \notin I$ such that $f(\frac{u^{i^*}}{t^{i^*}} + \alpha u^j) < f(\frac{u^{i^*}}{t^{i^*}})$ for some $\alpha > 0$, then set $p(K) = -\infty$. Otherwise, set

$$\beta(K) = \min \{f(\frac{u^i}{t^i}) : i \in I\} .$$

Proposition VII.16. *The above bounding operation is finitely consistent.*

Proof. Since we start with $\mathbb{R}_+^{n+1}$ or with a cone in $\mathbb{R}_+^{n+1}$ containing D, it is easily seen that any cone $K = \text{con}(y^1,y^2,\dots,y^r)$ obtained in the above subdivision process must have $I \neq \emptyset$. Furthermore, since the infimum of a concave function f(x) over a halfline either is $-\infty$ or else is attained at the origin of the halfline, the number $\beta(K)$ computed above actually yields a lower bound for min $f(\pi(K) \cap D)$. As in the bounded case (cf. the proof of Proposition VII.12), every trivial cone is fathomed, hence any unfathomed cone is capable of further subdivision. By Proposition VII.15 (consequence 1) it then follows that any nested sequence of cones generated by the branch and bound process must be finite. ■

Corollary VII.8. *The BB procedure using the subdivision and bounding methods defined above along with the selection rule*

$$\mathcal{R}_k = \{K_k\}, \quad K_k \in \arg\min \{\beta(K) : K \in \mathcal{R}_k\}$$

is finite.

Remark VII.10. Incidentally, we have obtained a procedure for generating all the vertices and extreme directions of a given polyhedron D.

4. RECTANGULAR ALGORITHMS

A standard method for lower bounding in branch and bound algorithms for concave minimization is to use convex underestimators (cf. IV.4.3). Namely, if φ is the convex envelope of the objective function f taken over a set M, then the number min $\varphi(M)$ (which can be computed by convex programming methods) yields a lower bound for min f(M). The fact that the convex envelope of a concave function f taken over a simplex is readily computable (and is actually linear), gave rise to the conical and simplicial subdivisions most commonly used in concave minimization branch and bound algorithms.

Another case where the convex envelope is easy to compute is when the function f(x) is separable, while the set M is a rectangle (parallelepiped). In fact, by

Theorem IV.8, if $f(x) = \sum_{j=1}^{n} f_j(x_j)$, then the convex envelope of f(x) taken over a rectangle $M = \{x: r_j \le x_j \le s_j \ (j=1,...,n)\}$ is equal to the sum of the convex envelopes of the functions $f_j(t)$ taken over the line segments $r_j \le t \le s_j$ $(j=1,...,n)$.

Moreover, the convex envelope of a concave function $f_j(t)$ (of one variable) taken over a line segment $[r_j,s_j]$ is simply the affine function that agrees with f_j at the endpoints of this segment, i.e., the function

$$\varphi_j(t) = f_j(r_j) + \frac{f_j(s_j)-f_j(r_j)}{s_j-r_j}(t - r_j) . \tag{37}$$

Therefore, when the function f(x) is separable or can be made separable by an affine transformation of the variables, then rectangular subdivisions might be conveniently used in branch and bound procedures.

A branch and bound algorithm of this type for separable concave programming was developed by Falk and Soland in 1969. A variant of this algorithm was discussed in Horst (1977).

The same idea of rectangular subdivision was used for minimizing a concave quadratic function subject to linear constraints by Kalantari (1984), Kalantari and Rosen (1987), Pardalos (1985), Rosen and Pardalos (1986), Phillips and Rosen (1988). An extension to the case of indefinite quadratic objective function is given in Pardalos, Glick and Rosen (1987).

In this section, a "normal rectangular method" is introduced that includes previous rectangular procedures. We present here in detail the Falk–Soland method and the approach of Kalantari–Rosen. The Rosen–Pardalos procedure that has been successfully applied for typical large–scale problems will be presented in connection with the decomposition methods of the next chapter.

4.1. Normal Rectangular Algorithm

We first define the concept of a normal rectangular subdivision, which is similar to that of a normal simplicial subdivision.

Consider the separable concave programming problem

(SCP) $\quad$ minimize $f(x) := \sum_{j=1}^{n} f_j(x_j)$

subject to $x \in D$,

where D is a polytope contained in $\{x: c \le x \le d\} \subset \mathbb{R}^n$ and each $f_j(t)$ is a concave function continuous on the interval $[c_j, d_j]$.

Let $M = \{x: r \le x \le s\}$ be a rectangle with $c \le r$, $s \le d$ (in the sequel by "rectangle" we shall always mean a set of this type). Let $\varphi_M(x) = \sum_j \varphi_{Mj}(x_j)$ be the convex envelope of the function f(x) taken over M. Denote by $\omega(M)$ and $\beta(M)$ a basic optimal solution and the optimal value, respectively, of the linear program

LP(M,D) $\quad$ min $\varphi_M(x)$ $\quad$ s.t. $x \in D \cap M$.

Now consider a rectangular subdivision process, i.e., a subdivision process in which a rectangle is subdivided into subrectangles by means of a finite number of hyperplanes parallel to certain facets of the orthant $\mathbb{R}^n_+$. If the initial rectangle is $M_0 = \{x: r^0 \le x \le s^0\}$, then such a process generates a family of rectangles which can be represented by a tree with root M_0 and such that a node is a successor of another one if and only if it represents an element of the partition of the rectangle corresponding to the latter node. An infinite path in this tree corresponds to an infinite nested sequence of rectangles M_h , h=0,1,... For each h let $\omega^h = \omega(M_h)$, $\varphi_h(x) = \varphi_{M_h}(x)$.

Definition VII.7. *A nested sequence M_h, $h=0,1,\dots$, is said to be* ***normal*** *if*

$$\lim_{h\to\infty} |f(\omega^h) - \beta(M_h)| = 0 . \tag{38}$$

A rectangular subdivision process is said to be ***normal*** *if any infinite nested sequence of rectangles that it generates is normal.*

We shall later discuss how to construct a normal rectangular subdivision (NRS) process.

The importance of this concept is that it provides a sufficient condition for the convergence of branch and bound algorithms operating with rectangular subdivisions. To be specific, suppose we have an NRS process. Then, using this subdivision process in conjunction with the lower bounding defined above we can construct the following branch and bound algorithm.

Algorithm VII.6 (Normal Rectangular Algorithm).

Select a tolerance $\varepsilon \geq 0$ and an NRS process.

Initialization:

Choose a rectangle M_0 containing D. Let x^0 be the best feasible point available. Set $\mathcal{M}_0 = \mathcal{N}_0 = \{M_0\}$.

Iteration k = 0,1,... :

1) For each $M \in \mathcal{N}_k$ compute the affine function $\varphi_M(x)$ that agrees with f(x) at the vertices of M and solve the linear program

LP(M,D) $\quad \min \varphi_M(x) \quad$ s.t. $x \in D \cap M$

to obtain a basic optimal solution $\omega(M)$ and the optimal value $\beta(M)$.

2) Delete all $M \in \mathcal{M}_k$ such that $\beta(M) \geq f(x^k)-\varepsilon$. Let $\mathcal{R}_k$ be the remaining

collection of rectangles.

3) If $\mathscr{R}_k = \emptyset$, terminate: x^k is an ε–optimal solution of (SCP). Otherwise,

4) Select $M_k \in \operatorname{argmin} \{\beta(M): M \in \mathscr{R}_k\}$ and subdivide M_k according to the NRS process.

5) Update the incumbent, setting x^{k+1} equal to the best of the feasible solutions known so far.

Let $\mathscr{N}_{k+1}$ be the collection of subrectangles of M_k provided by the subdivision in Step 5, and let $\mathscr{M}_{k+1} = (\mathscr{R}_k \setminus \{M_k\}) \cup \mathscr{N}_{k+1}$. Set $k \leftarrow k+1$ and return to Step 1.

Theorem VII.11. *The normal rectangular algorithm can be infinite only if $\varepsilon = 0$, and in this case it generates a sequence $\{x^k\}$ every accumulation point of which is a global optimal solution of (SCP).*

The proof of this theorem is exactly the same as that of Theorem VII.8 on convergence of the normal simplicial algorithm.

4.2. Construction of an NRS Process

Let us examine the question of how to construct an NRS process.

The selection of M_k implies that $\beta(M_k) < f(x^k)-\varepsilon$; hence, setting $\omega^k = \omega(M_k)$, $\varphi_k(x) = \varphi_{M_k}(x)$, $\varphi_{k,j}(x_j) = \varphi_{M_{k,j}}(x_j)$, we have

$$f(\omega^k) - \varphi_k(\omega^k) > 0 . \tag{39}$$

Let $M_k = \{x: r^k \le x \le s^k\}$.

Choose an index $j_k \in \{1,2,...,n\}$ and a number $w^k \in (r^k_{j_k}, s^k_{j_k})$, and, using the hyperplane $x_{j_k} = w^k$, subdivide M_k into two rectangles

$$M_{k,1} = \{x \in M_k: x_{j_k} \le w^k\}, \quad M_{k,2} = \{x \in M_k: x_{j_k} \ge w^k\}.$$

Proposition VII.17. *The above subdivision rule (i.e., the rule for selecting j_k and w^k) generates an NRS process if it satisfies either of the following conditions:*

(i) $\quad j_k \in \underset{j}{\operatorname{argmax}} |f_j(\omega_j^k) - \varphi_{kj}(\omega_j^k)|$ *and* $w^k = \omega_j^k$;

(ii) $\quad j_k \in \underset{j}{\operatorname{argmax}}\, \sigma_{kj}$ *and* $w^k = (r_{j_k}^k + s_{j_k}^k)/2$, *where* σ_{kj} *is such that*

$$f_j(\omega_j^k) - \varphi_{kj}(\omega_j^k) \le \sigma_{kj}, \ \sigma_{kj} \to 0 \ \text{ if } s_j^k - r_j^k \to 0. \tag{40}$$

Proof. Let M_{k_h}, h=1,2,..., be any infinite nested sequence of rectangles generated by the rule. To simplify the notation let us write h for k_h. By taking a subsequence if necessary, we may assume that $j_h = j_0$, for example $j_h = 1$, for all h. It suffices to show that

$$\sigma_{h1} \to 0 \quad (h \to \infty), \tag{41}$$

where in case (i) we set $\sigma_{h1} = f_1(\omega_1^h) - \varphi_{h1}(\omega_1^h)$. From the definition of j_h it then follows that $f_j(\omega_j^h) - \varphi_{hj}(\omega_j^h) \to 0 \ \forall j$ and hence $f(\omega^h) - \varphi_h(\omega^h) \to 0$, which is the desired normality condition.

Clearly, if rule (ii) is applied, then the interval $[r_1^h, s_1^h]$ shrinks to a point as $h \to \infty$, and this implies (41), in view of (40). Now consider the case when rule (i) is applied. Again taking a subsequence if necessary, we may assume that $\omega_1^h \to \bar{\omega}_1$ as $h \to \infty$. If $c_1 < \bar{\omega}_1 < d_1$, then, since the concave function $f_1(t)$ is Lipschitzian in any bounded interval contained in (c_1, d_1) (cf. Rockafellar (1970), Theorem 10.4), for all h sufficiently large we have

$$|f_1(s_1^h) - f_1(r_1^h)| \le \eta |s_1^h - r_1^h|,$$

where η is a positive constant. Hence, using formula (37) for $\varphi_{hj}(t)$, we obtain

$$|\varphi_{h1}(\omega_1^h) - \varphi_{h1}(\omega_1^{h-1})| \le \eta\,|(\omega_1^h - \omega_1^{h-1}| \longrightarrow 0 \text{ as } h \longrightarrow \infty .$$

Since ω_1^{h-1} is one of the endpoints of the interval $[r_1^h, s_1^h]$, it follows that $\varphi_{h1}(\omega_1^{h-1}) = f_1(\omega_1^{h-1})$. Therefore,

$$|f_1(\omega_1^h)-\varphi_{h1}(\omega_1^h)| \le |f_1(\omega_1^h)-f_1(\omega_1^{h-1})| + |\varphi_{h1}(\omega_1^h)-\varphi_{h1}(\omega_1^{h-1})| \le$$

$$\le 2\eta|\omega_1^h-\omega_1^{h-1}| \longrightarrow 0,$$

proving (41).

On the other hand, if $\bar{\omega}_1$ coincides with an endpoint of $[c_1,d_1]$, for example $\bar{\omega}_1 = c_1$, then we must have $r_1^h = c_1$, $s_1^h = \omega_1^{h-1}$ $\forall h$, and hence, $s_1^h \longrightarrow \bar{\omega}_1 = c_1$ as $h \longrightarrow \infty$. Noting that $\varphi_{h1}(\omega_1^h) \ge \min\{f_1(c_1), f_1(s_1^h)\}$, we then conclude that

$$|f_1(\omega_1^h)-\varphi_{h1}(\omega_1^h)| \le f_1(\omega_1^h)-\min\{f_1(c_1),f_1(s_1^h)\} \longrightarrow 0 \text{ as } h \longrightarrow \infty,$$

by the continuity of $f_1(t)$.

Thus, (41) holds in any case, completing the proof of the proposition. ■

Corollary VII.9. *If the subdivision in Step 5 is performed according to either of the rules (i), (ii) described in Proposition VII.17, then Algorithm VII.6 converges (in the sense of Theorem VII.11).*

Let us call a subdivision according to rule (i) in Proposition VII.17 a ***bisection***, and a subdivision according to rule (ii) ***ω–subdivision***.

Algorithm VII.6 with the ω–subdivision rule was proposed by Falk and Soland (1969) as a relaxed form of an algorithm which was finite but somewhat more complicated (see VI.4.6). In contrast to the "complete" algorithm, the "relaxed" algorithm may involve iterations in which there is no concave polyhedral function agreeing with $\varphi_M(x)$ on each rectangle M of the current partition. For this reason, the argument used to prove finite convergence of the "complete" algorithm cannot be extended to the "relaxed" algorithm.

Intuitively, it can be expected that, as in the case of conical and simplicial algorithms, variants of rectangular algorithms using ω-subdivision should converge more rapidly than those using bisection, because ω-subdivision takes account of the solution of the current approximating subproblem (this idea will be illustrated by an example in VII.4.4). But because of the separable structure, a new feature is that, while ω-subdivision has to be properly combined with bisection to produce convergent conical and simplicial algorithms, no such combination is necessary for rectangular algorithms.

4.3. Specialization to Concave Quadratic Programming

As seen earlier, an important property of a quadratic function is that it can be transformed to separable form by an affine transformation. It seems natural, then, that branch and bound algorithms with rectangular subdivisions should be used for concave quadratic programming. In fact, such algorithms have been developped by Kalantari and Rosen (1987), Rosen and Pardalos (1986), etc. (cf. the introduction to Section IV). We discuss the Kalantari–Rosen procedure here as a spezicalised version of Algorithm VII.6, whereas the Rosen–Pardalos method will be treated in Chapter VIII (large-scale problems).

Consider the concave quadratic programming problem

$$\text{(CQP)} \qquad \text{minimize } f(x) := px - \tfrac{1}{2}x(Cx) \qquad \text{subject to } x \in D ,$$

where $p \in \mathbb{R}^n$, C is a symmetric positive definite $n \times n$ matrix, and D is the polytope in $\mathbb{R}^n$ defined by the linear inequalities

$$Ax \le b , \tag{42}$$

$$x \geq 0 \tag{43}$$

with $b \in \mathbb{R}^n$ and A an m×n matrix.

If $U = [u^1,u^2,...,u^n]$ is a matrix formed by n C–conjugate vectors, so that $U^TCU = \text{diag}(\lambda_1,\lambda_2,...,\lambda_n) > 0$, then, as shown in Section VI.3.3, after the affine transformation $x = Uy$ this problem can be rewritten as

$$\min F(y):= \sum_{j=1}^{n} F_j(y_j) \quad \text{s.t.} \quad y \in \Omega , \tag{44}$$

where $F_j(y_j) = q_jy_j - \frac{1}{2}\lambda_jy_j^2$ with $q = U^Tp$, and $\Omega = \{y: Uy \in D\}$.

In this separable form the problem can be solved by the normal rectangular algorithm. To specialize Algorithm VII.6 to this case, we need the following properties of separable concave quadratic functions.

Proposition VII.18. *Let $M = \{y: r \leq x \leq s\}$ be a rectangle. Then:*

1) The convex envelope of $F(y)$ over M is the linear function

$$\psi_M(y) = \sum_{j=1}^{n} \psi_{M,j}(y_j) \text{ , with}$$

$$\psi_{M,j}(y_j) = q_jy_j - \frac{1}{2}\lambda_j(r_j + s_j)y_j + \frac{1}{2}\lambda_jr_js_j \tag{45}$$

2) $$\max\{|F_j(y_j) - \psi_{M,j}(y_j)| : y \in M\} \leq \frac{1}{8}\lambda_j(s_j-r_j)^2 \tag{46}$$

Proof. 1) By Theorem IV.8, the convex envelope of F(y) over M is equal to $\psi_M(y) = \Sigma\psi_{Mj}(y_j)$, where $\psi_{Mj}(t)$ is the convex envelope of $F_j(t)$ over the interval $[r_j,s_j]$. Let $G_j(t) = -\frac{1}{2}\lambda_jt^2$. Since $F_j(t) = q_jt + G_j(t)$, it follows that

$$\psi_{Mj}(t) = q_jt + \gamma_j(t) ,$$

where $\gamma_j(t)$ is the convex envelope of $G_j(t)$ over the interval $[r_j,s_j]$. But $\gamma_j(t)$ is an affine function that agrees with $G_j(t)$ at the endpoints of this interval; hence, after an easy computation we obtain

$$\gamma_j(t) = -\frac{1}{2}\lambda_j r_j^2 - \frac{1}{2}\lambda_j(r_j + s_j)(t - r_j)$$

$$= -\frac{1}{2}\lambda_j(r_j + s_j)t + \frac{1}{2}\lambda_j r_j s_j . \qquad (47)$$

This proves the first assertion in the theorem.

2) From (47) we have:

$$F_j(y) - \psi_{M,j}(y) = G_j(y_j) - \gamma_j(y_j)$$

$$= -\frac{1}{2}\lambda_j y_j^2 + \frac{1}{2}\lambda_j(r_j + s_j)y_j - \frac{1}{2}\lambda_j r_j s_j$$

$$= \frac{1}{2}\lambda_j(y_j - r_j)(s_j - y_j) . \qquad (48)$$

Since the two numbers $y_j - r_j$ and $s_j - y_j$ have a constant sum (equal to $s_j - r_j$), their product is maximum when they are equal, and then their product is equal to $\frac{1}{2}(s_j - r_j)^2$, from which (46) follows. ■

With these results in mind, it is now easy to specialize Algorithm VII.6 to the problem (44).

According to Corollary VII.8, for the subdivision operation in Step 5 we can use either of the rules given in Proposition VII.17. Rule (i) is easy to apply, and corresponds to the method of Falk and Soland. Rule (ii) requires us to choose numbers σ_{kj} satisfying (40). Formula (46) suggests taking

$$\sigma_{k,j} = \frac{1}{8}\lambda_j(s_j^k - r_j^k)^2 . \qquad (49)$$

For each rectangle $M = \{y: r \le y \le s\}$ to be investigated in Step 1, the objective function of the linear program LP(M,Ω) is computed according to formula (45), while the constraints are

$$AUy \le b,\ Uy \ge 0,\ r \le y \le s .$$

In certain cases, the constraints of the original polytope have some special structure which can be exploited in solving the subproblems. Therefore, it may be more advantageous to work with the original polytope D (in the x–space), rather than with the transformed polytope Ω (in the y–space). Since $x = Uy$ and $\lambda_j y_j = u^j(Cx)$, a rectangle $M = \{y: r \le y \le s\}$ is described by inequalities

$$\lambda_j r_j \le u^j(Cx) \le \lambda_j s_j \quad (j=1,2,...,n) ,$$

while the convex envelope $\varphi_M(x)$ of f(x) over M is

$$\varphi_M(x) = px - \frac{1}{2}\Sigma\,(r_j+s_j)\,u^j(Cx) + \frac{1}{2}\Sigma\,\lambda_j r_j s_j \,. \tag{50}$$

We can thus state the following algorithm of Kalantari and Rosen (1987) for solving (CQP):

Algorithm VII.7.

Initialization:

Select $\varepsilon \ge 0$. Solve the 2n linear programs

$$\min\,\{u^j(Cx): x \in D\},\ \max\,\{u^j(Cx): x \in D\}$$

obtaining the basic optimal solutions x^{0j}, $\bar{x}^{0j}$ and the optimal values η_j, $\bar{\eta}_j$ of these programs, respectively.

Clearly, $D \subset M_0 = \{x: \lambda_j\eta_j \le u^j(Cx) \le \lambda_j\bar{\eta}_j\ , j=1,2,...,n\}$. Set $\mathscr{M}_1 = \mathscr{N}_1 = \{M_0\}$, $x^0 = \text{argmin}\{f(x^{0j}), f(\bar{x}^{0j}), j=1,2,...,n\}$.

Iteration k = 1,2,... :

1) For each $M \in \mathscr{N}_k$ compute $\varphi_M(x) = \sum_j \varphi_{M,j}(x_j)$ according to (50) and solve the linear program

LP(M,D) $\quad \min\ \varphi_M(x)$ s.t.

$$x \in D,\ \lambda_j r_j \le u^j(Cx) \le \lambda_j s_j \quad (j=1,2,...,n)$$

to obtain a basic optimal solution $\omega(M)$ and the optimal value $\beta(M)$.

2) Update the incumbent by setting x^k equal to the best among all feasible solutions so far encountered: x^{k-1} and all $\omega(M)$, $M \in \mathcal{N}_k$.
Delete all $M \in \mathcal{M}_k$ such that $\beta(M) \geq f(x^k)-\varepsilon$. Let $\mathcal{R}_k$ be the remaining collection of rectangles.

3) If $\mathcal{R}_k = \emptyset$, terminate: x^k is a global ε-optimal solution of (CQP). Otherwise,

4) Select $M_k \in \operatorname{argmin} \{\beta(M): M \in \mathcal{R}_k\}$.

5) Let $j_k \in \operatorname{argmax} \{\sigma_{kj}: j=1,2,...,n\}$, where σ_{kj} is given by (49) (r^k, s^k are the vectors that define M_k), $w^k = \frac{1}{2}(r^k_{j_k} + s^k_{j_k})$.
Subdivide M_k into two subrectangles M_{k1}, M_{k2} by means of the hyperplane $u^j(Cx) = \frac{1}{2}\lambda_j(r^k_{j_k} + s^k_{j_k})$.

6) Let $\mathcal{N}_{k+1} = \{M_{k1}, M_{k2}\}$, $\mathcal{M}_{k+1} = (\mathcal{R}_k \setminus \{M_k\}) \cup \mathcal{N}_{k+1}$.
Set $k \leftarrow k+1$ and return to Step 1.

Remarks VII.11. (i) As already mentioned, in Step 5 instead of bisection one could also use ω-subdivision. If we denote $\omega^k = \omega(M_k)$, $\pi^k = U^{-1}\omega^k$, then, since $\lambda_j \pi^k_j = (u^j)^T C\omega^k$, we have from (48)

$$2[F_j(\pi^k)-\psi_{Mj}(\pi^k)] = [u^j(C\omega^k)-\lambda_j r^k_j]\,[\lambda_j s^k_j - u^j(C\omega^k)]\;/\;\lambda_j. \qquad (51)$$

Therefore, if one uses the ω-subdivision, then one should choose the index j_k that maximizes (51), and divide the parallelepiped M_k by the hyperplane $u^{j_k}(C(x-\omega^k)) = 0$.

(ii) Another way of improving the algorithm is to use a concavity cut from time to time (cf. Section V.1) to reduce the feasible polytope. Kalantari (1984) reported computational experiments showing the usefulness of such cuts for convergence of the algorithm. Of course, deeper cuts specially devised for concave quadratic programming, such as those developed by Konno (cf. Section V.4), should provide even better results.

(iii) Kalantari and Rosen (1987), and also Rosen and Pardalos (1986), Pardalos and Rosen (1987), actually considered linearly constrained concave quadratic programming problems in which the number of variables that enter the nonlinear part of the objective function is small in comparison with the total number of variables. In the next chapter these applications to the large-scale case will be discussed in a more general framework.

4.4. Example VII.2.

The following simple example is taken from Kalantari and Rosen (1987), where one can also find details of computational results with Algorithm VII.7.

Let $f(x) = -\frac{1}{2}(2x_1^2 + 8x_2^2)$, and let the polytope D be given by the constraints:

$$x_1 + x_2 \le 10,\ x_1 + 5x_2 \le 22,\ -3x_1 + 2x_2 \le 2\ ,$$

$$-x_1 - 4x_2 \le -4,\ x_1 - 2x_2 \le 4\ .$$

We have $u^i = e^i$ (i–th unit vector, i=1,2) and these are normalized eigenvectors of C, $\lambda_1 = 2$, $\lambda_2 = 8$.

It can easily be checked that

$$M_1 = \{x\colon 0 \le x_1 \le 8,\ 0 \le x_2 \le 4\},\ x_0 = (8,2),\ f(x^0) = -80\ .$$

(Below β_k, ω^k stand for $\beta(M_k)$, $\omega(M_k)$ respectively)

Algorithm VII.7 using bisection (Kalantari–Rosen).

Iteration 1:

$\beta_0 = -104,\ \omega_0 = (7,3)$,
$x_1 = (7,3),\ f(x^1) = -85$.
$M_1 = M_0$,
$f(\omega^1) - \varphi_{M_1} = 19$. Divide M_1 into
M_{11} and M_{12} $(j_1 = 1)$.

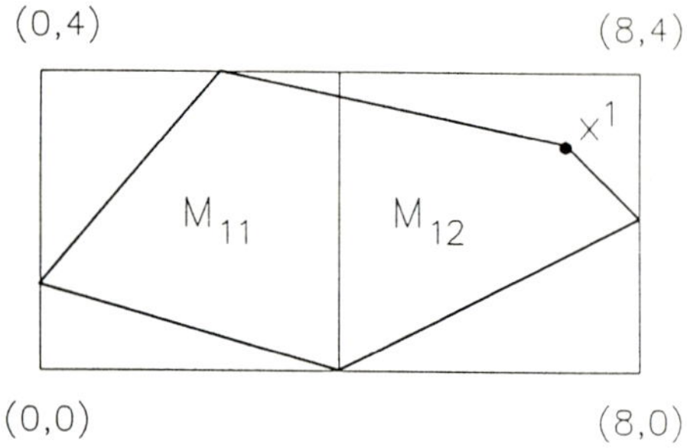

Fig. a

Iteration 2:

$\beta_{11} = -73.6,\ \omega^{11} = (4,3.6)$,
$\beta_{12} = -100.\ ,\ \omega^{12} = (7,3)$,
$x^2 = (7,3),\ f(x^2) = -85$.
M_{11} fathomed; $M_2 = M_{12}$
$f(\omega^2) - \varphi_{M_2}(\omega^2) = 15$. Divide M_2 into
M_{21} , M_{22} $(j_2 = 2)$.

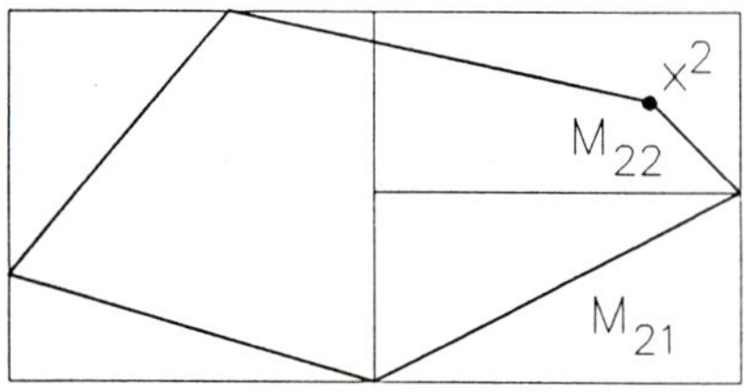

Fig. b

Iteration 3:

$\beta_{21} = -80\ ,\ \omega^{21} = (8,2)$,
$\beta_{22} = -90\ ,\ \omega^{22} = (7,3)$,
$x^3 = (7,3),\ f(x^3) = -85$.
M_{21} fathomed; $M_3 = M_{22}$,
$f(\omega^3) - \varphi_{M_3}(\omega^3) = 7$. Divide M_3 into
M_{31} , M_{32} $(j_3 = 1)$

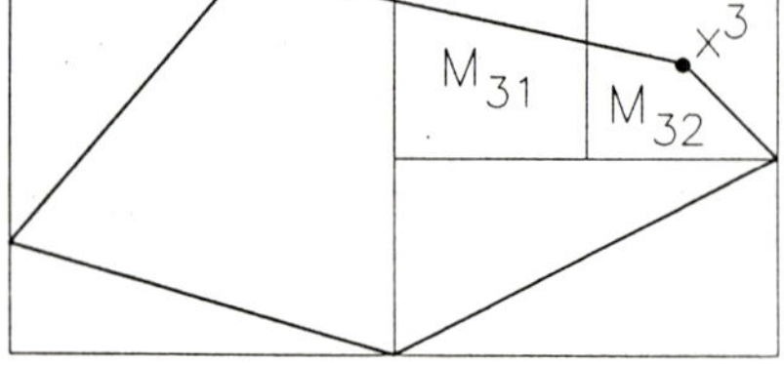

Fig. c

Iteration 4:

$\beta_{31} = -80.8$, $\omega^{31} = (6, 3.2)$,
$\beta_{32} = -90.$, $\omega^{32} = (7,3)$,
$x^4 = (7,3)$, $f(x^4) = -85$.
M_{31} is fathomed; $M_4 = M_{32}$,
$f(\omega^4) - \varphi_{M_4}(\omega^4 = 5$. Divide M_4 into
M_{41}, M_{42} $(j_4 = 2)$

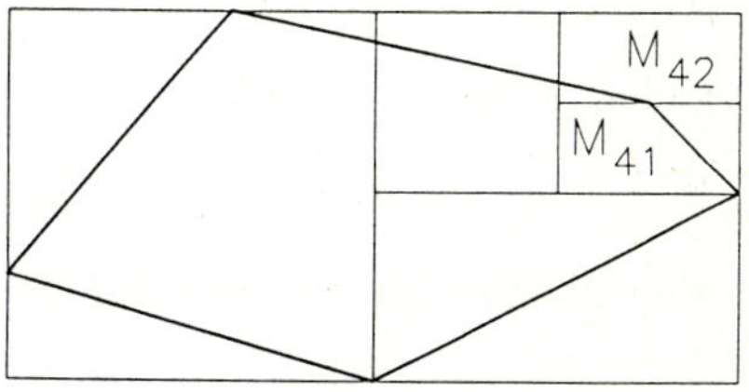

Fig. d

Iteration 5:

$\beta_{41} = -86$, $\omega^{41} = (7,3)$,
$\beta_{42} = -86$, $\omega^{42} = (7,3)$,
$x^5 = (7,3)$, $f(x^5) = -85$.
$M_5 = M_{42}$,
$f(\omega^5) - \varphi_{M_5}(\omega^5) = 1$. Divide M_5 into
M_{51} , M_{52} $(j_5 = 1)$.

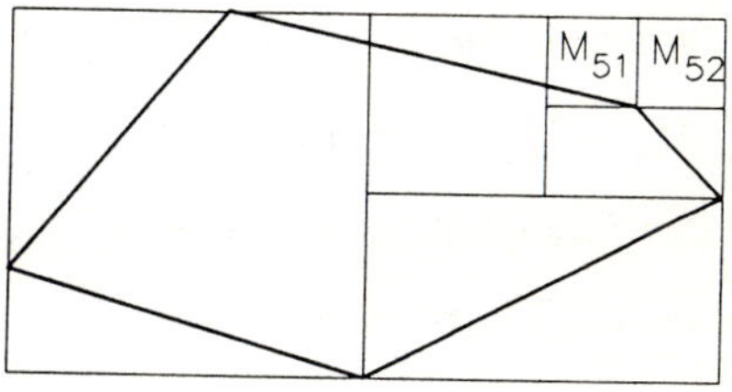

Fig. e

Iteration 6:

$\beta_{51} = -85$, $\omega^{51} = (7,3)$,
$\beta_{52} = -85$, $\omega_{52} = (7,3)$,
$x^5 = (7,3)$, $f(x^5) = -85$.
M_{51} , M_{52} are fathomed; $M_6 = M_{41}$,
$f(\omega^6) - \varphi_{M_6} = 1$. Divide M_6 into
M_{61} , M_{62} $(j_6 = 1)$.

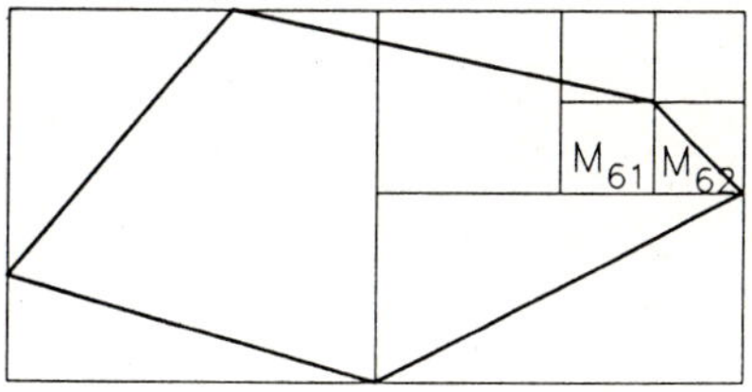

Fig. f

Iteration 7.

$\beta_{61} = \beta_{62} = -85$, $x^7 = (7,3)$, $f(x^7) = -85$. Hence $\mathcal{R}_7 = \emptyset$ and $x^7 = (7,3)$ is a global optimal solution.

Note that the solution (7,3) was already encountered in the first iteration, but the algorithm had to go through 6 more iterations to prove its optimality.

Algorithm using ω–subdivision (Falk–Soland).

Applying rule (ii) in Proposition VII.17 we have in Iteration 1:
$\omega^1 = (7,3)$, $j_1 = 1$ and M_1 is divided into M_{11} , M_{12} as in Fig. a.
In Iteration 2, since $\beta_{11} = \beta_{12} = -97$, we can take $M_2 = M_{12}$; then $\omega^2 = (7,3)$, $j_2 = 2$ and M_2 is divided into M_{21} , M_{22} as in Fig. b. In iteration 3, $\beta_{21} = \beta_{22} = -85$, while the incumbent is $x^3 = (7,3)$ with $f(x^3) = -85$, so M_{21} , M_{22} are fathomed, and $M_3 = M_{11}$. Then $\omega^3 = (7,3)$, $j_3 = 2$, and M_3 is divided into M_{31} , M_{32} , which will be fathomed in iteration 4, because $\beta_{31} = \beta_{32} = -85$. Thus, the algorithm will terminate after only 4 iterations.

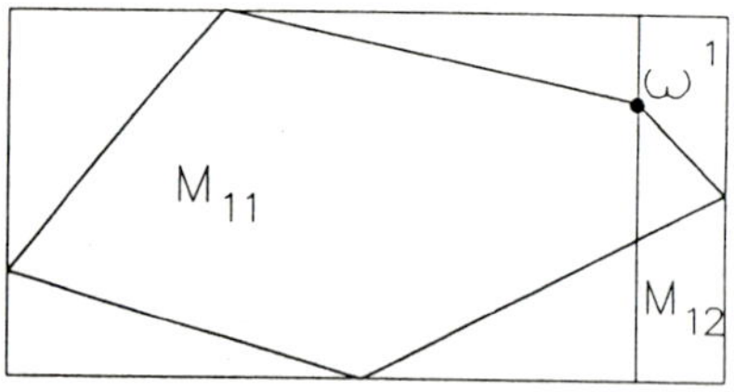

Fig. a

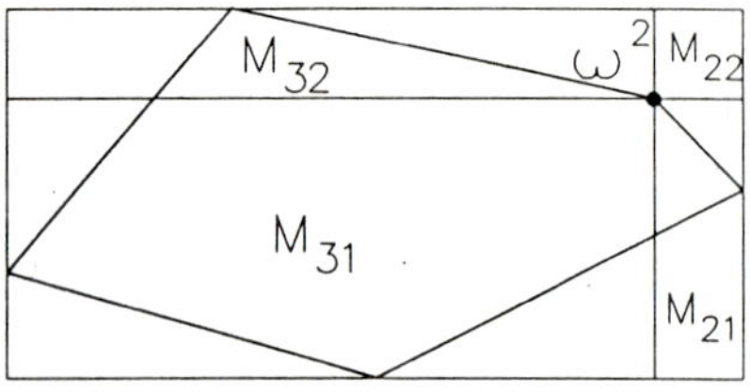

Fig. b

CHAPTER VIII

DECOMPOSITION OF LARGE SCALE PROBLEMS

In many problems of large size encountered in applications, the constraints are linear, while the objective function is a sum of two parts: a linear part involving most of the variables of the problem, and a concave part involving only a relatively small number of variables. More precisely, these problems have the form

$$\text{(P)} \qquad \text{minimize } f(x) + dy \quad \text{subject to } (x,y) \in \Omega \subset \mathbb{R}^n \times \mathbb{R}^h ,$$

where $f: \mathbb{R}^n \longrightarrow \mathbb{R}$ is a concave function, Ω is a polyhedron, d and y are vectors in $\mathbb{R}^h$, and n is generally much smaller than h.

In solving these problems it is essential to consider methods which take full advantage of this specific structure in order to save computational effort.

In this chapter, different approaches to the decomposition of problem (P) are presented within a general framework. These approaches include branch and bound techniques, polyhedral underestimation and outer approximation methods. Important special cases such as separable concave minimization and concave minimization on networks are discussed.

1. DECOMPOSITION FRAMEWORK

Denote by D the projection of Ω on the x–space, i.e.,

$$D = \{x \in \mathbb{R}^n : \exists y \text{ such that } (x,y) \in \Omega\} . \tag{1}$$

We shall assume that for every fixed $x \in D$ the linear function dy attains a minimum over the set of all y such that $(x,y) \in \Omega$. Define the function

$$g(x) = \min \{dy: (x,y) \in \Omega\} . \tag{2}$$

Proposition VIII.1. *D is a polyhedron, and g(x) is a convex polyhedral function with dom g = D.*

Proof. D is a polyhedron because it is the image of a polyhedron under a linear transformation from $\mathbb{R}^{n+1}$ to $\mathbb{R}^n$ (see e.g. Rockafellar (1970), Theorem 19.3). Since any $(x,y) \in \Omega$ can be written as

$$(x,y) = \sum_{i=1}^{s} \lambda_i(u^i,v^i), \ \sum_{i=1}^{r} \lambda_i = 1, \ \lambda_i \geq 0 \ (\forall i) , \tag{3}$$

where $r \leq s$, and since (u^i,v^i) for $i \leq r$ are the extreme points of Ω, while (u^i,v^i) for $i > r$ are the extreme directions of Ω, we have $g(x) = \inf \sum_{i=1}^{s} \lambda_i dv^i$, where the infimum is taken over all choices of λ_i satisfying (3). That is, g(x) is finitely generated, and hence convex polyhedral (see Rockafellar (1970), Corollary 19.1.2). Furthermore, from (1) and (2) it is obvious that $g(x) < +\infty$ if and only if $x \in D$. ■

Proposition VIII.2. *Problem (P) is equivalent to the problem*

(H) *minimize* $f(x) + g(x)$ *subject to* $x \in D \subset \mathbb{R}^n$.

Specifically, if $(\bar{x},\bar{y})$ *solves (P), then* $\bar{x}$ *solves (H), and if* $\bar{x}$ *solves (H), then* $(\bar{x},\bar{y})$ *solves (P), where* $\bar{y}$ *is the point satisfying* $g(\bar{x}) = d\bar{y}$, $(\bar{x},\bar{y}) \in \Omega$.

Proof. If $(\bar{x},\bar{y})$ solves P, then $\bar{x} \in D$ and $f(\bar{x}) + d\bar{y} \le f(\bar{x}) + dy$ for all y such that $(\bar{x},y) \in \Omega$; hence $d\bar{y} = g(\bar{x})$, and we have $f(\bar{x}) + g(\bar{x}) = f(\bar{x}) + d\bar{y} \le f(x) + dy$ for all $(x,y) \in \Omega$. This implies that $f(\bar{x}) + g(\bar{x}) \le f(x) + g(x)$ for all $x \in D$, i.e., $\bar{x}$ solves (H).

Conversely, suppose that $\bar{x}$ solves (H) and let $\bar{y} \in \operatorname{argmin} \{dy: (\bar{x},y) \in \Omega\}$. Then $f(\bar{x}) + d\bar{y} = f(\bar{x}) + g(\bar{x}) \le f(x) + g(x)$ for all $x \in D$ and hence $f(\bar{x}) + d\bar{y} \le f(x) + dy$ for all $(x,y) \in \Omega$, i.e., $(\bar{x},\bar{y})$ solves (P). ∎

Thus, solving (P) is reduced to solving (H), which involves only the x variables and might have a much smaller dimension than (P). Difficulties could arise for the following two reasons:

1) the function g(x) is convex (therefore, f(x) + g(x) is neither convex nor concave, but is a difference of two convex functions);

2) the function g(x) and the set D are not defined explicitly.

To cope with these issues, a first method is to convert the problem (H) to the form

$$(\tilde{H}) \qquad \text{minimize } f(x) + t \quad \text{subject to } x \in D,\ g(x) \le t, \tag{4}$$

and to deal with the implicitly defined constraints of this concave program in the same way as it is done in Benders' decomposition procedure.

We shall see, however, that in many cases the two mentioned points can merely be bypassed by employing appropriate methods such as branch and bound or concave underestimation.

We shall discuss these methods in the next sections, where, for the sake of simplicity, we shall assume that the set D is bounded.

2. BRANCH AND BOUND APPROACH

One of the most appropriate techniques for solving (H) without explicit knowledge of the function g(x) and the set D is the branch and bound approach (cf. Horst and Thoai (1989)).

In fact, in this approach all that we need to solve the problem is:

1) the construction of a simple polyhedron M_0 in the x–space which contains D, to be used for initialization;

2) a practical procedure for computing a lower bound for the minimum of f(x) + g(x) over any partition set M, such that the lower bounding is consistent (see Chapter IV).

Quite often, the construction of the starting polyhedron M_0 is straightforward. In any event, by solving the linear programs

$$\min \{x_j: (x,y) \in \Omega\} \ , \ \max \{x_j: (x,y) \in \Omega\} \quad (j=1,\dots,n)$$

one can always determine the smallest rectangle containing D.

As lower bounding, it should not present any particular difficulty, in view of the following result.

Proposition VIII.3. *For any polyhedron $M \subset M_0$, if $\psi_M(x)$ is a linear underestimator of f(x) on M, then the optimal value of the linear program*

$$(\mathrm{LP}(\mathrm{M},\Omega)) \qquad \min\ [\psi_M(x) + dy] \ \ s.t.\ \ x \in M,\ (x,y) \in \Omega \tag{5}$$

yields a lower bound for $\min \{f(x) + g(x): x \in M \cap D\}$.

Proof. Indeed, by Proposition VII.2, problem (5) is equivalent to

$$\min \{\psi_M(x) + dy: x \in M \cap D\} .$$

■

Thus, to compute a lower bound for f(x) + g(x) over $M \cap D$, it suffices to solve a linear program of the form (5). In this manner, any branch and bound algorithm originally devised for concave minimization over polytopes that *uses linear underestimators for lower bounding*, can be applied to solve the reduced problem (H). In doing this, there is no need to know the function f(x) and the set D explicitly. The convergence of such a branch and bound algorithm when extended to (H) can generally be established in much the same way as that of its original version.

Let us examine how the above decomposition scheme works with the branch and bound algorithms previously developed for (BCP).

2.1. Normal Simplicial Algorithm

When minimizing a concave function f(x) over a polytope $D \subset \mathbb{R}^n$, the normal simplicial algorithm (Algorithm VII.3) starts with an initial n–simplex $M_0 \supset D$, and proceeds through successive simplicial subdivision of M_0. For each n–subsimplex M of M_0, a lower bound for f(x) over $M \cap D$ is taken to be the optimal value of the linear program

$$\min \{\varphi_M(x): x \in M \cap D\} , \tag{6}$$

where $\varphi_M(x)$ is the linear underestimator of f(x) given simply by the affine function that agrees with f(x) at the n+1 vertices of M. In order to extend this algorithm to problem (H), it suffices to consider the linear program (5) in place of (6). We can thus formulate the following algorithm for solving (P).

Assume that the polyhedron Ω is defined by the inequalities

$$Ax + By \leq c, \quad x \geq 0, \quad y \geq 0 .$$

Then for any simplex $M = [v^1, v^2, \ldots, v^{n+1}]$ in x–space, since $x \in M$ is a convex combination of the vertices $v^1,v^2,...,v^{n+1}$, the lower bounding subproblem (5) can also be written as:

$$\begin{aligned} \min \quad & \Sigma \lambda_i f(v^i) + dy \\ \text{s.t.} \quad & \Sigma \lambda_i A v^i + By \leq c , \\ & \Sigma \lambda_i = 1 , \lambda_i \geq 0 \ \forall i, y \geq 0 \end{aligned} \tag{7}$$

(cf. Section VII.2.4).

Algorithm VIII.1 (Normal Simplicial Algorithm for (P)).

Select a tolerance $\varepsilon \geq 0$, and an exhaustive simplicial subdivision process (cf. Section VII.1.3 and VII.1.6)

Initialization:

Construct an n–simplex M_0 such that $D \subset M_0 \subset \mathbb{R}^n_+$. Let (x^0,y^0) be the best feasible solution available. Set $\mathcal{M}_1 = \mathcal{N}_1 = \{M_0\}$.

Iteration k = 1,2,...:

1) For each $M \in \mathcal{N}_k$ solve the linear program (7) to obtain a basic optimal solution $(\omega(M),y(M))$ and the optimal value $\beta(M)$ of $LP(M,\Omega)$.

2) Update the incumbent by setting (x^k,y^k) equal to the best among all feasible solutions known so far: (x^{k-1},y^{k-1}) and $(\omega(M),y(M))$ for all $M \in \mathcal{N}_k$.
Delete all $M \in \mathcal{M}_k$ for which $\beta(M) \geq f(x^k) + dy^k - \varepsilon$. Let $\mathcal{R}_k$ be the remaining collection of simplices.

3) If $\mathcal{R}_k = \emptyset$, terminate: (x^k,y^k) is an ε–optimal solution of (P). Otherwise, continue.

4) Select $M_k \in \operatorname{argmin}\{\beta(M): M \in \mathcal{R}_k\}$ and subdivide it according to the chosen

exhaustive process.

5) Let $\mathscr{N}_{k+1}$ be the partition of M_k and $\mathscr{M}_{k+1} = (\mathscr{R}_k \setminus \{M_k\}) \cup \mathscr{N}_{k+1}$. Set $k \leftarrow k+1$ and return to Step 1.

Theorem VIII.1. *Algorithm VIII.1 can be infinite only if $\varepsilon = 0$, and in this case, any accumulation point of the generated sequence $\{(x^k,y^k)\}$ is a global optimal solution of (P).*

Proof. Consider the algorithm as a branch and bound procedure applied to problem (H), where for each simplex M, the value $\alpha(M) = f(\omega(M)) + d(y(M))$ is an upper bound for min f(M), and $\beta(M)$ is a lower bound for min f(M) (cf. Section IV.1). It suffices to show that the bounding operation is consistent, i.e., that for any infinite nested sequence $\{M_q\}$ generated by the algorithm we have $\lim(\alpha_q - \beta_q) = 0$ $(q \to \infty)$, where $\alpha_q = \alpha(M_q)$, $\beta_q = \beta(M_q)$ (then the proof can be completed in the same way as for Theorem VII.8).

But if we denote $\psi_q(\cdot) = \psi_{M_q}(\cdot)$, $\omega^q = \omega(M_q)$, then clearly $\alpha_q - \beta_q = f(\omega^q) - \psi_q(\omega^q)$. Now, by the exhaustiveness of the subdivision process, M_q shrinks to a point as $q \to \infty$. Since the function $\psi_q(x)$ agrees with f(x) at the vertices of M_q, we have $\min f(M_q) \le \psi_q(\omega^q) \le f(\omega^q)$. Furthermore, by the continuity of f, $[\min f(M_q) - f(\omega^q)] \to 0$. Hence, $f(\omega^q) - \psi_q(\omega^q) \to 0$, as was to be proved. ∎

We recall from the discussion in Section VII.1.6 that in practice, instead of an exhaustive rule one can often use the following simple rule for simplicial subdivision

Choose $\gamma > 0$ sufficiently small. At iteration k, let $\lambda^k = (\lambda_1^k,...,\lambda_n^k)$ be an optimal solution of the linear program (7) for $M = M_k$. Use ω–subdivision if $\min \{\lambda_i^k : \lambda_i^k > 0\} \ge \gamma$, and bisection otherwise.

2.2. Normal Rectangular Algorithm

Suppose that the function f(x) is separable:

$$f(x) = \sum_{j=1}^{n} f_j(x_j) .$$

As previously shown (Theorem IV.8), the convex envelope of such a function over a rectangle $M = \{x \in \mathbb{R}^n: r \le x \le s\}$ is equal to

$$\psi_M(x) = \Sigma \, \psi_{M,j}(x_j) , \tag{8}$$

where each $\psi_{M,j}(t)$ is the affine function of one variable which agrees with $f_j(t)$ at the endpoints of the interval $[r_j,s_j]$. Hence, a lower bound for f(x) + g(x) over M can be obtained by solving the linear program (5), with $\psi_M(x)$ defined by (8). This leads to the following extension of Algorithm VII.6 for solving problem (P).

Algorithm VIII.2 (Normal Rectangular Algorithm for (P)).

Select a tolerance $\varepsilon \ge 0$.

Initialization:

Construct a rectangle $M_0 = \{x \in \mathbb{R}^n: r^0 \le x \le s^0\}$ such that $D \subset M_0 \subset \mathbb{R}^n_+$. Let (x^0,y^0) be the best feasible solution available. Set $\mathcal{M}_1 = \mathcal{N}_1 = \{M_0\}$.

Iteration k = 1,2,...:

1) For each member $M = \{x \in \mathbb{R}^n: r \le x \le s^0\}$ of $\mathcal{N}_k$ compute the function $\psi_M(x)$ according to (8) and solve the linear programm

LP(M,Ω) $\quad\quad$ $\min \psi_M(x) + dy \quad$ s.t. $r \le x \le s, \ Ax + By \le c$

Let $(\omega(M),y(M))$ and $\beta(M)$ be a basic optimal solution and the optimal value of LP(M,Ω).

2) Update the incumbent by setting (x^k,y^i) equal to the best feasible solution

among (x^{k-1},y^{k-1}) and all $\omega(M),y(M))$, $M \in \mathcal{N}_k$.

Delete all $M \in \mathcal{M}_k$ for which $\beta(M) \geq f(x^k) + dy^k - \varepsilon$. Let $\mathcal{R}_k$ be the remaining collection of rectangles.

3) If $\mathcal{R}_k = \emptyset$, terminate: (x^k,y^k) is a global ε–optimal solution of (P). Otherwise, continue.

4) Select $M_k \in \operatorname{argmin}\{\beta(M): M \in \mathcal{R}_k\}$. Let $\omega^k = \omega(M_k)$, and let $\psi_{kj}(x) = \psi_{M_kj}(x)$.

5) Select $j_k \in \operatorname{argmax}\{|f_j(\omega_j^k) - \psi_{kj}(\omega_j^k)|: j=1,\ldots,n\}$. Divide M_k into two subrectangles by the hyperplane $x_{j_k} = \omega_{j_k}^k$.

6) Let $\mathcal{N}_{k+1}$ be the partition of M_k, $\mathcal{M}_{k+1} = (\mathcal{R}_k \setminus \{M_k\}) \cup \mathcal{N}_k$. Set $k \leftarrow k+1$ and return to 1).

Theorem VIII.2. *Algorithm VIII.2 can be infinite only if* $\varepsilon = 0$ *and in this case, any accumulation point of* $\{(x^k,y^k)\}$ *is a global optimal solution of (P).*

Proof. As with Algorithm VIII.1, all that needs to be proved is that $f(\omega^q) - \psi_q(\omega^q) \longrightarrow 0$ as $q \longrightarrow \infty$ for any infinite nested sequence $\{M_q\}$ generated by the algorithm (the notation is the same as in the proof of Theorem VIII.1). But this follows from Proposition VII.17(i). ■

Remark VIII.1. The above algorithm of course can be applied to concave quadratic programming problems, since a concave quadratic function f(x) can always be made separable by means of an affine transformation of the variables. An alternative algorithm can also be obtained by a similar extension of Algorithm VII.7 which is left to the reader.

2.3. Normal Conical Algorithm

Strictly speaking, the normal conical algorithm (Section VII.1) is not a branch and bound procedure using linear underestimators for lower bounding. However, it can be extended in a similar way to solve problems of the form (P), with $d = 0$ (i.e., where the objective function does not depend upon y):

$$\text{minimize } f(x) \quad \text{subject to } Ax + By \le c,\ x \ge 0,\ y \ge 0. \tag{9}$$

Obviously, by introducing an additional variable t, problem (P) can also be written as

$$\text{minimize } f(x) + t \quad \text{subject to } dy \le t,\ (x,y) \in \Omega\ ,$$

i.e., a problem of the form (9), with $(x,t) \in \mathbb{R}^{n+1}$ in the role of x.

As before, let $D = \{x \in \mathbb{R}^n\colon \exists y \ge 0 \text{ such that } Ax + By \le c,\ x \ge 0\}$. Then (9) becomes a BCP problem to which Algorithm VII.1* can be applied. The fact that D is not known explicitly does not matter, since the linear program

$$\text{LP}(Q;D) \qquad \max\ eQ^{-1}x \quad \text{s.t. } x \in D,\ Q^{-1}x \ge 0$$

that must be solved in Step 2 of Algorithm VII.1* is simply

$$\max\ eQ^{-1}x \quad \text{s.t. } Ax + By \le c,\ Q^{-1}x \ge 0\ ,\ y \ge 0$$

(we assume con $Q \subset \mathbb{R}^n_+$, so the constraint $x \ge 0$ is implied by $Q^{-1}x \ge 0$).

Of course, the same remark applies to Algorithm VII.1. Thus, after converting problem (P) into the form (10), the application of the normal conical algorithm is straightforward.

Note that in all the above methods, the function g(x) and the set D are needed only conceptually and are not used in the actual computation.

3. POLYHEDRAL UNDERESTIMATION METHOD

Another approach which allows us to solve (H) without explicit knowledge of the function g(x) and the set D uses polyhedral underestimation.

3.1. Nonseparable Problems

The polyhedral underestimation method for solving (BCP) is based on the following property which was derived in Section VI.4 (cf. Proposition VI.10):

To every finite set X_k in $\mathbb{R}^n$ such that conv $X_k = M_1 \supset D$ one can associate a polyhedron

$$S_k = \{q,q_0) \in \mathbb{R}^n \times \mathbb{R}: qx - q_0 \geq f(x) \ \ \forall x \in X_k\}$$

such that the function

$$\varphi_k(x) = \min \{qx - q_0: (q,q_0) \in \text{vert } S_k\} \tag{10}$$

is the lowest concave function which agrees with f(x) at all points of X_k (so, in particular, $\psi_k(x)$ underestimates f(x) over D).

Now suppose we start with an n–simplex $M_1 \supset D$ and the grid $X_1 = V(M_1)$ (vertex set of M_1). Then S_1 has a unique vertex which can easily be determined. At iteration k we have a finite grid X_k of M_1 such that conv $X_k = M_1$, together with the associated polyhedron S_k and its vertex set $\mathcal{M}_k$. By solving the relaxed problem

$$(P_k) \qquad \min \varphi_k(x) + dy \quad \text{s.t.} \quad (x,y) \in \Omega$$

with $\varphi_k(x)$ defined by (10), we obtain a lower estimate for min $\{f(x)+dy: (x,y) \in \Omega\}$. Therefore, if (x^k,y^k) is a basic optimal solution of (P_k) and (x^k,y^k) satisfies $\psi_k(x^k) = f(x^k)$, then $\psi_k(x^k) + dy^k = f(x^k) + dy^k$; hence (x^k,y^k) solves (P). Otherwise, $\psi_k(x^k) < f(x^k)$, which can happen only if $x^k \in D \setminus X_k$. We then consider the new grid $X_{k+1} = X_k \cup \{x^k\}$. Since

$$S_{k+1} = S_k \cap \{(q,q_0): qx^k - q_0 \ge f(x^k)\} ,$$

the vertex set of S_{k+1} can be derived from that of S_k by any of the procedures discussed in Section II.4.2. Once this vertex set is known, the procedure can be repeated, with X_{k+1} in place of X_k.

In this manner, starting from a simple grid X_1 (with conv $X_1 = M_1$), we generate a sequence of expanding grids

$$X_1 \subset X_2 \subset \dots \subset X_k \subset \dots \quad \text{satisfying} \quad X_{k+1} \setminus X_k = \{x^k\} , \tag{11}$$

a sequence of polyhedral concave functions

$$\varphi_1(x) \le \varphi_2(x) \le \dots \varphi_k(x) \le \dots \le f(x) \quad \forall\, x \in M_1 ,$$

and a sequence of points $(x^k,y^k) \in \Omega$. Since (x^k,y^k) is a vertex of Ω (optimal solution of (P_k)) and since, by (11), the sequence (x^k,y^k) never repeats, it follows that the number of iterations is bounded above by the number of vertices of Ω. Therefore, the procedure will terminate after finitely many iterations at a global optimal solution (x^k,y^k).

We are thus led to the following extension of Algorithm VI.5 (we allow restarting in Step 4 in order to avoid having $|\mathcal{M}_k|$ too large).

Algorithm VIII.3.

0) Set $X_1 = \{v^1,\dots,v^{n+1}\}$ (where $[v^1,\dots,v^{n+1}]$ is a simplex in $\mathbb{R}^n$ which contains D). Let $(\tilde{x}^0,\tilde{y}^0)$ be the best feasible solution known so far; set $S_1 = \{(q,q_0): q_0 - qv^i \ge f(v^i), i=1,\dots,n+1\}$. Let $\mathcal{M}_1$ be the vertex set of S_1, i.e., the singleton $\{[f(v^1),\dots,f(v^{n+1})]Q_1^{-1}\}$, where Q_1 is the matrix with n+1 columns $\begin{bmatrix} -v^j \\ 1 \end{bmatrix}$ (j=1,...,n+1). Set $\mathcal{N}_1 = \mathcal{M}_1$, k=1.

1) For every $(q,q_0) \in \mathcal{N}_k$ solve the linear program

$$\text{minimize } (q_0 - qx) + dy \quad \text{subject to } (x,y) \in \Omega \text{ ,}$$

obtaining a basic optimal solution $\omega(q,q_0)$ and the optimal value $\beta(q,q_0)$.

2) Compute

$$(q^k,q_0^k) \in \arg\min \{\beta(q,q_0): (q,q_0) \in \mathcal{M}_k\}$$

and let $(x^k,y^k) = \omega(q^k,q_0^k)$, $\beta^k = \beta(q^k,q_0^k)$.

3) Update the current best solution by taking the point $(\tilde{x}^k,\tilde{y}^k)$ with the smallest value of $f(x) + dy$ among $(\tilde{x}^{k-1},\tilde{y}^{k-1})$ and all points (q,q_0) in $\mathcal{M}_k$.

4) If $f(\tilde{x}^k) + d\tilde{y}^k = \beta^k$, then terminate: $(\tilde{x}^k,\tilde{y}^k)$ is a global optimal solution.

5) Otherwise, set $S_{k+1} = S_k \cap \{(q,q_0): q_0 - qx^k \geq f(x^k)\}$, and compute the vertex set $\mathcal{M}_{k+1}$ of S_{k+1}. Set $\mathcal{N}_{k+1} = \mathcal{M}_{k+1} \setminus \mathcal{M}_k$ and go to iteration k+1.

3.2. Separable Problems

When the function f(x) is separable

$$f(x) = \sum_{j=1}^{n} f_j(x_j) \text{ ,} \tag{12}$$

it is more convenient to start with a rectangle $M_1 = \{x: r^1 \leq x \leq s^1\}$ and to construct the grids X_k so as to determine rectangular subdivisions of M_1, i.e., $X_k = \prod_{j=1}^{n} X_{kj}$, where each X_{kj} consists of k+1 points on the x_j–axis. The polyhedral concave function $\varphi_k(x)$ corresponding to X_k is then easy to determine. In fact:

$$\varphi_k(x) = \sum_{j=1}^{n} \varphi_{kj}(x_j) \text{ ,} \tag{13}$$

where each $\varphi_{kj}(t)$ is a piecewise affine function defined by a formula of the type

((37), Section VII.4) in each of the k subintervals that the grid X_{kj} determines in the segment $[r_j^1,s_j^1]$.

Theoretically, the method looks very simple. From a computational point of view, however, for large values of k the functions $\varphi_{kj}(t)$ are not easy to manipulate. To cope with this difficulty, some authors propose using mixed integer programming, as in the following method of Rosen and Pardalos (1986) for concave quadratic minimization.

Consider the problem (P) where f(x) is concave quadratic. Without loss of generality we may assume that f(x) has the form (12) with

$$f_j(x_j) = q_j x_j - \frac{1}{2}\lambda_j x_j^2 \ , \ \lambda_j > 0 \ , \tag{14}$$

(cf. Section VI.3.3), and that Ω is the polytope defined by the linear inequalities

$$Ax + By \geq c \ , \ x \geq 0 \ , \ y \geq 0 \ . \tag{15}$$

As seen in Section VI.3.3, we can easily construct a rectangular domain $M = \{x \in \mathbb{R}^n: 0 \leq x_j \leq \beta_j \ (j=1,...,n)\}$ which contains D (recall that D is the projection of Ω on the x–space).

Suppose we partition each interval $I_j = [0,\beta_j]$ (j=1,...,n) into k_j equal subintervals of length $\delta_j = \beta_j/k_j$. Let $\varphi_j(t)$ be the lowest concave function that agrees with $f_j(t)$ at all subdivision points of the interval I_j, and let

$$\varphi(x) = \sum_{j=1}^{n} \varphi_j(x_j) \ .$$

If $(\hat{x},\hat{y})$ is an optimal solution of the approximating problem

$$\min \{\varphi(x) + dy: (x,y) \in \Omega\} \ , \tag{16}$$

then $(\hat{x},\hat{y})$ is an approximate optimal solution of (P) with an accuracy that obviously depends on the number k_j (j=1,...,n) of subintervals into which each interval I_j is divided. It turns out that we can select k_j (j=1,...,n) so as to ensure any prescribed level of accuracy (cf. Pardalos and Rosen (1987)).

Let $\Psi(x,y) = f(x) + dy$ and denote by ψ^* the global minimum of $\Psi(x,y)$ over Ω. The error at $(\hat{x},\hat{y})$ is given by $\Psi(\hat{x},\hat{y}) - \Psi^*$. Since

$$\Psi((\hat{x}) + d\hat{y} \leq \Psi^* \leq f((\hat{x}) + d\hat{y} = \Psi(\hat{x},\hat{y}) ,$$

we have $\Psi(\hat{x},\hat{y}) - \Psi^* \leq f(\hat{x}) - \varphi(\hat{x})$. So the error is bounded by

$$E(\hat{x}) := f(\hat{x}) - \varphi(\hat{x}) .$$

We now give a bound for $E(\hat{x})$ relative to the range of $f(x)$ over M. Let $f_{max} = \max_{x \in M} f(x)$, $f_{min} = \min_{x \in M} f(x)$. Then the range of $f(x)$ over M is

$$\Delta f = f_{max} - f_{min} .$$

We first need a lower bound for Δf.

Assume (without loss of generality) that

$$\lambda_1 \beta_1^2 \geq \lambda_j \beta_j^2 \quad (j=1,\ldots,n) . \tag{17}$$

Define the ratios

$$\rho_j = (\lambda_j \beta_j^2) / (\lambda_1 \beta_1^2) \leq 1 \quad (j=1,\ldots,n) . \tag{18}$$

The function $f_j(t)$ attains its unconstrained maximum at the point $\bar{x}_j = q_j/\lambda_j$. Define

$$\eta_j = \min \left\{1, \left|\frac{2\bar{x}_j}{\beta_j}\right| - 1\right\} \quad (j=1,\ldots,n) \tag{19}$$

Note that $0 \leq \eta_j \leq 1$.

Lemma VII.1. $\Delta f \geq \frac{1}{8} \lambda_1 \beta_1^2 \sum_{j=1}^{n} \rho_j (1 + \eta_j)^2$.

Proof. Denote

$$\Delta f_j = \max f_j(t) - \min f_j(t) ,$$

where the maximum and minimum are taken over the interval $[0,\beta_j]$.

There are four cases to consider:

(i) $0 \le \bar{x}_j \le \frac{1}{2}\beta_j$:

Then $f_j(\beta_j) = \lambda_j\beta_j(\bar{x}_j - \frac{1}{2}\beta_j) \le 0$, and since the minimum of the concave function $f_j(t)$ over the interval $0 \le t \le \beta_j$ is attained at an endpoint, we have $\min f_j(t) = f_j(\beta_j)$. On the other hand, $\max f_j(t) = \frac{1}{2}\lambda_j\bar{x}_j^2$. From (19) we have $\bar{x}_j = \frac{1}{2}\beta_j(1-\eta_j)$. Hence,

$$\Delta f_j = \frac{1}{2}\lambda_j\bar{x}_j^2 - \lambda_j\beta_j\bar{x}_j + \frac{1}{2}\lambda_j\beta_j^2 = \frac{1}{2}\lambda_j(\beta_j-\bar{x}_j)^2 = \frac{1}{8}\lambda_j\beta_j^2(1+\eta_j)^2 .$$

(ii) $\frac{1}{2}\beta_j \le \bar{x}_j \le \beta_j$:

Then $f_j(\beta_j) \ge 0$, so that $\min f_j(t) = f_j(0) = 0$. From (19) we have $\bar{x}_j = \frac{1}{2}\beta_j(1+\eta_j)$, hence

$$\Delta f_j = \frac{1}{2}\lambda_j\bar{x}_j^2 = \frac{1}{8}\lambda_j\beta_j^2(1+\eta_j)^2 .$$

(iii) $\bar{x}_j \le 0$:

This implies $q_j \le 0$, $f_j(t) \le -\frac{1}{2}\lambda_j t_j^2$ $(0 \le t \le \beta_j)$, so that $\max f_j(t) = 0$, $\min f_j(t) \le -\frac{1}{2}\lambda_j\beta_j^2$. Therefore, since $\eta_j = 1$,

$$\Delta f_j \ge \frac{1}{2}\lambda_j\beta_j^2 = \frac{1}{8}\lambda_j\beta_j^2(1+\eta_j)^2 .$$

(iv) $\bar{x}_j \ge \beta_j$:

Then $\min f_j(t) = f_j(0) = 0$, and since $\eta_j = 1$,

$$\Delta f_j = \max f_j(t) = f_j(\beta_j) = \lambda_j\beta_j(\bar{x}_j-\beta_j/2) \ge \frac{1}{2}\lambda_j\beta_j^2 = \frac{1}{8}\lambda_j\beta_j^2(1+\eta_j)^2 .$$

Finally, we have

$$\Delta f = \sum_{j=1}^{n} \Delta f_j \geq \frac{1}{8} \sum_{j=1}^{n} \lambda_j \beta_j^2 (1+\eta_j)^2 \geq \frac{1}{8} \sum_{j=1}^{n} \lambda_1 \beta_1^2 \rho_j (1+\eta_j)^2 .$$ ■

Theorem VIII.3. *We have*

$$\frac{\Psi(\hat{x},\hat{y}) - \Psi^*}{\Delta f} \leq \frac{E(\hat{x})}{\Delta f} \leq \frac{\sum_{j=1}^{n} (\rho_j / k_j^2)}{\sum_{j=1}^{n} \rho_j (1+\eta_j)^2} . \tag{20}$$

Proof. As we saw before,

$$\Psi(\hat{x},\hat{y}) - \Psi^* \leq f(\hat{x}) - \varphi(\hat{x}) = E(\hat{x}) . \tag{21}$$

Since f_j is concave and $\varphi_j(t)$ interpolates it at the points $t = i\delta_j$ $(i=0,1,...,k_j)$, it can easily be shown that

$$0 \leq f_j(\hat{x}_j) - \varphi_j(\hat{x}_j) \leq \frac{1}{8} \lambda_j \delta_j^2 = \frac{1}{8} \lambda_j (\beta_j / k_j)^2 \quad (j=1,...,n) .$$

(cf. Proposition VII.18).

Hence

$$f(\hat{x}) - \varphi(\hat{x}) = \sum_{j=1}^{n} \{f_j(\hat{x}_j) - \varphi_j(\hat{x}_j)\} \leq \frac{1}{8} \sum_{j=1}^{n} \lambda_j \frac{\beta_j^2}{k_j^2} \leq \frac{1}{8} \lambda_1 \beta_1^2 \sum_{j=1}^{n} \frac{\rho_j}{k_j^2} ,$$

and the inequality (20) follows from Lemma VII.1 and (21). ■

Corollary VIII.1. *Let $\varphi(x)$ be the convex envelope of $f(x)$ taken over M and let (x^0,y^0) be an optimal solution of the problem*

$$\min \{\varphi(x) + dy: (x,y) \in \Omega\} .$$

Then

$$\frac{\Psi(x^0,y^0) - \Psi^*}{\Delta f} \leq \frac{\sum_{j=1}^{n} \rho_j}{\sum_{j=1}^{n} \rho_j (1+\eta_j)^2} =: \sigma(\rho,\eta) \tag{22}$$

Proof. Take $k_j = 1$ (j=1,...,n) in Theorem VIII.3. ■

Note that $\sigma(\rho,\eta) \in [\frac{1}{4},1]$, and furthermore, $\sigma(\rho,\eta) < 1$, unless $\bar{x}_j = \frac{\beta_j}{2}$ for every j. In particular, if $\bar{x}_j \notin (0,\beta_j)$ $\forall j$ then $\sigma(\rho,\eta) = \frac{1}{4}$.

Corollary VIII.2. *If for each j=1,...,n:*

$$k_j \geq \left(\frac{n}{\alpha}\rho_j\right)^{1/2}, \text{ where } \alpha = \varepsilon \sum_{j=1}^{n} \rho_j (1+\eta_j)^2 \tag{23}$$

then an optimal solution $(\hat{x},\hat{y})$ *of the approximation problem (16) satisfies*

$$\Psi(\hat{x},\hat{y})) - \Psi* \leq \varepsilon\Delta f.$$

Proof. From (23) it immediately follows that

$$\sum_{j=1}^{n} \frac{\rho_j}{k_j^2} \leq \alpha = \varepsilon \sum_{j=1}^{n} \rho_j(1+\eta_j)^2 .$$ ■

Thus, in the case of a quadratic function f(x), we can choose the numbers k_j (j=1,...,n) so that the solution to the approximation problem (16) that corresponds to the subdivision of each interval $[0,\beta_j]$ into k_j subintervals will give an ε–approximate global optimal solution to the original problem (P) ($\varepsilon > 0$ is a prescribed tolerance).

In general, for small ε, the numbers k_j must be large, and the explicit construction of the functions $\varphi_j(t)$ may be cumbersome. Rosen and Pardalos (1986) suggest reformulating the problem (16) as a zero–one mixed integer programming problem as follows.

Let us introduce new variables ω_{ij} such that

$$x_j = \delta_j \Sigma_{i=1}^{k_j} \omega_{ij} \quad (j=1,...,n) . \tag{24}$$

The variables ω_{ij} are restricted to $\omega_{ij} \in [0,1]$ and furthermore, $\bar{\omega}_j = (\omega_{1j},...,\omega_{k_j j})$ is restricted to have the form $\bar{\omega}_j = (1,...,1,\omega_{\ell j},0,...,0)$. Then there will be a unique vector $\bar{\omega}_j$ representing any $x_j \in [0,\beta_j]$ and it is easy to see that

$$\varphi_j(x_j) = \sum_{i=1}^{k_j} \Delta f_{ij}\omega_{ij} \text{ , with}$$

$$\Delta f_{ij} = f_j(i\delta_j) - f_j((i-1)\delta_j) \text{ , } \quad i=1,\dots,k_j$$

where $\bar{\omega}_j$ is determined by (24). We can therefore reformulate (16) as

(MI)
$$\min \sum_{j=1}^{n} \sum_{i=1}^{k_j} \Delta f_{ij}\omega_{ij} + dy$$

$$\text{s.t.} \sum_{j=1}^{n} \delta_j a_j \sum_{i=1}^{k_j} \omega_{ij} + By \ge c \text{ ,}$$

$$0 \le \omega_{ij} \le 1 \text{ , } y \ge 0 \text{ ,}$$

$$\omega_{i+1,j} \le z_{ij} \le \omega_{ij} \quad (i=1,\dots,k_j-1,\ j=1,\dots,n) \text{ ,}$$

$$z_{ij} \in \{0,1\} \text{ ,}$$

where a_j is the j–th column of A in (15).

Algorithm VIII.4 (The Rosen–Pardalos Algorithm for (P)).

1) For the objective function, compute the eigenvalues λ_i and the corresponding orthonormal eigenvectors u^i, i=1,2,...,n. Construct a rectangular domain $M = \{x: 0 \le x_j \le \beta_j\}$ containing the projection D of Ω onto $\mathbb{R}^n$. Evaluate $\Psi(x,y)$ at every vertex of Ω encountered.

2) Choose the incumbent function value (IFV). Construct $\varphi(x),\Delta f,\rho_j,\bar{x}_j$ (j=1,...,n).

3) Solve the approximating problem (16) to obtain $(\hat{x},\hat{y})$. If $\Psi(\hat{x},\hat{y}) <$ IFV, reset IFV $= \Psi(x,y)$.

4) If IFV $- \varphi(\hat{x}) - d\hat{y} \le \varepsilon\Delta f$, then stop: IFV is an ε–approximate solution.

5) Otherwise, construct the piecewise affine functions $\varphi_j(t)$ corresponding to the

given tolerance ε.

Solve (MI) to obtain an ε–approximate solution, using the incumbent to accelerate pruning.

Remarks VIII.2. (i) The rectangular domain M is constructed by solving 2n linear programs:

$$\max \{x_j: (x,y) \in \Omega\} , \min \{x_j: (x,y) \in \Omega\} \quad (j=1,2,...,n) .$$

In the process of solving these multi–cost–row problems, each time a vertex of Ω is encountered, the corresponding value of $\Psi(x,y)$ is calculated. The vertex with the minimum Ψ is the incumbent and gives IFV in Step 2.

(ii) The formulation of (MI) requires a total of $N = \sum_{j=1}^{n} k_j - n$ (0–1) integer variables. If all possible combinations of the integer variables were allowed, this problem would have a maximum of 2^n possibilities to be evaluated. However, it can be shown that, because of the specific structure of the problem, the maximum number of possibilities is only $k_1 \times ... \times k_n$. In addition, the number N of 0–1 integer variables never exceeds

$$n \left[\frac{\sigma(\rho,\eta)}{\varepsilon} \right]^{1/2}$$

(cf. Rosen and Pardalos (1986)).

4. DECOMPOSITION BY OUTER APPROXIMATION

In this section we discuss a decomposition approach based on outer approximation of problem $(\tilde{H})$ (i.e., (4)), which is equivalent to (H). This method (cf. Tuy (1985, 1987) and Thieu (1989)) can also be regarded as an application of Benders' partitioning approach to concave minimization problems.

4.1. Basic Idea

As before, we shall assume that the constraint set Ω is given in the form

$$Ax + By \geq c \, , \, x \geq 0 \, , \, y \geq 0 \ . \tag{25}$$

The problem considered is

(P) minimize $f(x) + dy$ subject to (25).

As shown in Section VIII.1, this problem is equivalent to the following one in the space $\mathbb{R}^n \times \mathbb{R}$:

$(\tilde{H})$ minimize $[f(x) + t]$ subject to $g(x) \leq t \, , \, x \in D \, ,$

where

$$g(x) = \inf \{dy: -By \leq Ax - c, \, y \geq 0\} \, ,$$

$$D = \{x \geq 0: \exists \, y \geq 0 \text{ such that } Ax + By \leq c\} \, .$$

Since, by Proposition VIII.1, D is a polyhedron and g(x) is a convex polyhedral function with dom g = D, it follows that the constraint set G of $(\tilde{H})$ is a polyhedron; and hence problem $(\tilde{H})$ can be solved by the outer approximation method described in Section VI.1.

Note that the function g(x) is continuous on D, because any proper convex polyhedral function is closed and hence continuous on any polyhedron contained in its domain (Rockafellar, Corollary 19.1.2 and Theorem 10.2). Hence, if we assume that D is bounded, then g(x) is bounded on D: $\alpha \leq g(x) \leq \beta \ \forall x \in D$. Since obviously any optimal solution of $(\tilde{H})$ must satisfy $g(x) = t$, we can add the constraint $\alpha \leq g(x) \leq \beta$ to $(\tilde{H})$. In other words, we may assume that the constraint set G of $(\tilde{H})$ is contained in the bounded set $D \times [\alpha,\beta]$.

Under these conditions, an outer approximation method for solving $(\tilde{H})$ proceeds according to the following scheme.

Start with a polytope T_0 in $\mathbb{R}^n \times \mathbb{R}$ which contains G. At iteration k one has on hand a polytope T_k in $\mathbb{R}^n \times \mathbb{R}$ which contains M. Solve the relaxed problem

$$(H_k) \qquad \min \{f(x) + t: (x,t) \in T_k\} ,$$

and let (x^k,t^k) be an optimal solution. If (x^k,t^k) happens to be feasible (i.e., belong to G), then it solves (H). Otherwise, one constructs a linear constraint on G that is violated by (x^k,t^k). Adding this constraint to T_k, one defines a new polytope T_{k+1} that excludes (x^k,t^k) but contains G. Then the process is repeated, with T_{k+1} in place of T_k. Since the number of constraints on G is finite, the process will terminate at an optimal solution of (P) after finitely many iterations.

Clearly, to carry out this scheme we must carefully examine two essential questions:

(i) How can one check whether a point (x^k,t^k) is feasible for $(\tilde{H})$?

(ii) If (x^k,t^k) is infeasible, how does one construct a constraint on G that is violated by this point?

These questions arise, because, although we know that G is a polytope, we do not know its constraints explicitly.

In the next section we shall show how these two questions can be resolved, without, in general, having to generate explicitly all of the constraints of G.

4.2. Decomposition Algorithm

Let $(x^0,t^0) \in \mathbb{R}^n \times \mathbb{R}$. Recall that, by definition, $g(x^0)$ is the optimal value of the linear program

$$(C(x^0)) \qquad \min \{dy: -By \geq Ax^0 - c, y \geq 0\} .$$

Consider the dual program

$(C^*(x^0))$ $\max \{(Ax^0-c)w: B^Tw \ge -d, w \ge 0\}$.

Since the function g(x) is bounded on D, C(x) is either infeasible or else has a finite optimal value.

Proposition VIII.4. *The point (x^0,t^0) is feasible for $(\tilde{H})$, i.e., $x^0 \in D$ and $g(x^0) \le t^0$, if and only if both programs $C(x^0)$ and $C^*(x^0)$ are feasible and their common optimal value does not exceed t^0.*

Proof. Since the point (x^0,t^0) is feasible for $(\tilde{H})$ if and only if $C(x^0)$ has an optimal value that does not exceed t^0, the conclusion follows immediately from the duality theory of linear programming. ∎

Observe that, since C(x) has a finite optimal value ($g(x) > -\infty$), $C^*(x)$ must be feasible, i.e., the polyhedron

$$W = B^Tw \ge -d, \quad w \ge 0$$

is nonempty.

Now suppose that (x^0,t^0) is infeasible. By the proposition just stated, this can happen only in the following two cases:

I. $C(x^0)$ is infeasible.

Then, since $C^*(x^0)$ is always feasible (W is nonempty), it follows that $C^*(x^0)$ must be unbounded. That is, in solving $C^*(x^0)$ we must find an extreme direction v^0 of W such that $(Ax^0 - c)v^0 > 0$.

Proposition VIII.5. *The linear inequality*

$$(Ax^0 - c)v^0 \le 0 \tag{26}$$

is satisfied by all $x \in D$, but is violated by x^0.

Proof. Indeed, for any $x \in D$, $C(x)$ has a finite optimal value. Hence, $C^*(x)$ has a finite optimal value, too. This implies that $(x - c)v \leq 0$ for any extreme direction v of W. ■

II. $C(x^0)$ is feasible but its optimal value exceeds t^0.

Then $C^*(x^0)$ has an optimal solution w^0 with $(Ax^0 - c)w^0 = g(x^0) > t^0$.

Proposition VIII.6. *The linear inequality*

$$(Ax^0 - c)w^0 \leq t \tag{27}$$

is satisfied by all (x,t) such that $x \in D$, $g(x) \leq t$, but it is violated by (x^0,t^0).

Proof. For any $x \in D$, since $C(x)$ has $g(x)$ as its optimal value, it follows that $C^*(x)$, too, has $g(x)$ as its optimal value. Hence,

$$g(x) \geq (Ax - c)w^0 ,$$

which implies (27) if $g(x) \leq t$. Since $t^0 < g(x^0) = (Ax^0 - c)w^0$, the proposition follows. ■

Thus, given any (x^0,t^0), by solving $C(x^0)$ and $C^*(x^0)$ we obtain all the information needed to check feasibility of (x^0,t^0) and, in case of infeasibility, to construct the corresponding inequality that excludes (x^0,t^0) without excluding any feasible (x,t).

We are thus led to the following algorithm for solving (P).

Let $\alpha \leq g(x) \leq \beta \ \forall x \in D$.

Algorithm VIII.5

Initialization:

Construct a polytope $T_0 \subset \mathbb{R}^{n+1}$ containing $D \times [\alpha,\beta]$. Set $k = 0$.

Iteration k = 0,1,...:

1) Solve the relaxed problem

(H_k) $\min \{f(x) + t: (x,t) \in T_k\}$

obtaining an optimal solution (x^k,t^k) of it.

2) Solve the linear program $C^*(x^k)$.

3) If a basic optimal solution w^k of $C^*(x^k)$ is found with $(Ax^k - c)w^k \leq t^k$, then terminate: (x^k,y^k), with y^k a basic optimal solution of $C(x^k)$, is a global optimal solution of (P).

If a basic optimal solution w^k of $C^*(x^k)$ is found with $(Ax^k - c)w^k > t^k$, then form

$$T_{k+1} = T_k \cap \{x: (Ax^k - c)w^k \leq t\}$$

and go to 1) with $k \leftarrow k+1$.

4) Otherwise, an extreme direction v^k of the cone $B^Tv \geq 0$, $v \geq 0$, is found such that $(Ax^k - c)v^k > 0$. In this case form

$$T_{k+1} = T_k \cap \{x: (Ax^k - c)v^k \leq 0\}$$

and go to 1) with $k \leftarrow k+1$.

Theorem VIII.4. *Algorithm VIII.5 terminates after finitely many iterations.*

Proof. Each w^k is a vertex of the polyhedron $B^Tw \geq -d$, $w \geq 0$, while each v^k is an extreme direction of the cone $B^Tv \geq 0$, $v \geq 0$. But each T_{k+1} is obviously a proper subset of T_k. Hence, in each sequence $\{w^k\}$, $\{v^k\}$ there cannot be any repetition. Since the set of vertices and extreme directions of the polyhedron $B^Tw \geq -d$, $w \geq 0$ is finite, the finiteness of the algorithm follows. ■

Remarks VIII.3. (i) All of the linear programs $C^*(x^k)$ have the same constraints: $B^T w \geq -d$, $w \geq 0$, while the linear programs $C(x^k)$ have constraints that differ only on the right hand side: $-B^T y \geq Ax^k - c$, $y \geq 0$. This feature greatly simplifies the process of solving the auxiliary linear subproblems and is particularly valuable when the matrix B has some special structure which allows efficient procedures for solving the corresponding subproblems. We shall return to this matter in Section VIII.4.4.

(ii) If we know bounds α, β for g(x) over D, then we can take $T_0 = S \times [\alpha,\beta]$, where S is an n–simplex or a rectangle in $\mathbb{R}^n$ which contains D. Obviously, a lower bound for g(x) over D is

$$\alpha = \min \{dy: Ax + By \leq c, x \geq 0, y \geq 0\}$$

$$= \min \{g(x): x \in D\} .$$

It may be more difficult to compute an upper bound β. However, for the above algorithm to work it is not necessary that T_0 be a polytope. Instead, one can take T_0 to be a polyhedron of the form $S \times [\alpha,+\infty)$, where S is a polytope containing D. Indeed, this will suffice to ensure that any relaxed problem (H_k) will have an optimal solution which is a vertex of T_k.

(iii) If $x^k \in D$, i.e., if $C^*(x^k)$ has an optimal solution w^k, then $f(x^k)+g(x^k) = f(x^k) + (Ax^k-c)w^k$ provides an upper bound for the optimal value Ψ^* in (P), and so

$$f(x^k) + t^k \leq \Psi^* \leq f(x^k) + g(x^k) .$$

On the basis of this observation, the algorithm can be improved by the following modification.

To start, set CBS (Current Best Solution) = $\hat{x}$, UBD (Upper Bound) = $f(\hat{x}) + g(\hat{x})$, where $\hat{x}$ is some best available point of D (if no such point is available, set CBS = $\emptyset$, UBD = $+\infty$). At iteration k, in Step 1, if $f(x^k) + t^k \geq \text{UBD}-\varepsilon$, then stop: CBS is a global ε–optimal solution. In Step 3 (entered with an optimal solution w^k

of $C^*(x^k)$, i.e., $x^k \in D$), if $f(x^k) + (Ax^k - c)w^k < UBD$, reset $UBD = f(x^k) + g(x^k)$, $CBS = x^k$.

(iv) As in all outer approximation procedures, the relaxed problem (H_k) differs from (H_{k-1}) by just one additional constraint. Therefore, to solve the relaxed problems one should use an algorithm with restart capability. For example, if Algorithm VI.1 is used, then at the beginning the vertex set T_0 is known, and at iteration k the vertex set of T_k is computed using knowledge of the vertex set of T_{k-1} and the newly added constraint (see Section III.4). The optimal solution (x^k,t^k) is found by comparing the values of $f(x) + t$ at the vertices.

Example VIII.1. Consider the problem:

$$\begin{aligned} &\text{Minimize } f(x) + dy \\ &\text{s.t. } \quad Ax + By \le c,\ x \ge 0,\ y \ge 0, \end{aligned}$$

where $x \in \mathbb{R}^2$, $y \in \mathbb{R}^5$, $f(x) := -(x_1-1)^2 - (x_2+1)^2$, $d = (1,-1,2,1,-1)^T$,

$$A = \begin{bmatrix} 1 & 1 \\ 2 & 1 \\ 1 & 1 \end{bmatrix}, \quad B = \begin{bmatrix} 0 & 5 & 0 & -7 & 0 \\ 1 & 1 & 1 & 1 & 2 \\ 0 & 0 & 1 & 0 & 1 \end{bmatrix}, \quad c = \begin{bmatrix} -10 \\ -4 \\ -3 \end{bmatrix}$$

Initialization:

$\alpha = \min \{dy: Ax + By \le c, x \ge 0, y \ge 0\} = -3$.

$T_0 = S \times [\alpha,+\infty)$, $S = \{x: x_1 \ge 0, x_2 \ge 0, x_1 + x_2 \le 3\}$.

Iteration 0:

The relaxed problem is

$$(H_0) \qquad \min \{-(x_1-1)^2 - (x_2+1)^2 + t: x_1 \ge 0, x_2 \ge 0, x_1 + x_2 \le 3, t \ge -3\}$$

with the optimal solution $x^0 = (0,3)$, $t^0 = -3$.

Solving $C^*(x^0)$ yields $w^0 = (0,1,0)$ with $(Ax^0 - c)w^0 = -1 > t^0$.

Form (H_1) by adding the constraint $(Ax - c)w^0 = 2x_1 + x_2 - 4 \le t$.

Iteration 1:

Optimal solution of (H_1): $x^1 = (0,3)$, $t^1 = -1$.
Solving $C^*(x^1)$ yields $w^1 = (0,1,0)$, with $(Ax^1 - c)w^1 = -1 = t^1$.
The termination criterion in Step 3 is satisfied. Hence $(x^1;y^1) = (0,3;0,1,0,0,0)$ is the desired global optimal solution of (P).

4.3. An Extension

Algorithm VIII.5 can easily be extended to the case where

$$\Omega = \{(x,y) \in \mathbb{R}^n_+ \times \mathbb{R}^h \colon Ax + By \leq 0,\ x \in X,\ y \geq 0\}\ , \tag{28}$$

with X a convex polyhedron in $\mathbb{R}^n$, and the projection D of Ω on $\mathbb{R}^n$ is not necessarily bounded.

As above, define

$$g(x) = \inf\{dy\colon (x,y) \in \Omega\}\ ,$$

$$G = \{(x,t) \in \mathbb{R}^n \times \mathbb{R}\colon x \in D,\ g(x) \leq t\}\ ,$$

$$W = \{w \in \mathbb{R}^m\colon B^T w \geq -d,\ w \geq 0\}\ ,$$

and assume that $g(x) \geq \alpha\ \forall x \in X$; hence $W \neq \emptyset$ (this is seen by considering the dual linear programs $C(x)$, $C^*(x)$ for an arbitrary $x \in D$). Denote the set of vertices and the set of extreme directions of W by vert W and extd W, respectively.

Proposition VIII.7. *A vector $(x,t) \in X \times R$ belongs to M if and only if it satisfies*

$$(Ax - c)v \leq 0 \quad \forall v \in \mathit{extd}\ W \tag{29}$$

$$(Ax - c)w \leq t \quad \forall w \in \mathit{vert}\ W\ . \tag{30}$$

Proof. The proofs of Propositions VIII.5 and VIII.6 do not depend upon the hypotheses that $X = \mathbb{R}^n_+$ and D is bounded. According to these propositions, for any $x \in X$, if $(x,t) \in G$, then (x,t) violates at least one of the inequalities (29), (30). On the other hand, these inequalities are satisfied by all $(x,t) \in G$. ■

Thus, the feasible set G can be described by the system of constraints: $x \in X$, (29) and (30).

In view of this result, checking the feasibility of a point $(x^k,t^k) \in X \times R$ and constructing the constraint that excludes it when it is infeasible proceeds exactly as before, i.e., by solving $C^*(x^k)$.

The complication now is that, since the outer approximating polyhedron T_k may be unbounded, the relaxed problem (H_k) may have an unbounded optimal solution. That is, in solving (H_k) we may obtain an extreme direction (z^k,s^k) of T_k on which the function $f(x) + t$ is unbounded from below. Whenever this situation occurs, we must check whether this direction belongs to the recession cone of the feasible set G, and if not, we must construct a linear constraint on G that excludes this direction without excluding any feasible point.

Corollary VIII.3. *Let z be a recession direction of X. Then $(z,x) \in \mathbb{R}^n \times \mathbb{R}$ is a recession direction of G if and only if*

$$(Az)v \leq 0 \quad \forall v \in \text{extd } W \tag{31}$$

$$(Az)w \leq s \quad \forall w \in \text{vert } W . \tag{32}$$

Proof. Indeed, since (x,t) is an arbitrary point of G, the recession cone of G consists of all (z,s) such that for all $\lambda \geq 0$, $(A(x + \lambda z))v \leq 0 \quad \forall v \in \text{extd } W$ and $(A(x + \lambda z))w \leq s \quad \forall w \in \text{vert } W$. This is equivalent to (31), (32). ■

It follows from this corollary that to check whether a vector $(z^k,s^k) \in \mathbb{R}^n \times \mathbb{R}$ belongs to the recession cone of G it suffices to solve

$(S(z^k))$ $\max \{(Az^k)w: w \in W\}$.

If this linear program has a basic optimal solution w^k with $(Az^k)w^k \le s^k$, then (z^k,s^k) is a recession direction of G. If not, i.e., if $(Az^k)w > s^k$, then the constraint (30) corresponding to $w = w^k$ will exclude (z^k,s^k) from the recession cone. If $S(z^k)$ has an unbounded optimal solution with direction v^k, then the constraint (29) corresponding to $v = v^k$ will exclude (z^k,s^k) from the recession cone.

On the basis of the above results one can propose the following modification of Algorithm VIII.5 for the case where X is an arbitrary convex polyhedron and D may be unbounded:

Initialization:

Construct a polyhedron T_0 such that $G \subset T_0 \subset S \times [\alpha,+\infty)$. Set $k = 0$.

Iteration k $= 0,1,\dots$:

1) Solve the relaxed problem

(H_k) $\min \{f(x) + t: (x,t) \in T_k\}$.

If a finite optimal solution (x^k,t^k) of (H_k) is obtained, then go to 2). If an unbounded optimal solution with direction (z^k,s^k) is obtained, then go to 5).

2) – 4): as in Algorithm VIII.5.

5) Solve $S(z^k)$.

6) If a basic optimal solution w^k of $S(z^k)$ is found with $(Az^k)w^k \le s^k$, then terminate: $f(x)+t$ is unbounded from below over the direction (z^k,s^k).

If a basic optimal solution (z^k,s^k) of $S(z^k)$ is found with $(Az^k)w^k > s^k$, then form

$$T_{k+1} = T_k \cap \{x: (Ax - c)w^k \le t\}$$

and go to 1) with $k \leftarrow k+1$.

7) Otherwise, a direction $v^k \in \text{extd } W$ is found such that $(Az^k)v^k > 0$, Then form

$$T_{k+1} = T_k \cap \{x: (Ax - c)v^k \leq 0\}$$

and go to 1) with $k \leftarrow k+1$.

It is clear that the modified Algorithm VIII.5 will terminate in finitely many steps.

Remarks VIII.4. (i) If Ω is given in the form $Ax + By = c$, $x \in X$, $y \geq 0$, then the constraints of the linear program $C(x)$ (for $x \in X$) are $-By = Ax - c$, $y \geq 0$, Hence the nonnegativity constraint $w \geq 0$ in $C^*(x)$ is dropped, i.e., we have $W = \{w: B^T w \geq -d\}$.

(ii) A further extension of the algorithm can be made to the case when the nonnegativity constraint $y \geq 0$ is replaced by a constraint of the form $Ey \leq p$, where E is an $\ell \times h$ matrix and p an ℓ-vector. Then the problems $C^*(x)$, $S(z)$ become

$$(C^*(x)) \qquad \max \{(Ax - c)w - pu: (w,u) \in W\}$$

$$(S(z)) \qquad \max \{(Az)w: (w,u) \in W\} ,$$

where $W = B^T w + E^T u = -d$, $w \geq 0$, $u \geq 0$.
Setting

$$L = \{w: \exists\, u \text{ with } (w,u) \in \text{vert } W\},\ K = \{v: \exists\, v \text{ with } (v,u) \in \text{extd } W\} ,$$

$$\alpha(w) = \min \{(B^T w + d)y: Ey \leq p\},\ \beta(v) = \min \{(B^T v)y: Ey \leq p\} ,$$

one can extend Proposition VIII.7 as follows:

A vector $(x,t) \in X \times \mathbb{R}$ belongs to G if and only if:

$$(Ax - c)v + \alpha(v) \leq 0 \quad \forall v \in K ,$$

$$(Ax - c)w + \beta(w) \leq t \quad \forall w \in L .$$

(for the details, see Tuy (1985, 1987) and Thieu (1989)). From these results it is apparent how the algorithm should be modified to handle this case.

4.4. Outer Approximation Versus Successive Partition

As we saw in Chapter II and Section VI.1, a major difficulty with outer approximation methods is the rapid growth of the number of constraints in the relaxed problems (H_k). Despite this difficulty, there are instances when outer approximation methods appear to be more easily applicable than other methods. In the decomposition context discussed in this chapter, an obvious advantage of the outer approximation approach is that all the linear subproblems involved in it have the same constraint set. This is in contrast with branch and bound methods in which each linear subproblem has a different constraint set.

To illustrate this remark, let us consider a class of two level decision problems which are sometimes encountered in production planning. They have the following general formulation

$$(*) \qquad \begin{array}{l} \min\ [f(x) + dy] \\ \text{s.t. } x \in X\ , \\ By = b,\ Cy = x,\ y \geq 0\ . \end{array} \qquad (33)$$

Specifically, x might denote a production program to be chosen (at the first decision level) from a set X of feasible programs, and y might denote some transportation–distribution program that is to be determined once the production program x has been chosen in such a way that the requirements (33) are met. The objective function f(x) is the production cost, which is assumed to be concave, and dy is the transportation–distribution cost. Often, the structure of the constraints (33) is such that highly efficient algorithms are currently available for solving linear programs with these constraints. This is the case, for example, with the so–called plant location problem, in which $y = \{y_{ij}, i=1,...,m, j=1,...,n\}$ and the constraints (33) are:

$$\begin{aligned} &\sum_i y_{ij} = b_j \quad (j=1,...,m) , \\ &\sum_j y_{ij} = x_i \quad (i=1,...,n) , \\ &\quad y_{ij} \ge 0 \quad (i=1,...,n;\ j=1,...,m) . \end{aligned} \tag{34}$$

Clearly, branch and bound methods do not take advantage of the specific structure of the constraints (33): each linear subproblem corresponding to a partition set involves additional constraints which totally destroy the original structure. In contrast, the decomposition approach by outer approximation allows this structure to be fully exploited: the linear programs involved in the iterations are simply

$$(C(x)) \qquad \min \{dy: By = b, Cy = x, y \ge 0\} \tag{35}$$

and their duals

$$(C^*(x)) \qquad \max \{wx + ub: C^T w + B^T u \le d\} . \tag{36}$$

This is particularly convenient in the case of the constraints (34), for then each C(x) is a classical transportation problem, and the dual variables w, u are merely the associated potentials.

When specialized to the problem under consideration, Algorithm VIII.5, with the modifications described in Section VIII.4.3, goes as follows (cf. Thieu (1987)). For the sake of simplicity, assume that X is bounded and the constraints (33) are feasible for every fixed $x \in X$.

Algorithm VIII.6

Initialization:

Estimate a lower bound t^0 for the transportation–distribution cost, i.e., compute a number $t^0 \leq \inf\{dy: x \in X, By = b, Cy = x, y \geq 0\}$. Set $T_0 = \{x \in X, t^0 \leq t\}$. Let x^0 be an optimal solution of the problem

$$(H_0) \qquad \min\{f(x): x \in X\}\ .$$

Iteration k = 0,1,... :

1) Solve the linear program $C(x^k)$, obtaining the optimal transportation–distribution cost of x^k: $g(x^k) = w^k x^k + u^k b$, where (w^k, u^k) is a basic optimal solution of the dual $C^*(x^k)$.

2) If $g(x^k) \leq t^k$ (the optimal transportation–distribution costs do not exceed the current estimate t^k), then terminate: x^k yields a global optimal solution of the problem (*). Otherwise, continue.

3) Form T_{k+1} by adding the constraint

$$w^k x + u^k b \leq t$$

to T_k. Solve the new relaxed problem

$$(H_{k+1}) \quad \min\{f(x) + t: (x,t) \in T_{k+1}\}\ ,$$

obtaining an optimal solution (x^{k+1}, t^{k+1}) (the new production program together with the new estimate of transportation–distribution cost).

Go to iteration k+1.

Example VIII.2. Solve the plant location problem:

$$\begin{aligned} \text{minimize } & \sum_i f_i(x_i) + \sum_i\sum_j d_{ij} y_{ij} \\ \text{s.t. } & \sum_i x_i = \sum_j b_j\ , \quad x_i \geq 0 \quad (i=1,\dots,n)\ , \\ & \sum_i y_{ij} = b_j \quad (j=1,\dots,m)\ , \\ & \sum_j y_{ij} = x_i \quad (i=1,\dots,n)\ , \\ & y_{ij} \geq 0 \quad (i=1,\dots,n;\ j=1,\dots,m)\ , \end{aligned} \tag{37}$$

with the following data:

$f_i(x_i) = 0$ if $x_i = 0$, $f_i(x_i) = r_i + s_ix_i$ if $x_i > 0$ $(i=1,...,n)$

(r_i is a setup cost for plant i; see Section I.2.2) ;

$n = 3$, $m = 5$, $r = (1,88,39)$, $s = (1.7, 8.4, 4.7)$.

$b = (62, 65, 51, 10, 15)$,

$$d = (d_{ij}) = \begin{bmatrix} 6 & 66 & 68 & 81 & 4 \\ 40 & 20 & 34 & 83 & 27 \\ 90 & 22 & 82 & 17 & 8 \end{bmatrix} .$$

Applying Algorithm VIII.6 with X defined by (37) we obtain the following results.

Initialization:

$t^0 = 3636$, $x^0 = (203, 0, 0)$.

Iteration 0:

Optimal value of $C(x^0)$: $9000 > t^0$.

Potentials: $w^0 = (0, -46, -64)$, $u^0 = (6, 66, 68, 81, 4)$.

New constraint: $-46x_2 - 64x_3 + 9000 \leq t$.

Optimal solution of $H(_1)$: $x^1 = (119.1875, 0, 83.8125)$, $t^1 = 3636$

Iteration 1:

Optimal value of $C(x^1)$: $5535.25 > t^1$

Potentials: $w^1 = (-106, -140, -102)$, $u^1 = (112, 124, 174, 119, 110)$

New constraint: $-106x_1 - 140x_2 - 102x_3 + 26718 \leq t$

Optimal solution of (H_2): $x^2 = (104.7, 51.5, 46.8)$, $t^2 = 3636$

Iteration 2:

Optimal value of $C(x^2)$: $4651.44407 > t^2$

Potentials: $w^2 = (-60, -94, -92)$, $u^2 = (66, 114, 128, 109, 64)$

New constraint: $-60x_1 - 94x_2 - 92x_3 + 20080 \leq t$

Optimal solution of (H_3): $x^3 = (73.176, 54.824,75)$, $t^3 = 3636$.

Iteration 3:

Optimal value of $C(x^3)$: $3773.647 > t^3$

Potentials: $w^3 = (-48, -46, -44)$, $u^3 = (54, 66, 80, 61, 52)$

New constraint: $-48x_1 - 46x2 - 44x_3 + 13108 \le t$

Optimal solution of (H_4): $x^4 = (77, 51, 75, 3766)$, $t^4 = 3766$

Iteration 4:

Optimal value of $C(x^4)$: $3766 = t^4$.

Thus, a global optimal solution of the plant location problem is $x^4 = (77, 51, 75)$, with corresponding transportation program

$$(y_{ij}) = \begin{bmatrix} 62 & 0 & 0 & 0 & 15 \\ 0 & 0 & 51 & 0 & 0 \\ 0 & 65 & 0 & 10 & 0 \end{bmatrix}.$$

4.5. Outer Approximation Combined with Branch and Bound

As discussed in connection with Algorithm VIII.5, in order to solve the relaxed problem (H_k) one can use any procedure, as long as it is capable of being restarted. One such procedure is the method of Thieu– Tam–Ban (Algorithm VI.1), which relies on the inductive computation of the vertex set V_k of the approximating polytope T_k. However, for relatively large values of n, this method encounters serious computational difficulties in view of the rapid growth of the set V_k which may attain a prohibitive size.

One possible way of avoiding these difficulties is to solve the relaxed problems by a restart branch and bound procedure (Section IV.6). This idea is similar to that underlying Horst–Thoai– Benson's method (1988) for concave minimization under convex constraints (Section VII.1.9.).

In fact, solving the relaxed problems in Algorithm VIII.5 by the restart normal conical algorithm amounts to applying Algorithm VII.2 (normal conical algorithm for CP) to the problem $(\tilde{H})$:

$$(\tilde{H}) \qquad \min \{f(x) + t: x \in D, g(x) \leq t\} .$$

For the sake of simplicity, as before we assume that D is bounded; as seen above (in Section VIII.4.2), this implies that g(x) is bounded on D, and for any $x \in \mathbb{R}^n$ the linear program

$$(C^*(x)) \qquad \max \{(Ax - c)w: B^T w \geq -d, w \geq 0\}$$

is feasible.

Let $G = \{(x,t): x \in D, g(x) \leq t\}$ be the feasible set of $(\tilde{H})$. From the above results (Section VIII.4.2) it follows that for any point (x^k,t^k), one of the following three cases must occur:

a) $C^*(x^k)$ has an optimal solution w^k with $(Ax^k - c)w^k \leq t^k$. Then $(x^k,t^k) \in G$;

b) $C^*(x^k)$ has an optimal solution w^k with $(Ax^k - c)w^k > t^k$.

Then $x^k \in D$ but $(x^k,t^k) \notin G$ and the inequality

$$(Ax - c)w^k \leq t \tag{38}$$

excludes (x^k,t^k) without excluding any point (x,t) of G.

c) $C^*(x^k)$ is unbounded, so that an extreme direction v^k of the polyhedron $W = \{w: B^T w \geq -d, w \geq 0\}$ can be found over which $(Ax^k - c)w^k$ is unbounded. Then $x^k \notin D$ and the inequality

$$(Ax - c)v^k \leq 0 \tag{39}$$

excludes x^k without excluding any point x of D.

Now, when applying Algorithm VII.2 to problem $(\tilde{H})$, let us observe that the feasible set G of $(\tilde{H})$ is a convex set of a particular kind, namely it is the epigraph of a convex function on D. In view of this particular structure of G, instead of using a conical subdivision of the (x,t)–space as prescribed by Algorithm VII.2, it is more convenient to subdivide the space into prisms of the form M×R, where M is an n–simplex in the x–space (such a prism can also be viewed as a cone with vertex at infinity, in the direction $t \longrightarrow +\infty$).

With this subdivision in mind, let T_k be a polyhedron containing G in the (x,t)–space, and let γ_k be the incumbent function value. For every prism $M \times \mathbb{R}$ which is in some partition of T_k, where $M = [s^1,...,s^{n+1}]$, we can compute the points (s^i,θ^i) on the verticals $x = s^i$ such that $f(s^i) + \theta^i = \gamma_k$ and consider the linear program

$$\max \{\Sigma \lambda_i \theta^i - t: (\Sigma \lambda_i s^i, t) \in T_k, \Sigma \lambda_i = 1, \lambda_i \geq 0 \ (i=1,...,n+1)\} \tag{40}$$

which is equivalent to

$$\max \{\psi_M(x) - t: (x,t) \in T_k , x \in M\} , \tag{41}$$

where $\psi_M(x)$ is the intersection point of the vertical through x with the hyperplane through $(s^1,\theta^1),...,(s^{n+1},\theta^{n+1})$. Let $\mu(M)$ be the optimal value, and let $z(M) = (x(M),t(M))$ be a basic optimal solution of (40). Clearly, if $\mu(M) \leq 0$, then the portion of G contained in $M \times \mathbb{R}$ lies entirely above our hyperplane. Hence, by the concavity of $\Phi(x,t) := f(x)+t$, we must have $f(x)+t \geq \gamma_k \ \forall (x,t) \in G \cap (M \times R)$, i.e., this prism can be fathomed. Otherwise, if $\mu(M) > 0$, this prism should be further investigated. In any case, x(M) is distinct from all vertices of M, so that M can be subdivided with respect to this point. In addition, the number

$$\gamma_k - \mu(M) = \min \{f(s^i) + \theta^i - \mu(M): i=1,2,...,n+1\}$$

clearly yields a lower bound for $\min \{f(x) + t: x \in M, (x,t) \in T_k\}$.

On the other hand, by solving $C^*(x(M))$ we can check whether z(M) belongs to G, and if not, construct a constraint to add to T_k to define the new polyhedron T_{k+1}. For every $x \in D$ denote $F(x) = f(x) + g(x)$. The above development leads to the following procedure:

Algorithm VIII.7.

Select $\varepsilon > 0$ and an exhaustive rule for simplicial subdivision.

1) Construct an n–simplex M_0 containing D in the x–space and let $T_0 = M_0 \times \mathbb{R}$. Choose a point $\bar{x}^0 \in D$ and let $\gamma_0 = F(\bar{x}^0)$. Set $\mathscr{M}_0 = \mathscr{P}_0 = \{M_0\}$, $k = 0$.

2) For each $M \in \mathscr{P}_k$ solve the linear program (40), obtaining the optimal value $\mu(M)$ and a basic optimal solution $(x(M),t(M))$ of (41).

3) Let $\mathscr{R}_k = \{M \in \mathscr{M}_k: \mu(M) > \varepsilon\}$. If $\mathscr{R}_k = \emptyset$, then terminate: $\bar{x}^k$ is a global ε–optimal solution of (H). Otherwise, go to 4).

4) Select $M_k \in \operatorname{argmax} \{\mu(M): M \in \mathscr{R}_k\}$, and subdivide it according to the exhaustive process that was chosen.
Let $\mathscr{P}_{k+1}$ be the resulting partition of M_k.

5) Let $x^k = x(M_k)$, $t^k = t(M_k)$. Solve $C^*(x^k)$. If case a) occurs, i.e., $(x^k,t^k) \in G$, then let $T_{k+1} = T_k$. Otherwise, form T_{k+1} by adding the new constraint (38), or (39), to T_k, according to whether case b) or case c) occurs.

6) Let $\bar{x}^{k+1}$ be the best (in terms of the value of F(x)) among $\bar{x}^k$, all x(M) for $M \in \mathscr{P}_{k+1}$, and $u(M_k)$ if $u(M_k)$ was used, and $u(M_k) \in D$. Let $\gamma_{k+1} = F(\bar{x}^{k+1})$, $\mathscr{M}_{k+1} = (\mathscr{R}_k \setminus \{M_k\}) \cup \mathscr{P}_{k+1}$. Set $k \leftarrow k+1$ and return to 2).

Theorem VIII.5. *Algorithm VIII.7 terminates after finitely many iterations*

Proof. By viewing a prism $M \times \mathbb{R}$ as a cone with vertex at infinity in the direction

$t \longrightarrow \infty$, the proof can be carried out in the same way as for Theorem VII.5 on the convergence of Algorithm VII.2. Since the subdivision process is exhaustive, the argument is even simpler. ∎

Remark VIII.4. Note the difference between the above algorithm and Algorithm VIII.1 (normal simplicial algorithm for (P), Section VIII.2.1). Although both algorithms proceed by simplicial subdivision of the x–space, the lower bounding subproblem LP(M,Ω) in Algorithm VIII.1 is much larger than the corresponding linear program (40) in Algorithm VIII.7. Of course, Algorithm VIII.7 requires us to solve an additional subproblem $C^*(x^k)$ (which is the dual to a linear program in y). However, since at least two new simplices appear at each iteration, in all, Algorithm VIII.7 should be less expensive.

5. DECOMPOSITION OF CONCAVE MINIMIZATION PROBLEMS OVER NETWORKS

A significant class of concave minimization problems relates to networks. These include problems in inventory and production planning, capacity sizing, location and network design which involve set–up charges, discounting, or economies of scale. Large scale problems of this class can often be treated by appropriate decomposition methods that take into account the specific underlying network structure.

5.1. The Minimum Concave Cost Flow Problem

Consider a (directed) graph G = (V, A), where V is the set of nodes and A is the set of arcs (an arc is an ordered pair of nodes). Suppose we are given a real number d(v) for each node $v \in V$ and two nonnegative numbers p_a, q_a ($p_a \le q_a$) for each arc

$a \in A$. A vector x with components $x(a) \geq 0$, $a \in A$, is called a *flow* in the network G (where the component x(a) is the flow value in the arc a). A flow is said to be feasible if

$$p_a \leq x(a) \leq q_a \quad \forall a \in A\ , \tag{42}$$

$$\sum_{a \in A^+(v)} x(a) - \sum_{a \in A^-(v)} x(a) = d(v) \quad \forall v \in V\ , \tag{43}$$

where $A^+(v)$ (resp. $A^-(v)$) denotes the set of arcs entering (resp. leaving) node v. The number d(v) expresses the "demand" at node v (if $d(v) < 0$, then node v is a "supply" node with supply $-d(v)$). The numbers p_a, q_a represent lower and upper bounds on the flow value in arc a. The relation (43) expresses flow conservation. It follows immediately from (43) that a feasible flow exists only if $\sum_{v \in V} d(v) = 0$.

Furthermore, to each arc we associate a concave function $f_a: \mathbb{R}_+ \longrightarrow \mathbb{R}_+$ whose value $f_a(t)$ at a given $t \geq 0$ represents the cost of sending an amount t of the flow through the arc a. The *minimum concave cost flow problem* (CF) is to find a feasible flow x with smallest cost

$$f(x) = \sum_{a \in A} f_a(x(a))\ . \tag{44}$$

When $p_a = 0$, $q_a = +\infty$ $\forall a \in A$, the problem is called the *uncapacitated minimum concave cost flow problem* (UCF). When, in addition, there is only one supply node (one node v with $d(v) < 0$), the problem is referred to as the *single source uncapacitated minimum concave cost flow problem* (SUCF). It is known that (SUCF) is NP–hard (cf.,e.g., Nemhauser and Wolsey (1988), Guisewite and Pardalos (1989)).

The (CF) problem has been studied by several authors. One of the earliest works is a paper of Zangwill (1968), where a dynamic programming method was developed for certain important special cases of (SUCF). The dynamic programming approach of Zangwill was further extended for (CF) in a recent study by Erickson et al. (1987) (send–and–split method).

Other methods using branch and bound concepts have been proposed by Soland

(1974), Gallo, Sandi and Sodini (1980), Konno (1988) and others (cf. Chapter I, Florian (1986) and references therein).

Denote by A^I the set of all arcs $a \in A$ for which the cost function f_a is affine, i.e., of the form $f_a(x(a)) = f_a(p_a) + c_a(x(a) - p_a)$ $(c_a \ge 0)$, and let $A^{II} = A \setminus A^I$. In many practical cases, $|A^{II}|$ is relatively small compared to $|A^I|$, i.e., the problem involves relatively few nonlinear variables. Then the minimum concave cost flow problem belongs to the class considered in this chapter and can be treated by the methods discussed above.

For every flow x let us write $x = (x^I, x^{II})$ where $x^I = (x(a), a \in A^I)$, $x^{II} = (x(a), a \in A^{II})$. The following Algorithm VIII.8 is a specialization to the concave cost flow problem of the normal rectangular algorithm for separable problems (Algorithm VIII.2).

Algorithm VIII.8.

Start with the rectangle $M_1 = \prod_{a \in A^{II}} [p_a, q_a]$. Set $x^0 = 0$, $\gamma_0 = +\infty$. Let $\mathcal{M}_1 = \mathcal{N}_1 = \{M_1\}$.

Iteration k=1,2,...:

1) For each rectangle $M \in \mathcal{N}_k$ with $M = \prod_{a \in A^{II}} [r_a, s_a]$ solve the linear problem:

$$\text{minimize } \sum_{a \in A^I} c_a x(a) + \sum_{a \in A^{II}} c_a^M x(a)$$

$$\text{(LP(M))} \quad \text{s.t. } p_a \le x(a) \le q_a \quad (a \in A^I) \,; \; r_a \le x(a) \le s_a \quad (a \in A^{II}) \,,$$

$$\sum_{a \in A^+(v)} x(a) - \sum_{a \in A^-(v)} x(a) = d(v) \quad (v \in V) \,,$$

where

$$c_a^M = \frac{f_a(s_a) - f_a(r_a)}{s_a - r_a}$$

(c_a^M is the slope of the affine function $\psi_a^M(t)$, which agrees with f(t) at the points $t = r_a$ and $t = s_a$).

If (LP(M)) is infeasible, then set $\beta(M) = +\infty$. Otherwise, let $\omega(M)$ be an optimal solution of (LP(M)) with components $\omega_a(M)$ and let $\beta(M) = f(\omega(M))$.

2) Define the incumbent by setting x^k equal to the best feasible solution among x^{k-1}, and all $\omega(M)$, $M \in \mathcal{N}_k$. Let $\gamma_k = f(x^k)$ if x^k exists, $\gamma_k = +\infty$ otherwise. Delete all $M \in \mathcal{M}_k$ for which $\beta(M) \geq f(x^k)$. Let $\mathcal{R}_k$ be the remaining collection of rectangles.

3) If $\mathcal{R}_k = \emptyset$, then terminate: if $\gamma_k < +\infty$, then x^k is an optimal solution (optimal flow), otherwise the problem is infeasible.

4) If $\mathcal{R}_k \neq \emptyset$, then choose $M_k \in \operatorname{argmin} \{\beta(M): M \in \mathcal{R}_k\}$. Let $\omega^k = \omega(M_k)$ and $\psi_{k,a}(t) = f_a(r_a) + c_a^{M_k}(t-r_a)$

5) Select an arc $a_k \in \operatorname{argmax} \{f_a(\omega_a^k) - \psi_{k,a}(\omega_a^k): a \in A^{II}\}$. Denote by M_k^1, M_k^2 the rectangles obtained from M_k by replacing $[r_{a_k}, s_{a_k}]$ by $[r_{a_k}, \omega_{a_k}^k]$ and $[\omega_{a_k}^k, s_{a_k}]$, respectively.

6) Let $\mathcal{N}_{k+1} = \{M_k^1, M_k^2\}$, $\mathcal{M}_{k+1} = (\mathcal{R}_k \setminus \{M_k\}) \cup \mathcal{N}_{k+1}$. Set $k \leftarrow k+1$ and return to 1).

It is easily seen that $\gamma_k < +\infty \ \forall k > 1$.

Convergence of Algorithm VIII.8 can be deduced from Theorem VIII.2, since Algorithm VIII.8 is a specialization to the concave cost flow problem of Algorithm VIII.2 which handles more general separable problems. It follows that every accumulation point of the sequence $\{x^k\}$ is an optimal solution of (CF), and $f(x^k)$ converges to the optimal value of (CF) as $k \rightarrow \infty$.

5.2. The Single Source Uncapacitated Minimum Concave Cost Flow (SUCF) Problem

Now we consider the SUCF problem, i.e., the special case of (CF) when there are no capacity constraints, (i.e., $p_a = 0$, $q_a = +\infty$ $\forall a \in A$) and there is only one supply node (i.e., one node $v \in V$ with $d_v < 0$). Since (SUCF) is NP–hard, large scale SUCF problems cannot be expected to be effectively solved by general purpose methods. Fortunately, many SUCF problems encountered in practice have special additional structure that can be exploited to devise specialized algorithms. Examples include the concave warehouse problem, the plant location problem, the multi-product production and inventory models, etc. All of these problems and many others can be described as (SUCF) over a network which consists of several pairwise interconnected subnetworks like the one depicted in Fig. VIII.1.

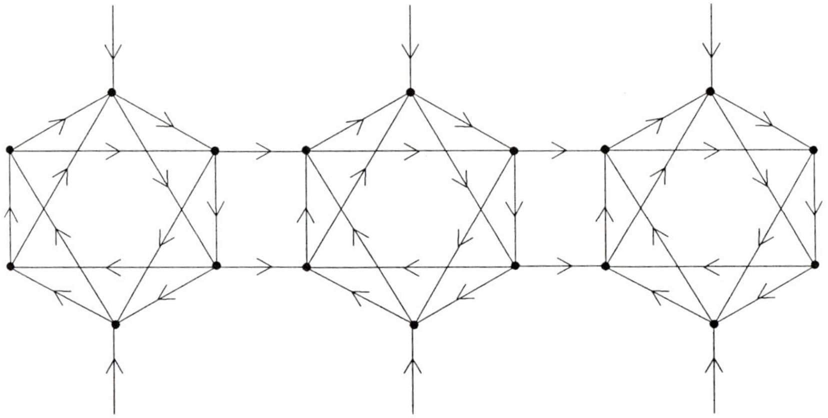

Fig. VIII.1 Typical structure of (SUCF)

For problems with such a reccuring special structure, a decomposition method can be derived that combines dynamic programming and polyhedral underestimation techniques. We follow the approach of Thach (1988).

Let us first restate the problem and its structure in a convenient form. A node v of a graph G = (V,A) is called a *source* if $A_v^+ = \emptyset$. Let S(G) be the set of all sources of G, and suppose that we are given a demand $d(v) \geq 0$ for each node $v \in V \setminus S(G)$ and a concave cost function $f_a: \mathbb{R}_+ \longrightarrow \mathbb{R}_+$ for each arc $a \in A$. Then we consider the problem

$$\text{minimize } f(x) = \sum_{a \in A} f_a(x(a)) \tag{45}$$

$$\text{s.t.} \quad \sum_{a \in A^+(v)} x(a) - \sum_{a \in A^-(v)} x(a) = d(v) \quad (v \in V \setminus S(G)) \tag{46}$$

$$x(a) \geq 0 \quad (a \in A)\,. \tag{47}$$

Clearly this problem becomes an (SUCF) problem when we add to the network a fictive supply node v^0 along with a fictive arc (v^0,v) for each $v \in S(G)$. Furthermore, we set $d(v) = 0 \;\; (v \in S(G))$, $d(v^0) := -\Sigma\,\{d(v): v \in V \setminus S(G)\}$.

Conversely, an (SUCF) problem in the original formulation corresponds to the case $|S(G)| = 1$. Therefore, from now on, by (SUCF) we shall mean the problem (45) – (47) with the fictive node v^0 and the fictive arcs (v^0,v) $(v \in S(G))$ as introduced above.

We shall further impose a certain structure on the network G = (V,A):

(*) There is a partition $\{V_1,\dots,V_n\}$ *of V such that for every i=1,...,n we have the following property:*

if the initial node of an arc belongs to V_i *, then its final node belongs to either* V_i *or* V_{i+1}.

In view of this condition, the coefficient matrix of the flow conservation constraints (46) has a staircase structure which we exploit in order to derive a decomposition method.

For each i=1,...,n we define

$$A_i = \{a \in A: a = (u,v),\ u \in V,\ v \in V_i\},$$

and for each v we let $A_i^+(v)$ (resp. $A_i^-(v)$) denote the set of arcs in A_i which enter (resp. leave) v.

Then $A_i \cap A_j = \emptyset$ $(i \neq j)$ and $A = \bigcup_{i=1}^{n} A_i$, i.e., the sets $A_1,\ldots,A_n$ form a partition of A. Furthermore, let G_i be the subgraph of G generated by A_i, i.e., $G_i = (W_i, A_i)$, where W_i is the set of all nodes incident to arcs in A_i. Then any flow x in G can be written as $x = (x_1,\ldots,x_n)$, where x_i is the restriction of x to A_i.
Setting $U_i = W_i \cap W_{i+1}$, we see that $v \in U_i$ if and only if v is the initial node of an arc going from V_i to V_{i+1}. Let $h_i\colon V \to \mathbb{R}_+$ be a function such that $h_i(v) = 0$ for $v \notin U_i$. Denote by X(G,d) the set of all feasible flows in G for the demand vector d with components d(v), $v \in V \setminus S(G)$, i.e., the set of all vectors $x = (x(a),\ a \in A)$ satisfying (46) and (47); similarly, denote by $X(G_i, d + h_i)$ the set of all feasible flows in G_i for the demands

$$d(v) + h_i(v) \quad (v \in W_i).$$

Proposition VIII.8. *We have*

$$X(G,d) = \{x=(x_1,\ldots,x_n)\colon x_1 \in X(G_1,d+h_1),\ldots,\ x_{n-1} \in X(G_{n-1},d+h_{n-1}),$$

$$x_n \in X(G_n,d),\ h_i(v) = \sum_{a \in A_{i+1}^-(v)} x_{i+1}(a) \quad \forall v \in U_i, i=1,\ldots,n-1\},$$

where $A_{i+1}^-(v)$ denotes the set of arcs $a \in A_{i+1}$ leaving v.

Proof. By virtue of assumption (*) and the definition of A_i we have

$$A^+(v) = A_i^+(v) \quad \forall v \in V_i$$

$$A^-(v) = \begin{cases} A_i^-(v) \cup A_{i+1}^-(v) & \text{if } v \in U_i \\ A_i^-(v) & \text{if } v \in V_i \setminus U_i \end{cases},$$

from which the assertion follows. ■

In a network $G = (V,A)$ a ***path*** is a finite sequence of distinct arcs such that (except for the last arc) the final node of any arc in the sequence is the initial node of the next arc.

We shall assume that for any node $v \in V \setminus S(G)$ there is at least one path going from a source to v. This amounts to requiring that $X(G,d)$ is nonempty for any vector d of nonnegative demands $d(v)$.

A subgraph $T = (V,B)$ of G is called a ***spanning forest*** of G if for any node $v \in V \setminus S(G)$ there is exactly one path in T going from a source to v.

An ***extreme flow*** in G is an extreme point of the polyhedron $X(G,d)$.

Proposition VIII.9. *A feasible flow x is an extreme flow if and only if there exists a spanning forest $T = (V,B)$ such that $\{a \in A: x(a) > 0\} \subset B$.*

We omit the proof, which can be found in textbooks on elementary network flow theory (see also Zangwill (1968)).

Now consider the function

$$\Phi(d) = \inf \{f(x): x \in X(G, d)\} .$$

Proposition VIII.10. *The function $\Phi(d)$ is concave on the set of all $d = (d(v))$, $v \in V \setminus S(G)$, $d(v) \geq 0$.*

Proof. For $d = \lambda d' + (1-\lambda)d''$, $0 \leq \lambda \leq 1$, let $\bar{x} \in \operatorname{argmin} \{f(x): x \in X(G,d)\}$. We may assume that $\bar{x}$ is an extreme flow (because $f(x)$ is concave). Then there exists a spanning forest $T = (V,B)$ such that $B \supset \{a \in A: \bar{x}(a) > 0\}$. Since T is a forest, there exist a unique $x' \in X(T,d')$ and a unique $x'' \in X(T,d'')$. Clearly $\lambda x' + (1-\lambda)x'' \in X(T,d)$ and hence by uniqueness, $\bar{x} = \lambda x' + (1-\lambda)x''$. But we have $\Phi(d) = f(\bar{x}) \geq \lambda f(x') + (1-\lambda) f(x'') \geq \lambda\Phi(d') + (1-\lambda)\Phi(d'')$, which proves the concavity of Φ. ■

Next, consider the functions

$$F_1(h_1) = \inf \{ \sum_{a \in A_1} f_a(x_1(a)) \colon x_1 \in X(G_1, d + h_1) \} ,$$

$$F_2(h_2) = \inf \{ F_1(h_1) + \sum_{a \in A_2} f_a(x_2(a)) \colon x_2 \in X(G_2, d + h_2) ,$$

$$h_1(u) = \sum_{a \in A_2^-(u)} x_2(a) \quad \forall u \in U_1 \}$$

....

$$F_{n-1}(h_{n-1}) = \inf \{ F_{n-2}(h_{n-2}) + \sum_{a \in A_{n-1}} f_a(x_{n-1}(a)) \colon$$

$$x_{n-1} \in X(G_{n-1}, d + h_{n-1}), h_{n-2}(u) = \sum_{a \in A_{n-1(u)}^-} x_{n-1}(a) \quad \forall u \in U_{n-2} \} .$$

Proposition VIII.11. *The functions $F_i(\cdot)$ are concave on their domains of definition. If $\varphi_i(x_i)$ denotes the function obtained from $F_{i-1}(h_{i-1})$ by replacing h_{i-1} by the vector $\sum_{a \in A_i^-(u)} x_i(a)$, $u \in U_{i-1}$, then $\varphi_i(x_i)$ is concave.*

Proof. The concavity of $F_1(\cdot)$ follows from Proposition VIII.10. Since the mapping $x_2 \longrightarrow (\sum_{a \in A_2^-(u)} x_2(a), u \in U_1)$ is linear in x_2, and since $F_1(\cdot)$ is concave, we deduce that $\varphi_2(x_2)$ is concave. But

$$F_2(h_2) = \inf \{ \varphi_2(x_2) + \sum_{a \in A_2} f_a(x_2(a)) \colon x_2 \in X(G_2, d + h_2) \}.$$

Therefore, again using Proposition VIII.10, we see that $F_2(\cdot)$ and $\varphi_3(\cdot)$ are concave. The proof can be completed by induction. ∎

With the above background, we can describe the basic idea of the decomposition method proposed in Thach (1988).

Consider the subproblem

$$(P_n) \qquad \text{minimize } (F_{h-1}(h_{n-1}) + \sum_{a \in A_n} f_a(x_n(a)))$$

$$\text{s.t. } x_n \in X(G_n,d),\ h_{n-1}(u) = \sum_{a \in A_n^-(u)} x_n(a) \quad \forall u \in U_{n-1} .$$

Let $(\bar{x}_n,\bar{h}_{n-1})$ denote an optimal solution of (P_n), and consider the next subproblem

$$(P_{n-1}) \qquad \text{minimize } (F_{n-2}(h_{n-2}) + \sum_{a \in A_{n-1}} f_a(x_{n-1}(a)))$$

$$\text{s.t. } x_{n-1} \in X(G_{n-1},d+\bar{h}_{n-1}),\ h_{n-2}(u) = \sum_{a \in A_{n-1}^-(u)} x_{n-1}(a) \quad \forall u \in U_{n-2}$$

Continuing in this way, we can successively define the subproblems $P_{n-2},\dots,P_2$, where P_i is the problem

$$(P_i) \qquad \text{minimize } (F_{i-1}(h_{i-1}) + \sum_{a \in A_i} f_a(x_i(a)))$$

$$\text{s.t. } x_i \in X(G_i,\ d + \bar{h}_i),\ h_{i-1}(u) = \sum_{a \in A_i^-(u)} x_i(a) \quad \forall u \in U_{i-1} ,$$

in which $\bar{h}_i$ is obtained from an optimal solution $(\bar{x}_{i+1},\ \bar{h}_i)$ of P_{i+1} .

Finally, let $\bar{x}_1$ be an optimal solution of the subproblem

$$(P_1) \qquad \text{minimize } \sum_{a \in A_1} f_a(x_1(a)), \quad \text{s.t. } x_1 \in X(G_1,\ d + \bar{h}_1) .$$

Note that, if we agree to set $\varphi_1(\cdot) \equiv 0$, then, by Proposition VIII.11 each problem (P_i) is equivalent to minimizing the concave function $\varphi_i(x_i) + \sum_{a \in A_i} f_a(x_i(a))$ subject to $x_i \in X(G_i,\ d + \bar{h}_i)$.

Theorem VIII.6. *Let α be the optimal value of the objective function in (P_n). Then α is the optimal value of the objective function in (P), and $\bar{x} = (\bar{x}_1,\dots,\bar{x}_n)$, with $\bar{x}_i$ $(i=1,\dots,n)$ as defined above, is an optimal solution of (P).*

Proof. Replacing the functions F_i $(i=n-1,\dots,1)$ in the problems (P_i) $(i=n-1,\dots,1)$ by their defining expressions we can easily see that α is the optimal value of the objective function and $\bar{x}$ is a minimizer of the function $\sum_{a\in A} f_a(x(a))$ over the domain

$$\{x = (x_1,\dots,x_n)\colon x_1 \in X(G_1,\ d + h_1),\dots,x_{n-1} \in X(G_{n-1},\ d + h_{n-1})\ ,$$

$$x_n \in X(G_n,\ d) \text{ and } h_i(v) = \sum_{a\in A^-_{i-1}(v)} x_{i+1}(a) \quad \forall v \in U_i\ ,\ i=1,\dots,n-1\}\ .$$

From Proposition VIII.8 it follows that this is just problem (P). ■

Thus, to obtain an optimal flow $\bar{x}$, it suffices to solve $(P_n),(P_{n-1}),\dots(P_1)$ successively (note that by solving (P_i) we obtain $(\bar{x}_i,\ \bar{h}_{i-1})$, and then $\bar{h}_{i-1}$ is used to define P_{i-1}). Since each (P_i) is a concave minimization over $X(G_i,\ d + \bar{h}_i)$, the original problem (P) decomposes into n subproblems of the same type but much smaller size. We next discuss how to solve these subproblems.

5.3. Decomposition Method for (SUCF)

When solving the subproblems (P_i), the difficulty is that the functions $F_{i-1}(h_{i-1})$ which occur in the objective functions of these subproblems are defined only implicitly. One way to overcome this difficulty is as follows: first approximate the functions $F_i(\cdot)$ (which are nonnegative and concave by Proposition VIII.11) with certain polyhedral concave underestimators ψ_i^0, and for $i=n,n-1,\dots,1$ solve the approximate subproblems (P_i^0) obtained from (P_i) by substituting ψ_i^0 for F_{i-1}. Then use the solutions of $(P_n^0),\dots,(P_1^0)$ to define functions ψ_i^1 that are better approximations to F_i than ψ_i^0, and solve the new approximate subproblems $(P_n^1),\dots,(P_1^1)$, and so on. It turns out that with an appropriate choice of approximating functions $\psi_i^0,\psi_i^1,\dots,\psi_i^k,\dots$, this iterative procedure will generate an optimal flow $\bar{x}$ after finitely many iterations.

Recall that for any real valued function $\psi(h)$ defined on a set dom ψ, the hypograph of $\psi(h)$ is the set

$$\text{hypo } \psi = \{(h,t) \in \text{dom } \psi \times R: t \le \psi(h)\} .$$

Clearly, a concave function is determined by its hypograph.

Denote by $\mathbb{R}^U$ the space of all vectors $h = (h(u), u \in U)$ (i.e., all functions $h: U \longrightarrow \mathbb{R}$), and denote by $\mathbb{R}_+^U$ the set of all $h \in \mathbb{R}^U$ such that $h(u) \ge 0 \ \ \forall u \in U$.

Algorithm VIII.9 (Decomposition Algorithm for SUCF).

Initialization:

For each $i=1,...,n-1$ let $\psi_i^0 \equiv 0$. Set $k = 0$.

Iteration k=0,1,2,...:

This iteration is entered with knowledge of ψ_i^k , $i=1,...,n-1$. Define $h_n^k \equiv 0$, $\psi_0^k \equiv 0$. Set $i=n$.

k.1. Solve

$$(P_i^k) \qquad \text{minimize } \psi_{i-1}^k(h_{i-1}) + \sum_{a \in A_i} f_a(x_i(a))$$

$$\text{s.t. } x_i \in X(G_i,d + h_i^k), \ h_{i-1}(u) = \sum_{a \in A_{i-1}^-(u)} x_i(a) \ \ \forall u \in U_{i-1} \qquad (48)$$

obtaining an optimal solution (x_i^k, h_{i-1}^k) and the optimal value t_i^k of (P_i^k).

If $i \ge 2$, set $i \longleftarrow i-1$ and return to k.1. Otherwise, go to k.2.

k.2. If $t_i^k \le \psi_i^k(h_i^k)$ for all $i=1,...,n-1$, then stop: $x^k = (x_1^k,...,x_n^k)$ is an optimal flow. Otherwise, go to k.3.

k.3. Construct a new concave underestimator ψ_i^{k+1} for each F_i such that the hypograph of ψ_i^{k+1} is the convex hull of the set obtained by adjoining the point (h_i^k,t_i^k) to hypo ψ_i^k , i.e.,

$$\text{hypo } \psi_i^{k+1} = \text{conv } \{\text{hypo } \psi_i^k , (h_i^k,t_i^k)\} \tag{49}$$

Go to iteration k+1.

Remarks VIII.5. Before discussing the convergence of the algorithm, we make some remarks on how to construct the functions ψ_i^{k+1} and how to solve the subproblem (P_i^k).

Construction of the functions ψ_i^{k+1} (step k.3):

In view of (49) and the relation $\psi_0^i \equiv 0$ we can write

$$\begin{aligned}\text{hypo } \psi_i^{k+1} &= \text{conv } \{\text{hypo } \psi_i^0 , (h_i^j,t_i^j) \quad j=0,\dots,k\} \\ &= \text{conv } \{\mathbb{R}_+^{U_i} \times \mathbb{R}_- , (h_i^j,t_i^j) \quad j=0,\dots,k\} .\end{aligned}$$

Thus, for any $h_i \in \mathbb{R}_+^{U_i}$ we have

$$\begin{aligned}\psi_i^{k+1}(h_i) &= \sup \{t: (h_i,t) \in \text{hypo } \psi_i^{k+1}\} \\ &= \sup \{t: t \le \sum_{j=1}^k s_j t_i^j , \sum_{j=1}^k s_j h_i^j(u) \le h_i(u) \quad \forall u \in U_i , \\ &\qquad \sum_{j=1}^k s_j \le 1 , s_j \ge 0 \quad \forall j=0,\dots,k\} \\ &= \sup \{ \sum_{j=1}^k s_j t_i^j : \sum_{j=1}^k s_j h_i^j(u) \le h_i(u) \quad \forall u \in U_i , \\ &\qquad \sum_{j=1}^k s_j \le 1 , s_j \ge 0 \quad \forall j=0,\dots,k\} .\end{aligned} \tag{50}$$

For a given $h_i \in \mathbb{R}_+^{U_i}$ the value of $\psi_i^{k+1}(h_i)$ is equal to the optimal value of the linear program (50).

On the other hand, since hypo ψ_i^{k+1} is a polyhedral convex set, ψ_i^{k+1} can be expressed as a pointwise minimum of affine functions. The graphs of these affine functions are hyperplanes through the nonvertical facets of hypo ψ_i^{k+1}. Starting with the fact that ψ_i^0 has a unique nonvertical facet, namely $\mathbb{R}^{U_i}\times\{0\}$, and using formula (49), we can inductively determine the nonvertical facets of $\psi_i^1,\psi_i^2,\ldots,\psi_i^{k+1}$ by the polyhedral annexation technique (see Section VI.4.3). In the present context this procedure works in the following way.

Consider the dual problem of (50):

$$\text{minimize} \sum_{u\in U_i} r(u)h_i(u) + t \tag{51}$$

$$\text{s.t.} \sum_{u\in U_i} r(u)h_i^j(u) + t \ge t_i^j \quad (j=0,1,\ldots,k) , \tag{52}$$

$$r \in \mathbb{R}_+^{U_i} , \; t \ge 0 . \tag{53}$$

Since $\psi_i^{k+1}(h_i)$ is equal to the optimal value in (50), it is also equal to the optimal value in (51)–(53). Furthermore, since $\psi_i^{k+1}(h_i)$ is finite, the optimal value in (51)–(53) must be achieved at least at one vertex of the polyhedral convex set Z_i^k defined by (52) and (53). Let E_i^k denote the vertex set of Z_i^k. Then E_i^k is finite, and we have

$$\psi_i^{k+1}(h_i) = \min \{ \sum_{u\in U_i} r(u)h_i(u) + t\colon (r,t) \in E_i^k\} . \tag{54}$$

Therefore, in order to obtain an explicit expression for ψ_i^{k+1} we compute the vertex set E_i^k of Z_i^k. Clearly, $E_i^0 = \{0\}$ for all $i=1,\ldots,n-1$, and Z_i^{k+1} differs from Z_i^k by just one additional linear constraint. Hence, we can compute the vertex set of Z_i^k by the methods discussed in Section II.4.2. Since $Z_i^k \subset \mathbb{R}^{U_i}\times\mathbb{R}$, the above procedure is practical if $|U_i|$ is relatively small.

Solving the subproblems (P_i^k) (step k.1):

From (54) it is easily seen that (P_i^k) can be rewritten as

$$\min_{(r,t)\in E_{i-1}^{k-1}} \{t + \min P_i^k(r)\} ,$$

where $\min P_i^k(r)$ denotes the optimal value of the problem

$$(P_i^k(r)) \qquad \min \sum_{u \in U_{i-1}} r(u)h_{i-1}(u) + \sum_{a \in A_i} f_a(x_i(a))$$

$$\text{s.t. } x_i \in X(G_i, d + h_i^k),\ h_{i-1}(u) = \sum_{a \in A_i^-(u)} x_i(a) \ \ \forall u \in U_{i-1} .$$

Recalling now that $G_i = (W_i, A_i)$ and using an artificial node w, let us define the network $\overline{G}_i = (\overline{W}_i, \overline{A}_i)$, where $\overline{W}_i = W_i \cup \{w\}$, $\overline{A}_i = A_i \cup \{(w,u)\colon u \in U_{i-1}\}$ and to each arc $a = (w,u)$ $(u \in U_{i-1})$ one assigns the linear cost function $f_a(x) = r(u)x$. Then $P_i^k(r)$ becomes an uncapacitated minimum concave cost flow problem on $\overline{G}_i$ which can be solved by several currently available algorithms (for example, Algorithm VIII.8). Note that for $r' \neq r$ the problem $P_i^k(r')$ differs from $P_i^k(r)$ only in the linear cost functions on the artificial arcs $a = (w,u)$ $(u \in U_{i-1})$.

Moreover, it follows from the above discussion that each (P_i^k) has an optimal solution (x_i^k, h_{i-1}^k) such that x_i^k is an extreme flow in G_i. Hence, we may assume that for every k, $x^k = (x_1^k, \dots, x_n^k)$ is an extreme flow in G.

In order to show convergence of the above algorithm, we first formulate the following propositions.

Proposition VIII.12. *For any $i=1,\dots,n-1$ and any $k=0,1,2,\dots$ we have $\psi_i^k(h_i) \le F_i(h_i)$ $\forall h_i \in \mathbb{R}_+^{U_i}$ (i.e., ψ_i^k is actually an underestimator of F_i).*

Proof. Since F_i is nonnegative on $\mathbb{R}_+^{U_i}$ $(i=1,\dots,n-1)$, it is immediate that

$$\psi_i^0(h_i) \le F_i(h_i) \quad \forall h_i \in \mathbb{R}_+^{U_i} \quad (i=1,\dots,n-1) . \tag{55}$$

Now, for any k we have

$$t_1^k = \inf (P_1^k) = \inf \{ \sum_{a \in A_1} f_a(x_1(a))\colon x_1 \in X(G_1, d + h_1^k)\} = F_1(h_1^k) .$$

Furthermore, since $\psi_1^0(h_1) \le F_1(h_1)$, we further have hypo $\psi_1^0 \subset$ hypo F_1. Therefore, by (55), hypo $\psi_1^k \subset$ hypo F_1 $\forall k$, or, equivalently, $\psi_1^k(h_1) \le F_1(h_1)$ $\forall k$. Since

$$t_2^k = \inf\{\psi_1^k(h_1) + \sum_{a\in A_2} f_a(x_2(a)) : x_2 \in X(G_2, d+h_2^k),$$

$$h_1(u) = \sum_{a\in A_2^-(u)} x_2(a) \;\; \forall u \in U_1\}$$

$$\le \inf\{F_1(h_1) + \sum_{a\in A_2} f_a(x_2(a)) : x_2 \in X(G_2, d + h_2^k),$$

$$h_1(u) = \sum_{a\in A_2^-(u)} x_2(a) \;\; \forall u \in U_1\}$$

$$= F_2(h_2^k) ,$$

and hypo $\psi_2^0 \subset$ hypo F_2, it follows from (49) that $\psi_2^k(h_2) \le F_2(h_2)$ for all k.

By the same argument we have

$$\psi_i^k(h_i) \le F_i(h_i) \quad (i=1,\ldots,n-1,\ k=0,1,2,\ldots) .$$

■

Proposition VIII.13. *If at some iteration k*

$$t_i^k \le \psi_i^k(h_i^k) \quad \forall\ i=1,\ldots,n-1 , \tag{56}$$

or, equivalently,

$$\psi_i^k = \psi_i^{k+1} \quad \forall\ i=1,\ldots,n-1 , \tag{57}$$

then $x^k = (x_1^k,\ldots,x_n^k)$ *is an optimal flow.*

Proof. First we prove that if (56) (or (57)) holds, then

$$\psi_i^k(h_i^k) = F_i(h_i^k) \tag{58}$$

for all i=1,...,n–1. Indeed, for i=1 it is obvious that

$$\psi_1^k(h_1^k) = \psi_1^{k+1}(h_1^k) \ge t_1^k = \inf(P_1^k)$$

$$= \inf\{\sum_{a\in A_1} f_a(x_1(a)) : x_1 \in X(G_1, d + h_1)\} = F_1(h_1^k) .$$

In view of Proposition VIII.12, we then deduce that $\psi_1^k(h_1^k) = F_1(h_1^k)$.

Now assuming that (58) is true for i–1, let us prove (58) for i. We have

$$F_i(h_i^k) \geq \psi_i^k(h_i^k) = \psi_i^{k+1}(h_i^k) \geq t_i^k = \inf (P_i^k)$$

$$= \psi_{i-1}^k(h_{i-1}^k) + \sum_{a\in A_i} f_a(x_i^k(a))$$

$$= F_{i-1}(h_{i-1}^k) + \sum_{a\in A_i} f_a(x_i^k(a))$$

$$= \inf \{F_{i-1}(h_{i-1}) + \sum_{a\in A_i} f_a(x_i^k(a)) \colon x_i \in X(G_i, d + h_i^k) ,$$

$$h_{i-1}(u) = \sum_{a\in A_{i-1}^-(u)} x_i(a) \ \ \forall u \in U_{i-1}\} = F_i(h_i^k) .$$

Therefore, (58) holds for all i=1,...,n–1.

To complete the proof it remains to show that (x_i^k, h_{i-1}^k) is an optimal solution of

$$(P_i) \qquad \min \{F_{i-1}(h_{i-1}) + \sum_{a\in A_i} f_a(x_i(a)) \colon x_i \in X(G_i, d + h_i),$$

$$h_{i-1}(u) = \sum_{a\in A_{i-1}^-(u)} x_i(a) \ \ \forall u \in U_{i-1}\}$$

(where we agree to set $h_n^k = 0$ and $F_0 = 0$). We have

$$F_{i-1}(h_{i-1}^k) + \sum_{a\in A_i} f_a(x_i^k(a)) = \psi_{i-1}^k(h_{i-1}^k) + \sum_{a\in A_i} f_a(x_i^k(a))$$

$$= \inf \{\psi_{i-1}^k(h_{i-1}) + \sum_{a\in A_i} f_a(x_i(a)) \colon x_i \in X(G_i, d + h_i^k) ,$$

$$h_{i-1}(u) = \sum_{a\in A_{i-1}^-(u)} x_i(a) \ \ \forall u \in U_i\}$$

$$\leq \inf \{F_{i-1}(h_{i-1}) + \sum_{a\in A_i} f_a(x_i(a)) \colon x_i \in X(G_i, d + h_i^k) ,$$

$$h_{i-1}(u) = \sum_{a\in A_{i-1}^-(u)} x_i(a) \ \ \forall u \in U_i\} ,$$

i.e., (x_i^k, h_{i-1}^k) is an optimal solution of (P_i).

Finally, from Theorem VIII.6 it follows that $x^k = (x_1^k,...,x_n^k)$ is an optimal solution of (P). ■

Theorem VIII.7. *Algorithm VIII.9 terminates after finitely many iterations at an optimal solution of (SUCF).*

Proof. We first show that for any fixed $i \in \{0,1,...,n-1\}$ there is a finite collection $\mathscr{H}_i$ of functions such that $\psi_i^k \in \mathscr{H}_i$ for all k=0,1,2,.... Indeed, this is obvious for i=0 since $\psi_0^k \equiv 0 \ \forall k$. Arguing by induction, suppose that the claim is true for $i = p-1$ $(p \geq 1)$ and consider the case $i = p$. For $k = 0$, we have $\psi_p^0 \equiv 0$, while for $k \geq 1$

$$\text{hypo}\ \psi_p^k = \text{conv}\ \{\text{hypo}\ \psi_p^{k-1}, (h_p^{k-1}, t_p^{k-1})\}\ , \tag{59}$$

where

$$t_p^{k-1} = \psi_{p-1}^{k-1}(h_{p-1}^{k-1}) + \sum_{a \in A_p} f_a(x_p^{k-1}(a))\ . \tag{60}$$

Since every $x^{k-1} = (x_1^{k-1},...,x_n^{k-1})$ is an extreme flow in G and the number of extreme flows is finite, x_p^{k-1} must belong to some finite set X_p. Moreover, since the quantities h_i^{k-1} (i=1,...,n−1) are uniquely determined by x^{k-1}, they must belong to certain finite sets H_i (i=1,...,n−1). Therefore,

$$h_{p-1}^{k-1} \in H_{p-1}\ , \ x_p^{k-1} \in X_p\ ,$$

and since $\psi_{p-1}^{k-1} \in \mathscr{H}_{p-1}$ and $\mathscr{H}_{p-1}$ is finite by assumption, it follows from (60) that t_p^{k-1} belongs to some finite set $T_p \subset \mathbb{R}_+$. But

$$(h_p^{k-1}, t_p^{k-1}) \in H_p \times T_p\ .$$

By virtue of (59), this implies that any ψ_p^k is the convex hull of the union of the hypograph of ψ_p^0 and some subset of the finite set $H_p \times T_p$. Hence, ψ_p^k itself belongs to some finite family $\mathscr{H}_p$.

We have thus proved that for any i=0,1,...,n−1, $\mathscr{H}_i$ is finite. Since for any k

$$\psi_i^k \le \psi_i^{k+1} \quad (i=0,1,...,n-1) ,$$

and both ψ_i^k and ψ_i^{k+1} belong to the finite set $\mathscr{H}_i$, there must exist a k such that

$$\psi_i^k = \psi_i^{k+1} \quad (i=0,1,...,n-1) .$$

In other words, the algorithm must stop at some iteration k. By Proposition VIII.13, x^k is then an optimal flow. ■

Example VIII.3. Consider the network G in Fig. VIII.2, with cost functions of the form

$$f_a(t) = c(a)t + b(a)\delta(t) ,$$

where $\delta(t) = 1$ if $t > 0$, $\delta(t) = 0$ if $t = 0$ (i.e., b(a) is a fixed cost).

The data of the problem are as follows:

Node v:	2	3	4	5	6	7	9	10	11	12
Demand d(v):	3	29.2	1.5	2.5	0	0	20.3	39.5	1.5	30.5
Arc a:	1	2	3	4	5	6	7	8		
Fixed cost b(a):		25.1	15.7	15.7	14.9	14.9	14.8	30	29	
c(a):	3.2	3.3	3.3	2.8	2.8	2.7	5.5	5.4		
Arc a:	9	10	11	12	13	14	15	16		
Fixed cost b(a):		15.7	15.7	15.5	15.7	50.5	41.5	55	41.5	
c(a):		3.3	3.3	3.2	3.3	9.5	5.7	8.5	5.7	
Arc a:		17	18	19						
Fixed cost b(a):		15.7	15.7	41.5						
c(a):		3.3	3.3	5.7						

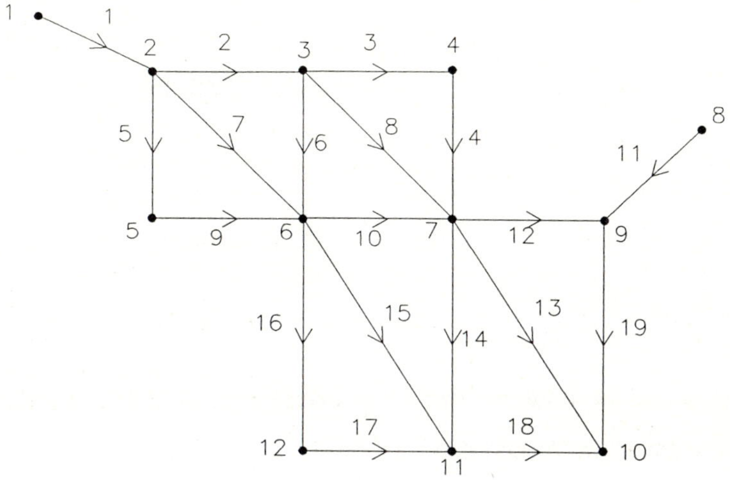

Fig. VIII.2

Clearly, the set of nodes V can be partitioned into two subsets $V_1 = \{1,2,3,4,5,6,7\}$ and $V_2 = \{8,9,10,11,12\}$ satisfying assumption (*).

Applying the above method we obtain the following results ($U_1 = \{6,7\}$).

Iteration 0. (Here we write x_j^0 rather than $x^0(j)$ to denote the value of the flow x^0 in arc j).

0.1: Solving (P_2^0):

$x_{11}^0 = 59.8$, $x_{12}^0 = 0$, $x_{13}^0 = 0$, $x_{14}^0 = 0$, $x_{15}^0 = 0$, $x_{16}^0 = 32$, $x_{17}^0 = 1.5$, $x_{18}^0 = 0$, $x_{19}^0 = 39.5$.

Solving (P_1^0):

$x_1^0 = 68.2$, $x_2^0 = 30.7$, $x_3^0 = 1.5$, $x_4^0 = 0$, $x_5^0 = 2.5$, $x_6^0 = 0$, $x_7^0 = 32$, $x_8^0 = 0$, $x_9^0 = 0$, $x_{10}^0 = 0$.

Optimal value of (P_1^0): $t_1^0 = 599.3$.

0.2.: $t_1^0 = 599.3 > 0 = \psi_1^0(h_1^0)$ (stopping criterion not satisfied).

0.3.: (From step 0.1 we have $h_1^0(6) = 32$, $h_1^0(7) = 0$)

hypo $\psi_1^1 = \text{conv}\,\{\mathbb{R}_+^2 \times R,\ (32,0;599.3)\} = \{(\theta_1,\theta_2,t):\ t \le \min\,(599.3,\ 18.728\theta_1),$ $\theta_1 \ge 0,\ \theta_2 \ge 0\}$.

Hence, $\psi_1^1(h_1) = \psi_1^1(h_1(6),\ h_1(7)) = \min\,\{599.3,\ 18.728h_1(6)\}$.

Iteration 1.

1.1. Solving (P_2^1) and (P_1^1), we obtain $x^1 = x^0$, $t_1^1 = 599.3$

1.2. $t_1^1 = 599.3 = \psi_1^1(h_1^1)$: stop. Optimal value: 1317.36.

In this example the total number N of extreme flows is 162. Further numerical results are reported in Thach (1988).

5.4. Extension

If we set $X_i(h_i) := X(G_i,\ d + h_i)$ $(i=1,\dots,n)$, $H_i(x_{i+1}) := (\sum_{a \in A_{i+1}^-(u)} x_{i+1}(a),$ $u \in U_i)$ $(i=1,\dots,n-1)$, $f_i(x_i) := \sum_{a \in A_i} f_a(x_i(a))$ $(i=1,\dots,n)$, then it is easily seen that (SUCF) is a special case of the following more general problem:

$$\text{minimize} \sum_{i=1}^{n} f_i(x_i) \tag{61}$$

(P)

$$\text{s.t. } x_i \in X_i(h_i) \quad (i=1,\dots,n)\ , \tag{62}$$

$$h_i = H_i(x_{i+1}) \quad (i=1,\dots,n-1)\ , \tag{63}$$

$$h_n \equiv 0\ . \tag{64}$$

Here, $x_i \in \mathbb{R}_+^{m_i}$, $h_i \in \mathbb{R}_+^{k_i}$ and it is assumed that:

(i) each $f_i(\cdot)$ is a concave function on $\mathbb{R}_+^{m_i}$;

(ii) each $X_i(h_i)$ is a convex polyhedron, and the point–to–set mapping $h_i \mapsto X_i(h_i)$

is affine, i.e.,

$$X_i(\lambda h_i' + (1-\lambda)h_i'') = \lambda X_i(h_i') + (1-\lambda)X_i(h_i'')$$

for any $h_i', h_i'' \in \mathbb{R}_+$ and $0 \le \lambda \le 1$;

(iii) each $H_i: \mathbb{R}_+^{m_i+1} \longrightarrow \mathbb{R}_+^{k_i}$ is a linear mapping.

It is obvious that (SUCF) satisfies (i) and (iii). To see that (SUCF) also satisfies (iii), it suffices to show that any extreme point $\bar{x}_i$ of $X_i(\lambda h_i' + (1-\lambda)h_i'') = X(G_i, d + \lambda h_i' + (1-\lambda)h_i'')$ is of the form $\bar{x}_i = \lambda x_i' + (1-\lambda)x_i''$ with $x_i' \in X_i(h_i')$, $x_i'' \in X_i(h_i'')$. But, since $\bar{x}_i$ is an extreme flow in $G_i = (V_i, A_i)$, there exists (by Proposition VIII.9) a spanning forest $T_i = (V_i,B)$ such that $\{a \in A_i: \bar{x}_i(a) > 0\} \subset B$. Let x_i', x_i'' be feasible flows in $T_i = (V_i,B)$ for the demands $d + h_i'$ and $d + h_i''$, respectively. Then $x_i' \in X_i(h_i')$, $x_i'' \in X_i(h_i'')$, and hence $\lambda x_i' + (1-\lambda)x_i'' \in X_i(\lambda h_i' + (1-\lambda)h_i'')$. Since there is a unique feasible flow in T_i for the demand $d + \lambda h_i' + (1-\lambda)h_i''$ we conclude that $\bar{x}_i = \lambda_i' + (1-\lambda)x_i''$.

Note that several problems of practical interest can be formulated as special cases of problem (P). Consider for example the following *concave cost production and inventory model* (see, e.g., Zangwill (1968)).

$$\text{minimize} \sum_{i=1}^{n} (p_i(y_i) + q_i(h_i)) \tag{69}$$

$$\text{s.t.} \quad h_{i-1} + y_i = h_i + d_i \quad (i=1,\ldots,n) , \tag{70}$$

$$y_i \ge 0 , h_i \ge 0 \quad (i=1,\ldots,n) , h_0 = h_n = 0 \tag{71}$$

Here $d_i > 0$ is the given market demand for a product in period i, y_i is the amount to be produced and h_i is the inventory in that period. The function $p_i(y_i)$ is the production cost, and $q_i(h_i)$ is the inventory holding cost in period i (where both $p_i(\cdot)$ and $q_i(\cdot)$ are assumed to be concave functions).

Setting $x_i = (y_i,h_{i-1})$ (i=1,...,n), and

$$x_i \in X_i(h_i) \Longleftrightarrow x_i = (y_i, h_{i-1}),\ h_{i-1} + y_i = h_i + d_i\ ,$$

$$h_i = H_i(x_{i+1}) \Longleftrightarrow x_{i+1} = (y_{i+1}, h_i)\ ,$$

$$f_i(x_i) = p_i(y_i) + q_i(h_i)\ ,$$

we see that we have a special case of problem (P).

A feasible solution $x = (x_1,...,x_n)$ for (P) can be generated in the following way: first, choose $x_n \in X_n(0)$ and compute $h_{n-1} = H_{n-1}(x_n)$; then choose $x_{n-1} \in X_{n-1}(h_{n-1})$ and compute $h_{n-2} = H_{n-2}(x_{n-1})$, and so on; finally, choose $x_1 \in X_1(h_1)$.

Therefore, the problem can be considered as a multistage decision process. This suggests decomposing (P) into a sequence of smaller problems which corresponds to different stages of the process.

For each j=1,...,n consider the problem

$$\text{minimize} \sum_{i=1}^{j} f_i(x_i) \tag{65}$$

$$P_j(h_j) \qquad \text{s.t.}\ \ x_i \in X_i(h_i) \quad (i=1,...,j)\ , \tag{66}$$

$$h_i = H_i(x_{i+1}) \quad (i=1,...,j-1)\ , \tag{67}$$

$$h_j \text{ given.} \tag{68}$$

Let $F_j(h_j)$ denote the optimal value of $P_j(h_j)$ as a function of h_j.

Proposition VIII.14. *The function $F_j(\cdot)$ is concave.*

Proof. Denote by $\Omega_j(h_j)$ the set of all $(x_1,...,x_j)$ that are feasible for $P_j(h_j)$, i.e., for each $(x_1,...,x_j) \in \Omega_j(h_j)$ there exist $h_1,...,h_{j-1}$ satisfying (66) (67) (68). By induction, it can easily be shown that the point–to–set mapping $h_j \longrightarrow \Omega_j(h_j)$ is affine. Now let $h_j = \lambda h_j' + (1-\lambda)h_j''$ $(0 \le \lambda \le 1)$, and let $(x_1,...,x_j) \in \Omega_j(h_j)$ satisfy $(x_1,...,x_j) = \lambda(x_1',...,x_j') + (1-\lambda)(x_1'' ,...,x_j'')$ with $(x_1',...,x_j') \in \Omega_j(h_j')$ and

$(x_1'',...,x_j'') \in \Omega_j(h_j'')$. Since $f_i(\cdot)$ is concave, we have $f_i(x_i) \geq \lambda f_i(x_i') + (1-\lambda)f_i(x_i'')$. Hence, if $(x_1,...,x_j)$ is an optimal solution of $P_j(h_j)$, then $F_j(h_j) = \sum_{i=1}^{j} f_i(x_i) \geq \lambda \sum_{i=1}^{j} f_i(x_i') + (1-\lambda) \sum_{i=1}^{j} f_i(x_i'') \geq \lambda F_j(h_j') + (1-\lambda)F_j(h_j'')$.

This proves the concavity of $F_j(\cdot)$. ∎

Obviously the optimal value of (P) is simply $F_n(0)$, and we can write the *recursive equations*:

$$F_1(h_1) = \min \{f_1(x_1): x_1 \in X_1(h_1)\} .$$

$$F_j(h_j) = \min \{F_{j-1}(h_{j-1}) + f_j(x_j): h_{j-1} = H_{j-1}(x_j), x_j \in X_j(h_j)\}$$
$(j=2,...,n)$.

Because of Proposition VIII.14, these subproblems are concave minimization problems under linear constraints. Hence, in these problems one can replace the polyhedra $X_j(h_j)$ by their vertex sets. Since the latter sets are finite, one can compute the function $F_1(h_1)$ (i.e., the tableau of its values corresponding to different possible values of h_1). Then, using the recursive equation, one can find $F_2(h_2)$ from $F_1(h_1)$,..., and finally $F_n(0)$ from $F_{n-1}(h_{n-1})$. However, this method is practical only if n and the vertex sets of $X_j(h_j)$ are small.

Under more general conditions it is easy to see that the same method that was developed in Section VIII.5.3 for the SUCF problem can be applied to solve (P), provided that the dimensions of the variables h_i are relatively small (but the dimensions of the variables x_i may be fairly large, as in (SUCF)).

CHAPTER IX

SPECIAL PROBLEMS OF CONCAVE MINIMIZATION

Many nonconvex optimization problems can be reduced to concave minimization problems of a special form and can be solved by specialized concave minimization methods. In this chapter we shall study some of the most important examples of these problems. They include bilinear programming, complementarity problems and certain parametric concave minimization problems. An important subclass of parametric concave minimization which we will study is linear programming subject to an additional reverse convex constraint.

1. BILINEAR PROGRAMMING

A number of situations in engineering design, economic management, operations research (e.g., constrained bimatrix games (Mangasarian (1964)), three dimensional assignment (Frieze (1974)), multicommodity network flow, production scheduling, rectilinear distance location–allocation (e.g., Konno (1971a), Soland (1974, Sherali and Shetty (1980a), etc.) can be modeled by the following general mathematical formulation, often called the **bilinear programming problem**:

(BLP) minimize $F(x,y) := px + y(Cx) + qy$,

subject to $x \in X$, $y \in Y$,

where X, Y are nonempty polyhedra given by

$$X = \{x \in \mathbb{R}^n: Ax \leq a , x \geq 0\} ,$$

$$Y = \{y \in \mathbb{R}^{n'}: By \leq b , y \geq 0\} ,$$

with $a \in \mathbb{R}^m$, $b \in \mathbb{R}^{m'}$, $p \in \mathbb{R}^{n'}$, $q \in \mathbb{R}^n$, and C,A,B are matrices of dimension $n' \times n$, $m \times n$, $m' \times n'$, respectively. Let V(X) and V(Y) denote the vertex set of X and Y, respectively.

This problem has been extensively studied in the literature for more than twenty years. The earliest solution methods were proposed by Mangasarian (1964), Mangasarian and Stone (1964) and Altman (1968). Subsequently, a number of other algorithms were developed, including those of Falk (1973), Konno (1976), Vaish and Shetty (1976 and 1977), Gallo and Ulkücü (1977), Mukhamediev (1982), Sherali and Shetty (1980), Thieu (1980 and 1988), Czochralska (1982 and 1982a), Al–Khayyal (1986), etc. (cf. Section I.2.4).
We shall focus on methods that are directly related to concave minimization.

1.1. Basic Properties

The key property which is exploited in most methods for bilinear programming is the equivalence of (BLP) with a polyhedral concave minimization problem (see Section I.2.4). Recall that the problem (BLP) can be rewritten as

$$\min \{f(x): x \in X\} , \tag{1}$$

where

$$f(x) = \inf \{F(x,y): y \in Y\}$$

$$= px + \inf \{(q + Cx)y : y \in Y\} . \tag{2}$$

If Y has at least one vertex, and if we denote the vertex set of Y by V(Y), then the hypograph P of f is a polyhedron

$$P = \{(x,t) \in \mathbb{R}^n \times \mathbb{R}: px + (q + Cx)y \geq t \quad \forall y \in V(Y)\} . \tag{3}$$

Hence, f(x) is a polyhedral concave function. In particular, dom f:= $\{x \in \mathbb{R}^n: f(x) > -\infty\}$ is a polyhedron (note that dom f = $\mathbb{R}^n$ if Y is bounded). Moreover, the function f(x) is piecewise affine on dom f.

As a consequence of this equivalence, any method of concave minimization, when specialized to problem (1), will produce an algorithm for solving problem (BLP). It is the aim of this section to examine some of the methods for which this specialisation is not trivial. But before turning to the methods, let us mention some general properties of the optimal solutions of (BLP).

Proposition IX.1. *If problem (BLP) has a finite optimal value (e.g., if X and Y are bounded), then an optimal solution $(\bar{x},\bar{y})$ exists such that $\bar{x} \in V(X)$, $\bar{y} \in V(Y)$.*

Proof. Indeed, the infimum of f(x) over X, if it is finite, must be attained at some $\bar{x} \in V(X)$. Furthermore, from (3) we see that $f(\bar{x}) = F(\bar{x},\bar{y})$ for some $\bar{y} \in V(Y)$ (cf. Theorem I.3). ∎

Of course, by interchanging the roles of x and y, we obtain another equivalent form of (BLP):

$$\min \{g(y): y \in Y\} ,$$

where $g(y) = qy + \inf \{(p + C^T y)x: x \in X\}$.

In view of this symmetric structure, it is clear that a necessary condition for a pair $(\bar{x},\bar{y})$ with $\bar{x} \in V(X)$, $\bar{y} \in V(Y)$ to be an optimal solution of the problem is that

$$\min_{x \in X} F(x,\bar{y}) = F(\bar{x},\bar{y}) = \min_{y \in Y} F(\bar{x},y) . \tag{4}$$

However, this condition is not sufficient. We only have

Proposition IX.2. *Let $(\bar{x},\bar{y})$ satisfy (4). If $\bar{y} = \arg\min_{y \in Y} F(\bar{x},y)$ (i.e., $\bar{y}$ is the unique minimizer of $F(\bar{x},.)$ over Y), then $\bar{x}$ is a local optimal solution of (1).*

Proof. By hypothesis, $F(\bar{x},\bar{y}) < F(\bar{x},y)$ for all $y \in Y$ satisfying $y \neq \bar{y}$. Therefore, by continuity, for each $y \in Y$, $y \neq \bar{y}$, there exists an open neighbourhood U_y of $\bar{x}$ satisfying $F(\bar{x},\bar{y}) < F(x,y)$ for all $x \in U_y$. Let $U = \cap \{U_y: y \in V(Y), y \neq \bar{y}\}$. Then for all $x \in U$ we have

$$F(\bar{x},\bar{y}) < F(x,y) \quad \forall y \in V(Y), y \neq \bar{y} .$$

But $F(\bar{x},\bar{y}) = \min_{x \in X} F(x,\bar{y})$. Hence, for all $x \in U$:

$$F(\bar{x},\bar{y}) \leq \min \{F(x,y): y \in V(Y)\} = \min_{y \in Y} F(x,y) ,$$

i.e., $f(\bar{x}) \leq f(x)$. ■

To find a pair $(\bar{x},\bar{y})$ satisfying (4), one can use the following "mountain climbing" procedure (e.g., Konno (1976)).

Assume that X,Y are bounded. Let $x^0 \in V(X)$. Set $h = 0$.

1) Solve the linear program $\min \{(q + Cx^h)y: y \in Y\}$ to obtain a vertex y^h of Y such that $F(x^h,y^h) = \min_{y \in Y} F(x^h,y)$.
2) Solve the linear program $\min \{(p + C^T y^h)x: x \in X\}$ to obtain a vertex x^{h+1} of X such that $F(x^{h+1},y^h) = \min_{x \in X} F(x,y^h)$.
   If $F(x^{h+1},y^h) = F(x^h,y^h)$, then stop. Otherwise, set $h \leftarrow h+1$ and go to 1).

In view of the finiteness of the set $V(X) \times V(Y)$, the situation $F(x^{h+1},y^h) < F(x^h,y^h)$ cannot occur infinitely many times. Therefore, the above procedure must terminate after finitely many steps with a pair (x^h,y^h) such that $\min_{y \in Y} F(x^h,y)) = F(x^h,y^h) = F(x^{h+1},y^h) = \min_{x \in X} F(x,y^h)$.

1.2. Cutting Plane Method

As seen above, problem (BLP) is equivalent to each of the following concave minimization problems:

$$\min \{f(x): x \in X\} , \; f(x) = \inf \{F(x,y): y \in Y\} , \tag{5}$$

$$\min \{g(y): y \in Y\} , \; g(y) = \inf \{F(x,y): x \in X\} . \tag{6}$$

To solve (BLP) we can specialize Algorithm V.1 to either of these problems. However, if we use the symmetric structure of the bilinear programming problem, a more efficient method might be to alternate the cutting process for problem (5) with the cutting process for problem (6) in such a way that one uses the information obtained in the course of one process to speed up the other.

Specifically, consider two polyhedra $X_0 \subset X$, $Y_0 \subset Y$ with vertex sets $V(X_0)$, $V(Y_0)$, respectively; and let

$$f_0(x) = \min \{F(x,y): y \in Y_0\} , \; g_0(y) = \min \{F(x,y): x \in X_0\} .$$

By the "mountain climbing" procedure we can find a pair $(x^0,y^0) \in V(X_0) \times V(Y_0)$ such that

$$f_0(x^0) = F(x^0,y^0) = g_0(y^0) . \tag{7}$$

Let $\varepsilon > 0$ be a tolerance number, $\alpha = f_0(x^0) - \varepsilon$.

Assuming x^0 to be a nondegenerate vertex of X_0, denote by d^j (j=1,...,n) the directions of the n edges of X_0 emanating from x^0. Then, as shown in Section III.1, an α–valid cut for the concave program min $f_0(X^0)$ is given by a vector $\pi_{X_0}(Y_0)$ such that

$$\pi_{X_0}(Y_0) = (\frac{1}{\theta_1},...,\frac{1}{\theta_n})Q^{-1} , \; Q = (d^1,...,d^n) , \tag{8}$$

$$\theta_j = \max \{\lambda: f_0(x^0 + \lambda d^j) \geq \alpha\} \quad (j=1,...,n) . \tag{9}$$

Therefore, setting

$$\Delta_{X_0}(Y_0) = \{x \in X^0: \pi_{X_0}(Y_0)(x - x^0) < 1\} , \tag{10}$$

we have $f_0(x) \geq \alpha$ for all $x \in \Delta_{X_0}(Y_0)$, i.e., any candidate $(x,y) \in X_0 \times Y_0$ with $F(x,y) < \alpha$ must lie in the region $X_1 \times Y_0$, where

$$X_1 = X_0 \backslash \Delta_{X_0}(Y_0) .$$

Thus, if $X_1 = \emptyset$, then

$$\alpha \leq \min \{F(x,y): x \in X_0 , y \in Y_0\} .$$

Otherwise, $X_1 \neq \emptyset$, and we can consider the problem

$$\min \{F(x,y): x \in X_1 , y \in Y_0\} .$$

In an analogous manner, but interchanging the roles of x and y, we can construct a vector $\pi_{Y_0}(X_1)$ which determines an α-valid cut for the concave program min $g_1(Y_0)$, where $g_1(y) = \min \{F(x,y): x \in X_1\}$. (Note that this is possible, provided y^0 is a nondegenerate vertex of Y^0, because $\alpha = g_0(Y^0) - \varepsilon \leq g_1(y^0) - \varepsilon$). Setting

$$\Delta_{Y_0}(X_1) = \{y \in Y_0: (\pi_{Y_0}(X_1)(y - y^0) < 1\} , \tag{11}$$

we have $g_1(y) \geq \alpha$ for all $y \in \Delta_{Y_0}(X_1)$, so that any candidate $(x,y) \in X_1 \times Y_0$ with $F(x,y) < \alpha$ must lie in the region $X_1 \times Y_1$, where

$$Y_1 = Y_0 \backslash \Delta_{Y_0}(X_1) .$$

Thus, the problem to be considered now is $\min \{F(x,y): x \in X_1, y \in Y_1\}$. Of course, the same operations can be repeated with X_1, Y_1 in place of X_0, Y_0. We are led to the following procedure (Konno (1976)).

Algorithm IX.1.

Initialization: Let $X_0 = X$, $Y_0 = Y$, $\alpha = +\infty$.

Step 1: Compute a pair (x^0,y^0) satisfying (7). If $F(x^0,y^0) - \varepsilon < \alpha$, then reset $\alpha \leftarrow F(x^0,y^0) - \varepsilon$.

Step 2: Construct the cut $\pi_{X_0}(Y_0)$ defined by (8), (9), and let $\Delta_{X_0}(Y_0)$ denote the set (10). If $X_1 = X_0 \setminus \Delta_{X_0}(Y_0) = \emptyset$, then stop: (x^0,y^0) is a global ε–optimal solution of the problem (BL).

Otherwise, go to Step 3.

Step 3: Construct the cut $\pi_{Y_0}(X_1)$, and define the set (11). If $Y_1 = Y_0 \setminus \Delta_{Y_0}(X_1) = \emptyset$, then stop: (x^0,y^0) is a global ε–optimal solution.

Otherwise, go to Step 4.

Step 4: Set $X_0 \leftarrow X_1$, $Y_0 \leftarrow Y_1$ and go to Step 1.

Denote by π_X^k and π_Y^k the cuts generated in Steps 2 and 3 of iteration k, respectively. From Theorem V.2 we know that the cutting plane algorithm just described will converge if the sequences $\{\pi_X^k\}$, $\{\pi_Y^k\}$ are bounded. However, in the general case the algorithm may not converge. To ensure convergence one could, from time to time, insert a facial cut as described in Section V.2; but this may be computationally expensive.

For the implementation of the algorithm, note that, because of the specific structure of the function $f_0(x)$, the computation of the numbers θ_j defined by (9) reduces simply to solving linear programs. This will be shown in the following proposition.

Proposition IX.3. *Let $X_0 = X$. Then θ_j equals the optimal value of the linear program*

$$\text{minimize } (px^0 - \alpha)s_0 + (q + Cx^0)s$$

$$\text{s.t. } (pd^j)s_0 + (Cd^j)s = -1$$

$$-bs_0 + Bs \le 0$$

$$s_0 \ge 0, \quad s=(s_1,\dots,s_m)^T \ge 0.$$

Proof. Define $\varphi_j(\lambda) = \lambda pd^j + \min \{(q + Cx^0 + \lambda Cd^j)y\colon By \le b,\ y \ge 0\}$. From (1) we have $f(x^0) + \lambda d^j) = px^0 + \varphi_j(\lambda)$, so that

$$\theta_j = \max \{\lambda\colon \varphi_j(\lambda) \ge \alpha - px^0\} .$$

By the duality theorem of linear programming we have

$$\varphi_j(\lambda) = \lambda pd^j + \max \{-bu\colon -B^Tu \le q + C(x^0 + \lambda d^j)\ ,\ u \ge 0\} .$$

Hence,

$$\theta_j = \max \lambda$$

$$\text{s.t. } \lambda pd^j - bu \ge \alpha - px^0$$

$$\lambda Cd^j + B^Tu \ge -q-Cx^0$$

$$u \ge 0 .$$

The assertion follows by passing to the dual of this linear program. ■

Konno (1976) has also indicated a procedure for constructing an α-valid cut which is usually deeper than the concavity cut $\pi_{X_0}(Y_0)$ defined by (8), (9).

Note that the numbers θ_j in (9) depend upon the set Y_0. If we write $\theta_j(Y_0)$ to emphasize this dependence, then for $Y_1 \subset Y_0$ and Y_1 smaller than Y_0 we have

$\theta_j(Y_1) \geq \theta_j(Y_0)$ $\forall j$, usually with strict inequality for at least one j; i.e., the cut $\pi_{X_0}(Y_1)$ is usually deeper than $\pi_{X_0}(Y_0)$. Based on this observation, Konno's cut improvement procedure consists in the following.

Construct $\pi_{X_0}(Y_0)$, then $\pi_{Y_0}(X_0)$ for $X_1 = X_0 \setminus \Delta_{X_0}(Y_0)$ and $\pi_{X_0}(Y_1)$ for $Y_1 = Y_0 \setminus \Delta_{Y_0}(X_1)$.

The cut $\pi_{X_0}(Y_1)$ is also an α–valid cut at x^0 for the concave program $\min f_0(X_0)$, and it is generally deeper than the cut $\pi_{X_0}(Y_0)$. Of course, the process can be iterated until successive cuts converge within some tolerance.

This cut improvement procedure seems to be particularly efficient when the problem is symmetric with respect to the variables x,y, as it happens, e.g., in the bilinear programming problem associated with a given quadratic minimization problem (see Section V.4).

Example IX.1 (Konno (1976).

Consider the problem

$$\begin{aligned}
\text{minimize } F(x,y) = \; & x_1 - x_2 - y_1 + (x_1 - x_2)(y_2 - y_1) \\
\text{s.t.} \quad & x_1 + 4x_2 \leq 8 , \qquad 2y_1 + y_2 \leq 8 , \\
& 4x_1 + x_2 \leq 12 , \qquad y_1 + 2y_2 \leq 8 , \\
& 3x_1 + 4x_2 \leq 12 , \qquad y_1 + y_2 \leq 5 , \\
& x_1 \geq 0 , \; x_2 \geq 0 , \qquad y_1 \geq 0 , \; y_2 \geq 0 .
\end{aligned}$$

Applying Algorithm IX.1, where Step 3 is omitted (with $Y_1 = Y_0$), we obtain the following results ($\varepsilon = 0$):

1st iteration: $x^0 = P_1 , y_0 = Q_1$ (see Fig. IX.1); $\alpha = -10$

cut: $\frac{1}{20} x_1 - \frac{1}{20} x_2 \geq 1$

$X_1 \neq \emptyset$ (shaded region)

2nd iteration: $x^1 = P_4 , y^1 = Q_4$; $\alpha = -13$

$$\text{cut: } -\frac{1}{4.44}x_1 - \frac{1}{1.45}x_2 \geq 1$$

$X_2 = \emptyset$. Optimal solution.

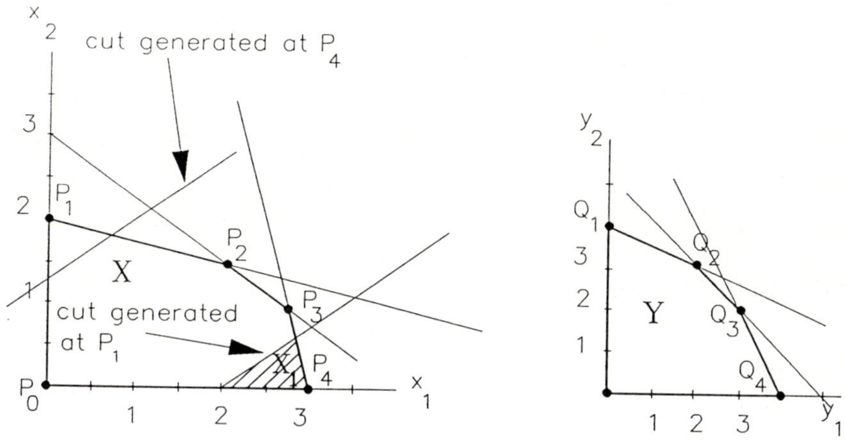

Fig. IX.1

1.3. Polyhedral Annexation

Polyhedral annexation type algorithms for bilinear programming have been proposed by several authors (see, e.g., Vaish and Shetty (1976), Mukhamediev (1978)). We present here an algorithm similar to that of Vaish and Shetty (1976), which is a direct application of the PA algorithm (Algorithm VI.3) to the concave minimization problem (1).

Because of the special form of the concave function f(x), the α-extension of a point with respect to a given vertex x^0 of X such that $f(x^0) \geq \alpha$ can be computed by solving a linear program. Namely, by Proposition IX.3, if $x^0 = 0$, then the value

$$\theta = \max \{t: f(tx) \geq \alpha\}$$

is equal to the optimal value of the linear program

$$\text{minimize } -\alpha s_0 + qs$$

$$\text{s.t. } (px)s_0 + (Cx)s = -1 \ ,$$
$$- b s_0 + B^T s \le 0 \ ,$$
$$s_0 \ge 0 \ , \quad s = (s_1,\dots,s_m)^T \ge 0 \ .$$

We can now give the following polyhedral annexation algorithm (assuming X,Y to be bounded).

Algorithm IX.2 (PA Algorithm for (BLP)).

Compute a point $z \in X$. Set $X_0 = X$.

0) Starting with z, search for a vertex x^0 of X_0 which is a local minimizer of f(x) over X_0. Let $\alpha = f(x^0)$. Translate the origin to x^0, and construct a cone K_0 containing X_0 such that for each i the i–th edge of K_0 contains a point $y^{0i} \neq 0$ satisfying $f(y^{0i}) \ge \alpha$. Construct the α–extension z^{0i} of y^{0i} (i=1,...,n), and find the (unique) vertex v^1 of

$$S_1 = \{x: z^{0i}x \le 1 \ (i=1,\dots,n)\} \ .$$

Let $V_1 = \{v^1\}$, $V_1^* = V_1$. Set k = 1

1) For each $v \in V_k^*$ solve the linear program

$$\max \{vx: x \in X_0\}$$

to obtain the optimal value $\mu(v)$ and a basic optimal solution $\omega(v)$. If for some $v \in V_k^*$ we have $f(w(v)) < \alpha$, then set $z \leftarrow w(v)$,

$$X_0 \leftarrow X_0 \cap \{x: v^1x \ge 1\}$$

where v^1 was defined in Step 0, and return to Step 0. Otherwise, go to 2.

2) Select $v^k \in \operatorname{argmax} \{\mu(v): v \leftarrow V_k^*\}$. If $\mu(v^k) \le 1$, then stop: an optimal solution of (BLP) is (x^0,y^0), where $y^0 \in \operatorname{argmin} \{(q + Cx^0)y: y \in Y\}$. Otherwise, go to 3).

3) Construct the α-extension z^k of v^k and form S_{k+1} by adjoining to S_k the constraint

$$z^k x \le 1 .$$

Compute the vertex set V_{k+1} of S_{k+1}, and let $V_{k+1}^* = V_{k+1} \setminus V_k$. Set $k \leftarrow k+1$ and return to 1.

It follows from Theorem VI.3 that this algorithm must terminate after finitely many steps at an optimal solution of problem (BLP).

1.4. Conical Algorithm

A cone splitting algorithm to solve problem (BLP) was proposed by Gallo and Ülkücü (1977). However, it has been shown subsequently that this algorithm can be considered as a specialization of the cut and split algorithm (Algorithm V.3) to the concave program (5), which is equivalent to problem (BLP) (cf. Thieu (1980)). Though the latter algorithm may work quite successfully in many circumstances, we now know that its convergence is not guaranteed (see Sections V.3.3 and VII.1.6). In fact, an example by Vaish (1974) has shown that the algorithm of Gallo and Ülkücü may lead to cycling.

However, from the results of Chapter VII it follows that a convergent conical algorithm for solving (BLP) can be obtained by specializing Algorithm VII.1 to problem (5).

Recall that for a given x the value f(x) is computed by solving the linear program

$$\min \{(q + Cx)y: By \le b,\ y \ge 0\}\ ,$$

while the α-extension of x with respect to a vertex x^0 of X (when $f(x^0) < \alpha$, $x \neq x^0$ and $f(x) \ge \alpha$) is computed according to Proposition IX.3. More precisely, the number

$$\theta = \max \{\lambda: f(x^0 + \lambda(x-x^0)) \ge \alpha\}$$

is equal to the optimal value of the linear program

$$\begin{aligned} &\max \lambda \\ \text{s.t.}\quad & -\lambda p(x-x^0) + bu \le px^0 - \alpha\ , \\ & -\lambda C(x-x^0) - B^T u \le Cx^0 + q\ , \\ & u \ge 0\ , \end{aligned}$$

or of its dual

$$\begin{aligned} &\min\ (px^0 - \alpha)s_0 + (q + Cx^0)s \\ \text{s.t.}\quad & (p(x-x^0))s_0 + (C(x-x^0))s = -1\ , \\ & -bs_0 + B^T s \le 0 \\ & s_0 \ge 0\ ,\ s = (s_1,\ldots,s_m)^T \ge 0\ . \end{aligned}$$

Assuming X and Y to be bounded, we thus can specialize Algorithm VII.1 to problem (5) as follows.

Algorithm IX.3.

Select $\varepsilon \ge 0$ and an NCS rule (see VII.1.4). Compute a point $z \in X$.

0) Starting with z, find a vertex x^0 of X such that $f(x^0) \le f(z)$. Let $\bar{x}$ be the best among x^0 and all the vertices of X adjacent to x^0. Let $\gamma = f(\bar{x})$.

1) Let $\alpha = \gamma - \varepsilon$. Translate the origin to x^0, and construct a cone $K_0 \supset X$ such that for each i the i-th edge of K_0 contains a point $y^{0i} \neq x^0$ satisfying $f(y^{0i}) \ge \alpha$.

Let $Q_0 = (z^{01},z^{02},...,z^{0n})$, where each z^{0i} is the α–extension of y^{0i}. Let $\mathcal{M} = \mathcal{P} = \{Q_0\}$.

2) For each $Q = (z^1,z^2,...,z^n) \in \mathcal{P}$, solve the linear program

$$\text{LP}(Q,X) \quad \max \Sigma\lambda^i \quad \text{s.t. } \Sigma\, \lambda_i A z^i \le \alpha,\ \lambda^i \ge 0 \quad (i=1,...,n)$$

to obtain its optimal value $\mu(Q)$ and basic optimal solution $\omega(Q)$. If $f(\omega(Q)) < \gamma$ for some Q, then return to 0) with

$$z \leftarrow \omega(Q)\ ,\ X \leftarrow X \cap \{x\colon eQ_0^{-1}x \ge 1\}\ ,$$

where Q_0 is the matrix in Step 1. Otherwise go to 3.

3) In $\mathcal{M}$ delete all Q with $\mu(Q) \le 1$. Let $\mathcal{R}$ be the collection of remaining matrices. If $\mathcal{R} = \emptyset$, then terminate: $\bar{x}$ is a global ε–optimal solution of (5). Otherwise,

4) Select $Q_* \in \operatorname{argmax} \{\mu(Q)\colon Q \in \mathcal{R}\}$ and subdivide Q_* according to the chosen NCS role.

5) Let $\mathcal{P}_*$ be the partition of Q_*. For each $Q \in \mathcal{P}_*$ reset $Q = (z^1,z^2,...,z^n)$ with z^i such that $f(z^i) = \alpha$.

Return to 2) with $\mathcal{P} \leftarrow \mathcal{P}_*$, $\mathcal{M} \leftarrow \{\mathbb{R}^n\}\{Q_*\}\} \cup \mathcal{P}_*$.

By Theorem VII.1, this algorithm terminates at a global ε–optimal solution after finitely many steps whenever $\varepsilon > 0$.

Example IX.2 (Gallo and Ülkücü, 1977).

Consider the problem

$$\text{minimize} \quad F(x,y) = -2x_1 - y_2 + (x_1 - x_2)(y_2 - y_1)$$

$$\begin{array}{llll} \text{s.t.} & x_1 + x_2 \leq 5 , & & y_1 + 2y_2 \leq 8 , \\ & 2x_1 + x_2 \leq 7 , & & 3y_1 + y_2 \leq 14 , \\ & 3x_1 + x_2 \leq 6 , & & 2y_1 \leq 9 , \\ & x_1 - 2x_2 \leq 1 , & & y_2 \leq 3 , \\ & x_1 \geq 0 , \; x_2 \geq 0 , & & y_1 \geq 0 , \; y_2 \geq 0 , \end{array}$$

Applying Algorithm IX.3 with $\varepsilon = 0$ and the NCS rule (*) of Section VII.1.6, where N is very large and ρ is very close to 1,leads to the following calculations.

0) Choose $x^0 = (0;0)$. The neighbouring vertices are (1;0) and (0;5). The values of f(x) at these three points: –3, –13/2, –18. Hence, $\bar{x} = (0;5)$ and $\gamma = -18$.

Iteration 1:

1) Choose $Q_0 = \{z^{01},z^{02}\}$, with $z^{01} = (36/13;0)$, $z^{02} = (0;5)$. $\mathcal{M}_1 = \mathcal{P}_1 = \{Q_0\}$,

2) Solve $LP(Q_0,X)$: $\mu(Q_0) = 119/90 > 1$, $\omega(Q_0) = (2;3)$ with $f(\omega(Q_0)) = -10 > \gamma$.

3) $\mathcal{R}_1 = \{Q_0\}$.

4) $Q_* = Q_0$. Subdivide Q_0 with respect to $\omega_0 = (2;3)$. The α–extension of ω^0 is $\hat{\omega}^0 = (30/17;45/7)$.

$\mathcal{P}_2 = \{Q_{01},Q_{02}\}$ with $Q_1 = (z^{01},\hat{\omega}^0)$, $Q_{02} = (z^{02},\hat{\omega}^0)$.

Iteration 2:

1) Solve $LP(Q_{01},X)$, i=1,2: $\mu(Q_{01}) = 707/900 < 1$; $\mu(Q_{02}) = 1$.

2) Both Q_{01}, Q_{02} are deleted. $\mathcal{R}_2 = \emptyset$, and hence the global optimal solution of (BLP) is $\bar{x} = (0,5)$, $\bar{y} = (0,3)$.

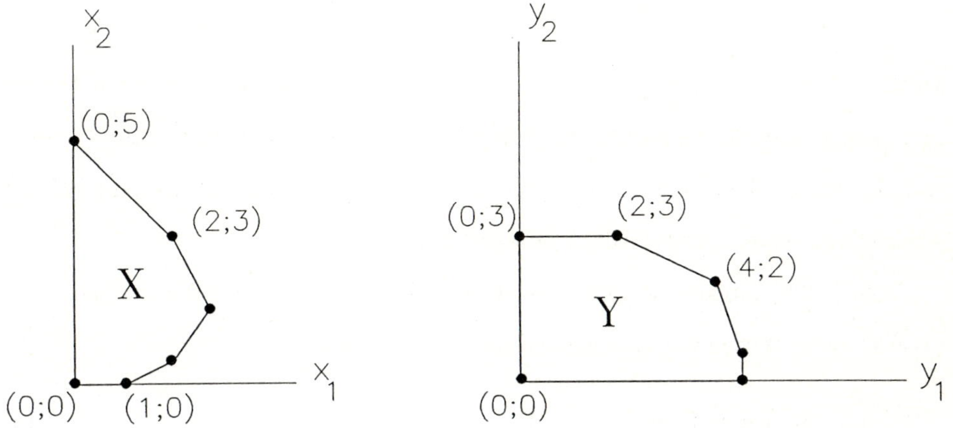

Fig. IX.2

1.5. Outer Approximation Method

In the previous methods we have assumed that the polyhedra X, Y are bounded. We now present a method, due to Thieu (1988), which applies to the general case, when X and (or) Y may be **unbounded**.

This method is obtained by specializing the outer approximation method for concave minimization (Section VI.1) to problem (5) (or (6)).

A nontrivial question that arises here is how to check whether the function f(x) is bounded from below over a given halfline $\Gamma = \{x^0 + \theta u: 0 \le \theta < +\infty\}$ (note that f(x) is not given in an explicit form, as assumed in the conventional formulation of problem (BCP)). A natural approach would be to consider the parametric linear program

$$\min \{F(x^0 + \theta u,y): y \in Y\} , \quad 0 \le \theta \le \infty , \tag{12}$$

or, alternatively, the subproblem

$$\max \{\tau\colon f(x^0 + \theta u) \geq f(x^0), \quad \theta \geq 0\}, \tag{13}$$

which can be shown to be a linear program. However, this approach is not the best one. In fact, solving the parametric linear program (12) is computationally expensive and the subproblem (12) or (13) depends upon x^0, and this means that different subproblems have to be solved for different points x^0.

Thieu (1988) proposed a more efficient method, based on the following fact. Let f(x) be a concave function defined by

$$f(x) = \inf_{y \in J} \{r(y)x + s(y)\}, \tag{14}$$

where $r(y) \in \mathbb{R}^n$, $s(y) \in \mathbb{R}$ and J is an arbitrary set of indices (in the present context, $J = Y$, $r(y) = p + C^T y$, $s(y) = qy$). Let Γ be the halfline emanating from a point $x^0 \in \mathbb{R}^n$ in the direction u.

Proposition IX.4. *The function f(x) is bounded from below on Γ if and only if*

$$\rho\,(u) := \inf_{y \in J} r(y)u \geq 0\,. \tag{15}$$

Proof. Suppose that $\rho(u) \geq 0$. From (14) we have, for every $\theta \geq 0$,

$$\begin{aligned} f(x^0 + \theta u) &= \inf_{y \in J} \{r(y)(x^0 + \theta u) + s(y)\} \\ &\geq \inf_{y \in J} \{r(y)x^0 + s(y)\} + \theta \inf_{y \in J} r(y) \\ &= f(x^0) + \theta \rho u \geq f(x^0)\,. \end{aligned}$$

This shows that f(x) is bounded from below on Γ. In the case $\rho(u) < 0$, let $y^0 \in J$ be such that $\gamma = r(y^0)u < 0$. Then from (4) we see that

$$\begin{aligned} f(x^0 + \theta u) &\leq r(y^0)(x^0 + \theta u) + s(y^0) \\ &= r(y^0)x^0 + s(y^0) + \theta\gamma \longrightarrow -\infty \text{ as } \theta \longrightarrow +\infty\,. \end{aligned}$$

Therefore, f(x) is unbounded from below on Γ. ■

In the general case, when the index set J is arbitrary, the value $\rho(u)$ defined by (15) might not be easy to determine. For example, for the formula

$$f(x) = \inf \{x^*x - f^*(x^*): x^* \in \operatorname{dom} f^*\}$$

where $f^*(x^*)$ is the concave conjugate of f(x) (cf. Rockafellar (1970)), computing $\inf \{x^*u: x^* \in \operatorname{dom} f^*\}$ may be difficult. But in our case, because of the specific structure of f(x), the determination of $\rho(u)$ is very simple. Indeed, since J = Y, $r(y) = p + C^Ty$, $s(y) = qy$, the inequality (15) reduces to

$$\rho(u) = \inf_{y \in Y} (p + C^Ty)u$$

$$= pu + \inf_{y \in Y} (Cu)y \geq 0 .$$

Thus, in order to check whether f(x) is bounded from below over Γ it suffices to solve the linear subprogram

$$\text{minimize } (Cu)y \quad \text{subject to } y \in Y . \tag{16}$$

Note that all of these subproblems have the same constraint set Y, and their objective function depends only on u, but not on x^0.

Let us now consider the concave minimization problem (5) (which is equivalent to (BLP)) in explicit form:

$$\text{minimize } \; f(x) := px + \min \{(q + Cx)y: y \in Y\} ,$$

$$\text{subject to } \; A_ix \leq a_i \quad (i=1,\dots,m) ,$$

$$x_j \geq 0 \quad (j=1,\dots,n) .$$

For the sake of convenience we shall assume that both polyhedra X, Y are nonempty.

On the basis of the above results, we can give the following outer approximation method for solving the problem (BLP):

Algorithm IX.4.

Initialization:

Set $X_1 = \mathbb{R}^n_+$. Let $V_1 = \{0\}$ (vertex set of X_1), $U_1 = \{e^1,...,e^n\}$ (extreme direction set of X_1), where e^i is the i–th unit vector of $\mathbb{R}^n$. Set $I_1 = \{1,...,m\}$.

Iteration k = 1,2,...:

1) For each $u \in U_k$ compute $\rho(u) = pu + \inf\{(Cu)y: y \in Y\}$ by solving (16). If $\rho(u^k) < 0$ for some $u^k \in U_k$, then:

a) If $\forall i \in I_k$ $A_i u^k \leq 0$, stop: problem (5) has no finite optimal solution, and f(x) is unbounded from below on any halfline in X parallel to u^k. In this case (BLP) is unsolvable.

b) Otherwise, select

$$i_k \in \operatorname{argmax}\{A_i u^k : i \in I_k\}$$

and go to 3).

2) If $\rho(u) \geq 0$ $\forall u \in U_k$, then select

$$x^k \in \operatorname{argmin}\{f(x): x \in V_k\}.$$

a) If $A_i x^i \leq a_i$ $\forall i \in I_k$, then terminate: x^k is a global optimal solution of (5). Compute

$$y^k \in \operatorname{argmin}\{(q + Cx^k)y: y \in Y\}$$

(this linear program must be solvable, because $f(x^k) = px^k + \inf\{(q + C^T x^k)y: y \in Y\}$ is finite). Then (x^k,y^k) is an optimal solution of (BLP).

b) If $A_i x^k > a_i$ for some $i \in I_k$, then select

$$i_k \in \operatorname{argmax} \{A_i x^k - a_i : i \in I_k\}$$

and go to 3).

3) Form

$$X_{k+1} = X_k \cap \{x: A_{i_k} x \le a_{i_k}\} .$$

Determine the vertex set V_{k+1} and the extreme direction set U_{k+1} of X_{k+1} from V_k and U_k (see Section II.4.2). Set $I_{k+1} = I_k \setminus \{i_k\}$ and go to iteration k+1.

Remark IX.1. In the worst case, the above algorithm might stop when $k = m$. Then the algorithm enumerates not only all of the vertices and extreme directions of X, but also the vertices and extreme directions of intermediate polyhedra X_k generated during the procedure. However, computational experiments reported in Thieu (1988) suggest that this case generally can be expected not to occur, and that the number of linear programs to be solved is likely to be substantially less than the total number of vertices and extreme directions of X.

Example IX.2.

Minimize $F(x,y) := 3x_1 + 5x_2 - 2x_1y_1 + (x_1-x_2)y_2 + (2x_1+x_2)y_3 + 2y_1 - y_2 + y_3$

$$\text{s.t.} \quad \begin{array}{rlrl} -x_1 + x_2 \le & 5 \ , & y_1 + y_2 + y_3 \le & 6 \ , \\ x_1 - 4x_2 \le & 2 \ , & y_1 - y_2 + y_3 \le & 2 \ , \\ -3x_1 + x_2 \le & 1 \ , & -y_1 + y_2 + y_3 \le & 2 \ , \\ -3x_1 - 5x_2 \le & -23 \ , & -y_1 - y_2 + y_3 \le & -2 \ . \end{array}$$

The sets X and Y are depicted in the following figures.

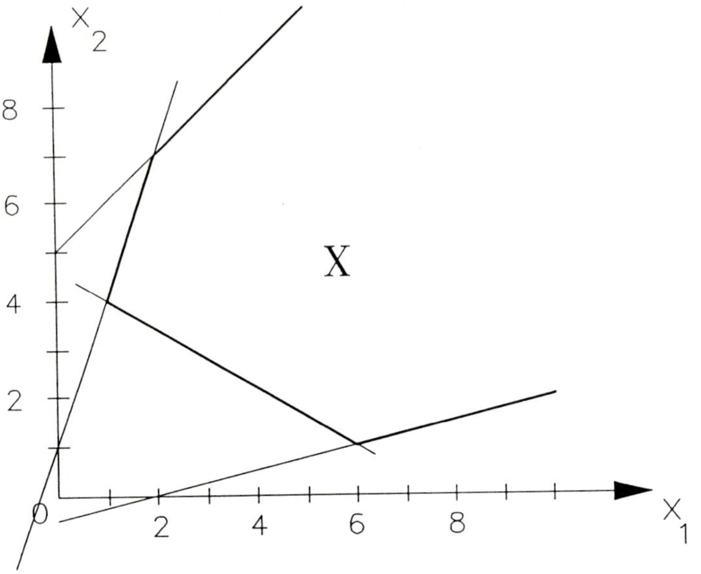

Fig. IX.3

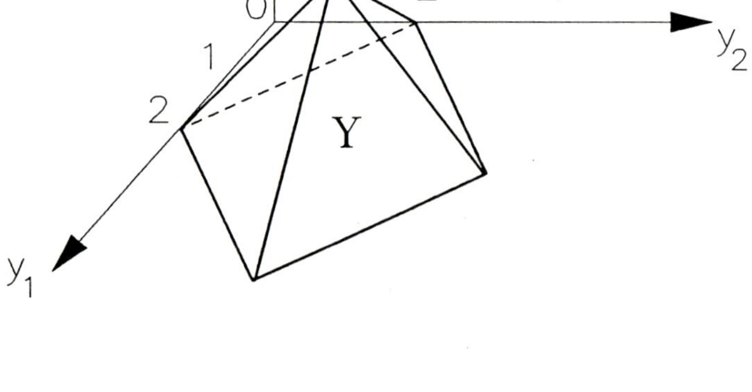

Fig. IX.4

The algorithm starts with $X_1 = \mathbb{R}^2_+$, $V_1 = \{0\}$, $U_1 = \{e^1,e^2\}$, $I_1 = \{1,2,3,4\}$.

Iteration 1.

$\rho(e^1) = pe^1 + \min \{(Ce^1)y: y \in Y\} = 3 + 0 = 3 > 0.$

$\rho(e^2) = 5 - 10 = -5 < 0.$

$\max \{A_ie^2: i \in I_1\} = \max \{1,-4,1,-5\} = 1 > 0$. This maximum is achieved for $i = 1$.

Hence $i_1 = 1$.

Form $X_2 = X_1 \cap \{x: -x_1 + x_2 \leq 5\}$.

Then $V_2 = \{0,v^2\}$ with $v^2 = (0,5)$; $U_2 = \{e^1,u^3\}$ with $u^3 = (1,1)$.

$I_2 = I_1 \setminus \{i_1\} = \{2,3,4\}$.

Iteration 2.

$\rho(e^1) = 3 > 0$, $\rho(u^3) = 0$.

$\min \{f(x): x \in X_2\} = -19$. This minimum is achieved at $x^2 = (0,5) = v^2$.

$\max \{A_ix^2 - a_i: i \in I_2\} = \max \{-22,4,-2\} = 4 > 0$. This maximum is attained when $i = 3$. Hence $i_2 = 3$.

Form

$$X_3 = X_2 \cap \{x: -3x_1 + x_2 \leq 1\} .$$

Then $V_3 = \{0,v^3,v^4\}$ with $v^3 = (0,1)$, $v^4 = (2,7)$, while $U_3 = U_2 = \{e^1,u^3\}$.

$I_3 = I_2 \setminus \{i_2\} = \{2,4\}$.

Iteration 3.

$\min \{f(x): x \in X_3\} = -19$. This minimum is achieved at $x^3 = (2,7) = v^4$.

$\max \{A_ix^3 - a_i: i \in I_3\} = -18 < 0$. Hence, x^3 is a global optimal solution of the concave minimization problem $\min \{-f(x): x \in X\}$.

By solving the problem $\min \{(q + Cx^3)y: y \in Y\}$, we then obtain $y^3 = (4;2;0)$. Thus, (x^3,y^3) is a global optimal solution of the BLP problem under consideration.

Note that the polytope Y has five vertices: (2,2,2), (2,0,0), (0,2,0), (4,2,0), (2,4,0).

2. COMPLEMENTARITY PROBLEMS

Complementarity problems form a class of nonconvex optimization problems which play an important role in mathematical programming and are encountered in numerous applications, ranging from mechanics and engineering to economics (see Section I.2.5). In this section, we shall consider the *concave complementarity problem* (CCP), which can be formulated as follows:

Given a concave mapping $h: \mathbb{R}^n \longrightarrow \mathbb{R}^n$, i.e., a mapping $h(x) = (h_1(x),\dots,h_n(x))$, such that each $h_i(x)$ is a concave function, find a point $x \in \mathbb{R}^n$ satisfying

$$x \geq 0 \,, \quad h(x) \geq 0 \,, \quad \sum_{i=1}^{n} x_i h_i(x) = 0 \,. \tag{17}$$

Note that in the literature problem (17) is sometimes called a convex complementarity problem.

When h is an affine mapping, i.e., $h(x) = Mx + q$ (with $q \in \mathbb{R}^n$ and $M \in \mathbb{R}^{n\times n}$), the problem is called the *linear complementarity problem* (LCP): find a point x satisfying

$$x \geq 0 \,, \quad Mx + q \geq 0 \,, \quad \sum_{i=1}^{n} x_i(M_i x + q_i) = 0 \,, \tag{18}$$

where M_i denotes the i–th row of the matrix M.

Over the past 25 years, numerous methods have been devised to solve this problem. The best known of these methods – Lemke's complementarity pivot method (Lemke (1965), Tomlin (1978)) – and other pivoting methods due to Cottle and Dantzig (1968), Murty (1974), and Van der Heyden (1980), are guaranteed to work only under restrictive assumptions on the structure of the problem matrix M. Recently, optimization methods have been proposed to solve larger classes of linear complementarity problems (Mangasarian (1976, 1978 and 1979), Cottle and Pang (1978), Cheng (1982), Cirina (1983), Ramarao and Shetty (1984), Al–Khayyal (1986, 1986a and 1987), Pardalos (1988b), Pardalos and Rosen (1988) (see also the book of Murty (1988)).

In the sequel we shall be concerned with the global optimization approach to complementarity problems, as initiated by Thoai and Tuy (1983) and further developed in Tuy, Thieu and Thai (1985) for the convex complementarity problem and in Pardalos and Rosen (1987) for the (LCP). An advantage of this approach is that it can be applied to concave complementarity problems, and in the case of (LCP) it does not depend upon any special properties of the problem matrix M. This advantage, however, has to be paid for by a greater computational cost.

2.1. Basic Properties

As shown in Section I.2.5, by setting

$$f(x) = \sum_{i=1}^{n} \min \{x_i, h_i(x)\}$$

one can reduce the concave complementarity problem (17) to the concave minimization problem

$$\text{minimize } f(x) \quad \text{s.t.} \quad x \geq 0 \, , \, h(x) \geq 0 \, . \tag{19}$$

Proposition IX.5. *A vector $\bar{x}$ is a solution to the complementarity problem (17) if and only if it is a global optimal solution of the concave minimization problem (19) with $f(\bar{x}) = 0$.*

Proof. This follows from Theorem I.5, where $g(x) \equiv x$. ■

An immediate consequence of this equivalence between (17) and (19) is that:

Proposition IX.6. *If the concave complementarity problem (17) is solvable, then at least one solution is an extreme point of the convex set defined by $x \geq 0$, $h(x) \geq 0$. In particular, either (LCP) has no solution, or else at least one solution of (LCP) is a vertex of the polyhedron defined by*

$$x \geq 0 \, , \, Mx + q \geq 0 \, .$$

Many solution methods for (LCP) are based on this property of (LCP). From the equivalence between (17) and (19) it also follows that, in principle, any method of solution for concave minimization problems gives rise to a method for solving concave complementarity problems. In practice, however, there are some particular features of the concave minimization problem (19) that should be taken into account when devising methods for solving (17):

1) The objective function f(x) is nonnegative on the feasible domain D and must be zero at an optimal solution.

2) The feasible domain D, as well as the level sets of the objective function f(x), may be unbounded.

Furthermore, what we want to compute is not really the optimal value of (19), but rather a point $\bar{x}$ (if it exists) such that

$$\bar{x} \in D , f(\bar{x}) = 0 .$$

In other words, if we denote

$$D = \{x: x \geq 0 , h(x) \geq 0\} , \quad G = \{x: f(x) > 0\} ,$$

then the problem is to *find a point* $\bar{x} \in D \setminus G$ *or else establish that* $D \setminus G = \emptyset$ *(i.e.,* $D \subset G$*).*

Observe that, since the functions h(x) and f(x) are concave, both of the sets D, G are convex. Thus, the complementarity problem is a special case of the following general "geometric complementarity problem":

Given two convex sets D and G, find an element of the complement of G with respect to D.

In Chapters VI and VII we saw that the concave minimization problem is also closely related to a problem of this form (the "(DG) problem", cf. Section VI.2). It turns out that the procedures developed in Sections VI.2 and VII.1 for solving the (DG) problem can be extended to the above geometric complementarity problem.

2.2. Polyhedral Annexation Method for the Linear Complementarity Problem

Consider the linear complementarity problem (18)

(LCP) $\quad x \in D \ , \ f(x) = 0 \ ,$

where $D = \{x: x \geq 0 \ , Mx + q \geq 0\}$, $f(x) = \sum_{i=1}^{n} \min \{x_i, M_i x + q_i\}$.

Let x^0 be a vertex of the polyhedron D. If $f(x^0) = 0$ then x^0 solves (LCP). Otherwise, we have $f(x^0) > 0$.

We introduce the slack variables $x_{n+i} = M_i x + q_i$ $(i=1,...,n)$, and express the basic variables (relative to the basic solution x^0) in terms of the nonbasic ones. If we then change the notation, we can rewrite (LCP) in the form

$$y \geq 0, \ Cy + d \geq 0 \ , \ \tilde{f}(y) = 0 \ , \tag{20}$$

where y is related to x by a linear equation $x = x^0 + Uy$ (with U an n×n–matrix), $\tilde{f}(y) = f(x^0 + Uy)$ is a concave function with $\tilde{f}(0) = f(x^0) > 0$, $y = 0$ is a vertex of the polyhedron

$$\tilde{D} = \{y: y \geq 0 \ , Cy + d \geq 0\} \ ,$$

and $\tilde{f}(y) \geq 0$ for all $y \in \tilde{D}$. Setting

$$\tilde{G} = \{y: \tilde{f}(y) > 0\} \ ,$$

we see that $0 \in \text{int} \ \tilde{G}$, and all of the conditions assumed in the $(\tilde{D}\tilde{G})$ problem as formulated in Section VI.2 are fulfilled, except that $\tilde{G}$ is an open (rather than closed) set and $\tilde{D}$ and $\tilde{G}$ may be unbounded. Therefore, with some suitable modifications, the polyhedral annexation algorithm (Section VI.2.4) can be applied to solve (20), and hence (LCP).

In Section VI.2.6 it was shown how this algorithm can be extended in a natural way to the case when $\tilde{D}$ and $\tilde{G}$ may be unbounded, but $\tilde{D}$ contains no line and inf $\tilde{f}(\tilde{D}) > -\infty$ (the latter conditions are fulfilled here, because $\tilde{D} \subset \mathbb{R}^n_+$ and $\tilde{f}(y) \geq 0$ for all $y \in \tilde{D}$). On the other hand, it is straightforward to see that, when $\tilde{G}$ is open, the stopping criterion in Step 2 of the polyhedral annexation algorithm should be $\mu(\nu^k) < 1$ rather than $\mu(v^k) \le 1$. We can thus give the following

Algorithm IX.5 (PA Algorithm for LCP).

Using a vertex x^0 of the polyhedron D, rewrite (LCP) in the form (20).

0) For every i=1,...,n compute

$$\theta_i = \sup \{t: \tilde{f}(te^i) \ge 0\} \quad (i=1,...,n) ,$$

where e^i is the i–th unit vector of $\mathbb{R}^n$ ($\theta_i > 0$ because $\tilde{f}(0) > 0$ and $\tilde{f}$ is continuous). Let

$$S_1 = \{y: y_i \le \frac{1}{\theta_i} \quad (i=1,...,n)\}$$

(with the usual convention that $\frac{1}{\infty} = 0$), and let $v^1 = (\frac{1}{\theta_1}, ..., \frac{1}{\theta_n})$ (the unique vertex of S_1).

Set $V_1 = \{v^1\}$, $V_1^* = V_1$, $k = 1$ (for $k > 1$, V_k^* is the set of new vertices of S_k).

k.1. For each $v \in V_k^*$ solve the linear program

$$\text{LP}(v,\tilde{D}) \qquad \max \{vy: y \in \tilde{D}\}$$

to obtain the optimal value $\mu(v)$ and a basic optimal solution $\omega(v)$ (when $\mu(v) = +\infty$, $\omega(v)$ is an extreme direction of $\tilde{D}$ over which $vx \longrightarrow +\infty$). If for some $v \in V_k^*$ the point $\omega(v)$ satisfies $\tilde{f}(\omega(v)) = 0$, then stop. Otherwise, go to k.2.

k.2. Select $v^k \in \text{argmax} \{\mu(v): v \in V_k\}$. If $\mu(v^k) < 1$, then stop: (LCP) has no solution. Otherwise, go to k.3.

k.3. Let $\omega^k = \omega(v^k)$,

$$\tau_k = \sup \{t: \tilde{f}(t\omega^k) \ge 0\}$$

(if ω^k is an extreme direction of $\tilde{D}$, then necessarily $\tau_k = +\infty$ because $\tilde{f}(x) \geq 0$ $\forall x \in \tilde{D}$).

Form the polyhedron

$$S_{k+1} = S_k \cap \{y: \omega^k y \leq \frac{1}{\tau_k}\} .$$

Compute the vertex set V_{k+1} of S_{k+1}, and let $V_{k+1}^* = V_{k+1} \setminus V_k$. Set $k \leftarrow k+1$ and return to k.1.

Theorem IX.1. *The above algorithm terminates after finitely many steps, either yielding a solution to (LCP) or else establishing that (LCP) has no solution.*

Proof. Denote by P_k the polar set of S_k. It is easily verified that P_1 is the convex hull of $\{0,u^1,...,u^n\}$, where u^i is the point $\theta_i e^i$ if $\theta_i < +\infty$, and the direction e^i if $\theta_i = +\infty$. Similarly, P_{k+1} is the convex hull of $P_k \cup \{z^k\}$, where z^k is the point $\tau_k \omega^k$ if $\tau_k < +\infty$ and the direction ω^k if $\tau_k = +\infty$ (see Section VI.2.6). Since each $\omega^k = \omega(v^k)$ is a vertex or an extreme direction of $\tilde{D}$ which does not belong to P_k, there can be no repetition in the sequence $\{\omega^k\}$. Hence, the procedure must terminate after finitely many steps. If it terminates at a step k.1., then a solution $\bar{y}$ of (20) (and hence a solution $\bar{x} = U\bar{y} + x^0$ of (LCP)) is obtained. If it terminates at a step k.2 (because $\mu(v^k) < 1$), then $\mu(v) < 1$ for all $v \in V_k$, and hence $\tilde{D} \subset \tilde{G}$, i.e., $\tilde{f}(y) > 0$ for all $y \in \tilde{D}$. This implies that $f(x) > 0$ for all $x \in D$. ■

Note that the set V_k may increase quickly in size as the algorithm proceeds. Therefore, to alleviate storage problems and other difficulties that can arise when V_k becomes too large, it is recommended to restart the algorithm from some $\omega(v) \in V_k^*$, i.e., to return to Step 0 with $x^0 \leftarrow \omega(v)$ and

$$D \leftarrow \tilde{D} \cap \{y: v^1 y \geq 1\} , f \leftarrow \tilde{f} ,$$

where v^1 is the vertex computed in Step 0 of the current cycle of iterations. At each restart the feasible domain is reduced, while the starting vertex x^0 changes; therefore, the chance for successful termination will increase.

2.3. Conical Algorithm for the Linear Complementarity Problem

Like the polyhedral annexation algorithm, the conical (DG) procedure in Section VII.1.2 can be extended to solve the problem (20). This extension, however, requires a careful examination of the following circumstance.

In the conical (DG) procedure, when the sets D and G are not bounded (as assumed in Section VII.1.2), the optimal value $\mu(Q)$ of the linear program LP(Q,D) associated with a given cone $K = \mathrm{con}(Q)$ might be $+\infty$. In this case, the procedure might generate an infinite nested sequence of cones $K_s = \mathrm{con}(Q_s)$ with $\mu(Q_s) = +\infty$. To avoid this difficulty and ensure convergence of the method, we need an appropriate subdivision process in conjunction with an appropriate selection rule in order to prevent such sequences of cones from being generated when the problem is solvable.

To be specific, consider (LCP) in the formulation (20). Let e^i be the i–th unit vector of $\mathbb{R}^n$, and let H_0 be the hyperplane passing through $e^1,\dots,e^n$, i.e., $H_0 = \{x\colon \sum_{i=1}^{n} x_i = 1\}$. Then for any cone K in $\mathbb{R}^n_+$ we can define a nonsingular $n \times n$ matrix $Z = (v^1,\dots,v^n)$ whose i–th column v^i is the intersection of H_0 with the i–th edge of K. Since $\tilde{f}(0) > 0$ and the function $\tilde{f}$ is continuous, we have

$$\theta_i = \sup\,\{t\colon \tilde{f}(tv^i) \ge 0\} > 0 \quad (i=1,\dots,n)\,. \tag{21}$$

Consider the linear program

$$\mathrm{LP}(Z,\tilde{D}) \qquad \max \sum_{i=1}^{n} \frac{\lambda_i}{\theta_i} \quad \text{s.t. } CZ\lambda + d \ge 0\,,\ \lambda \ge 0 \tag{22}$$

(here, as before, it is agreed that $\frac{\lambda_i}{\theta_i} = 0$ if $\theta_i = +\infty$).

Denote by $\mu(Z)$ and $\hat{\lambda} = (\hat{\lambda}_1,\dots,\hat{\lambda}_n)^T$ the optimal value and a basic optimal solution of this linear program (when $\mu(Z) = +\infty$, $\hat{\lambda}$ denotes an extreme direction of the polyhedron $CZ\lambda + d \geq 0$, $\lambda \geq 0$ over which $\sum_i \frac{\lambda_i}{\theta_i} \rightarrow +\infty$).

Proposition IX.7. *Let $\ell(y) = 1$ be the equation of the hyperplane passing through $z^i = \theta_i v^i$ $(i \in I)$ which is parallel to the directions v^j $(j \notin I)$, where $I = \{i\colon \theta_i < +\infty\}$. Then $\mu(Z)$ and $\omega(Z) = Z\hat{\lambda}$ are the optimal value and a basic optimal solution of the linear program*

$$\max \ell(y) \quad s.t. \quad y \in \tilde{D} \cap K .$$

Proof. Since Z is a nonsingular n×n matrix, we can write $y = Z\lambda = \sum_{i=1}^{n} \lambda_i v^i = \sum_{i \in I} \frac{\lambda_i}{\theta_i} z^i$, with $\lambda = Z^{-1}y$. Hence, noting that $\ell(z^i) = 1$ $(i \in I)$, we see that:

$$\ell(y) = \sum_{i \in I} \frac{\lambda_i}{\theta_i} \ell(z^i) = \sum_{i \in I} \frac{\lambda_i}{\theta_i} .$$

On the other hand, we have $\tilde{D} \cap K = \{y\colon Cy + d \geq 0,\ Z^{-1}y \geq 0\} = \{y = Z\lambda\colon CZ\lambda + d \geq 0,\ \lambda \geq 0\}$, from which the assertion immediately follows. ∎

Corollary IX.1. *If $\mu(Z) < 1$, then $\tilde{f}(y) > 0$ for all $y \in \tilde{D} \cap K$.*

Proof. The hyperplane $\ell(y) = \mu(Z)$ passes through $z^{*i} = \theta_i^* v^i$ $(i \in I)$ and is parallel to the directions v^j $(j \notin I)$, where $0 < \theta_i^* < \theta_i$. Since $\tilde{D} \cap K \subset \{y \in K\colon \ell(y) \leq \mu(Z)\}$, any $y \in \tilde{D} \cap K$ can be represented as $y = \sum_{i \in I} t_i z^{*i} + \sum_{j \notin I} t_j v^j$, where $t_i \geq 0$ $(i=1,\dots,n)$ and $\sum_{i \in I} t_i \leq 1$. But clearly $f(z^{*i}) > 0$ $(i \in I)$ and $\tilde{f}(tv^j) > 0$ $(j \notin I)$ for all $t > 0$. Hence $\tilde{f}(y) > 0$ by the concavity of $\tilde{f}(y)$. ∎

Now for each cone K with matrix $Z = (v^1,\dots,v^n)$ define the number

$$\theta(Z) = \min \{\theta_i : i = 1,\dots,n\} ,$$

where the θ_i are computed according to (21). Let $\mathcal{R}$ be the collection of matrices that remain for study at a given stage the a conical procedure. It turns out that the following selection rule coupled with an exhaustive subdivision process will suffice to ensure convergence of this procedure:

A cone K is said to be *of the first category* if the corresponding matrix $Z = (v^1,...,v^n)$ is such that $\theta_i < +\infty$ $\forall i$ and $\mu(Z) < +\infty$; K is said to be *of the second category* otherwise. If there exists at least one cone of the first category in $\mathcal{R}$, then choose a cone of the first category with maximal $\mu(Z)$ for further subdivision; otherwise, choose a cone with minimal $\theta(Z)$.

The corresponding algorithm can be described as follows.

Algorithm IX.6 (Conical Algorithm for LCP).

Assume that a vertex x^0 of the polyhedron

$$x \geq 0 \ , \ Mx + q \geq 0 \ ,$$

is available such that $f(x^0) > 0$. Using x^0, rewrite (LCP) in the form (20). Select an exhaustive cone subdivision process.

0) Let $Z_1 = (e^1,e^2,...,e^n)$, where e^i is the i–th unit vector of $\mathbb{R}^n$, and let $\mathcal{P}_1 = \mathcal{M}_1 = \{Z_1\}$. Set k = 1.

k.1. For each $Z \in \mathcal{P}_k$ solve the linear program LP($Z,\tilde{D}$) (see (22)). Let $\mu(Z)$ and $\hat{\lambda}$ be the optimal value and a basic optimal solution of LP($Z,\tilde{D}$), and let $\omega(Z) = Z\hat{\lambda}$. If $\tilde{f}(\omega(Z)) = 0$ for some $Z \in \mathcal{P}_k$, then terminate. Otherwise ($\tilde{f}(\omega(Z)) > 0$ $\forall Z \in \mathcal{P}_k$), go to k.2.

k.2. In $\mathcal{M}_k$ delete all $Z \in \mathcal{P}_k$ satisfying $\mu(Z) < 1$. Let $\mathcal{R}_k$ be the collection of the remaining elements of $\mathcal{M}_k$. If $\mathcal{R}_k = \emptyset$, then terminate: $\tilde{f}(y) > 0$ $\forall y \in \tilde{D}$ (i.e., (LCP) has no solution). Otherwise, go to k.3.

k.3. Let $\mathcal{R}_k^{(1)} = \{Z \in \mathcal{P}_k\colon Z = (v^1,...,v^n),\ \theta_i < +\infty\ \forall i,\ \mu(Z) < +\infty\}$, where θ_i is defined by (21). If $\mathcal{R}_k^{(1)} \neq \emptyset$, then select

$$Z_k \in \operatorname{argmax} \{\mu(Z): Z \in \mathcal{R}_k^{(1)}\} .$$

Otherwise, select

$$Z_k \in \operatorname{argmin} \{\theta(Z): Z \in \mathcal{R}_k\} .$$

Subdivide the corresponding cone according to the chosen exhaustive cone subdivision rule.

k.4. Let $\mathcal{P}_{k+1}$ be the partition of Z_k, and let $\mathcal{M}_{k+1}$ be the collection obtained from $\mathcal{R}_k$ by replacing Z_k with $\mathcal{P}_{k+1}$. Set $k \leftarrow k+1$ and return to k.1.

To establish convergence of this algorithm, let us first observe the following

Lemma IX.1. *For every point v of the simplex $[e^1,...,e^n]$ denote by $\Gamma(v)$ the halfline from 0 through v. If $\Gamma(v^*) \cap \tilde{D}$ is a line segment $[0,y^*]$, then for all $v \in [e^1,...,e^n]$ sufficiently close to $v*$, $\Gamma(v) \cap \tilde{D}$ is a line segment $[0,y]$ satisfying $y \longrightarrow y^*$ as $v \longrightarrow v^*$.*

Proof. Since $0 \in \tilde{D}$, we must have $d \geq 0$. Consider a point v^* such that $\Gamma(v^*) \cap \tilde{D}$ is a line segment $[0,y^*]$. First Suppose that $y^* \neq 0$. Then, since y^* is a boundary point of $\tilde{D}$, there exists an i such that $C_iy^* + d_i = 0$ but $\lambda C_iy^* + d_i < 0$ for all $\lambda > 1$, i.e., $d_i > 0$. Hence, $J = \{i: d_i > 0\} \neq \emptyset$. Define

$$g(u) = \max \{ -\frac{C_iu}{d_i} : i \in J\} .$$

Obviously, $g(y^*) = 1$, and if $y^* = \theta v^*$ then $1 = g(\theta v^*) = \theta g(v^*)$. It follows that $g(v^*) > 0$ and $y^* = \frac{v^*}{g(v^*)}$. Since $g(\cdot)$ is continuous, we see that for all v sufficiently close to v^* we still have $g(v) > 0$. Then $y = \frac{v}{g(v)}$ will satisfy $g(y) = \frac{g(v)}{g(v)} = 1$, and this implies that y is a boundary point of $\tilde{D}$, and $\Gamma(v) \cap \tilde{D} = [0,y]$. Clearly, as $v \longrightarrow v^*$ we have $y = \frac{v}{g(v)} \longrightarrow \frac{v^*}{g(v^*)} = y^*$.

On the other hand, if $y^* = 0$, then $\Gamma(v^*) \cap \tilde{D} = \{0\}$. Therefore, for all v sufficiently close to v^*, we have $\Gamma(v) \cap \tilde{D} = [0,y]$ with $y = 0$. This completes the proof of the lemma. ∎

Lemma IX.2. *If K_s, $s \in \Delta \subset \{1,2,...\}$, is an infinite nested sequence of cones of the first category, then* $\bigcap_{s\in\Delta} K_s = \Gamma$ *is a ray such that* $\Gamma \cap \tilde{D} = [0,y*]$ *with*

$$y* = \lim_{\substack{s\in\Delta \\ s\to\infty}} \omega(Z_s) .$$

Proof. Recall that K_s denotes the cone to be subdivided at iteration s. From the exhaustiveness of the subdivision process it follows that the intersection Γ of all K_s, $s \in \Delta$, is a ray. Now, since K_s is of the first category, the set $K_s \cap \{y: \ell_s(y) \le \mu(Z_s)\}$ (which contains $K_s \cap \tilde{D}$) is a simplex ($\ell_s(y) = 1$ is the equation of the hyperplane through $z^{si} = \theta_{si} v^{si}$, i=1,...,n, where $Z_s = (v^{s1},...,v^{sn})$, $\theta_{si} = \sup \{t: f(tv^{si}) \ge 0\}$). Hence $K_s \cap \tilde{D}$ is bounded, and consequently $\Gamma \cap \tilde{D}$ is bounded, because $\Gamma \cap \tilde{D} \subset K_s \cap D$. That is, $\Gamma \cap \tilde{D}$ is a line segment $[0,y^*]$. If v^* and v^s denote the intersections of the rays through y^* and $\omega(Z_s)$, respectively, with the simplex $[e^1,...,e^n]$, then, as $s \to \infty$, $s \in \Delta$, we have $v^s \to v^*$, and hence, by Lemma IX.1, $\omega(Z_s) \to y^*$. ∎

Lemma IX.3. *If the cone K_s chosen for further subdivision is of the second category at some iteration s, then $\tilde{f}(y) > 0$ for all $y \in \tilde{D}$ in the simplex $T_s = \{y \in \mathbb{R}^n_+ : \sum_{i=1}^{n} y_i \le \theta(Z_s)\}$.*

Proof. The selection rule implies that $\mathcal{R}_s^{(1)} = \emptyset$ and $\theta(Z_s) \le \theta(Z)$ for all $Z \in \mathcal{R}_s$. Now if $y \in \tilde{D} \cap T_s$, then y either belongs to a cone already deleted at an iteration $k \le s$ (in which case $\tilde{f}(y) > 0$), or else belongs to a cone with matrix $Z \in \mathcal{R}_s$. In the latter case, since $\theta(Z_s) \le \theta(Z)$, we have $\sum_{i=1}^{n} y_i \le \theta(Z)$, and it follows from the definition of $\theta(Z)$ that $\tilde{f}(y) > 0$. ∎

Lemma IX.4. *If the algorithm generates infinitely many cones K_s of the second category $s \in \Delta \subset \{1,2,...\}$, then $\theta(Z_s) \longrightarrow \infty$ as $s \longrightarrow \infty$, $s \in \Delta$.*

Proof. Among the cones in the partition of $K_1 = \mathbb{R}^n_+$ there exists a cone, say K_{s_1}, that contains infinitely many cones of the second category generated by the algorithm. Then among the cones in the partition of K_{s_1} there exists a cone, say K_{s_2}, that contains infinitely many cones of the second category. Continuing in this way, we find an infinite nested sequence of cones K_{s_ν}, $\nu=1,2,...$, each of which contains infinitely many cones of the second category. Since the subdivision is exhaustive, the intersection $\bigcap_{\nu=1}^{\infty} K_{s_\nu} = \Gamma$ is a ray. If Γ contains a point y such that $f(y) < 0$, then, since $y \notin \tilde{D}$, there exists a ball U around y, disjoint from $\tilde{D}$, such that $f(u) < 0$ for all $u \in U$. Then for ν sufficiently large, any ray contained in K_{s_ν} will meet U. This implies that for all k such that $K_k \subset K_{s_\nu}$, we have $\theta_{ki} < \infty$ (i=1,...,n) and $\mu(Z_k) < \infty$. That is, all subcones of K_{s_ν} generated by the algorithm will be of the first category. This contradicts the above property of K_{s_ν}. Therefore, we must have $f(y) \geq 0$ for all $y \in \Gamma$. Since f is concave and $f(0) > 0$, it follows that $f(y) > 0$ for all $y \in \Gamma$.

For any given positive number N, consider a point $c \in \Gamma$ and a ball W around c such that $\sum_{i=1}^{n} y_i > N$ and $f(y) > 0$ for all $y \in W$. When ν is sufficiently large, say $\nu \geq \nu_0$, the edges of K_{s_ν} will meet W at points $y^{s_\nu,i}$ such that $f(y^{s_\nu,i}) > 0$ (i=1,...,n). Since $\sum_{j=1}^{n} y_j^{s_\nu,i} > N$, it follows that $\theta_{s_\nu,i} > N$ (i=1,...,n), and hence $\theta(Z_{s_\nu}) > N$. Now for any $s \in \Delta$ such that $s \geq s_{\nu_0}$, the cone K_s must be a subcone of some cone in $\mathcal{R}_{s_{\nu_0}}$. Therefore, $\theta(Z_s) \geq \theta(Z_{s_{\nu_0}}) > N$. ■

Theorem IX.2. *If Algorithm IX.5 generates infinitely many cones of the second category, then the LCP problem has no solution. Otherwise, beginning at some iteration, the algorithm generates only cones of the first category. In the latter case, if the algorithm is infinite, then the sequence* $\omega^k = \omega(Z_k)$ *has at least one accumulation point and each of its accumulation points yields a solution of (LCP).*

Proof. Suppose that the algorithm generates an infinite number of cones K_s of the second category, $s \in \Delta \subset \{1,2,...\}$. By Lemma IX.3, the problem has no solution in the simplices $T_s = \{y \in \mathbb{R}^n_+ : \sum_{i=1}^{n} y_i \leq \theta(Z_s)\}$. On the other hand, by Lemma IX.4, $\theta(Z_s) \longrightarrow \infty$ as $s \longrightarrow \infty$, $s \in \Delta$. Therefore, the problem has no solution.
Now suppose that for all $k \geq k_0$, $\mathcal{P}_k$ consists only of cones of the first category. If the algorithm is infinite, then it generates at least one infinite nested sequence of cones K_s, $s \in \Delta \subset \{1,2,...\}$. Since all K_s, $s \geq k_0$, are of the first category, it follows from Lemma IX.2 that $\lim_{s\to\infty} \omega(Z_s) = y^*$, where y^* is such that $[0,y^*] = \tilde{D} \cap \bigcap_{s=1}^{\infty} K_s$.
Now consider any accumulation point $\bar{y}$ of $\{\omega(Z_k)\}$, for example, $\bar{y} = \lim_{r\to\infty} \omega(Z_{k_r})$. Reasoning as in the beginning of the proof of Lemma IX.4, we can find an infinite nested sequence of cones K_s, $s \in \Delta' \subset \{1,2,...\}$, such that each K_s, $s \in \Delta'$, contains infinitely many members of the sequence $\{K_{k_r}, r=1,2,...\}$. Without loss of generality, we may assume that $K_{k_s} \subset K_s$ $(s \in \Delta')$. Since the subdivision process is exhaustive, the intersection of all of the K_s, $s \in \Delta'$, is the ray passing through $\bar{y}$. If $f(\bar{y}) > 0$, then around some point c of this ray there exists a ball U such that $[0,\bar{y}] \cap U = \emptyset$ and $f(u) > 0 \ \ \forall u \in U$. Then for all sufficiently large $s \in \Delta'$, we have $[0,\omega(Z_{k_s})] \cap U = \emptyset$.

On the other hand, the i–th edge of the cone K_{k_s} meets U at some point $u^{k_s,i}$. Since $f(u^{k_s,i}) > 0$, it follows that $u^{k_s,i}$ will lie on the line segment $[0,z^{k_s,i}]$. Consequently, since $\mu(Z_{k_s}) \geq 1$, the line segment $[0,\omega(Z_{k_s})]$ meets the simplex

$[u^{k_s,1},\ldots,u^{k_s,n}]$ at some point u^{k_s}. Then we have $u^{k_s} \notin U$ (because $u^{k_s} \in [0,\omega(Z_{k_s})] \subset \mathbb{R}^n \setminus U$), while $u^{k_s} \in [u^{k_s,1},\ldots,u^{k_s,n}] \subset U$. This contradiction shows that $f(\bar{y}) \le 0$, and hence $f(\bar{y}) = 0$, since $\bar{y} \in \tilde{D}$. ■

Corollary IX.2. *For any $\varepsilon > 0$ and any $N > 0$, Algorithm IX.5 either finds an ε-approximate solution of (20) (i.e., a point $y \in \tilde{D}$ such that $\tilde{f}(y) < \varepsilon$) after finitely many iterations or else establishes that the problem has no solution in the ball $\|y\| < N$.*

Proof. If the first alternative in the previous theorem holds, then, for all k such that K_k is of the second category and the simplex $T_k = \{y \in \mathbb{R}^n_+ : \sum_{i=1}^n y_i \le \theta(Z_k)\}$ contains the ball $\|y\| < N$, it follows that we have $f(y) > 0$ for all $y \in \tilde{D}$ with $\|y\| < N$.

If the second alternative holds, then for sufficiently large k we have $\tilde{f}(\omega(Z_k)) < \varepsilon$. ■

Remarks IX.2. (i) As in the case of Algorithm IX.4, when $\mathcal{R}_k$ becomes too large, it is advisable to restart (i.e., to return to step 0), with $x^0 \leftarrow \omega(Z)$ and

$$D \leftarrow \tilde{D} \cap \{y: \ell_1(y) \ge 1\} \ , \ f \leftarrow \tilde{f} \ ,$$

where Z is the matrix of a cone such that $\omega(Z)$ is a vertex of $\tilde{D}$ satisfying $f(\omega(Z)) > 0$, and where $\ell_1(y) = 1$ is the equation of the hyperplane in Proposition IX.7 for Z_1 constructed in Step 0.

(ii) When the problem is known to be solvable, the algorithm can be made finite as follows. At each iteration k, denote by $\mathcal{P}_k^{(1)}$ the set of cones of the first category in $\mathcal{P}_k$, and let $\bar{y}^{-1} = \omega(Z_1)$, $y^k \in \operatorname{argmin} \{f(\bar{y}^{k-1}), f(\omega(Z)) \ \forall Z \in \mathcal{P}_k^{(1)}\}$ (i.e., y^k is the best point of $\tilde{D}$ known up to iteration k). If y^k is a vertex of $\tilde{D}$, let $\bar{y}^k = y^k$; otherwise, find a vertex $\bar{y}^k$ of $\tilde{D}$ such that $\tilde{f}(\bar{y}^k) \le f(y^k)$. Since $f(\bar{y}^k) \to 0$ $(k \to \infty)$ and the vertex set of $\tilde{D}$ is finite, we have $f(\bar{y}^k) = 0$ after finitely many steps.

(iii) To generate an exhaustive subdivision process one can use the rules discussed in Section VII.1.6, for example, the rule (*) in Section VII.1.6, which generates mostly ω-subdivisions. Note that when $\omega(Z_k)$ is a direction (i.e., $\mu(Z_k) = +\infty$), the ω-subdivision of Z_k is the subdivision with respect to the point where the ray in the direction $\omega(Z_k)$ intersects the simplex $[Z_k]$ defined by the matrix Z_k.

(iv) Algorithm IX.6 can be considered as an improved version of an earlier conical algorithm for (LCP) proposed by Thoai and Tuy (1983). The major improvement consists in using a more efficient selection rule and an exhaustive subdivision process which involves mostly ω-subdivisions instead of pure bisection.

2.4. Other Global Optimization Approaches to (LCP)

The above methods for solving (LCP) are based on the reduction of (LCP) to the concave minimization problem

$$\text{minimize} \sum_{i=1}^{n} \min\{x_i, Mx_i + q_i\} \quad \text{s.t. } x \geq 0,\ Mx + q \geq 0 .$$

Since $\min\{x_i, Mx_i + q_i\} = x_i + \min\{0, M_i x - x_i + q_i\}$, by introducing the auxiliary variables w_i we can rewrite this concave minimization problem in the separable form

$$\text{minimize} \sum_{i=1}^{n} \{x_i + \min(0, w_i)\}$$

$$\text{s.t. } x \geq 0 \ ,\quad Mx + q \geq 0 \ ,\quad w - Mx + x = q .$$

Bard and Falk (1982) proposed solving this separable program by a branch and bound algorithm which reduces the problem to a series of linear programs with constraint set $D = \{x\colon x \geq 0,\ Mx + q \geq 0\}$ and cost functions

$$\sum_{i=1}^{n} \{x_i + \alpha_i(M_i x - x_i + q_i)\} ,$$

where α_i is a parameter with value 0, $\frac{1}{2}$ or 1, depending on the stage of the algorithm.

A more direct approach to (LCP) is based on the equivalence of (LCP) and the quadratic programming problem

$$\text{minimize } xMx + qx \quad \text{s.t. } x \geq 0 \, , \ Mx + q \geq 0 \, . \tag{23}$$

Here M can be replaced by the symmetric matrix $\overline{M} = \frac{1}{2}(M + M^T)$. When $\overline{M}$ is positive semidefinite, it is a convex quadratic program and can be solved by efficient procedures. When $\overline{M}$ is negative semidefinite, this is a concave quadratic program and can be treated by the methods in Sections V.4.2., VI.3.3, VII.4.3, and VIII.3.2. In the general case when $\overline{M}$ is indefinite, (23) becomes a harder d.c. optimization problem (see Chapter X).

Pardalos and Rosen (1988) showed that (LCP) is equivalent to the following mixed zero–one integer program:

$$\text{(MIP)} \quad \begin{array}{ll} & \text{maximize } \alpha \\ \text{s.t.} & 0 \leq M_i y + q_i \alpha \leq 1 - z_i \quad (i=1,\ldots,n) \\ & 0 \leq y_i \leq z_i \quad (i=1,\ldots,n) \\ & z_i \in \{0,1\} \quad (i=1,\ldots,n) \, , \ 0 \leq \alpha \leq 1 \end{array}$$

Of course, here we assume that $q_i < 0$ for at least one i (otherwise $x = 0$ is an obvious solution of (LCP)).

Proposition IX.8. *If (MIP) has an optimal solution $(\overline{\alpha},\overline{y},\overline{z})$ with $\overline{\alpha} > 0$, then $\overline{x} = \overline{y} / \overline{\alpha}$ solves (LCP). If the optimal value of (MIP) is $\overline{\alpha} = 0$, then (LCP) has no solution.*

Proof. (MIP) always has the feasible solution $y = 0$, $\alpha = 0$, and $z_i = 0$ or 1 (i=1,...,n). Since the constraint set is bounded, (MIP) has an optimal solution $(\overline{\alpha},\overline{y},\overline{z})$. Suppose that $\overline{\alpha} > 0$, and let $\overline{x} = \overline{y} / \overline{\alpha}$. Then $M\overline{y} + \overline{\alpha}q = \overline{\alpha}(M\overline{x} + q) \geq 0$, and hence $M\overline{x} + q \geq 0$. Furthermore, for each i, either $\overline{z}_i = 0$ (and hence $\overline{x}_i = 0$) or else $\bar{z}_i = 1$ (and hence $M_i\bar{y} + q_i\bar{\alpha} = 0$, i.e., $M_i\bar{x} + q_i = 0$). Therefore, $\bar{x}$ solves (LCP).

Now suppose that $\bar{\alpha} = 0$. If (LCP) has a solution x, then we have max $\{x_i, M_ix + q_i \ (i=1,...,n)\} > 0$. Denote by α the reciprocal of this positive number. Then a feasible solution of (MIP) is α, $y = \alpha x$, $z_i = 0$ if $x_i = 0$, $z_i = 1$ if $x_i > 0$. Hence we have $\alpha \le \bar{\alpha} = 0$, a contradiction. Therefore, (LCP) has no solution. ■

Using this result, Pardalos and Rosen suggested the following method of solving (LCP):

1. Solve the quadratic program (23) by a local optimization method. If $\bar{x}$ is a solution and $f(\bar{x}) = 0$, then stop. Otherwise, go to 2).

2. Choose n orthogonal directions u^i (i=1,...,n), and solve the linear programs min $\{c^Tx: x \in D\}$ with $c = u^i$ or $c = -u^i$ (i=1,...,n). This will generate $k \le 2n$ vertices x^j of D $(j \in J)$. If $f(x^j) = 0$ for some j, then stop. Otherwise, go to 3).

3. Starting from the vertex x^j $(j \in J)$ with smallest $f(x^j)$, solve the quadratic program (23) (by a local optimization method) to obtain a Kuhn–Tucker point $\bar{x}^j$. If $f(\bar{x}^j) = 0$, then stop. Otherwise, go to 4)

4. Solve (MIP) by a mixed integer programming algorithm.

In this approch, (MIP) is used only as a last resort, when local methods fail. From the computational results reported by Pardalos and Rosen (1987), it seems that the average complexity of this algorithm is $0(n^4)$.

Let us also mention another approach to (LCP), which consists in reformulating (23) as a bilinear program:

$$\text{minimize } xw \quad \text{s.t. } x \ge 0,\ w \ge 0,\ -Mx + w = q\ . \tag{24}$$

Since the constraints of this program involve both x and w, the standard bilinear programming algorithms discussed in Section IX.1 cannot be used. The problem can, however, be handled by a method of Al–Khayyal and Falk (1983) for jointly constrained biconvex programming (cf. Chapter X). For details of this approach, we refer to Al–Khayyal (1986a).

2.5. The Concave Complementarity Problem

Now consider the concave complementarity problem (17)

$$x \geq 0 \,, \quad h(x) \geq 0 \,, \quad \sum_{i=1}^{n} x_i h_i(x) = 0 \,,$$

where $h: \mathbb{R}^n \longrightarrow \mathbb{R}^n$ is a given concave mapping such that $h_i(0) < 0$ for at least one $i=1,...,n$. Setting

$$D = \{x: x \geq 0, h(x) \geq 0\} \,, \; G = \{x: \sum_{i=1}^{n} \min \{x_i \,, h_i(x)\} > 0\} \,,$$

we saw that the sets D and G are convex, and the problem is to find a point $x \in D \setminus G$.

By translating the origin to a suitable point $x \in G$ and performing some simple manipulations, we can reformulate the problem as follows.

Given an open convex set $\tilde{G}$ containing the origin 0 and a closed convex set $\tilde{D} \subset \mathbb{R}^n_+ \cap \text{cl}\, \tilde{G}$ (where $\text{cl}\, \tilde{G}$ denotes the closure of $\tilde{G}$), find a point $x \in \tilde{D} \setminus \tilde{G}$.

Note that the assumptions in this reformulation imply that, if the problem is solvable, then a solution always exists on the part of the boundary of $\tilde{D}$ that consists of points x such that $x = \theta y$ with $y \in \mathbb{R}^n_+ \setminus \{0\}$, $\theta = \sup \{t: ty \in \tilde{D}\}$. Therefore, replacing $\tilde{D}$ by the convex hull of $\tilde{D} \cup \{0\}$ if necessary, we may assume that $0 \in \tilde{D}$. The following algorithm can then be deduced from the normal conical algorithm for (CP) (Algorithm VII.2).

Algorithm IX.7.

Select an exhaustive subdivision process.

0) Let $Z_1 = (e^1,e^2,\dots,e^n)$, where e^i is the i–th unit vector of $\mathbb{R}^n$, and let $\mathscr{P}_1 = \mathscr{M}_1 = \{Z_1\}$. Construct a polyhedron $D_1 \supset \mathrm{conv}\,\{\tilde{D} \cup \{0\}\}$. Set k = 1.

k.1. For each $Z \in \mathscr{P}_k$ solve the linear program

$$\mathrm{LP}(Z, D_1) \qquad \max \sum_{i=1}^{n} \frac{\lambda_i}{\theta_i} \quad \text{s.t. } Z\lambda \in D_1 ,$$

where $Z = (v^1,\dots,v^n)$, $\theta_i = \sup\,\{t\colon tv^i \in \tilde{G}\}$ $(i=1,\dots,n)$, $\lambda = (\lambda_1,\dots,\lambda_n)^T$ (as usual, $\frac{\lambda_i}{\theta_i} = 0$ if $\theta_i = +\infty$). Let $\mu(Z)$ and $\hat{\lambda}$ be the optimal value and a basic optimal solution of this linear program, and let $\omega(Z) = Z\hat{\lambda}$. If we have $\omega(Z) \in \tilde{D} \setminus \tilde{G}$ for some $Z \in \mathscr{P}_k$, then terminate. Otherwise, ($\omega(Z) \in \tilde{G}$ or $\omega(Z) \notin \tilde{D}$ $\forall Z \in \mathscr{P}_k$), go to k.2.

k.2. In $\mathscr{M}_k$ delete all $Z \in \mathscr{P}_k$ satisfying $\mu(Z) < 1$. Let $\mathscr{R}_k$ be the collection of remaining elements of $\mathscr{M}_k$. If $\mathscr{R}_k = \emptyset$, then terminate: $\tilde{D} \subset \tilde{G}$ (the problem has no solution). Otherwise, go to k.3.

k.3. Let $\mathscr{R}_k^{(1)} = \{Z \in \mathscr{R}_k\colon Z = (v^1,\dots,v^n),\ \theta_i < +\infty\ \forall i,\ \mu(Z) < +\infty\}$. If $\mathscr{R}_k^{(1)} \neq \emptyset$, select

$$Z_k \in \mathrm{argmax}\,\{\mu(Z)\colon Z \in \mathscr{R}_k^{(1)}\} .$$

Otherwise, select

$$Z_k \in \mathrm{argmin}\,\{\theta(Z)\colon Z \in \mathscr{R}_k\} .$$

where, for $Z = (v^1,\dots,v^n)$ we define

$$\theta(Z) = \min\,\{\theta_i\colon i=1,\dots,n\} = \min_i \sup\,\{t\colon tv^i \in \tilde{G}\} .$$

Subdivide (the cone generated by) Z_k according to the chosen exhaustive subdivision rule.

k.4. Denote $\omega^k = \omega(Z_k)$. If $\omega^k \in \tilde{D}$, then set $D_{k+1} = D_k$. Otherwise, take a vector p^k such that the halfspace $p^k(x - \bar{\omega}^k) \leq 0$ separates ω^k from $\tilde{D}$ (where $\bar{\omega}^k = \theta\omega^k$, $\theta = \sup \{t: t\omega^k \in \tilde{D}\}$) and set $D_{k+1} = D_k \cap \{x: p^k(x - \bar{\omega}^k) \leq 0\}$.

k.5. Let $\mathscr{P}_{k+1}$ be the partition of Z_k obtained in Step k.3., and let $\mathscr{M}_{k+1}$ be the collection that results from $\mathscr{M}_k$ by substituting $\mathscr{P}_{k+1}$ for Z_k. Set $k \leftarrow k+1$ and go to k.1.

As before, we shall say that a cone K with matrix $Z = (v^1,...,v^n)$ is of the first category if $\theta_i < +\infty \quad \forall i$ and $\mu(Z) < +\alpha$, and that it is of the second category otherwise.

Theorem IX.3. *If Algorithm IX.6 generates infinitely many cones of the second category, then the convex complementarity problem has no solution. Otherwise, beginning at some iteration, the algorithm generates only cones of the first category. In the latter case, if the algorithm is infinite, then the sequence $\omega^k = \omega(Z_k)$ has at least one accumulation point, and any of its accumulation points yields a solution of the problem.*

Proof. It is easily seen that Lemmas IX.3 and IX.4 are still valid ($\tilde{f}(y) > 0$ means that $y \in \tilde{G}$), and hence the first part of the theorem can be established in the same way as the first part of Theorem IX.2.

Now suppose that for all sufficiently large k, $\mathscr{P}_k$ consists only of cones of the first category. If the algorithm is infinite, then it generates at least one infinite nested sequence of cones K_s of the first category, $s \in \Delta \subset \{1,2,...\}$. For each s, let $Z_s = (v^{s,1},...,v^{s,n})$, $z^{s,i} = \theta_{s,i}v^{s,i}$, $\theta_{s,i} = \sup \{t: tv^{s,i} \in G\}$ $(i=1,...,n)$, and let $\ell_s(x) = 1$ be the equation of the hyperplane through $z^{s,1},...,z^{s,n}$ (so that $\omega^s \in \operatorname{argmax} \{\ell_s(x): x \in D_s \cap K_s\}$; see Proposition IX.7). Then, denoting the smallest index $s \in \Delta$ by s_1, we find that $\omega^s \in \{x \in K_{s_1} : \ell_{s_1}(x) \leq \mu(Z_{s_1})\}$ for all $s \in \Delta$, i.e., the sequence $\{\omega^s, s \in \Delta\}$ is bounded, and hence must have an accumulation point. Consider any accumulation point $\bar{x}$ of the sequence $\{\omega^k\}$ (for all k sufficiently large, ω^k is a point). For example, let $\bar{x} = \lim_{r\to\infty} \omega^{k_r}$. Reasoning as in the proof of Lemma IX.4, we can find an infinite nested sequence of cones K_s, $s \in \Delta' \subset \{1,2,...\}$, such that each K_s, $s \in \Delta'$, contains infinitely many members of the sequence $\{K_{k_r}, r=1,2,...\}$. Without loss of generality we may assume that $K_{k_s} \subset K_s$ $(s \in \Delta')$ and some K_{s_1} is of the first category. It is easily verified that all of the conditions of Theorem II.2 are fulfilled for the set $\tilde{D} \cap \{x \in K_{s_1}: \ell_{s_1}(x) \le \mu(Z_{s_1})\}$ and the sequences $\{\omega^{k_r}\}$, $\{p^{k_r}\}$. Therefore, by this theorem, we conclude that $\bar{x} \in \tilde{D}$.

On the other hand, the exhaustiveness of the subdivision process implies that the simplex $[Z_{k_r}] = [v^{k_r,1},...,v^{k_r,n}]$ shrinks to a point $\bar{v}$. Hence $z^{k_r,i} = \theta_{k_r,i} v^{k_r,i}$ converges to a point $\bar{z} = \bar{\theta}\bar{v}$ as $r \to \infty$. Since $z^{k_r,i} \in \partial\tilde{G}$ (the boundary of $\tilde{G}$), we must have $\bar{z} \in \partial\tilde{G}$. If z^{k_r} denotes the point where the halfline from 0 through ω^{k_r} meets the simplex $[z^{k_r,1},...,z^{k_r,n}]$, then obviously $z^{k_r} \to \bar{z}$. But $\omega^{k_r} = \mu(Z_{k_r})z^{k_r}$, and since $\mu(Z_{k_r}) \ge 1$, it follows that $\bar{x} = \lim_{r\to\infty} \omega^{k_r} \notin \tilde{G}$. Therefore we have $\bar{x} \in \tilde{D} \setminus \tilde{G}$. ■

3. PARAMETRIC CONCAVE PROGRAMMING

An important problem that arises in certain applications is the following

(PCP) Find the smallest value θ such that

$$\min \{f(x): x \in D, cx \le \theta\} \le \alpha , \tag{25}$$

where D is a polyhedron in $\mathbb{R}^n$, c is an n–vector, and $f: \mathbb{R}^n \to \mathbb{R}$ is a concave function.

We shall call this a ***parametric concave programming problem***, since the minimization problem in (25) is a concave program depending on the parameter θ.

In a typical interpretation of (PCP), D represents the set of all feasible production programs, while the inequality $cx \le \theta$ expresses a constraint on the amount of a certain scarce commodity that can be used in the production, and f(x) is the production cost of the program x. Then the problem is to find the least amount of the scarce commodity required for a feasible production program with a cost not exceeding a given level α.

In the literature, the PCP problem has received another formulation which is often more convenient.

Proposition IX.9. *The PCP problem is equivalent to*

$$\text{(LRCP)} \qquad \text{minimize } cx \quad \text{s.t.} \quad x \in D,\ f(x) \le \alpha \ . \tag{26}$$

Proof. We may of course assume that min f(D) $\le \alpha$, for otherwise both problems are infeasible. If θ^0 is optimal for (PCP) and x^0 is an optimal solution of the corresponding concave program, then obviously $cx^0 = \theta^0$, and x^0 is feasible for (LRCP), hence $\theta^0 \ge \theta^1 :=$ optimal value of (LRCP).

Conversely, if x^1 is optimal for (LRCP) and $cx^1 = \theta^1$, then θ^1 is feasible for (PCP), hence $\theta^1 \ge \theta^0$. Therefore, $\theta^1 = \theta^0$, and x^0 is optimal for (LRCP), while θ^1 is optimal for (PCP). ■

An inequality of the form $f(x) \le \alpha$, where f(x) is a concave function, is called a ***reverse convex inequality***, because it becomes convex when reversed (see Chapter I). If this inequality is omitted, then problem (26) is merely a linear program; therefore it is often referred to as a ***linear program with an additional reverse convex constraint***.

LRCP problems were first studied with respect to global solutions by Bansal and Jacobsen (1975 and 1975a), Hillestad (1975) and also Hillestad and Jacobsen (1980).

In Bansal and Jacobsen (1975 and 1975a) the special problem of optimizing a network flow capacity under economies of scale was discussed. Several methods for globally solving (LRCP) with bounded feasible domain have been proposed since then. Hillestad (1975) and Hillestad and Jacobsen (1980 and 1980a) developed methods based on the property that an optimal solution lies on an edge of the polyhedron D. These authors also showed how cuts that were originally devised for concave minimization problems can be applied to (LRCP). Further developments along these lines were given in Sen and Sherali (1985 and 1987), Gurlitz (1985) and Fulöp (1988). On the other hand, the branch and bound methods originally proposed for minimizing concave functions over polytopes have been extended to (LRCP) by Muu (1985), Hamami and Jacobsen (1988), Utkin, Khachaturov and Tuy (1988), Horst (1988).

For some important applications of (LRCP) we refer to the discussion in Section I.2.5.

3.1. Basic Properties

We shall make the following assumptions:

(a) D is nonempty and contains no lines;

(b) either D or $G := \{x: f(x) \geq \alpha\}$ is compact;

(c) $\min \{cx: x \in D\} < \min \{cx: x \in D, f(x) \leq \alpha\}$.

The last assumption simply means that the constraint $f(x) \leq \alpha$ is *essential* and (LRCP) does not reduce to the trivial linear program $\min \{cx: x \in D\}$. It follows that there exists a point w satisfying

$$w \in D,\ f(w) > \alpha,\ cw < cx \quad \forall x \in D \setminus G \tag{27}$$

(such a point is provided, for example, by an optimal solution of the linear program $\min \{cx: x \in D\}$).

The following property was first established by Hillestad and Jacobsen (1980) in the case when D is bounded, and was later extended to the general case by Tuy (1983) (see also Sen and Sherali (1985)).

Let $\overline{\text{conv}}$ A denote the closure of conv A.

Proposition IX.10. *The set $\overline{conv}(D \setminus int\ G)$ is a polyhedron whose extreme directions are the same as those of D and whose vertices are endpoints of sets of the form conv(E \ int G), where E is any edge of D.*

Proof. Denote by M the set of directions and points described in the proposition. Obviously, $M \subset D \setminus \text{int } G$ (note that, in view of assumption (b), any recession direction of D must be a recession direction of $D \setminus \text{int } G$). Hence, $\text{conv}M \subset \overline{\text{conv}}(D \setminus \text{int } G)$.

We now show the inverse inclusion. Suppose $x \in D \setminus \text{int } G$. Since G is convex, there is a hyperplane $H = \{x: hx = \eta\}$ separating x from G whose intersection with D is nonempty. Then, since $D \cap H$ contains no lines, x belongs to the convex hull of the set of extreme points and directions of $H \cap D$. But obviously any extreme direction of the latter polyhedron is a recession direction of D, while any extreme point must lie on an edge E of D such that $E \setminus \text{int } G$ is nonempty, and hence must belong to the convex hull of the endpoints of $\text{conv}(E \setminus \text{int } G)$. Consequently, $D \setminus \text{int } G$ is contained in conv M; and, since conv M is closed (M is finite), $\overline{\text{conv}}(D \setminus \text{int } G) \subset \text{conv } M$. Hence, $\overline{\text{conv}}(D \setminus \text{int } G) = \text{conv } M$. ■

Since minimizing a linear form over a closed set is equivalent to minimizing it over the closure of the convex hull of this set we have

Corollary IX.3. *(LRCP) or (PCP) is equivalent to the implicit linear program*

$$\min \{cx: x \in \overline{conv}(D \setminus int\ G)\} . \tag{28}$$

Here the term "implicit" refers to the fact that, although $\overline{\text{conv}}(D \setminus \text{int } G)$ is a polyhedron, its constraints are not given explicitly (this constitutes, of course, the main difficulty of the problem).

Proposition IX.11. *If (LRCP) has an optimal solution, then at least one of its optimal solutions of it lies on the intersection of the boundary ∂G of G with an edge of D.*

Proof. If (LRCP) is solvable, then at least one of its optimal solutions (i.e., an optimal solution of (28)), say x^0, is a vertex of $\overline{\text{conv}}(D \setminus \text{int } G)$, and hence is an endpoint of the set $\text{conv}(E \setminus \text{int } G)$, where E is some edge of D. If $f(x^0) < \alpha$, then x^0 must be a local (hence, global) minimum of cx over D, contrary to assumption (c). Hence, $f(x^0) = \alpha$, i.e., $x^0 \in \partial G$. ■

It follows from the above that in the search for an optimal solution of (LRCP) we can restrict ourselves to the set of intersection points of ∂G with the edges of D. Several earlier approaches to solving (LRCP) are based on this property (see, e.g., Hillestad and Jacobsen (1980), and Tuy and Thuong (1984)).

Another property which is fundamental in some recent approaches to the LRCP problem is the following.

Definition IX.1. *We say that the LRCP problem is regular if $D \setminus \text{int } G = cl(D \setminus G)$, i.e., if any feasible point is the limit of a sequence of points $x \in D$ satisfying $f(x) < \alpha$.*

Thus, if $D \setminus \text{int } G$ has isolated points, as in Fig. IX.5, then the problem is not regular. However, a regular problem may have a disconnected feasible set as in Fig. IX.6.

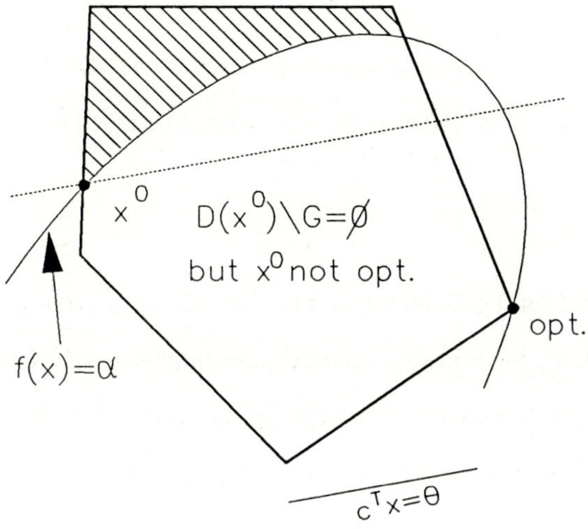

Fig. IX.5

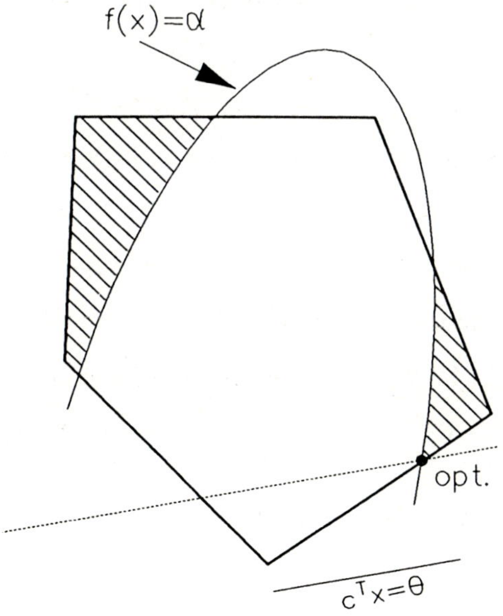

Fig. IX.6

For any point x^0 of D let us denote

$$D(x^0) = \{x \in D: cx \le cx^0\} \tag{29}$$

Theorem IX.4. *In order that a feasible solution x^0 be globally optimal for (LRCP), it is necessary and, if the problem is regular, also sufficient that*

$$D(x^0) \setminus G = \emptyset . \tag{30}$$

Proof. Suppose $D(x^0) \setminus G$ is not empty, i.e., there exists a point z of $D(x^0)$ such that $f(z) < \alpha$. Let x^1 be the point where the boundary ∂G of G intersects the line segment joining x and the point w satisfying (3). Then x^1 belongs to $D \setminus \text{int } G$, and, since $cw < cz$, it follows that $cx^1 < cz \le cx^0$; and hence x^0 is not optimal. Conversely, suppose that (30) holds and the problem is regular. If there were a feasible point x with $cx < cx^0$, in any neighbourhood of x we would find a point $x' \in D$ with $f(x') < \alpha$. When this point is sufficiently near to x, we would have $cx' < cx^0$, i.e., $x' \in D(x^0) \setminus G$, contrary to the assumption. Therefore, x^0 must be optimal. This completes the proof. ■

It is easy to give examples of problems where, because regularity fails, condition (30) holds without x^0 being optimal (see Fig. IX.5). However, the next result shows the usefulness of condition (30) in the most general case, even when we do not know whether a given problem is regular.

For each $k = 1,2,...$ let $\varepsilon_k: \mathbb{R}^n \to \mathbb{R}$ be a convex function such that $0 < \varepsilon_k(x)$ $\forall x \in D$, $\max \{\varepsilon_k(x): x \in D\} \to 0$ $(k \to \infty)$, and consider the *perturbed problem*

$$(\text{LRCP}_k) \quad \text{minimize } cx \quad \text{s.t. } x \in D,\ f(x) - \varepsilon_k(x) \le \alpha .$$

Denote $G_k = \{x: f(x) - \varepsilon_k(x) \ge \alpha\}$.

Theorem IX.5. *Let x^k be a feasible solution to $(LRCP_k)$ satisfying $D(x^k) \setminus G_k = \emptyset$. Then any accumulation point $\bar{x}$ of the sequence x^k, $k=1,2,\ldots$, is a global optimal solution of (LRCP).*

Proof. Clearly, because of the continuity of $f(x) - \varepsilon_k(x)$, $\bar{x}$ is feasible for (LRCP). For any feasible solution x of (LRCP), since $f(x) - \varepsilon_k(x) < f(x) \leq \alpha$, we have $x \notin G_k$. Therefore, the condition $D(x^k) \setminus G_k = \emptyset$ implies that $x \notin D(x^k)$. Since $x \in D$, we must have $cx \geq cx^k$, and hence $cx \geq c\bar{x}$. This proves that $\bar{x}$ is a global optimal solution of (LRCP). ■

In practice, the most commonly used perturbation functions are $\varepsilon(x) = \varepsilon$ or $\varepsilon(x) = \varepsilon(\|x\|^2 + 1)$ (for D bounded), where $\varepsilon \downarrow 0$.

Proposition IX.13. *For sufficiently small $\varepsilon > 0$, the problem*

$$(LRCP-\varepsilon) \qquad \min cx \quad \text{s.t.} \quad x \in D,\ f(x) - \varepsilon(\|x\|^2 + 1) \leq \alpha$$

is regular.

Proof. Since the vertex set V of D is finite, there exists $\varepsilon_0 > 0$ small enough so that $\varepsilon \in (0,\varepsilon_0)$ implies that $F(\varepsilon,x) := f(x) - \varepsilon(\|x\|^2 + 1) \neq \alpha \ \ \forall x \in V$. Indeed, if $V_1 = \{x \in V\colon f(x) > \alpha\}$ and ε_0 satisfies $\varepsilon_0(\|x\|^2 + 1) < f(x) - \alpha \ \forall x \in V_1$, then whenever $0 < \varepsilon < \varepsilon_0$, we have for all $x \in V \setminus V_1$: $f(x) - \varepsilon(\|x\|^2 + 1) \leq \alpha - \varepsilon < \alpha$, while for all $x \in V_1$: $f(x) - \varepsilon(\|x\|^2 + 1) > f(x) - \varepsilon_0(\|x\|^2 + 1) > \alpha$. Also note that the function $F(\varepsilon,x) := f(x) - \varepsilon(\|x\|^2 + 1)$ is strictly concave in x. Now consider the problem $(LRCP-\varepsilon)$, where $0 < \varepsilon < \varepsilon_0$, and let $x \in D$ be such that $F(\varepsilon,x) \leq \alpha$. If x is not a vertex of D, then x is the midpoint of a line segment $\Delta \subset D$, and, because of the strict concavity of $F(\varepsilon,x)$, any neighbourhood of x must contain a point x' of Δ such that $F(\varepsilon,x') < \alpha$. On the other hand, if x is a vertex of D, then $F(\varepsilon,x) < \alpha$, and any point x' of D sufficiently near to x will satisfy $F(\varepsilon,x') < \alpha$. Thus, given any $x \in D$ such that $F(\varepsilon,x) \leq \alpha$, there exists a point x' arbitrarily close to x such that

$x' \in D$, $F(\varepsilon,x') < \alpha$. This means that the problem (LRCP$-\varepsilon$) is regular. ■

It follows from the above result that a LRCP problem can always be regularized by a slight perturbation. Moreover, this perturbation makes the function f(x) strictly concave, a property which may be very convenient in certain circumstances.

3.2. Outer Approximation Method for (LRCP)

To simplify the presentation of the methods, in the sequel instead of (b) we shall assume a stronger condition:

(b') D is a bounded polyhedron (a polytope).

With suitable modifications, most of the results below can be extended to the case when D is unbounded.

Under assumptions (a), (b'), (c), if w is a basic optimal solution of the linear program $\min \{cx: x \in D\}$, then, by transforming the problem to a space of nonbasic variables relative to this basic solution, we can always arranged it so that:

1) $w = 0$ is a vertex of D;
2) $f(0) > \alpha$ and $\min \{cx: x \in D\} > 0$ (see (27));
3) D is defined by constraints of the form $Ax \le b$, $x \ge 0$.

One of the most natural approaches to solving the LRCP problem is outer approximation (cf. Forgo (1988), Hillestad and Jacobsen (1980a), and Fulöp (1988); see also Bulatov (1977) for a related discussion). This approach is motivated by the following simple observation.

Proposition IX.14. *Let x^0 be a basic optimal solution of the linear program*

$$\min \{cx: x \in D\} .$$

Suppose that $f(x^0) > \alpha$, i.e., x^0 is not feasible for (LRCP). If $\pi(x-x^0) \ge 1$ is an *α–valid cut constructed at x^0 for the concave program*

$$\min \{f(x): x \in D\} ,$$

then the inequality

$$\ell(x):= \pi(x-x^0) - 1 \geq 0$$

excludes x^0 without excluding any feasible solution of (LRCP).

Proof. The proof is trivial, since, by definition, an α–valid cut at x^0 is an inequality $\ell(x) \geq 0$ which excludes x^0 without excluding any point $x \in D$ such that $f(x) \leq \alpha$. ■

It follows from this fact that concavity cuts (see Chapter III) can be used in outer approximation methods to solve (LRCP).

Specifically, to solve (LRCP) by the outer approximation approach one constructs a nested sequence of polytopes $S^0 \supset S_1 \supset ... \supset S_k \supset ...$ in the following way. One starts with $S_0 = D$. When $S_0,...,S_k$ have been constructed, one solves the linear program

$$\min \{cx: x \in S_k\} ,$$

obtaining a basic optimal solution, x^k. If x^k happens to be feasible for (LRCP), then the procedure terminates: x^k solves (LRCP), since S_k contains the feasible set of (LRCP). Otherwise, one generates a concavity cut $\pi^k(x-x^k) \geq 1$ to exclude x^k and forms S_{k+1} by adding this constraint to S_k. The procedure is then repeated with S_{k+1} in place of S_k (see Fig. IX.7, page 486).

Although this method is conceptually simple, and, as reported in Hillestad and Jacobsen (1980a), it may sometimes help solve problems which otherwise would be difficult to attack, its convergence is not guaranteed. For example, Gurlitz (1985) has shown that, when applied to the 5–dimensional problem

$$\min \{x_1: 0 \leq x_1 \leq 1 \ (i=1,...,5), \ \Sigma x_i^2 \geq 4.5\} ,$$

the outer approximation method using such cuts will generate a sequence x^k which converges to an infeasible solution.

To overcome the difficulty and ensure finiteness of the procedure, Fulöp (1988) proposed combining concavity cuts with facial cuts, similar to those introduced by Majthay and Whinston (1974) (see Section V.2). His method involves solving a set covering subproblem in certain steps.

Note that by Corollary III.4, a sufficient convergence condition is that the sequence $\{\pi^k\}$ be bounded. A promising approach suggested by this condition is to combine cutting with partitioning the feasible domain by means of cone splitting. This leads to conical algorithms, which will be discussed later in this section.

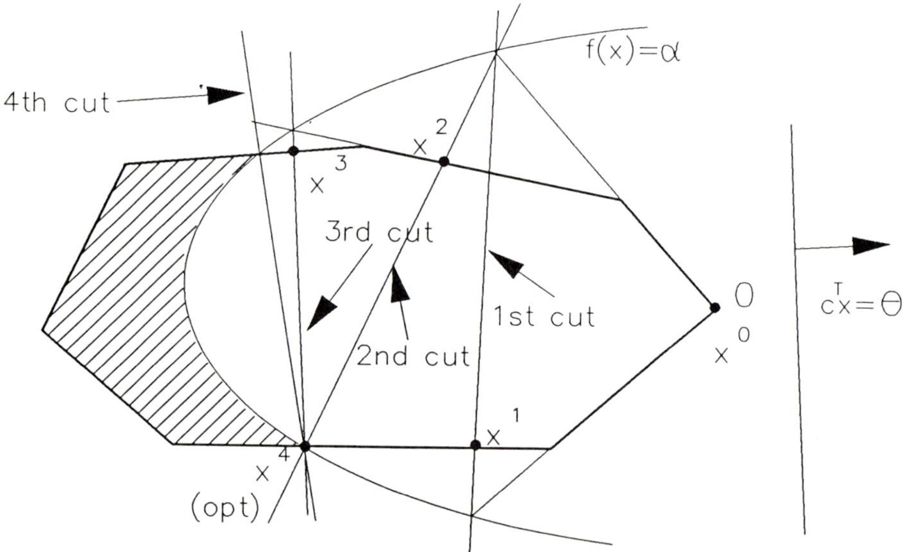

Fig. IX.7

3.3. Methods Based on the Edge Property

Since, by Proposition IX.11, an optimal solution must exist on some edge of D intersecting ∂G, and since the number of such edges is finite, one can hope to solve the problem by a suitable edge search procedure. The first method along these lines is due to Hillestad (1975). Another method, proposed by Hillestad and Jacobsen (1980a), is based on a characterization of these edges which can contain an optimal solution in term of the best feasible vertex of the polytope D.

In addition to (a), (b'), (c), assume that f(x) is strictly concave and the problem is regular (these additional assumptions are innocuous, by virtue of Proposition IX.4).

Typically, a method based on the edge property alternates between steps of two kinds: "forward" and "backward".

First, starting with a vertex s^0 of D such that $f(s^0) < \alpha$, we try to decrease the objective function value cx, while moving *forward* to the surface $f(x) = \alpha$. To do this, it suffices to apply the simplex procedure to the linear program

$$\min \{cx: x \in D, cx \le cs^0\} .$$

In view of assumption (c), s^0 cannot be a minimizer of cx over D, so at least one neighbouring vertex u to s^0 satisfies $cu < cs^0$. If $f(u) < \alpha$, we perform a simplex pivot to move from s^0 to u. This pivoting process is continued until we find a pair of vertices u,v of D such that $f(u) < \alpha$, $f(v) \ge \alpha$ (this must occur, again because of assumption (c)). Then we move along the edge [u,v] of D to the point x^0 where this edge meets the surface $f(x) = \alpha$ (due to the strict concavity of f(x), x^0 is uniquely determined). At this stage x^0 is the best feasible point obtained thus far, so for further investigation we need only consider

$$D(x^0) = \{x \in D: cx \le cx^0\} .$$

Since we are now stopped by the "wall" $f(x) = \alpha$, we try to move *backward* to the

region $f(x) < \alpha$, while keeping the objective function value at the lowest level already attained. This can be done by finding a vertex s^1 of $D(x^0)$ such that $f(s^1) < \alpha$ which is as far as possible from the surface $f(x) = \alpha$ (intuitively, the further we can move backward, the more we will gain in the next forward step).
If such a point s^1 can be found, then another forward step can be performed from s^1, and the whole process can be repeated with s^1 and $D(x^0)$ replacing s^0 and D.
On the other hand, if such an s^1 does not exist, this means that $D(x^0) \setminus G$ is empty. By Theorem IV.1 and the regularity of the problem, this implies that x^0 is a global optimal solution of (LRCP).

The most difficult part of this forward–backward scheme of course lies in the backward steps: given the polyhedron $D(x^0)$, how do we find a vertex s^1 such that $f(s^1) > \alpha$?
Hillestad and Jacobsen (1980) suggested a combinatorial procedure for the backward step which involves enumerating the vertices of certain polytopes. Unfortunately, in certain cases this procedure may require us to solve by a rather expensive method (vertex enumeration) a subproblem which is almost as difficult as the original one (finding a feasible point s^1 better than the current best feasible solution x^0).
However, a more systematic way to check whether $D(x^0)$ has a vertex s such that $f(s) < \alpha$, and to find such a vertex if one exists, is to solve the concave program

$$\min \{f(x): x \in D(x^0)\} .$$

Therefore, Thuong and Tuy (1985) proposed that one solves this concave program in the backward step. With this approach, the algorithm can be summarized as follows:

Algorithm IX.8.

Initialization:

If a vertex s^0 of D is available such that $f(s^0) < \alpha$, set $D_0 = D$, $k = 0$ and go to 1).

Otherwise, apply any finite algorithm to the concave program min $\{f(x): x \in D\}$, until a vertex s^0 of D is found such that $f(s^0) \le \alpha$ (if such a vertex cannot be found, the problem has no feasible solution). Set $D_0 = D$, $k = 0$. Go to 1) if $f(s^0) < \alpha$, go to 2) (with $x^0 = s^0$) if $f(s^0) = \alpha$.

Iteration k = 1,2,...:

1) Starting from s^{k-1} pivot by means of the simplex algorithm for solving the linear program

$$\min \{cx: x \in D_{k-1}\} \tag{31}$$

until a pair of vertices u,v of D_{k-1} is found so that $f(u) < \alpha$, $f(v) \ge \alpha$, and $cv < cu \le cs^{k-1}$. Let x^k be the (unique) point of the line segment $[u,v]$ such that $f(x^k) = \alpha$. Go to 2).

2) Form $D_k = \{x \in D: cx \le cx^k\}$ and solve the concave program

$$\min \{f(x): x \in D_k\} , \tag{32}$$

obtaining an optimal vertex solution s^k.

a) If $f(s^k) = \alpha$, terminate.

b) Otherwise, $f(s^k) < \alpha$, set $k \leftarrow k+1$ and go to 1).

Theorem IX.6. *Assume that (a), (b'), (c) hold, and that moreover, f(x) is strictly concave and the LRCP problem is regular. If the problem has a feasible solution, then the above algorithm terminates at Step 2a) after finitely many iterations, yielding a global optimal solution.*

Proof. If the algorithm terminates at Step 2a), then min $\{f(x): x \in D_k\} = \alpha$, and hence, $D(x^k) \setminus G = D_k \setminus G = \emptyset$. Therefore, by Theorem IX.1 and the regularity assumption, x^k is a global optimal solution. Now, by construction, $[u^k,v^k]$ is an edge of $D_{k-1} = D(x^{k-1})$. Since $cv^k < cu^k \le cs^{k-1} \le cx^{k-1}$, $[u^k,v^k]$ cannot be contained in the face $\{x: cx = cx^{k-1}\}$ of D_{k-1}. Hence, $[u^k,v^k]$ is contained in an edge of D.

Let M denote the set of all $x \in D$ such that $f(x) = \alpha$ and x is contained in some edge of D. M is finite, since the number of edges of D is finite, and the strictly concave function f(x) can assume the value 0 on each edge of D at most at two distinct points. Finiteness of the algorithm then follows from finiteness of M and the fact that each iteration generates a point $x^k \in M$ satisfying $cx^k < cx^{k-1}$. ■

The algorithm involves an alternating sequence of linear programming steps (31) and concave programming steps (32).

An important feature of the concave programming subproblems

$$\min \{f(x): x \in D, cx \le cx^k\}$$

is that the subproblem in each iteration differs from the one in the previous iteration only in the right hand side of the constraint $cx \le cx^k$. To increase efficiency, the solution method chosen for the subproblems should take advantage of this structure. For example, if D is bounded and the outer approximation algorithm (Algorithm VI.1) is used for the concave programs (32), then the algorithm could proceed as follows:

Algorithm IX.8*.

Initialization:

Construct a polytope $D_0 \supset D$, with a known (and small) vertex set V_0. Let s^0 be a vertex of D_0 such that $f(s^0) < \alpha$. Set $k = 0$.

Iteration k = 1,2,...:

1) Starting from s^{k-1}, pivot by means of the simplex algorithm for solving the linear program $\min \{cx: x \in D_{k-1}\}$ until a pair of vertices u,v of D_{k-1} is found so that $f(u) < \alpha$, $f(v) \ge \alpha$, and $cv < cu \le cs^{k-1}$. Let x^k be the intersection of [u,v]

with the surface $f(x) = \alpha$. Go to 2).

2) If $x^k \in D$, set $D_k = D_{k-1} \cap \{x: cx \le cx^k\}$. Otherwise, set $D_k = D_{k-1} \cap \{x: \ell_k(x) \le 0\}$, where $\ell_k(x) \le 0$ is the constraint of D that is the most violated by x^k. Compute the vertex set V_k of D_k (from knowledge of V_{k-1}).
Let $s^k \in \text{argmin}\ \{f(x): x \in V_k\}$.

a) If $f(s^k) = \alpha$, then terminate: $\bar{x}^k \in \text{argmin}\ \{cx^i: x^i \in D, i=0,1,...,i\}$ is a global optimal solution of (LRCP).

b) If $f(s^k) < \alpha$, set $k \leftarrow k+1$ and return to 1).

c) If $f(s^k) > \alpha$, terminate: (LRCP) is infeasible.

Theorem IX.7. *Under the same assumptions as in Theorem IX.6, Algorithm IX.8* terminates at Step 2a) or 2c) after finitely many iterations.*

Proof. Since the number of constraints on D is finite, either all of these constraints are generated (and from then on Theorem IX.6 applies), or else the algorithm terminates before that. In the latter case, if 2a) occurs, since $\{x \in D_k: cx \le c\bar{x}^k\} \setminus G = \emptyset$, we have $D(\bar{x}^k) \setminus G = \emptyset$. Hence, $\bar{x}^k$ is optimal for (LRCP). Similarly, if 2c) occurs, then $\{x \in D_k: cx \le c\bar{x}^k\} \setminus \text{int}\ G = \emptyset$. Hence $x^i \notin D$ $(i=0,1,...,k)$, and therefore $D \setminus \text{int}\ G = \emptyset$. ■

Example IX.3.

$$\begin{aligned} \text{Minimize}\quad & -2x_1 + x_2 \\ \text{s.t.}\quad & x_1 + x_2 \le 10\ , \\ & -x_1 + 2x_2 \le 8\ , \\ & -2x_1 - 3x_2 \le -6\ , \\ & x_1 - x_2 \le 4\ , \\ & x_1 \ge 0\ ,\ x_2 \ge 0\ , \\ & -x_1^2 + x_1x_2 - x_2^2 + 6x_1 \le 0\ . \end{aligned}$$

Since the subproblems are only one–dimensional it will suffice to use Algo–

rithm IX.8. However, we shall also show Algorithm IX.8* for comparison.

Applying Algorithm IX.8, we start with the vertex $s^0 = (0;4)$.

Iteration 1. Step 1 finds $u^1 = (0,2)$, $v^1 = (3,0)$ and $x^1 = (0.4079356,1.728049)$. Step 2 solves the concave program $\min \{f(x): x \in D(x^1)\}$ and finds $s^1 = (2,5)$.

Iteration 2. Step 1 finds $u^2 = (4,6)$, $v^2 = (7;3)$ and $x^2 = (4.3670068,5.6329334)$. Since the optimal value of the concave program $\min \{f(x): x \in D(x^2)\}$ is 0, Step 2 concludes that x^2 is an optimal solution of the problem (see Fig. IX.8).

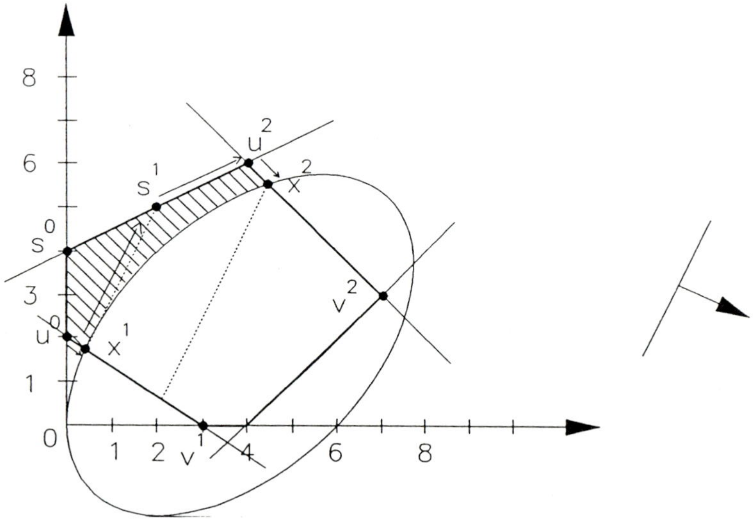

Fig. IX.8

Applying Algorithm IX.8*, we start with the simplex $D_0 = \{x_1 \geq 0,\ x_2 \geq 0,\ x_1 + x_2 \leq 10\}$ and its vertex $s^0 = (0,10)$.

Iteration 1: Step 1 finds $x^1 = (4.3670068,5.6329334)$. Since x^1 is feasible, $D_1 = D_0 \cap \{x: cx \leq cx^1\}$. Step 2 finds the vertex $s^1 = (10;0)$ which achieves the minimum of cx over D_1.

Iteration 2: Step 1 finds $x^2 = (7.632993,2.367007)$. Since this point is infeasible,

$D_2 = D_1 \cap \{x: x_1 - x_2 \leq 4\}$. Since the minimum of f(x) over D_2 is 0, Step 2 concludes that the best feasible solution so far obtained, i.e., x^1, is optimal.

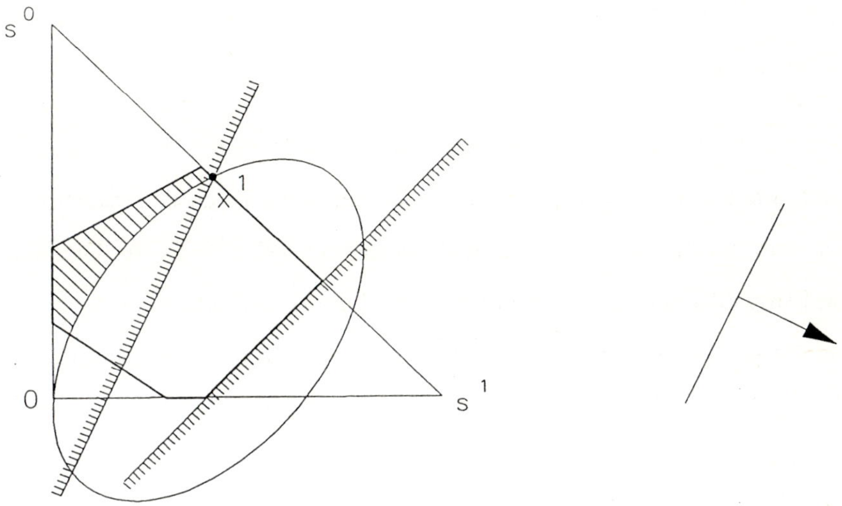

Fig. IX.9

Remark IX.3. The above method assumes that the function f(x) is strictly concave and the LRCP problem is regular. If these assumptions are not readily verifiable, we apply Algorithm IX.8 or IX.8* to the perturbed problem

$$\min cx \quad \text{s.t. } x \in D,\ f(x) - \varepsilon(\|x\|^1 + 1) \leq \alpha \tag{33}$$

which is regular and satisfies the strict concavity assumption. Clearly, if this perturbed problem is infeasible, then the original problem itself is infeasible. Otherwise, we obtain a global optimal solution $x(\varepsilon)$ of the perturbed problem, and, by Theorem IX.5, as $\varepsilon \downarrow 0$ any accumulation point of the sequence $\{x(\varepsilon)\}$ yields a global optimal solution of (LRCP).

3.4. Conical Algorithms for (LRCP)

It is also natural to extend also conical algorithms to (LRCP). The extension is based on the fact, that a basic subproblem of the LRCP problem is the following:

(∗) Given a feasible point x^k, find a point $y \in D(x^k) \setminus G$, or else establish that $D(x^k) \subset G$.

Indeed, whenever a point $y \in D(x^k) \setminus G$ exists, then, using inexpensive local methods, one can find a vertex s of $D(x^k)$ such that $f(s) \le f(y) < \alpha$. Therefore, the backward step in Algorithm 1 is reduced to solving the subproblem (∗).
Now recall that, because of the basic assumptions (a), (b'), (c), we can arrange it so that the conditions 1) – 3) at the beginning of Section IX.5.2 hold (in particular, $0 \in D \cap \text{int } G$ and $D \subset \mathbb{R}^n_+$). Then, setting $D_k = D(x^k)$, in (∗) we recognize the (D_kG) problem studied in Sections VI.2.2 and VII.1.1. If we then use the (D_kG) procedure (Section VII.1.2) in the backward step of each iteration, then we obtain an algorithm which differs from Algorithm IX.8 only in the way the backward step is carried out. However, just as with Algorithm IX.8, it is important to take advantage of the fact that D_k differs from D_{k-1} only in the right hand side of the constraint $cx \le cx^{k-1}$. This suggests integrating all of the (D_kG) procedures in successive iterations k = 1,2,... into a unified conical algorithm, as follows.

Algorithm IX.9 (Normal Conical Algorithm for (LRCP)).

Select an NCS rule for cone subdivision (see Sections VII.1.4 and VII.1.6). Set $\gamma_0 = +\infty$, k = 0.

0) Let $D_k = D \cap \{x: cx \le \gamma_k\}$. Select a cone K_0 such that $D_k \subset K_0 \subset \mathbb{R}^n_+$ and $\min \{cx: x \in K_0\} = 0$ (e.g., $K_0 = \mathbb{R}^n_+$ if $D_k = D$). Compute the points $z^{0i} \neq 0$ where the surface $f(x) = \alpha$ meets the i–th edge of K_0 (i=1,2,...,n). Set $Q_0 = (z^{01}, z^{02}, \dots, z^{0n})$, $\mathcal{M} = \mathcal{P} = \{Q_0\}$.

1) For each $Q \in \mathcal{P}$ with $Q = (z^1,z^2,\dots,z^n)$, $f(z^i) = \alpha$ $(i=1,2,\dots,n)$, solve the linear program

$$\text{LP}(Q,D_k) \quad \max\{\Sigma\lambda_i\colon \Sigma\lambda_i Az^i \le b, \Sigma\lambda_i cz^i \le \gamma_k, \lambda_i \ge 0 \ \forall i\} = $$
$$\max\{eQ^{-1}x\colon x \in D_k, Q^{-1}x \ge 0\}$$

obtaining the optimal value $\mu(Q)$ and a basic optimal solution $\omega(Q)$. If $f(\omega(Q)) < \alpha$ for some Q, then go to 5). Otherwise, go to 2).

2) In $\mathcal{M}$ delete all $Q \in \mathcal{P}$ such that $\mu(Q) \le 1$. Let $\mathcal{R}$ be the remaining collection of matrices. If $\mathcal{R} = \emptyset$, then terminate. Otherwise go to 3).

3) Select $Q_* \in \operatorname{argmax}\{\mu(Q)\colon Q \in \mathcal{R}\}$ and split it according to the NCS rule chosen.

4) Let $\mathcal{P}_*$ be the partition of Q_* obtained in this way.
Set $\mathcal{P} \leftarrow \mathcal{P}_*$, $\mathcal{M} \leftarrow (\mathcal{R} \setminus \{Q_*\}) \cup \mathcal{P}_*$ and go to 1).

5) Compute a vertex s^k of D_k such that $f(s^k) \le f(\omega(Q))$, and, starting from s^k, perform a forward step (Step 1 in Algorithm IX.8) to obtain a point x^{k+1} which lies on the intersection of the surface $f(x) = \alpha$ with an edge of D. Let $\gamma_{k+1} = cx^{k+1}$. Set $k \leftarrow k+1$ and go to 0).

Theorem IX.8. *Assume that the LRCP problem is regular. If Algorithm IX.9 is infinite, then some iteration $k \ge 1$ is infinite and x^k is a global optimal solution. If the algorithm terminates at iteration k with $\gamma_k < +\infty$ (or equivalently, $k \ge 1$), then x^k is a global optimal solution; if it terminates at iteration k with $\gamma_k = +\infty$ (or equivalently, $k = 0$), then the problem is infeasible.*

Proof. Before proving the theorem, observe that so long as Step 5) has not yet occurred, we are executing the (D_kG) procedure described in Section VII.1.2. An occurrence of Step 5) marks the end of an iteration k and the passage to iteration k+1, with $\gamma_{k+1} < \gamma_k$. Bearing this in mind, suppose first that the algorithm is infinite. Since each x^k lies on some edge E of D and achieves the minimum of cx over $E \setminus \text{int } G$, it follows that the number of all possible values of γ_k, and hence the number of iterations, is finite. That is, some iteration k must continue endlessly. Since this iteration is exactly the (D_k,G) procedure, by Proposition VII.2 the set $D_k \cap \partial G$ is nonempty, while $D_k \subset G$. If $k = 0$, i.e., $D_k = D$, this would mean that the problem is feasible, but has no feasible point x such that $f(x) < \alpha$, conflicting with the regularity assumption. Hence, $k \geq 1$, and then the fact that $D_k \subset G$ together with the regularity assumption imply that x^k is a global optimal solution. Now suppose that the algorithm terminates at Step 2 of some iteration k. Then, since the current set $\mathcal{R}$ is empty, no cone of the current partition of K_0 contains points of $D_k \setminus G$. Hence, if $\gamma_k < +\infty$, then by Theorem IX.4, x^k is a global optimal solution. On the other hand, if $\gamma_k = +\infty$ (i.e., $D_k = D$), then $D \subset G$; hence, since the problem is regular, it must be infeasible. ■

As with Algorithms IX.8 and IX.8*, the regularity assumption here is not too restrictive. If this assumption cannot be readily checked, the problem can always be handled by replacing f(x) with $f(x) - \varepsilon$, where $\varepsilon > 0$ is sufficiently small.
On the other hand, there exist variants of conical algorithms which do not require regularity of the problem (Muu (1985), Sen and Whiteson (1985)). However, these algorithms approach the global optimum from outside the feasible region and at any stage can generally guarantee only an infeasible solution sufficiently near to a global optimal solution.

Next we present an algorithm of this class which is an improved version of an earlier algorithm of Muu (1985). The main improvement consists in allowing any exhaustive subdivision process instead of a pure bisection process. This is possible due to the following lower bounding method.

Proposition IX.15. *Let $K = con(Q)$, $Q = (z^1,z^2,\dots,z^n)$, be a cone generated by n linearly independent vectors $z^i \in \mathbb{R}^n_+$ such that $f(z^i) = \alpha$. Then a lower bound for cx over the set $K \cap (D \setminus int\ G)$ is given by the optimal value $\beta(Q)$ of the linear program:*

$$min\ cx \quad s.t. \quad x \in D,\ eQ^{-1}x \geq 1,\ Q^{-1}x \geq 0, \tag{34}$$

i.e.,

$$min\ \{\Sigma\lambda_i cz^i\colon\ \Sigma\lambda_i Az^i \leq b,\ \Sigma\lambda_i \geq 1,\ \lambda_i \geq 0\ (i=1,...,n)\}\ . \tag{35}$$

If $cx > 0 \ \ \forall x \in K \setminus int\ G$, *then for an optimal solution* $\omega(Q)$ *of (34) either* $f(\omega(Q)) = \alpha$ *(and in this case* $\beta(Q)$ *equals the exact minimum of cx over the feasible portion contained in K), or else* $\omega(Q)$ *does not lie on any edge of K.*

Proof. Because of the convexity of G, $K \cap (D \setminus int\ G)$ is contained in $K \cap D \cap H$, where H is the halfspace not containing 0 with bounding hyperplane passing through $z^1,z^2,...,z^n$. This proves the first assertion (the formulation (35) follows from the fact that any point $x \in K \cap H$ can be represented as $x = \Sigma\lambda_i z^i$, with $\Sigma\lambda_i \geq 1$, $\lambda_i \geq 0$ $(i=1,...,n)$). The assumption $cx > 0 \ \ \forall x \in K \setminus int\ G$ implies that $cx(\lambda z^i) > cz^i$ $\forall \lambda > 1$, while the convexity of D implies that $[0,\omega(Q)] \subset D$. Therefore, if an optimal solution $\omega(Q)$ of (34) lies on the i–th edge of K, i.e., if it satisfies $\omega(Q) = \lambda z^i$ for some $\lambda \geq 1$, then necessarily $\omega(Q) = z^i$, and hence $f(\omega(Q)) = \alpha$. On the other hand, if $f(\omega(Q)) = \alpha$, then $\omega(Q) \in D \setminus int\ G$, and hence $\omega(Q)$ achieves the minimum of cx over $K \cap (D \setminus int\ G) \subset K \cap (D \cap H)$. ■

The algorithm we are going to describe is a branch and bound procedure similar to Algorithm IX.8*, in which branching is performed by means of conical subdivision and lower bounding is based on Proposition IX.15.

We start with a cone K_0 as in Algorithm X.9. Then for any subcone $K = con(Q)$ of this cone the condition $cx > 0 \ \forall x \in K \setminus int\ G$ in Proposition X.15 is satisfied. At any given stage, if $con(Q_k)$ denotes the cone chosen for further partition, then $\omega(Q_k)$ must be infeasible (otherwise, the exact minimum of cx over the feasible portion contained in this cone would be known, and Q would have been fathomed); hence, by Proposition X.15, $\omega(Q_k)$ does not lie on any edge of $con(Q_k)$ and can be used for further subdivision of $con(Q_k)$. We can thus state:

Algorithm IX.9*

Choose a rule for cone subdivision so as to generate an exhaustive process (cf. VII.1.4).

0) Construct a matrix $Q_0 = (z^{01},z^{02},...,z^{0n})$ as in Step 0 of iteration k = 0 of Algorithm X.9. Set $\mathcal{M}_0 = \{Q_0\}$, $\beta(Q_0) = -\infty$, $\gamma_0 = +\infty$ (or $\gamma_0 = cx^0$ if a feasible point x^0 is available). Set k = 0.

1) In $\mathcal{M}_k$ delete all Q such that $\beta(Q) \geq \gamma_k$. Let $\mathcal{R}_k$ be the remaining collection of matrices. If $\mathcal{R}_k = \emptyset$, terminate: x^k is a global optimal solution of (LRCP) if $\gamma_k < +\infty$; the problem is infeasible if $\gamma_k = +\infty$. Otherwise, if $\mathcal{R}_k$ is nonempty, go to 2).

2) Select $Q_k \in \operatorname{argmin} \{\beta(Q): Q \in \mathcal{R}_k\}$ and split it according to the subdivision process chosen.

3) Let $\mathcal{P}_k$ be the partition of Q_k so obtained. For each $Q \in \mathcal{P}_k$ solve (35) to obtain the optimal value $\beta(Q)$ and a basic optimal solution $\omega(Q)$ of (34).

4) Update the incumbent: set x^{k+1} equal to the best among: $x^k, u^k = u(Q_k)$ (if these points exist) and all $\omega(Q)$, $Q \in \mathcal{P}_k$, that are feasible. Set $\gamma_{k+1} = cx^{k+1}$. Set $\mathcal{M}_{k+1} = (\mathcal{R}_k \setminus \{Q\}) \cup \mathcal{P}_k$, $k \leftarrow k+1$ and go to 1).

Theorem IX.7*. *If Algorithm IX.9* is infinite, it generates an infinite sequence* $\omega^k = \omega(Q_k)$, *every accumulation point of which is a global optimal solution.*

Proof. If the algorithm is infinite, it generates at least one infinite nested sequence of cones $K_s = \operatorname{con}(Q_s)$, $s \in \Delta \subset \{0,1,...\}$, with $Q_s = (z^{s1},z^{s2},...,z^{sn})$ such that $f(z^{si}) = \alpha$ (i=1,2,...,n). By virtue of the subdivision rule, such a sequence shrinks to a ray; consequently, $z^{si} \longrightarrow x^*$ ($s \longrightarrow \infty$, $s \in \Delta$), for i=1,2,...,n.

Since ω^s belongs to the halfspace $\{x: eQ_s^{-1}x \geq 1\}$, the halfline from 0 through ω^s

meets the simplex $[z^{s1},...,z^{sn}]$ at some point v^s, and it meets the surface $f(x) = \alpha$ at some point y^s. Clearly, $v^s \to x^*$, $y^s \to x^*$. But we have $f(\omega^s) > \alpha$, for otherwise, by Proposition IX.15, ω^s would be feasible and $c\omega^s \geq \gamma_s$, i.e., $\beta(Q_s) \geq \gamma_s$, conflicting with the fact that $Q_s \in \mathcal{R}_s$.

On the other hand, since $f(\omega^s) \geq \alpha$, by the concavity of f(x), it follows that ω^s belongs to the line segment $[v^s,y^s]$. Hence, $\omega^s \to x^*$. Noting that $\omega^s \in D$ and $f(z^{si}) = \alpha$ $\forall s$, by letting $s \to \infty$ we then deduce that $x^* \in D$, $f(x^*) = \alpha$. Thus, x^* is feasible, proving that the lower bounding used in the algorithm is strongly consistent in the sense of Definition IV.7. It then follows by Corollary IV.3 that $\lim \beta(Q_s) = \lim \beta(Q_k) = \min \{cx: x \in D, f(x) \leq \alpha\}$, and, since $cx^* = \lim \beta(Q_s)$, we conclude that x^* is a global optimal solution.

(This can also be seen directly: for any feasible point x, if x belongs to one of the cones that have been deleted at some iteration $h \leq s$, then $cx \geq \gamma_h \geq \gamma_s > \beta(Q_s) = c\omega^s$, and hence $cx \geq cx^*$; if x belongs to some con(Q), $Q \in \mathcal{R}_s$, then $cx \geq \beta(Q) \geq \beta(Q_s) = c\omega^s$, and hence $cx \geq cx^*$). Now let $\bar{x}$ be an arbitrary accumulation point of the sequence $\{\omega^k = \omega(Q_k)\}$, e.g., $\bar{x} = \lim \omega^h$ $(h \to \infty, h \in H)$. It is easy to see that there exists a nested sequence $K_s = \text{con}(Q_s)$, $s \in \Delta$, such that any K_s contains infinitely many K_h, $h \in H$. Indeed, at least one of the cones con(Q), $Q \in \mathcal{R}_1$, contains infinitely many K_h: such a cone must be split at some subsequent iteration; let this cone be K_{s_1} for some $s_1 \geq 1$. Next, among the successors of K_{s_1} at least one contains infinitely many K_h: such a cone must, in turn, be split at some iteration $s_2 > s_1$; let this cone be K_{s_2}. Continuing in this way, we obtain an infinite nested sequence $\{K_s, s \in \Delta\}$ with the desired property, where $\Delta = \{s_1,s_2,...\}$. Since for every s there exists an h_s such that K_{h_s} is a descendant of K_s, because the sequence $\{K_s\}$ shrinks to a ray, it follows that $\omega(Q_{h_s}) \to x^*$, where $x^* = \lim \omega(Q_s)$ $(s \to \infty, s \in \Delta)$. That is, $\bar{x} = x^*$, and hence, by the above, $\bar{x}$ is a global optimal solution of (LRCP). ■

Remark IX.4. In the interest of efficiency of the procedure, one should choose a cone subdivision rule that involves mostly ω–subdivisions. It can be verified that the algorithm will still work if an arbitrary NCS rule is allowed.

Examples IX.4. Fig. IX.10 illustrates Algorithm IX.9 for a regular problem. In Step 1 a point ω^0 of D is found with $f(\omega^0) < \alpha$; hence the algorithm goes to 5). A forward step from s^0 then finds the optimal solution x*.

Fig. IX.11 illustrates Algorithm IX.9* for a nonregular problem. The algorithm generates a sequence of infeasible points $\omega^1, \omega^2, \ldots$ approaching the optimal solution x*, which is an isolated point of the feasible region.

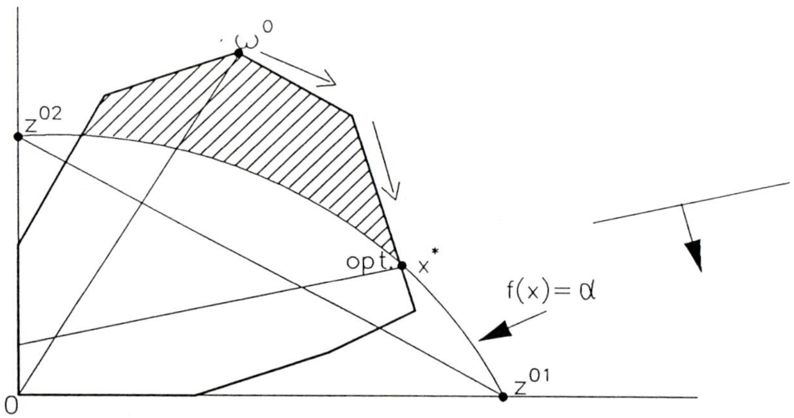

Fig. IX.10

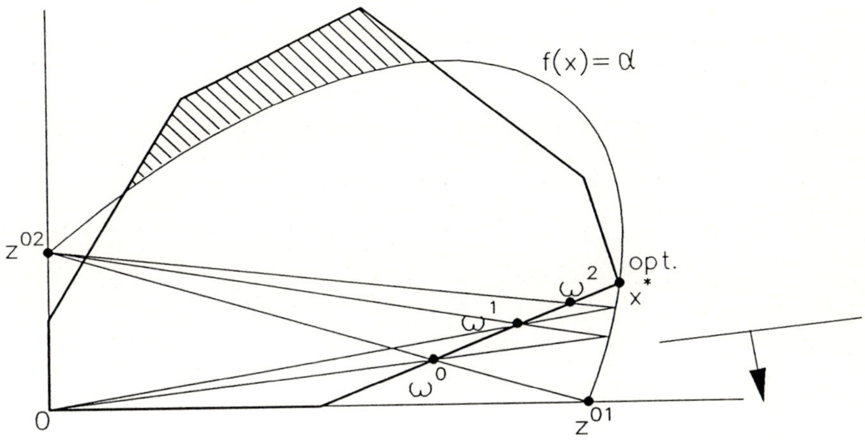

Fig. IX.11

PART C

GENERAL NONLINEAR PROBLEMS

Part C is devoted to the study of methods of solution for very general global optimization problems. Several outer approximation algorithms, branch and bound procedures and combinations thereof are developed for solving d.c. programming and Lipschitzian optimization problems. The "relief indicator method" may serve as a conceptual tool for even more general global problems. The applications that we discuss include design centering problems, biconvex programming, optimization problems with separable indefinite quadratic constraints and systems of equations and / or inequalities.

CHAPTER X

D.C. PROGRAMMING

In Chapter X, we continue the discussion of d.c. programming problems. First, a duality theory is developed between the objective and the constraints of a very general class of optimization problems. This theory allows one to derive several outer approximation methods for solving canonical d.c. problems and even certain d.c. problems that involve functions whose d.c. representations are not known. Then we present branch and bound methods for the general d.c. program and a combination of outer approximations and branch and bound. Finally, the design centering problem and biconvex programming are discussed in some detail.

1. OUTER APPROXIMATION METHODS FOR SOLVING THE CANONICAL D.C. PROGRAMMING PROBLEM

Recall from Chapter I that a d.c. programming problem is a global optimization problem of the form

$$\begin{aligned} &\text{minimize } f(x) \\ &\text{s.t. } x \in C,\ g_i(x) \le 0 \quad (i=1,\dots,m) \end{aligned} \tag{1}$$

where $C \subset \mathbb{R}^n$ is convex and all of the functions f, g_j are d.c. Suppose that C is defined by a finite system of convex inequalities $h_k(x) \le 0$, $k \in I \subset \mathbb{N}$. In Theorem I.9, it is shown that, by introducing at most two additional variables, every d.c. programming problem can be transformed into an equivalent ***canonical d.c. programming problem***

$$\text{(CDC)} \qquad \begin{array}{l} \text{minimize } cx \\ \text{s.t. } h(x) \le 0,\ g(x) \ge 0 \end{array}, \tag{2}$$

where $c \in \mathbb{R}^n$ (cx denotes the inner product), and where h and g: $\mathbb{R}^n \to \mathbb{R}$ are real-valued convex functions on $\mathbb{R}^n$.

In this section, an outer approximation method is presented for solving (CDC) that is based on Tuy (1987), Thoai (1988), Tuy and Thuong (1988). Moreover, it will be shown that the method can easily be extended to solve problems where in (2) the objective function is **convex** (cf. Tuy (1987).

1.1. Duality between the Objective and the Constraints

A general and simple duality principle allows one to derive an optimality condition for problem (CDC) (cf. Tichonov (1980)). We present a modification of the development given in Tuy (1987) and Tuy and Thuong (1988). Let D be an arbitrary subset of $\mathbb{R}^n$, and let f: $\mathbb{R}^n \to \mathbb{R}$, g: $\mathbb{R}^n \to \mathbb{R}$, $\alpha, \beta \in \mathbb{R}$. Consider the following pair of global optimization problems, in which the objective function of one is the constraint of the other, and vice versa:

$$(P_\beta) \qquad \inf\{f(x): x \in D,\ g(x) \ge \beta\}, \tag{3}$$

$$(Q_\alpha) \qquad \sup\{g(x): x \in D,\ f(x) \le \alpha\}. \tag{4}$$

Denote by inf P_β and sup Q_α the optimal values of (P_β) and (Q_α), respectively.

Definition X.1. *Problem (P_β) is said to be **stable** if*

$$\lim_{\beta' \to \beta+0} \inf P_{\beta'} = \inf P_\beta . \tag{5}$$

Similarly, problem (Q_α) *is stable if*

$$\lim_{\alpha' \to \alpha - 0} \sup Q_{\alpha'} = \sup Q_\alpha . \tag{6}$$

Lemma X.1. *(i) If* (Q_α) *is stable, then* $\alpha \le \inf P_\beta$ *implies* $\beta \ge \sup Q_\alpha$.
(ii) If (P_β) *is stable, then* $\beta \ge \sup Q_\alpha$ *implies* $\alpha \le \inf P_\beta$.

Proof. (i) Assume that (Q_α) is stable and $\alpha \le \inf P_\beta$. Then for all $\alpha' < \alpha$ the set $\{x \in D: g(x) \ge \beta, f(x) \le \alpha'\}$ is empty. Hence,

$$\sup Q_{\alpha'} := \sup \{g(x): x \in D, f(x) \le \alpha'\} \le \beta \quad \forall \alpha' < \alpha ,$$

and, letting $\alpha' \longrightarrow \alpha - 0$, we see from (6) that $\beta \ge \sup Q_\alpha$.

(ii) Similarly, if (P_β) is stable and $\beta \ge \sup Q_\alpha$, then for all $\beta' > \beta$ the set $\{x \in D: g(x) \ge \beta', f(x) \le \alpha\}$ is empty. Hence,

$$\inf P_{\beta'} := \inf \{f(x): x \in D, g(x) \ge \beta'\} \ge \alpha \quad \forall \beta' > \beta ,$$

and, letting $\beta' \longrightarrow \beta + 0$, we see from (5) that $\inf P_\beta \ge \alpha$. ■

Corollary X.1. *If both* (P_β) *and* (Q_α) *are stable, then*

$$\alpha \le \inf P_\beta \Longleftrightarrow \beta \ge \sup Q_\alpha .$$

Proposition X.1. *(i) If* (Q_α) *is stable and* $\alpha = \min P_\beta$, *then* $\beta = \sup Q_\alpha$.
(ii) If (P_β) *is stable and* $\beta = \max Q_\alpha$, *then* $\alpha = \inf P_\beta$.

Proof. (i) Since $\alpha = \min P_\beta$, there must exist an $\bar{x} \in D$ satisfying $f(\bar{x}) = \alpha$, $g(\bar{x}) \ge \beta$. It follows that $\beta \le \sup Q_\alpha$. But, by Lemma X.1, we know that $\beta \ge \sup Q_\alpha$.

(ii) Similarly, since $\beta = \max Q_\alpha$, we see that there exists $\bar{y} \in D$ satisfying $g(\bar{y}) = \beta$, $f(\bar{y}) \le \alpha$. Hence, $\inf P_\beta \le \alpha$, and Lemma X.1 shows that we must have $\inf P_\beta = \alpha$. ■

In order to apply the above results, it is important to have some criteria for checking stability of a given problem.

Lemma X.2. *If* $\inf P_\beta < +\infty$, *f is upper semicontinuous (u.s.c.) and* β *is not a local maximum of g over D, then* (P_β) *is stable. Similarly, if* $\sup Q_\alpha > -\infty$, *g is lower semicontinuous (l.s.c.) and* α *is not a local minimum of f over D, then* (Q_α) *is stable.*

Proof. We prove only the first assertion, since the second can be established in a similar way.

Suppose that $\inf P_\beta < +\infty$, f is u.s.c. and β is not a local maximum of g with respect to D. Then there is a sequence $\{x^k\} \subset D$ such that

$$g(x^k) \geq \beta, \quad f(x^k) \leq c_k, \quad c_k \downarrow \inf P_\beta ,$$

where $\{c_k\}$ is any sequence of real numbers having the above property. If, for some $k \in \mathbb{N}$ we have $g(x^k) > \beta$, then, obviously, for all $\beta' > \beta$ sufficiently close to β we also have $g(x^k) \geq \beta'$, and hence

$$\inf P_{\beta'} \leq f(x^k) \leq c_k.$$

It follows that

$$\lim_{\beta' \to \beta+0} \inf P_{\beta'} \leq c_k .$$

Therefore, since $c_k \downarrow \inf P_\beta$ we see that, if $g(x^k) > \beta$ holds for infinitely many k, then

$$\lim_{\beta' \to \beta+0} \inf P_{\beta'} \leq \inf P_\beta . \tag{7}$$

Since obviously $\inf P_{\beta'} \geq \inf P_\beta$ for all $\beta' \geq \beta$, we have equality in (7).

On the other hand, if $g(x^k) = \beta$ for all but finitely many k, then, since β is not a local maximum of g over D, it follows that for every k with $g(x^k) = \beta$ there is a sequence $x^{k,\nu} \xrightarrow[\nu]{} x^k$ such that $x^{k,\nu} \in D$, $g(x^{k,\nu}) > \beta$. Then for all β' sufficiently close to β we have $g(x^{k,\nu}) \geq \beta'$, and hence

$$\inf P_{\beta'} \leq f(x^{k,\nu})$$

and

$$\lim_{\beta' \to \beta+0} \inf P_{\beta'} \leq f(x^{k,\nu}) .$$

Letting $\nu \longrightarrow \infty$, we see from the upper semicontinuity of f that

$$\lim_{\beta' \to \beta+0} \inf P_{\beta'} \leq f(x^k) \leq c_k .$$

Finally, letting $k \longrightarrow \infty$, as above we see that

$$\lim_{\beta' \to \beta+0} \inf P_{\beta'} = \inf P_{\beta} .$$ ■

The following concept of regularity provides further insight into the notion of stability.

Definition X.2. *A feasible point $\bar{x}$ of (P_β) is said to be* **regular** *for (P_β) if every neighbourhood of $\bar{x}$ contains a point $x \in D$ satisfying $g(x) > \beta$ (i.e., there exists a sequence $x^k \longrightarrow \bar{x}$ such that $x^k \in D$, $g(x^k) > \beta$).*

Clearly, if the function g is l.s.c. (so that $\{x: g(x) > \beta\}$ is open), then every nonisolated point $\bar{x} \in D$ satisfying $g(\bar{x}) > \beta$ is regular for (P_β). Saying that β is not a local maximum of g over D (cf. Lemma X.2) is equivalent to saying that every point $\bar{x} \in D$ satisfying $g(\bar{x}) = \beta$ is regular.

The notion of a regular point for (Q_α) is defined in a similar way, replacing the condition $g(x) > \beta$ by $f(x) < \alpha$.

Note that the notion of a regular linear program with an additional reverse convex constraint in Definition IX.1 is related to Definition X.2 in an obvious way.

Proposition X.2. *If f is u.s.c. and if there exists at least one optimal solution of (P_β) that is regular for (P_β), then (P_β) is stable. Similarly, if g is l.s.c. and there exists at least one optimal solution of (Q_α) that is regular for (Q_α), then (Q_α) is stable.*

Proof. Let $\bar{x}$ be an optimal solution of (P_β) that is regular. Then there exists a sequence $\{x^k\} \subset D$ satisfying $g(x^k) > \beta$, $x^k \to \bar{x}$. For any fixed k we have $g(x^k) > \beta'$ for all β' sufficiently close to β; hence $\inf P_{\beta'} \le f(x^k)$. This implies that

$$\lim_{\beta' \to \beta+0} \inf P_{\beta'} \le f(x^k) .$$

But from the upper semicontinuity of f we know that

$$\overline{\lim}\, f(x^k) \le f(\bar{x}) = \inf P_\beta .$$

Therefore,

$$\lim_{\beta' \to \beta+0} \inf P_{\beta'} \le \inf P_\beta .$$

Since the reverse inequality is obvious, the first assertion in Proposition X.2 is proved.

The second assertion can be proved in an analogous way. ■

Now let us return to the canonical d.c. problem (CDC). Denote $G := \{x: g(x) \ge 0\}$, $H := \{x: h(x) \le 0\}$, and recall that g and h are convex functions. In (3) let $D = H$, $f(x) = cx$ and $\beta = 0$. Then problems (P_0) and (CDC) coincide. Assume that $H \cap G \neq \emptyset$.

Corollary X.2. *(i) Let H be bounded, and suppose that $g(x) \neq 0$ at every extreme point x of H. Then problem (CDC) is stable.*

(ii) If at least one optimal solution of (CDC) is regular, then problem (CDC) is stable.

Proof. (i): A local maximum of the convex function g with respect to H is always attained at an extreme point of H. Therefore, 0 cannot be a local maximum of g over H, and hence (CDC) is stable, by Lemma X.2.
(ii) Assertion (ii) follows from Proposition X.2. ■

As in Section I.3.4 assume that H is bounded, $H \cap G \neq \emptyset$, and the reverse convex constraint $g(x) \geq 0$ is essential, i.e., we have

$$\min \{cx: x \in H\} < \min \{cx: x \in H \cap G\} \tag{8}$$

(cf. Definition I.6). Then the following optimality criterion can be derived from the above considerations. Recall from Theorem I.10 that an optimal solution is attained on $\partial H \cap \partial G$.

Theorem X.1. *In problem (CDC) let H be bounded, $H \cap G \neq \emptyset$ and $g(x) \geq 0$ be essential.*
For a point $\bar{x} \in \partial G$ to be an optimal solution to problem (CDC) it is necessary that

$$\max \{g(x): x \in H, cx \leq c\bar{x}\} = 0 . \tag{9}$$

This condition is also sufficient if problem (CDC) is stable.

Proof. Let $f(x) = cx$, $D = H$, $\beta = 0$, $\alpha = \min \{cx: x \in H \cap G\}$, and consider the problems (Q_α) and $(P_0) = $ (CDC). The convex function g is continuous and it follows from (8) that α is not a local minimum of $f(x) = cx$ over H. Hence from the second part of Lemma X.2 we see that (Q_α) is stable. Therefore, if $\bar{x}$ is optimal, i.e., $c\bar{x} = \alpha$, then, by Lemma X.1(i),

$$0 \geq \max \{g(x): x \in H, cx \leq c\bar{x}\}. \tag{10}$$

But since $g(\bar{x}) = 0$, $\bar{x} \in H$, it follows that in (10) we have equality, i.e., (9) holds.

Conversely, if (CDC) is stable and (9) holds for $\bar{x} \in \partial G$, then, by Proposition X.1

(ii), we have

$$\alpha = c\bar{x} = \inf \{cx: x \in H \cap G\} = \min \{cx: x \in H \cap G\} ,$$

since (CDC) has an optimal solution. ■

1.2. Outer Approximation Algorithms for Canonical D.C. Problems

As above consider the problem (CDC)

$$\begin{array}{l} \text{minimize } cx \\ \text{s.t. } x \in H \cap G \end{array} , \tag{11}$$

where $c \in \mathbb{R}^n$, $H:= \{x: h(x) \le 0\}$, $G:= \{x: g(x) \ge 0\}$ with $h,g: \mathbb{R}^n \longrightarrow \mathbb{R}$ convex. Assume that

(a) $\quad$ $\text{int } H = \{x: h(x) < 0\} \neq \emptyset$,

(b) $\quad$ H is bounded,

(c) $\quad$ the reverse convex constraint is essential.

A straightforward application of the development presented in Section X.1.1. leads to a sequence of concave minimization problems. In addition to (a), (b) and (c) assume that $H \cap G \neq \emptyset$, so that an optimal solution to (11) exists. Let $w \in \text{int } H$ be a point satisfying

$$g(w) < 0 , \tag{12}$$

$$cw < \min \{cx: x \in H \cap G\} . \tag{13}$$

Note that w is readily available by assumptions (a), (b), (c) (cf. Section I.3.4). For example, solve the convex minimization problem

$$\begin{array}{ll} \text{minimize} & cx \\ \text{s.t.} & x \in H \end{array}$$

obtaining an optimal solution $\overline{w}$. If $\overline{w} \in H \cap G$, then assumption (c) above does not hold, and $\overline{w}$ is an optimal solution to (11). Hence, by assumption (c), we have $g(\overline{w}) < 0$, $c\overline{w} < \min \{cx: x \in H \cap G\}$, where $\overline{w}$ is a boundary point of H. Then $w \in \text{int } H$ can be found by a small perturbation of $\overline{w}$.

For every $x \in G$, let $\pi(x)$ denote the point where the line segment $[w,x]$ intersects the boundary ∂G of G. Since g is convex, and since $g(w) < 0$, while $g(x) \geq 0$, it is clear that

$$\pi(x) = tx + (1-t)w ,$$

with $t \in (0,1]$ is uniquely determined by an univariate convex minimization problem (cf. Chapter II.).

Note that for $x \in H \cap G$ it follows from (12) and (13) that $c\pi(x)=tcx+(1-t)cw<cx$.

Algorithm X.1.

Initialization:

Determine a point $x^1 \in H \cap \partial G$. Set $k \leftarrow 1$.

Iteration k=1,2,...:

Solve the subproblem

$$(Q(x^k)) \qquad \begin{array}{ll} \text{maximize} & g(x) \\ \text{s.t.} & x \in H, \ cx \leq cx^k \end{array} \tag{14}$$

obtaining an optimal solution z^k of $(Q(x^k))$. If $g(z^k) = 0$, then stop.
Otherwise, set $x^{k+1} = \pi(z^k)$, and go to iteration k+1.

Remarks X.1. (i) A point $x^1 \in H \cap \partial G$ can be determined by running an algorithm to solve the convex maximization problem max $\{g(x): x \in H\}$ until a point $y^1 \in H$ has been found satisfying $g(y^1) \geq 0$. Set $x^1 = \pi(y^1)$.

(ii) Also note that x^k satisfies $g(x^k) = 0$ for all k, so that $g(z^k) \geq 0$.

Proposition X.3. *Assume that problem (CDC) is stable. Then the following assertions hold.*

(i) If Algorithm X.1 stops at z^k, then x^k is an optimal solution of (CDC).

(ii) If Algorithm X.1 is infinite, then it generates a sequence $\{x^k\} \subset H \cap \partial G$, every accumulation point of which is an optimal solution to (CDC).

Proof. If the algorithm stops at z^k, then we have $g(z^k) = \max \{g(x): x \in H, cx \leq cx^k\} = 0$, and x^k satisfies the necessary and sufficient optimality condition (9) in Theorem X.1.

Suppose that Algorithm X.1 is infinite. Then, since $H \cap \partial G$ is compact, the sequence $\{x^k\} \subset H \cap \partial G$ has accumulation points in $H \cap \partial G$. Let $\{x^{k_q}\}$ be a subsequence of $\{x^k\}$ such that

$$\bar{x} = \lim_{q \to \infty} x^{k_q} .$$

Since the sequences $\{x^k\}$ and $\{z^k\}$ are bounded, we may, by considering a subsequence if necessary, assume that $x^{k_q+1} \xrightarrow[q]{} \tilde{x}$, $z^{k_q} \xrightarrow[q]{} z$. Clearly, since $c\pi(x) < cx$ for all $x \in H \cap G$, it follows that $cx^{k+1} < cx^k$ for all k. Moreover, for $x \in H$ satisfying $cx \leq c\bar{x}$, we have $cx \leq cx^k$, k=1,2,... But, by the definition of z^{k_q} in Algorithm X.1, it follows that $g(x) \leq g(z^{k_q})$; and hence, letting $q \longrightarrow \infty$: $g(x) \leq g(z)$. Since $z \in H$ and $cz \leq c\bar{x}$, we thus see that z is an optimal solution of the subproblem $(Q(\bar{x}))$.

Now suppose that $g(z) > 0$. Since $x^{k_q+1} = \pi(z^{k_q})$, it is easily seen (from the definition of $\pi(x)$) that $\tilde{x} = \pi(z)$. Thus,

$$\tilde{x} = tz + (1-t)w \quad , \quad 0 < t < 1 \; ;$$

hence, using (13),

$$c\tilde{x} = tcz + (1-t)cw < tcz + (1-t)cz = cz \; . \tag{15}$$

But

$$cx^{k_{q+1}} \le cx^{k_q+1} < cz^{k_q} \le cx^{k_q} , \tag{16}$$

where the strict inequality holds, by an argument similar to that used to derive (15). If we let $q \longrightarrow \infty$, then (16) yields

$$c\bar{x} \le c\tilde{x} \le cz \le c\bar{x} ,$$

contradicting (15). Therefore, $g(z) = 0$; hence (9) is satisfied, and Proposition X.4 is established by virtue of Theorem X.1. ■

Remark X.2. Theorem X.1 and Proposition X.3 remain true if we replace the linear function cx by a **convex function f(x)**. Hence, Algorithm X.1 can be used to minimize a convex function f(x) subject to convex and reverse convex constraints (cf. Tuy (1987)).

Note that each subproblem $(Q(x^k))$ is a difficult global optimization problem that cannot be assumed to be solved in a finite number of iterations, since its feasible set is not polyhedral.

Therefore, we propose a second algorithm that can be interpreted as an outer approximation method for solving

$$\max \{g(x) \colon x \in H \; , \; cx \le \alpha\} \; ,$$

where $\alpha = \min \{cx \colon x \in H \cap G\}$ is the unknown optimal value of (CDC).

Denote by $V(D_k)$ the vertex set of a polytope D_k, and let $\partial f(x)$ denote the set of subgradients of a convex function f at x.

Algorithm X.2.

Initialization:

Compute a feasible point $x^1 \in H \cap \partial G$. Set $\alpha_1 = cx^1$. Generate a polytope D_1 containing the compact convex set $\{x \in H: cx \leq \alpha_1\}$. Let $k \leftarrow 1$.

Iteration k = 1,2,...:

Solve the subproblem

$$(Q_k) \qquad \max \{g(x): x \in D_k\} = \max \{g(x): x \in V(D_k)\} .$$

Let z^k be an optimal solution for (Q_k).

If $g(z^k) = 0$, then stop.

Otherwise, determine the point y^k where the line segment $[w,z^k]$ intersects the surface

$$\max \{cx - \alpha_k, g(x)\} = 0 . \tag{17}$$

(a) If $y^k \in H$ (i.e., $h(y^k) \leq 0$), then set

$$\ell_k(x) = c(x - y^k). \tag{18}$$

(b) If $y^k \notin H$ (i.e., $h(y^k) > 0$), then choose $p^k \in \partial h(y^k)$ and set

$$\ell_k(x) = p^k(x - y^k) + h(y^k) . \tag{19}$$

Let

$$D_{k+1} = D_k \cap \{x: \ell_k(x) \leq 0\} . \tag{20}$$

Set

$$x^{k+1} = y^k, \text{ if } y^k \in H \text{ and } g(y^k) = 0,$$

$$x^{k+1} = x^k, \text{ otherwise.}$$

Set $\alpha_{k+1} = cx^{k+1}$ and go to iteration k+1.

We establish convergence of this algorithm under the assumptions (a), (b), (c) stated at the beginning of Section X.1.2.

Observe that for each k=1,2,..., x^k is the best feasible point obtained until step k, while $\alpha_k = cx^k$ is the corresponding objective function value, $\alpha_k \ge \alpha_{k+1}$.

Lemma X.3. *For every k, we have*

$$D_k \supset \{x \in H: cx \le \alpha_k\} .$$

Proof. Let $x \in H$, $cx \le \alpha_k$. Since $\alpha_k \le \alpha_1$ and $\{y \in H: cy \le \alpha_1\} \subset D_1$, it follows that $x \in D_1$. Furthermore, if $i < k$ and $y^i \in H$, then $\ell_i(x) = c(x - y^i) = cx - cy^i \le cx - \alpha_k \le 0$. If $y^i \notin H$, then $p^i \in \partial h(y^i)$, and, by the definition of a subgradient, we have $\ell_i(x) = p^i(x - y^i) + h(y^i) \le h(x) \le 0$. Therefore, $x \in D_k$. ■

Lemma X.4. *(i) If $g(z^k) = 0$, then*

$$0 = max \{g(x): x \in H, cx \le \alpha_k\} . \tag{21}$$

(ii) If the sequence $\{z^k\}$ has an accumulation point $\bar{z}$ satisfying $g(\bar{z}) = 0$, then

$$0 = max \{g(x): x \in H, cx \le \bar{\alpha}\} , \tag{22}$$

where $\bar{\alpha} = inf \{\alpha_k: k = 1,2,...\}$.

Proof. (i) From $\alpha_k \ge \alpha := \min \{cx: x \in H, g(x) \ge 0\}$ and Lemma X.3 we see that

$$D_k \supset \{x \in H: cx \le \alpha_k\} \supset \{x \in H: cx \le \alpha\} . \tag{23}$$

By Theorem X.1, we have

$$0 = \max \{g(x): x \in H, cx \le \alpha\} , \tag{24}$$

by hypothesis,

$$g(z^k) = \max \{g(x): x \in D_k\} = 0 ,$$

and (21) follows in view of (23).

(ii) To prove (22), observe that

$$g(z^k) \geq g(x) \quad \forall x \in D_k .$$

Hence, using the continuity of g, we have

$$g(\bar{z}) \geq \sup \{g(x): x \in \bigcap_{k=1}^{\infty} D_k\} . \tag{25}$$

But (23) implies that

$$\{x \in H: cx \leq \alpha\} \subset \{x \in H: cx \leq \bar{\alpha}\} \subset \bigcap_{k=1}^{\infty} D_k . \tag{26}$$

Since $g(\bar{z}) = 0$, the relations (24) – (26) together imply (22). ■

Lemma X.5. *Assume that problem (CDC) is stable, and let $\tilde{\alpha} \in \mathbb{R}$ satisfy*

$$0 = max \{g(x): x \in H, cx \leq \tilde{\alpha}\}. \tag{27}$$

Then

$$\tilde{\alpha} \leq \alpha := min \{cx: x \in H, g(x) \geq 0\} .$$

Proof. From (27) we see that for any number $\alpha' < \tilde{\alpha}$ we have

$$0 \geq \max \{g(x): x \in H, cx \leq \alpha'\} .$$

Since (CDC) is stable, it follows by Lemma X.1(ii) that $\alpha' \leq \alpha$, and hence $\tilde{\alpha} \leq \alpha$. ■

We need the last lemma to establish the convergence of Algorithm X.2. The next lemma gives a convergence principle for outer approximation methods which is similar to Theorem II.1.

Lemma X.6. *Let A be an arbitrary nonempty set in $\mathbb{R}^n$, and let $\{z^k\}$ be a bounded sequence in $\mathbb{R}^n$. Moreover, assume that for every k=1,2,... there is a continuous* *function* $\ell_k: \mathbb{R}^n \longrightarrow \mathbb{R}$ *such that the following conditions hold.*

(1) $\ell_j(z^k) \le 0$, *for all* $j < k$;

(2) $\ell_k(z^k) > 0$;

(3) *for each convergent subsequence* $\{z^r\} \subset \{z^k\}$ *satisfying* $z^r \xrightarrow[r]{} \bar{z}$, *there is a subsequence* $\{z^q\}$ *such that* $\ell_q(z^q) \xrightarrow[q]{} \ell(\bar{z})$, *where* $\ell: \mathbb{R}^n \longrightarrow \mathbb{R}$,

and

(4) $\ell(\bar{z}) = 0$ *implies that* $\bar{z} \in A$.

Then every accumulation point $\bar{z}$ *of* $\{z^k\}$ *is in* A.

Proof. Lemma X.6 can be verified in a manner similar to Theorem II.1. ■

Theorem X.2. *Assume that the conditions (a), (b), (c) are fulfilled and that problem (CDC) is stable. If Algorithm X.2 terminates at iteration k, then* x^k *is an optimal solution for problem (CDC).*
If the algorithm is infinite, then every accumulation point $\bar{x}$ *of the sequence* $\{x^k\}$ *is an optimal solution for problem (CDC).*

Proof. If $g(z^k) = 0$, then we see from Lemma X.4 (i) that

$$0 = \max \{g(x): x \in H, cx \le cx^k\} ,$$

and, by Theorem X.1, x^k is an optimal solution for problem (CDC).

In order to prove the second part of Theorem X.2, we shall show that any accumulation point $\bar{z}$ of the sequence $\{z^k\}$ satisfies $g(\bar{z}) = 0$. Then, in view of Lemma X.4 (ii), we have

$$0 = \max \{g(x): x \in H, cx \le c\bar{x}\} ,$$

and the equality $c\bar{x} = \alpha = \min \{cx: x \in H, g(x) \ge 0\}$ follows from Lemma X.5, since

$\bar{x}$ is feasible for (CDC).

The assertion $g(\bar{z}) = 0$ will be established by checking that all of the conditions of Lemma X.6 are fulfilled for

$$A = \partial G = \{z: g(z) = 0\} .$$

Condition (1) of Lemma X.6 obviously holds. To verify condition (2), observe that

$$z^k = y^k + \lambda_k(y^k - w) , \lambda_k > 0 . \tag{28}$$

If ℓ_k is of the form (18), then from (28) and the inequality $cw < \min \{cx: x \in H \cap G\}$ we deduce that

$$\ell_k(z^k) = c(z^k - y^k) = \lambda_k c(y^k - w) > 0 .$$

If ℓ_k is of the form (19), then observe that from the definition of a subgradient and the inequality $h(w) < 0$ we have

$$\ell_k(w) = p^k(w - y^k) + h(y^k) \le h(w) < 0 . \tag{29}$$

Using (28), (29) and $\ell_k(y^k) = h(y^k) > 0$, we see that

$$\ell_k(z^k) = \ell_k(y^k) + \lambda_k(\ell_k(y^k) - \ell_k(w)) > 0 .$$

Now let $\bar{z}$ be an accumulation point of $\{z^k\}$. Then without loss of generality we may assume that one of the following cases occurs for a subsequence $z^q \longrightarrow \bar{z}$.

(a) Suppose that $y^q \in H$ for all q. Then ℓ_q has the form (18); and, since $z^q, y^q \in D_1$ and D_1 is compact, we may (by passing to a subsequence if necessary) assume that $y^q \longrightarrow \bar{y}$, $\lambda_q \longrightarrow \bar{\lambda}$. Therefore, by (18) and (28), we have

$$\bar{z} = \bar{y} + \bar{\lambda} (\bar{y} - w) , \bar{\lambda} \ge 0 \tag{30}$$

and

$$\ell(x) = c(x - \bar{y}) .$$

Let

$$\ell(\bar{z}) = c(\bar{z} - \bar{y}) = \bar{\lambda}c(\bar{y} - w) = 0. \tag{31}$$

As above, from the definition of w we deduce that $c\bar{y} > cw$. Therefore, we have $\bar{\lambda} = 0$ and $\bar{z} = \bar{y}$. But for all q we have $g(z^q) \geq 0$, $g(y^q) \leq 0$, by the construction of the algorithm. Therefore, $\bar{z} = \bar{y}$ implies that $g(\bar{z}) = 0$.

(b) Suppose that $y^q \notin H$ for all q. Then ℓ_q has the form (19); and, as before, we may assume that $y^q \longrightarrow \bar{y}$, $\lambda_q \longrightarrow \bar{\lambda}$. Moreover, we may also assume that $p^q \longrightarrow \bar{p} \in \partial h(\bar{y})$ (cf. Rockafellar (1970), Chapter 24). It follows that

$$\ell(x) = \bar{p}(x - \bar{y}) + h(\bar{y}) .$$

Clearly, $\ell(w) = \bar{p}(w - \bar{y}) + h(\bar{y}) \leq h(w) < 0$ and $\ell(\bar{y}) = h(\bar{y}) \geq 0$, since $h(y^q) > 0$.

As above, from (30) we conclude that then

$$\ell(\bar{z}) = \ell(\bar{y}) + \bar{\lambda}(\ell(\bar{y}) - \ell(w)) = 0$$

is only possible for $\bar{\lambda} = 0$. But this implies that $\bar{z} = \bar{y}$ and $g(\bar{z}) = 0$. ■

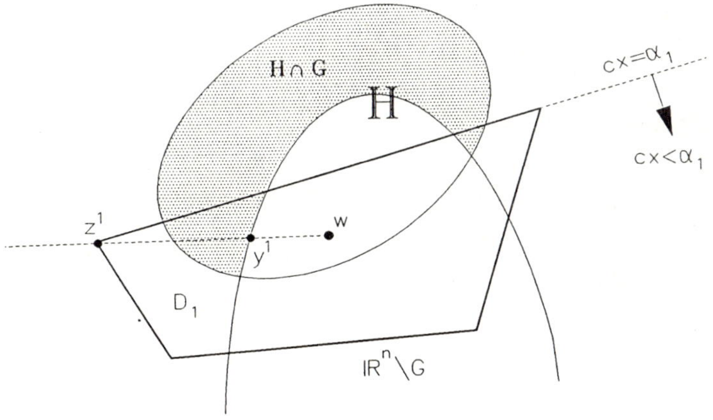

Fig. X.1. Algorithm X.2

Remark X.3. Algorithm X.2 is formulated in a manner that allows us to extent it in a straightforward way to the more general case of minimizing a **convex function** f(x) over $H \cap G$. For that purpose we only have to start with a polytope $D_1 \supset H$, and to replace equation (17) by $\max \{f(x) - \alpha_k, g(x)\} = 0$ and $\ell_k(x) = c(x-y^k)$ in (18) by $\ell_k) = t^k(x-y^k)$, where $t^k \in \partial f(y^k)$ (cf. Remark X.2 and Tuy (1987)).

Remark X.4. In the case of linear objective function (for which Algorithm X.2 is formulated) it is easy to say that certain simplifications can be built in. For example, if (18) occurs, i.e. if a cut $cx \leq cy^k$ is added, then obviously all previous cuts of the form $cx \leq cy^i$, $i < k$, are redundant and can be omitted. Moreover, if the initial polytope D_1 is contained in the halfspace $\{x \in \mathbb{R}^n: cx \leq \alpha_1\}$, i.e., if $cx \leq \alpha_1$ is among the constraints which define D_1, then equation (17) can be replaced by $g(x) = 0$.

Remark X.5. In practice, given a tolerance $\varepsilon > 0$, we terminate when

$$g(z^k) < \varepsilon \,. \tag{32}$$

Since any accumulation point $\bar{z}$ of the sequence $\{z^k\}$ satisfies $g(\bar{z}) = 0$, (32) must occur after finitely many iterations. Suppose that

$$\gamma = \max \{g(x): x \in H\} > 0 \,, \tag{33}$$

and $\varepsilon < \gamma$. Then we have

$$\max \{g(x): x \in H, cx \leq \alpha_k\} \leq g(z^k) < \varepsilon < \gamma \,,$$

which, because of (33), implies that $\alpha_k < +\infty$, i.e., there is a point $x^k \in H$ satisfying $g(x^k) = 0$, $cx^k = \alpha_k$. Furthermore, there is no $x \in H$ such that $cx \leq \alpha_k$, $g(x) \geq \varepsilon$. Hence,

$$cx^k = \alpha_k < \min \{cx: x \in H, g(x) \geq \varepsilon\} \,. \tag{34}$$

Therefore, with the stopping rule (32), where $\varepsilon < \gamma$, the algorithm is finite and provides an ε-optimal solution in the sense of (34). Note that this assertion is valid no matter whether or not the problem is stable.

In Thoai (1988), a modification of Algorithm X.2 is proposed which does not require the calculation neither of w nor of an initial feasible point $x^1 \in H \cap \partial G$ and which detects whether we have $H \cap G = \emptyset$. Moreover, preliminary numerical tests indicate that this modification is quite efficient.

Algorithm X.2'.

Initialization:

Generate a polytope $D_1 \supset H$ and its vertex set $V(D_1)$. Set $k \leftarrow 1$.

Iteration k=1,2,...:

1.) Compute $z^k \in \arg\min \{cx: x \in V(D_k)\}$.

1.a.) If $h(z^k) \le 0$ and $g(z^k) \ge 0$, then stop.

1.b.) If $h(z^k) > 0$ and $g(z^k) \ge 0$, then set $y^k = z^k$, go to Step 3.

2.) Compute $x^k \in \operatorname{argmax} \{g(x): x \in V(D_k)\}$.

2.a.) If $g(x^k) = 0$, and there is a feasible point v which was computed in Step 2.c. at some iteration k', k' < k, then stop.

2.b.) If $g(x^k) < 0$, then stop.

2.c.) Otherwise, compute the point u^k where the line segment $[z^k, x^k]$ intersects the boundary ∂G of G.

If $h(u^k) > 0$, then set $y^k = u^k$ and go to Step 3.

If $h(u^k) \le 0$ and $cu^k = cz^k$, then stop;

else, set

$$v = u^k \tag{35}$$

$$\ell_k(x) = cx - cv \tag{36}$$

and go to Step 4.

3.) Compute $p^k \in \partial h(y^k)$ and set

$$l_k(x) = p^k(x - y^k) + h(y^k) . \tag{37}$$

4.) Generate the polytope

$$D_{k+1} = D_k \cap \{x: l_k(x) \le 0\} ;$$

compute its vertex set $V(D_{k+1})$, set $k \leftarrow k+1$ and go to Step 1.

Proposition X.4. *If for some k Algorithm X.2' terminates at step 1.a, then z^k is an optimal solution of problem (CDC). If the algorithm terminates at step 2.c., then u^k is an optimal solution of problem (CDC).*

Proof. In a manner similar to the proof of Lemma X.3, we see that for $k \ge 1$ we have $D^1 \supset D^k \supset \{x \in H: cx \le \alpha\}$, where α is the optimal value of problem (CDC). Therefore, if $z^k \in H \cap G$, i.e., if z^k is feasible for problem (CDC), then z^k is an optimal solution.

If the algorithm terminates at step 2.c, then we have $cu^k = cz^k = \min \{cx: x \in D_k\} \le \alpha$ and u^k is feasible, and hence optimal. ∎

Proposition X.5. *If Algorithm X.2' is infinite, and if an infinite subsequence $\{z^q\}$ is generated such that $g(z^q) \ge 0$ for each q, then every cluster point $\bar{z}$ of $\{z^q\}$ is an optimal solution of problem (CDC).*

Proof. From the general theory of outer approximation methods as developed in Chapter II, and by reasons similar to those given in the proof of Lemma X.3, it follows that $\bar{z}$ solves the problem $\min \{cx: x \in H\}$. But $g(z^q) \ge 0$ for every q, and hence $g(\bar{z}) \geq 0$. This implies that $\bar{z}$ is an optimal solution of problem (CDC). ■

Proposition X.6. *Assume that problem (CDC) is stable.*

If for some k the algorithm terminates at Step 2.a, then the point v is an optimal solution of problem (CDC).

If an infinite subsequence $\{u^q\}$ is generated, then every cluster point $\bar{u}$ of $\{u^q\}$ is an optimal solution.

Proof. Proposition X.5 can be verified in a way similar to Theorem X.2. ■

Proposition X.7. *The algorithm terminates at Step 2.b for some k if and only if problem (CDC) has no solution.*

Proof. If $H \cap G = \emptyset$, then the algorithm must terminate after a finite number of iterations at Step 2.b. Otherwise, it would either generate a subsequence $\{z^q\}$ satisfying $g(z^q) \geq 0$ for every q, which, according to Proposition X.5, converges to a feasible point; or else it would generate a subsequence $\{u^q\}$ which converges to a feasible point by Proposition X.6.

Conversely, when the algorithm terminates at iteration k, Step 2.b, then we have $\max \{g(x): x \in D_k\} < 0$. This implies that problem (CDC) has no solution, because every feasible point $\tilde{x}$ must satisfy $\max \{g(x): x \in D_k\} \geq g(\tilde{x}) \geq 0$. ■

Example X.1. Consider the following d.c. problem in $\mathbb{R}^3$ (Thoai (1988)):

$$
\begin{aligned}
\text{minimize} \quad & (x_1^4 + x_2 + x_3) - (x_1 + x_2^2 - x_3) \\
\text{s.t.} \quad & x_1 \geq 1.4 , \\
& x_2 \geq 1.6 , \\
& x_3 \geq 1.8 , \\
& x_1 + x_2 + x_3 \leq 6.5 , \\
& (x_1 - x_2 - 1.2)^2 + x_3 \leq 4.4 .
\end{aligned}
$$

By introducing the additional variables x_4, x_5, we obtain an equivalent problem in $\mathbb{R}^5$:

$$\begin{aligned}
&\text{minimize } x_4 \\
&\text{s.t.} \quad h_1(x) = -x_1 + 1.4 \leq 0\,, \\
&\qquad h_2(x) = -x_2 + 1.6 \leq 0\,, \\
&\qquad h_3(x) = -x_3 + 1.8 \leq 0\,, \\
&\qquad h_4(x) = x_1 + x_2 + x_3 - 6.5 \leq 0\,, \\
&\qquad h_5(x) = (x_1 - x_2 - 1.2)^2 + x_3 - 4.4 \leq 0\,, \\
&\qquad h_6(x) = x_1^4 + x_2 + x_3 - x_5 \leq 0\,, \\
&\qquad g(x) = (x_1 + x_2^2 - x_3) + x_4 - x_5 \geq 0\,.
\end{aligned}$$

This is equivalent to the canonical form

$$\begin{aligned}
&\text{minimize } x_4 \\
&\text{s.t.} \quad h(x) \leq 0,\ g(x) \geq 0
\end{aligned}$$

where $h(x) = \max\{h_i(x)\colon i = 1,\dots,6\}$.

It is easily seen that the feasible set $H \cap G$ is contained in $\mathbb{R}^5_+$, and a solution exists. We start the algorithm with the simplex

$$D_1 = \{x \in \mathbb{R}^5_+\colon \sum_{i=1}^{5} x_i \leq 100\}\,.$$

Iteration 1:

$z^1 = (0.0,0.0,0.0,0.0,0.0)$; $h(z^1) = 1.8$; $g(z^1) = 0.0$;

cutting plane $l(x) = -x_3 + 1.8 \leq 0$.

Iteration 2:

$z^2 = (0.0, 0.0, 100.0, 0.0, 0.0)$; $h(z^2) = 100.0$; $g(z^2) = -100.0$;

$x^2 = (0.0, 98.2, 1.8, 0.0, 0.0)$; $g(x^2) = 9641.44$;

$u^2 = (0.0, 9.51, 90.49, 0.0, 0.0)$;

cutting plane $l(x) = -21.43\, x_1 + 21.43\, x_2 + x_3 - 93.45 \leq 0$.

After 9 iterations the feasible point $v = u^9 = (2.14, 1.6, 1.8, 45.78, 48.68)$ is found; cutting plane $l(x) = x_4 - 45.78 \le 0$.

An optimal solution is found after 81 iterations: $x^* = (1.4001, 1.8124, 1.800, 4.5719, 7.4569)$. The first three components of x^* constitute the solution vector of the original problem, and $x_4^* = 4.5719$ is the optimal value.

1.3. Outer Approximation for Solving Noncanonical D.C. Problems

Transforming a general d.c. problem to a canonical d.c. program (or to a similar program with convex objective function) requires the introduction of additional variables. Moreover, the functions h and g in a canonical d.c. program that is derived from a general d.c. problem can be quite complicated (cf. Section I.3.4).

Even more important, however, is the fact that, in order to apply one of the approaches to solve d.c. programs discussed so far, we must know a d.c. representation of the functions involved. Although such a representation is readily available in many interesting cases (cf. Section I.3), very often we know that a given function is d.c., but we are not able to find one of its d.c. representations.

A first attempt to overcome the resulting difficulties was made by Tuy and Thuong (1988), who derived a conceptual outer approximation scheme that is applicable for certain noncanonical d.c. problems, and even in certain cases where a d.c. representation of some of the functions involved is not available.

Consider the problem

$$(P) \qquad \begin{array}{ll} \text{minimize } f(x) & \\ \text{s.t. } g_i(x) \ge 0 & (i=1,\dots,m) \end{array}$$

where f is a finite convex function on $\mathbb{R}^n$ and g_i $(i=1,\dots,m)$ are continuous functions in a large class satisfying the assumption below.

Let us set

$$C_i = \{x \in \mathbb{R}^n : g_i(x) < 0\} \quad (i=1,\ldots,m) \; ; \tag{38}$$

$$g(x) = \min_{i=1,\ldots,m} g_i(x); \quad C = \{x \in \mathbb{R}^n : g(x) < 0\} = \bigcup_{i=1}^{m} C_i \, .$$

With this notation, problem (P) asks us to find the global minimum of f(x) over the complement of the set C.

Suppose that the following assumptions are fulfilled:

(i) *The finite convex function f has bounded level sets* $\{x \in \mathbb{R}^n : f(x) \le j\}$, *and the functions* g_i *are everywhere continuous;*

(ii) $\bar{\alpha} := \min \{f(x) : g(x) \ge 0\}$ *exists;*

(iii) *a point* $w \in \mathbb{R}^n$ *is available satisfying*

$$f(w) < \bar{\alpha} \; ; \; g(w) < 0 \; ;$$

(iv) *for any* $z \in \mathbb{R}^n$, *one can compute the point* $\pi(z)$ *nearest to* w *in the intersection of the line segment* $[w,z]$ *with the boundary* ∂C *of* C, *or else establish that such a point does not exist.*

Assumptions (i) and (ii) are self–explanatory. Assumption (iii) can be verified by solving the unconstrained convex program

$$\begin{array}{ll} \text{minimize} & f(x) \\ \text{s.t.} & x \in \mathbb{R}^n \end{array} .$$

If an optimal solution of this convex program exists satisfying $g(x) \ge 0$, then it will solve problem (P). Otherwise, a point w as required is available.

Assumption (iv) constitutes an essential restriction on the class of constraint functions that we admit. Since, by assumption (i), the functions $g_i(x)$ are continuous, the set of feasible points lying in a line segment $[w,z]$ is compact. This set does not contain w. Therefore, whenever nonempty, it must have an element nearest to w. It is easily seen that such a point must lie on ∂C, and is just $\pi(z)$. Assumption (iv) requires that an algorithm exist to compute $\pi(z)$.

For instance, if each of the functions $g_i(x)$ $(i=1,...,m)$ is convex, then $\pi(z)$ can be computed in the following way:

Initialization:

Set $w^1 = w$, $I_1 = \{1,...,m\}$.

Step k=1,2,...:

Choose $i_k \in \text{argmin}\,\{g_i(w^k): i \in I_k\}$.

If $g_{i_k}(z) < 0$, then stop: there is no feasible point on the line segment $[w,z]$ because $g_i(x) < 0\ \ \forall x \in [w,z]$.

Otherwise, compute the point $w^{k+1} \in [w,z]$ satisfying $g_{i_k}(w^{k+1}) = 0$. (Since C_{i_k} is convex, this point is unique and can be easily determined.)

Set $I_{k+1} = I_k \setminus \{i_k\}$. If $\min\,\{g_i(w^{k+1}): i \in I_{k+1}\} \geq 0$, then stop: $\pi(z) = w^{k+1}$.

Otherwise, go to Step k+1.

Clearly, after at most m steps this procedure either finds $\pi(z)$ or else establishes that $\pi(z)$ does not exist.

It is also not difficult to derive a procedure to determine $\pi(z)$ for other classes of functions $g_i(x)$, for example, when all of the $g_i(x)$ are piecewise affine (this will be left to the reader).

Proposition X.8. *Suppose that assumptions (i) – (iv) are satisfied. Then every optimal solution of (P) lies on the boundary ∂C of the set C.*

Proof. Suppose that z is a point satisfying $g(z) > 0$. Then $\pi(z)$ as defined in assumption (iv) exists. By assumption (iii), we have $f(w) < f(z)$, and hence

$$f(\pi(z)) = f(\lambda w + (1-\lambda)z) < \lambda f(w) + (1-\lambda)\, f(z) < f(z) \ ,$$

since f is convex and there is a number $\lambda \in (0,1)$ such that

$$\pi(z) = \lambda w + (1-\lambda)z \ . \qquad \blacksquare$$

Note that the duality theory discussed in the preceding sections also applies to problem (P). In particular, Theorem X.1 remains valid, i.e., we have the following corollary.

Corollary X.3. *For a point $\bar{x} \in \partial C$ to be an optimal solution of (P) it is necessary that*

$$\max \{g(x) \colon f(x) \le f(\bar{x})\} = 0 \ . \tag{39}$$

This condition is also sufficient if problem (P) is stable.

In addition to the conditions (i) – (iv) we shall assume that

(v) *the objective function f(x) is strictly convex.*

Assumption (v) is purely technical. If it is not satisfied (i.e., if f(x) is convex but not strictly convex), then we may replace f(x) by $f_\varepsilon(x) = f(x) + \varepsilon\|x\|^2$, which is obviously strictly convex. For an optimal solution $\bar{x}(\varepsilon)$ of the resulting problem we then have $f(\bar{x}(\varepsilon)) + \varepsilon\|\bar{x}\|^2 \le f(x) + \varepsilon\|x\|^2$ for all feasible points of the original problem (P). It follows that $f(\bar{x}(\varepsilon)) \longrightarrow \bar{\alpha}$ as $\varepsilon \longrightarrow 0$, since $\{\bar{x}(\varepsilon)\}$ is bounded, by assumption (i) (cf. Remark IX.3)..

The role of assumption (v) is to ensure that every supporting hyperplane of the level set $\{x \colon f(x) \le \alpha\}$ supports it at exactly one point. This property is needed in order to prevent the following algorithm from jamming.

Algorithm X.3.

Initialization:

Let $w \in \operatorname{argmin} \{f(x): x \in \mathbb{R}^n\}$ (or any point satisfying assumption (iii)). Compute a point $x^1 \in \partial C$.

Generate a polytope D_1 satisfying

$$\{x \in \mathbb{R}^n: x \in w + \bigcup_{\Theta \geq 1} \Theta\,(\partial C - w),\ f(x) \leq f(x^1)\} \subset D_1$$

while $f(z) > f(x^1)$ for any vertex z of D_1. Set $k = 1$.

Iteration k=1,2,...:

k.1.: Check whether $f(x^k) = \min f(D_k)$.

If $f(x^k) = \min f(D_k)$, then stop: x^k is an optimal solution of problem (P). Otherwise, continue.

k.2.: Solve the subproblem

$$(SP_k) \qquad \max \{f(x): x \in D_k\}$$

by a finite algorithm. Let z^k be an optimal solution of (SP_k) (z^k is a vertex of D_k since f is strictly convex).

k.3.: Compute $\pi(z^k)$. If $\pi(z^k)$ exists and $f(\pi(z^k)) < f(x^k)$, then set $x^{k+1} = \pi(z^k)$. Otherwise, set $x^{k+1} = x^k$.

k.4.: Let y^{k+1} be the point where the line segment $[w,z^k]$ intersects the surface $\{x: f(x) = f(x^{k+1})\}$. Compute $p^{k+1} \in \partial f(y^{k+1})$ ($\partial f(y^{k+1})$ denotes the subdifferential of f at y^{k+1}). Set

$$D_{k+1} = D_k \cap \{x: p^{k+1}(x-y^{k+1}) \leq 0\} .$$

Set $k \leftarrow k+1$ and go to Step k.1.

Remarks X.6. (i) In Step k.1, checking whether $f(x^k) = \min f(D_k)$ is easy, because f is convex and D_k is a polytope. It suffices to determine whether one of the standard first order optimality conditions holds. For example, x^k is optimal if

$$0 \in \partial f(x^k) + N_{D_k}(x^k) ,$$

where $N_{D_k}(x^k)$ denotes the outward normal cone to D_k at x^k, i.e., the cone which is generated by the normal vectors of the constraints of D_k that are binding (active) at x^k (cf., e.g., Rockafellar (1970)).

It will be shown that D_k contains all feasible points of (P) satisfying $f(x) \le f(x^k)$. Therefore, if $f(x^k) = \min f(D_k)$, then x^k must be an optimal solution of (P).

(ii) The subproblem (SP_k) is equivalent to the problem of globally minimizing the concave function (–f) over the polytope D_k, and it can be solved by any of the algorithms described in Part B. Since D_{k+1} differs from D_k by just one additional linear constraint, the algorithm for solving (SP_k) should have the capability of being restarted at the current solution of (SP_{k-1}) (cf., Chapter VII). However, most often one would proceed as discussed in Chapter II: start with a simple initial polytope D_1 (for example, a simplex) whose vertex set $V(D_1)$ is known and determine the vertex set $V(D_{k+1})$ of D_{k+1} from the vertex set $V(D_k)$ of D_k by one of the methods described in Chapter II. Since $\max f(D_k) = \max f(V(D_k))$, problem (SP_k) is then reduced to the problem of determining $V(D_{k+1})$ from $V(D_k)$.

The convergence of Algorithm X.3 is established through some lemmas. Let

$$\ell_k(x) := p^{k+1}(x - y^{k+1}) .$$

Lemma X.7. *For every k one has*

$$\{x \in D_1 : f(x) \le f(x^k)\} \subset D_k .$$

Proof. The assertion is obvious for k=1. Supposing that it holds for some k, we prove it for k+1.

If $x \in D_1$ and $f(x) \le f(x^{k+1})$, then $f(x) \le f(x^k)$, since $f(x^{k+1}) \le f(x^k)$; hence, $x \in D_k$ by the induction assumption. Furthermore, from the definition of a subgradient and the equality $f(y^{k+1}) = f(x^{k+1})$, we see that

$$p^{k+1}(x - y^{k+1}) \le f(x) - f(y^{k+1}) = f(x) - f(x^{k+1}) \le 0 ,$$

i.e., $\ell_k(x) \le 0$, and hence $x \in D_{k+1}$. ■

Lemma X.8. *(i) There exists a positive real number L such that*

$$|\ell_k(z) - \ell_k(x)| \le L\|z - x\| \quad \forall x,z \in D_1 .$$

(ii) There exists an $\alpha^* \in \mathbb{R}$ satisfying $f(x^k) \longrightarrow \alpha^*$ $(k \longrightarrow \infty)$. Moreover, whenever $z^{k_q} \longrightarrow \bar{z}$ and $l_{k_q}(z^{k_q}) \longrightarrow 0$ $(q \longrightarrow \infty)$ for some subsequence k_q, we have $f(\bar{z}) = \alpha^*$.

Proof. (i) Since $y^{k+1} \in D_1$ and D_1 is compact (it is a polytope), it follows from a well-known result of convex analysis (see, e.g., Rockafellar (1970)) that the sequence $\{p^{k+1}\}$, $p^{k+1} \in \partial f(y^{k+1})$, is bounded, i.e., $\|p^{k+1}\| < L$ for some $L > 0$. Hence, for every z and x we have

$$|\ell_k(z) - \ell_k(x)| = |p^{k+1}(z - x)| \le \|p^{k+1}\| \, \|z - x\| \le L\|z - x\| .$$

(ii) The sequence $\{f(x^k)\}$ is nonincreasing and bounded from below by f(w) (cf. assumption (iii)). Therefore, it must converge to some limit α^*.

Now consider a subsequence k_q such that $z^{k_q} \longrightarrow \bar{z}$ and $l_{k_q}(z^{k_q}) \longrightarrow 0$. In view of the boundedness of the sequences $\{p^{k+1}\}$ and $\{y^{k+1}\}$ we may assume (by passing to subsequences if necessary) that $p^{k_q+1} \longrightarrow \bar{p}$, $y^{k_q+1} \longrightarrow \bar{y}$. Then $\ell_{k_q}(z^{k_q}) = p^{k_q+1}(z^{k_q} - y^{k_q+1}) \longrightarrow \bar{p}(\bar{z} - \bar{y}) = 0$ (by hypothesis).

Since $\bar{p} \in \partial f(\bar{y})$ (cf., Rockafellar (1970)) and $f(\bar{y}) = \lim f(y^{k_q+1}) = \lim f(x^{k_q+1}) = \alpha^*$, we must have $\bar{p} \neq 0$. Otherwise, $0 = \bar{p}(w - \bar{y}) \leq f(w) - f(\bar{y}) = f(w) - \alpha^*$ would imply $f(\bar{y}) \leq f(w)$, which contradicts assumption (iii).

Furthermore, since $y^{k+1} \in [w,z^k]$, we see that $\bar{y} \in [w,\bar{z}]$, i.e., $\bar{z} - \bar{y} = -\lambda(w - \bar{y})$ for some $\lambda \geq 0$.

But from the above we know that $0 = \bar{p}(\bar{z} - \bar{y})$, and hence

$$0 = \bar{p}(\bar{z} - \bar{y}) = -\lambda\bar{p}(w - \bar{y}) \geq -\lambda(f(w) - f(\bar{y})) .$$

Finally, since $f(w) - f(\bar{y}) < 0$, this implies that $\lambda = 0$, i.e., $\bar{z} = \bar{y}$. Therefore, $f(\bar{z}) = f(\bar{y}) = x^*$. ■

Lemma X.9. *Every accumulation point $\bar{z}$ of the sequence $\{z^k\}$ satisfies $f(\bar{z}) = \alpha^*$.*

Proof. Lemma X.9 can easily be deduced from the preceding two lemmas and Theorem II.1. A simple direct proof is as follows:

Let $\bar{z} = \lim_{q\to\infty} z^{k_q}$. From Lemma X.8 (i) we see that

$$\ell_k(z^k) \leq \ell_k(\bar{z}) + L\|z^k - \bar{z}\| . \tag{40}$$

But from the construction of the algorithm it is clear that $\ell_k(z^j) \leq 0 \;\; \forall k < j$. Fixing k and setting $j = k_q \longrightarrow \infty$, we obtain $\ell_k(\bar{z}) \leq 0$. Inserting this in the above inequality yields

$$0 \leq \ell_{k_q}(z^{k_q}) \leq L\|z^{k_q} - \bar{z}\| \longrightarrow 0 ,$$

where the first inequality follows from the definition of z^{k_q} in the above algorithm ($\ell_k(w) < 0 \leq \ell_k(z^k) \;\; \forall k$).

The equality $f(\bar{z}) = \alpha^*$ then follows from Lemma X.8 (ii). ■

Lemma X.10. *Every point $\hat{x} \in D_1$ which satisfies $f(\hat{x}) = \alpha^*$ is the limit of some subsequence $\{y^{k_q}\}$ of $\{y^k\}$.*

Proof. Since $\alpha^* \le f(x^1)$, it follows from the construction of D_1 that $\hat{x}$ is not a vertex of D_1. Therefore, using the strict convexity of f(x), we see that there exists a point $u^q \in D_1$ such that $\|u^q - \hat{x}\| \le 1/q$ for any integer $q > 0$ and $f(u^q) > \alpha^*$. But the inclusion $z^k \in \text{argmax}\,\{f(z): z \in D_k\}$ and Lemma X.9 imply that max $\{f(z): z \in D_k\} \longrightarrow \alpha^*$ $(k \longrightarrow \infty)$. Hence, $u^q \notin D_{k_q}$ for some k_q, i.e., one has y^{k_q} and $p^{k_q} \in \partial f(y^{k_q})$ such that

$$p^{k_q}(u^q - y^{k_q}) > 0 . \tag{41}$$

By passing to subsequences if necessary, we may assume that $y^{k_q} \longrightarrow \bar{y}$ and $p^{k_q} \longrightarrow \bar{p}$ as $q \longrightarrow \infty$. Then we know from the proof of Lemma X.8 that $f(\bar{y}) = \alpha^*$ and $\bar{p} \in \partial f(\bar{y})$. We must have $\bar{p} \neq 0$, because $0 \in \partial f(\bar{y})$ would imply that $f(\bar{y}) = \alpha^* = \min f(\mathbb{R}^n)$, which contradicts assumption (iii).

Letting $q \longrightarrow \infty$, we see from (41) that $\bar{p}(\hat{x} - \bar{y}) \ge 0$, and since $f(\hat{x}) = f(\bar{y}) = \alpha^*$, from the strict convexity of f(x) again it follows that $\hat{x} = \hat{y}$, i.e., $\hat{x} = \lim_{q \to \infty} y^{k_q}$. ■

Proposition X.9. *Every accumulation point $\bar{x}$ of the sequence $\{x^k\}$ generated by Algorithm X.3 satisfies the condition*

$$0 = max\,\{g(x): f(x) \le f(\bar{x})\} . \tag{42}$$

If problem (P) is stable, then every accumulation point $\bar{x}$ of $\{x^k\}$ is an optimal solution of (P).

Proof. First note that, for every k the line segment $[w,y^k]$ contains at most one feasible point, namely either y^k, or no point at all. Indeed, either $y^k = x^k = \pi(z^{k-1})$ and, from the definition of $\pi(z^{k-1})$, y^k is the only feasible point in $[w,y^k]$, or else y^k satisfies $f(y^k) = f(x^{k-1})$ and there is no feasible point in $[w,y^k]$.

Now suppose that (42) does not hold. Then there is a point x satisfying

$f(x) \le f(\bar{x}) = \alpha^*$ and $g(x) > 0$. Since f and g are continuous and $f(w) < \alpha^*$, there is also a point $\tilde{x}$ satisfying $f(\tilde{x}) < \alpha^*$ and $g(\tilde{x}) > 0$. Let U denote a closed ball around $\tilde{x}$ such that $g(x) > 0$ and $f(x) < \alpha^*$ for all $x \in U$. Let $\hat{x} \in D_1$ be the point of the halfline from w through $\tilde{x}$ for which $f(\hat{x}) = \alpha^*$. From Lemma X.10 we know that there is a sequence $y^{k_q} \to \hat{x}$ $(q \to \infty)$. For sufficiently large q the line segment $[w,y^{k_q}]$ will intersect the ball U at a point x' satisfying $g(x') > 0$ and $f(x') < \alpha^*$. But since $f(y^{k_q}) = f(x^{k_q}) \ge \alpha^*$, this implies that $x \ne y^{k_q}$ $\forall q$, contradicting the above observation.

Therefore, relation (42) holds, and the second assertion follows from Corollary X.3. ■

Example X.2. Computing a global solution of a difficult multiextremal optimization problem is usually very expensive. Therefore, a reasonable approach is to transcendent local optimality by first computing a local minimum (or stationary point) x_{loc} by one of the standard nonlinear programming techniques, and then to apply a global optimization algorithm with the aim of locating a feasible solution x^k that is substantially better than x_{loc}, for example, such that

$$f(x^k) < f(x_{loc}) - \eta ,$$

where η is some prescribed positive number. As an example, consider the problem

$$\text{minimize } f(x) := (x_1 - 14)^2 + 2(x_2 - 10)^2 + 3(x_3 - 16)^2 + 4(x_4 - 8)^2$$

$$\text{s.t. } g_1(x) := 2(x_1-15)^2 + (x_2-9)^2 + 3(x_3-18)^2 + 2(x_4-10)^2 - 60 \ge 0,$$

$$g_2(x) := (x_1-10)^2 + 3(x_2-12)^2 + 2(x_3-14)^2 + (x_4-13)^2 - 50 \ge 0 .$$

The global minimum of f(x) over $\mathbb{R}^4$ is 0. Suppose that a feasible solution is given by

$$x^1 = (13.87481,\ 9.91058,\ 15.85692,\ 16.870087) \text{ with } f(x^1) = 314.86234.$$

We try to find a feasible solution $\bar{x}$ satisfying

$$f(\bar{x}) < f(x^1) - 300 = 14.86234$$

or to establish that such a feasible point does not exist.

We choose $w = (14, 10, 16, 8)$ with $f(w) = 0$, $g(w) = \min \{f_1(w), g_2(w)\}$: -37.

D_1 is the simplex in $\mathbb{R}^4$ with vertices $v^0 = 0$, $v^i = 10^3 \cdot e^i$ $(i=1,2,3,4)$, where e^i is the i–th unit vector in $\mathbb{R}^4$ $(i=1,2,3,4)$.

After eight iterations the algorithm finds

$x^9 = (14.44121, 10.48164, 14.30500, 7.15250)$ with $f(x^9) = 12.15073 < 14.86234$.

The intermediate results are:

$x^2 = (13.91441, 9.93887, 22.01549, 7.95109)$, $y^2 = x^2$, $f(x^2) = 108.58278$;

$x^3 = (13.92885, 15.03099, 15.91869, 7.95994)$, $y^3 = x^3$, $f(x^2) = 50.65313$;

$x^4 = x^3$, $y^4 = (13.92728, 15.03025, 15.91690, 8.07018)$;

$x^5 = x^4$, $y^5 = (13.92675, 15.03119, 16.06476, 7.95814)$;

$x^6 = (19.34123, 9.94583, 15.91333, 7.95667)$, $y^6 = x^6$, $f(x^6) = 28.56460$;

$x^7 = (19.32583, 10.06260, 15.91166, 7.95583)$, $y^7 = x^7$, $f(x^7) = 28.40348$;

$x^8 = (12.38792, 8.84852, 14.15763, 8.98799)$, $y^8 = x^8$, $f(x^8) = 19.33813$.

Now suppose that problem (P) is not stable. In this case, Algorithm X.3 is not guaranteed to converge to a global solution (cf. Proposition X.9). However, an ε–perturbation in the sense of the following proposition can always be used to handle unstable problems, no matter whether or not the perturbed problem is stable.

Proposition X.10. *Let $\bar{x}(\varepsilon)$ denote any accumulation point of the sequence $x^k(\varepsilon)$ generated by Algorithm X.3 when applied to the ε-perturbed problem*

$$(P(\varepsilon)) \qquad \begin{array}{l} \text{minimize } f(x) \\ \text{s.t. } g_i(x) + \varepsilon \geq 0 \quad (i=1,\dots,m) \end{array}$$

Then as $\varepsilon \longrightarrow 0$ every accumulation point of the sequence $\{\bar{x}(\varepsilon)\}$ is an optimal solution of problem (P).

Proof. Let $D(\varepsilon)$ denote the feasible set of problem $(P(\varepsilon))$. Clearly, $D(\varepsilon)$ contains the feasible set of (P) for all $\varepsilon > 0$. From Proposition X.9 we know that every accumulation point $\bar{x}(\varepsilon)$ of $\{x^k(\varepsilon)\}$ satisfies $g(\bar{x}(\varepsilon)) = -\varepsilon$, and

$$\max \{g(x): f(x) \leq f(\bar{x}(\varepsilon))\} = -\varepsilon.$$

This implies that $f(\bar{x}(\varepsilon)) < \min \{f(x): g(x) \geq -\varepsilon/2\}$. Therefore, as $\varepsilon \longrightarrow 0$ every accumulation point $\bar{x}$ of $\bar{x}(\varepsilon)$ satisfies $g(\bar{x}) = 0$, and $f(\bar{x}) \leq \min \{f(x): g(x) \geq 0\}$; hence, it is an optimal solution of (P). ■

Remark X.7. If we apply Algorithm X.3 to the above perturbed problem and stop as soon as

$$\max \{g(x): f(x) \leq f(x^k(\varepsilon))\} < 0 ,$$

then $x^k(\varepsilon)$ yields an approximate optimal solution of (P) in the sense that

$$g(x^k(\varepsilon)) + \varepsilon = 0, \; f(x^k(\varepsilon)) \leq \min \{f(x): g(x) \geq 0\}.$$

2. BRANCH AND BOUND METHODS

The development of BB methods presented in Chapter IV allows one to design various procedures for a variety of difficult global optimization problems.
We choose an appropriate type of partition set (e.g., simplices or rectangles) and an exhaustive subdivision (cf. Section IV.3.). From the examples in Section IV.4.5 we select a bounding operation in accordance with the given type of objective function which provides a lower bound $\beta(M)$ for min $f(D \cap M)$ or min $f(M)$, respectively. We apply a bound improving selection for the partition elements to be refined. Finally, if necessary, we choose from Section IV.5. the "deletion by infeasibility" rule that corresponds to the given feasible set, and we incorporate all of these elements into the prototype BB procedure described in Section IV.1.. Then the theory developed in Section IV guarantees convergence in the sense discussed here.

Note that, as shown in Section IV.4.5., whenever a lower bound $\beta(M)$ yields consistency or strong consistency, then any lower bound $\bar{\beta}(M)$ satisfying $\bar{\beta}(M) \geq \beta(M)$ for all partition sets M will, of course, also provide consistency or strong consistency, respectively. Hence, better and more sophisticated lower bounds than those discussed in Section IV.4.5. can be incorporated in the corresponding BB procedures without worrying about convergence.

As an example of the various possibilities, we present here an *algorithm for globally minimizing a d.c. function subject to a finite number of convex and reverse convex inequalities.*

Let $f_1: \mathbb{R}^n \to \mathbb{R}$ be a *concave* function, and let $f_2, g_i, h_j: \mathbb{R}^n \to \mathbb{R}$ be *convex* functions (i=1,...,m; j=1,...,r). Define the convex function

$$g(x) = \max \{g_i(x): i=1,...,m\} , \tag{43}$$

and the sets

$$D_1 = \{x \in \mathbb{R}^n: g(x) \leq 0\} , \tag{44}$$

$$D_2 = \{x \in \mathbb{R}^n: h_j(x) \geq 0 \quad (j=1,\dots,r)\} . \tag{45}$$

The problem to be considered is then

$$\begin{aligned} \text{minimize } f(x) := \ & f_1(x) + f_2(x) \\ \text{s.t. } x \in D := \ & D_1 \cap D_2 \end{aligned} \tag{46}$$

Assume that D is nonempty and compact and that a point y^0 satisfying $g(y^0) < 0$ is known.

Let min $f(S) = \infty$ whenever S is an empty set.

Algorithm X.4.

Step 0 (Initialization):

Construct an n–simplex $M_0 \supseteq D_1$ (cf. Chapter III) and its radial partition

$$\mathcal{M}_0 = \{M(i,y^0): i \in I_0\}$$

with respect to y^0 (cf. Section IV.3).

For each $M \in \mathcal{M}_0$:

Set $S_M = V(M) \cap D$ and $\alpha(M) = \min f(S_M)$, where V(M) denotes the vertex set of M.

Choose $v^* \in V(M)$, $p^* \in \partial f_2(v^*)$ and compute

$$\beta(M) = \min \{\phi_M(v) := f_1(v) + f_2(v^*) + p^*(v-v^*): v \in V(M)\} . \tag{47}$$

Choose $y(M) \in V(M)$ satisfying $\phi_M(y(M)) = \beta(M)$. (48)

Set $\alpha_0 = \min \{\alpha(M): M \in \mathcal{M}_0\}$, $\beta_0 = \min \{\beta(M): M \in \mathcal{M}_0\}$.
If $\alpha_0 < \infty$, then choose x^0 satisfying $f(x^0) = \alpha_0$.
If $\alpha_0 - \beta_0 = 0$ ($\leq \varepsilon$), then stop. x^0 is an (ε–)optimal solution.
Otherwise, choose $M_0 \in \mathcal{M}_0$ satisfying $\beta_0 = \beta(M_0)$ and set $y^0 = y(M_0)$.

Step k=1,2,...:

At the beginning of Step k we have the current partition $\mathcal{M}_{k-1}$ of a subset of M_0 still of interest.

Furthermore, for every $M \in \mathcal{M}_{k-1}$ we have $S_M \subseteq M \cap D$ (possibly $S_M = \emptyset$), and bounds $\beta(M)$, $\alpha(M)$ (possibly $\alpha(M) = \infty$) which satisfy

$$\beta(M) \le \min f(M \cap D) \le \alpha(M) \qquad \text{if M is known to be feasible,}$$

$$\beta(M) \le \min f(M) \qquad \text{if M is uncertain.}$$

Moreover, we have the current lower and upper bounds β_{k-1}, α_{k-1} (possibly $\alpha_{k-1} = \infty$) which satisfy

$$\beta_{k-1} \le \min f(D) \le \alpha_{k-1} ,$$

and a partition set M_{k-1} satisfying $\beta_{k-1} = \beta(M_{k-1})$.

Finally, we obtain a corresponding point $y^{k-1} = y(M_{k-1})$, and, if $\alpha_{k-1} < \infty$, we have $x^{k-1} \in D$ satisfying $f(x^{k-1}) = \alpha_{k-1}$.

k.1. Delete all $M \in \mathcal{M}_{k-1}$ satisfying $\beta(M) > \alpha_{k-1}$.

Let $\mathcal{R}_k$ be the collection of remaining members of $\mathcal{M}_{k-1}$.

k.2. Select a collection $\mathcal{P}_k \subset \mathcal{R}_k$ satisfying

$$M_{k-1} \in \mathcal{P}_k$$

and subdivide every member of $\mathcal{P}_k$ into a finite number of n–simplices by means of an exhaustive radial subdivision. Let $\mathcal{P}'_k$ be the collection of all new partition elements.

k.3. Delete every $M \in \mathcal{P}'_k$ for which the *deletion rule (DR2)* (cf. Section IV.5) applies or for which it is otherwise known that min f(D) cannot occur. Let $\mathcal{M}'_k$ be the collection of all remaining members of $\mathcal{P}'_k$.

k.4. For each $M \in \mathcal{M}'_k$:

Set $S_M = V(M) \cap D$ and $\alpha(M) = \min f(S_M)$.

Determine $\beta'(M) = \min \phi_M(V(M))$ and $y(M)$ according to (47), (48).

Set $\beta(M) = \max \{\beta'(M), \beta(M')\}$, where $M \subset M' \in \mathcal{M}_{k-1}$.

k.5. Set $\mathcal{M}_k = (\mathcal{R}_k \setminus \mathcal{P}_k) \cup \mathcal{M}'_k$.

Compute

$\alpha_k = \inf \{\alpha(M): M \in \mathcal{M}_k\}$, $\beta_k = \min \{\beta(M): M \in \mathcal{M}_k\}$.

If $\alpha_k < \infty$, then let $x^k \in D$ be such that $f(x^k) = \alpha_k$.

If $\alpha_k - \beta_k = 0$ ($\leq \varepsilon$), then stop. x^k is an (ε–)optimal solution.

Otherwise, choose $M_k \in \mathcal{M}_k$ satisfying $\beta_k = \beta(M_k)$ and set $y^k = y(M_k)$. Go to Step k+1.

The following Proposition shows convergence of the procedure ($\varepsilon = 0$).

Proposition X.11. (i) *If Algorithm X.4 does not terminate after a finite number of iterations, then every accumulation point of the sequence* $\{y^k\}$ *is an optimal solution of problem (46), and*

$$\lim_{k\to\infty} \beta_k = \min f(D). \tag{49}$$

(ii) *If* $S_M \neq \emptyset$ *for every partition element M that is never deleted, then the sequence* $\{x^k\}$ *has accumulation points, and every accumulation point is an optimal solution of problem (46) satisfying*

$$\lim_{k\to\infty} \beta_k = \min f(D) = \lim_{k\to\infty} \alpha_k = \lim_{k\to\infty} f(x^k) \tag{50}$$

Proof. Proposition X.11 follows from the theory presented in Chapter IV.. Since $\{\beta_k\}$ is a nondecreasing sequence bounded from above by min f(D), we have the existence of

$$\beta := \lim_{k\to\infty} \beta_k \leq \min f(D). \tag{51}$$

Now let $\bar{y}$ be an accumulation point of $\{y^k\}$, and denote by $\{y^r\}$ a subsequence of $\{y^k\}$ satisfying $y^r \to \bar{y}$. Then, since the selection is bound improving and the subdivision is exhaustive (cf., Definition IV.6 and Definition IV.10), we can use a standard argument on finiteness of the number of partition elements in each step (similarly to the argument in the proof of Theorem IV.3) to conclude that there exists a decreasing sequence $\{M_q\} \subset \{M_r\}$ of successively refined partiton elements satisfying

$$y^q \in M_q,\ \beta_q = \beta(M_q) \tag{52}$$

and

$$M_q \xrightarrow[q\to\infty]{} \{\bar{y}\}\ . \tag{53}$$

In Proposition IV.4, it was shown that deletion rule (DR2) is certain in the limit, and by Proposition IV.3 we have strong consistency of the bounding operation (cf., Definition VI.7). It follows that

$$\bar{y} \in D \tag{54}$$

and

$$\beta(M_q) \xrightarrow[q\to\infty]{} f(\bar{y})\ . \tag{55}$$

Considering (51), (52) and (55), we see that

$$\beta = \lim_{q\to\infty} \beta(M_q) = f(\bar{y}) \leq \min f(D)\ ,$$

and hence

$$\beta = f(\bar{y}) = \min f(D)\ ,$$

since $\bar{y} \in D$, and assertion (i) is verified.

In order to prove (ii), recall from Lemma IV.5 that under the assumptions of Proposition X.11(ii) the bounding operation is also consistent. The assertion then follows from Theorem IV.3 and Corollary IV.2, since f is continuous and D is compact. ■

Remark X.8. In addition to (47), several other bounding procedures are available. One possibility, for example, is to linearize the convex part f_2 of f at different vertices $v^* \in M$ and to choose the best bound obtained from (47) over all $v^* \in M$ considered.

Another method is to replace the concave part f_1 of f by its convex envelope φ over the simplex M (which is an affine function, cf. Section IV.4.3.) and to minimize the convex function $(\varphi + f_2)$ over M.

More sophisticated bounding operations have been applied to problems where additional structure can be exploited; examples include separable d.c. problems such as, e.g., the minimization of indefinite quadratic functions over polytopes where piecewise linearization has been used (Chapter IX).

3. SOLVING D.C. PROBLEMS BY A SEQUENCE OF LINEAR PROGRAMS AND LINE SEARCHES

The standard deterministic global optimization methods, such as outer approximation and branch and bound, were first investigated in order to solve the concave minimization problem and problems closely related to concave minimization (cf. Part B). In the preceding sections we saw that these basic methods can also be used to solve the more general and more difficult d.c. problem (general reverse convex programming problem).

Further developments in concave minimization, however, have led to certain combinations of outer approximation and branch and bound methods that involve only linear programming subproblems and line searches (Benson and Horst (1988), Horst, Thoai and Benson (1988)), cf. Algorithm VII.2 and the discussion in Section VII.1.9. The first numerical experiments indicate that, for concave minimization, these methods can be expected to be more efficient than pure outer approximation and pure branch and bound methods (cf. Horst, Thoai and Benson (1988) and the discussion in Chapter VII.). Therefore, some effort has been devoted to the extension of these approaches to the d.c. problem. The resulting procedure, which is presented below, can be viewed as an extension of Algorithm VII.2 to the d.c. problem which takes into account the more complicated nature of the latter problem by an appropriate deletion–by–infeasibility rule and a modified bounding procedure (cf. Horst et al. (1989)).

For other other possible extensions of the above mentioned linear programming –line search approaches for concave minimization to the d.c. problem we refer to Horst et al. (1989), Tuy (1989) and Horst et al. (1990).

Let us again consider the canonical d.c. problem

$$\text{(CDC)} \qquad \begin{array}{l} \text{minimize } f(x) := cx \\ \text{s.t. } \; h(x) \le 0 \,, \; g(x) \ge 0 \end{array} ,$$

where h, g are convex functions on $\mathbb{R}^n$ and $c \in \mathbb{R}^n$.

A modification of the algorithm which handles concave objective functions will be given below.

Let

$$\begin{aligned} &\Omega := \{x \in \mathbb{R}^n \colon h(x) \le 0\}, \\ &G := \{x \in \mathbb{R}^n \colon g(x) \ge 0\}, \;\; C = \{x \in \mathbb{R}^n \colon g(x) < 0\} = \mathbb{R}^n \setminus G \,, \\ &D := \Omega \cap G \,. \end{aligned} \tag{57}$$

(Note that Ω is the set which in Section X.1 was denoted by H.) Assume that *(a)* Ω *is bounded, and there is a polytope* T *containing* Ω*;*

(b) C *is bounded;*

(c) there is a point w *satisfying*

$$h(w) < 0,\ g(w) < 0,\ cw < \min\{cx: x \in \Omega \cap G\}\ .$$

Remarks X.9. (i) Assumptions (a), (b), (c) are quite similar to the standard assumptions in Section X.1.2. A polytope T satisfying (a) is often given as a rectangle defined by known bounds on the variables. Another possibility is to construct a simplex T satisfying $\Omega \subset T$ by one of the methods in Chapter II.

(ii) Assumption (b) is often not satisfied in formulations of (CDC) arising from applications. However, since Ω is bounded and a simple polytope $T \supset \Omega$ with known vertex set V(T) is at hand, we can always redefine

$$C \leftarrow C \cap \{x \in \mathbb{R}^n: \|x - w\| > \max\{\|v - w\|: v \in V(T)\}\}$$

without changing the problem.

(iii) Assumption (c) is fulfilled if Ω satisfies the Slater condition ($h(w) < 0$) and if the reverse convex constraint is essential, i.e., if it cannot be omitted (cf. Section X.1.2).

Initial conical partition and subdivision process

In order to simplify notation, let us assume that the coordinate system has been translated such that the point w in assumption (c) is the origin 0.

Let

$$S = \text{conv}\,\{v^1,...,v^{n+1}\}$$

denote an n–simplex with vertices $v^1,...,v^{n+1}$ which satisfies

$$0 \in \text{int}\, S\ .$$

For example, we can take $S = T$ if T is an n–simplex. Another example is given by $v^i = e^i$ $(i=1,\dots,n)$, $v^{n+1} = -e$, where e^i is the i–th unit vector in $\mathbb{R}^n$ and $e = (1,1,\dots,1)^T \in \mathbb{R}^n$.

Let F_i denote the facet of S opposite v^i, i.e., we have $v^i \notin F_i$ $(i=1,\dots,n+1)$. Clearly, F_i is an (n–1)–simplex.

Let $M(F_i)$ be the convex polyhedral cone generated by the vertices of F_i. Then we know from Chapter IV and Chapter VII that

$$\mathcal{M}_1 = \{M(F_i): i=1,\dots,n+1\}$$

is a conical partition of $\mathbb{R}^n$. This partition will be the initial partition of the algorithm.

The subdivision process for a cone $M = M(U)$, where U is an (n–1)–simplex contained in some facet F_i of S, is assumed to be induced by an *exhaustive radial subdivision* of U (cf. Chapters IV, and also VII where, however, a slightly different but equivalent matrix notation was used).

Lower bounds and deletion by infeasibility

Let U be an (n–1) simplex contained in some facet F_i of S, and let by $M = M(U)$ denote the cone generated by the vertices of U. Furthermore, suppose that α is the best objective function value attained at a feasible point known so far.

Let P be a polytope containing D, and consider the polytope

$$Q = P \cap \{x: cx \le \alpha\} .$$

The algorithm below begins with $P = T$, and then successively redefines P to include constraints generated by an outer approximation procedure.

We now present a method for calculating a lower bound of $f(x) = cx$ on the intersection $D \cap M \cap \{x: cx \le \alpha\}$ of the part of the feasible set still of interest with the cone M provided that $D \cap M \cap \{x: cx \le \alpha\} \neq \emptyset$. This method will also allow one to detect sufficiently many (but not necessarily all) cones M of a current partition which satisfy

$$D \cap M \cap \{x: cx \le \alpha\} = \emptyset .$$

These cones will be deleted from further consideration.

Let the above polytope Q be defined by

$$Q = \{x \in \mathbb{R}^n: Ax \le b\} ,$$

where A is a (p×n)–matrix and $b \in \mathbb{R}^p$.

Furthermore, for each i=1,...,n, let $y^i = y^i(M)$ denote the point where the i–th edge of the cone M intersects the boundary ∂C of C. (Note that y^i is uniquely determined by means of a univariate convex minimization (line search), since C is convex and bounded and $0 \in \text{int } C$.)

Suppose that an upper bound $\alpha \ge \min f(D)$ is known (usually α is the objective function value at the current best feasible point). Since $cw < \min \{cx: x \in D\}$, and since w was translated to the origin, we may assume that $\alpha \ge 0$. Then the cone M can be deleted from further consideration whenever we have

$$cy^i \ge \alpha \quad \forall i \in \{1,...,n\} . \tag{58}$$

To see this, consider the hyperplane $H = \{x \in \mathbb{R}^n: x = \sum_{i=1}^{n} \lambda_i y^i , \sum_{i=1}^{n} \lambda_i = 1\}$ which is uniquely determined by the linear independent points y^i (i=1,...,n). Let H^+ denote the closed halfspace generated by H that does not contain the origin. Then we have $M \cap D \subset H^+ \cap M$ because of the convexity of C. Since

$$H^+ \cap M = \{x \in \mathbb{R}^n: x = \sum_{i=1}^{n} \lambda_i y^i , \sum_{i=1}^{n} \lambda_i \ge 1, \ \lambda_i \ge 0 \ (i=1,...,n)\} ,$$

we see that (58) with $\alpha > 0$ implies that

$$cx = \sum_{i=1}^{n} \lambda_i cy^i \ge \sum_{i=1}^{n} \lambda_i \alpha \ge \alpha \quad \forall x \in H^+ \cap M ,$$

and hence $cx \geq \alpha$ $\forall x \in M \cap D$, i.e., the upper bound α cannot be improved in M.

Now let Y denote the (n×n)–matrix with columns y^i (i=1,...,n), and consider the linear programming problem

$$(LP) = (LP(M,Q)): \max \{ \sum_{i=1}^{n} \lambda_i : AY\lambda \leq b , \lambda \geq 0\} ,$$

where $\lambda = (\lambda_1,...,\lambda_n)^T \in \mathbb{R}^n$. Define $\mu(\lambda) = \sum_{i=1}^{n} \lambda_i$.

We recall the geometric meaning of (LP): consider the hyperplane $H = \{x \in \mathbb{R}^n: x = Y\lambda, \mu(\lambda) = 1\}$ defined above. Changing the value of $\mu(\lambda)$ results in a translation of H to a hyperplane parallel to H. The constraints in (LP) describe the set $M \cap Q$. Let $\lambda^*(M)$ and $\mu^* = \mu^*(M) = \mu(\lambda^*)$ denote an optimal solution and the optimal objective function value of (LP), respectively. Then $H^* = \{x \in \mathbb{R}^n: x = Y\lambda, \mu(\lambda) = \mu^*\}$ describes a hyperplane parallel to H that supports $Q \cap M$ at

$$x^* = x^*(M) = Y\lambda^* . \tag{59}$$

Let $z^i = \mu^* y^i$ denote the point where the i–th edge of M intersects H^* (i=1,...,n).

Lemma X.11. *Let $\lambda^* = \lambda^*(M)$, $\mu^* = \mu^*(M)$ and y^i, z^i (i=1,...,n) be defined as above.*

(i) If $\mu^* = \mu^*(M) < 1$, then $M \cap D \cap \{x: cx \leq \alpha\} = \emptyset$.

(ii) If $\mu^* = \mu^*(M) \geq 1$, then

$$\beta (M,Q):= \min \{cy^i, cz^i: i=1,...,n\}$$

is a lower bound for $f(x) = cx$ over $M \cap D \cap \{x: cx \leq \alpha\}$.

Proof. Let H^- and H^{*+} denote the closed halfspaces containing 0 generated by the hyperplanes H and H^* defined above, and let $\overset{o}{H}{}^-$ be the open halfspace int H^-. Since H is the hyperplane passing through the points $y^i \in \partial C$ (i=1,...,n), C is convex and $0 \in$ int C, it follows that the simplex $H^- \cap M = \text{conv}\,\{0,y^1,...,y^n\}$ is contained in $C \cup \partial C$; hence

$$D \cap \mathring{H}^- \cap M = \emptyset$$

(because $D \cap C = \emptyset$).

But from the definition of H^* it follows that

$$M \cap D \cap \{x: cx \le \alpha\} \subseteq M \cap Q \subseteq H^{*-} \cap M ,$$

and hence

$$M \cap D \cap \{x: cx \le \alpha\} \subseteq (H^{*-} \setminus \mathring{H}^-) \cap M . \tag{60}$$

Therefore, since $\mu^* < 1$ implies that $H^{*-} \subset \mathring{H}^-$, we see that assertion (i) holds.

Now consider the case $\mu^* \ge 1$. It is easily seen that $(H^{*-} \setminus \mathring{H}^-) \cap M$ is a polytope with vertex set $\{y^i, z^i \ (i=1,...,n)\}$. It follows from the linearity of the objective function cx and from (60) that assertion (ii) holds. ■

Remark X.10. When $\mu^*(M) < 1$ occurs, then the cone M is deleted.

Algorithm X.5.

Initialization:

Let $\mathscr{M}_1 = \{M(F_i): i=1,...,n+1\}$ be the initial conical partition as defined above. Determine the intersection points y^i $(i=1,...,n+1)$ of the rays emanating from 0 and passing through the vertices of the simplex S with the boundary ∂C of C. Determine

$$\alpha_1 := \min \{cy^i: y^i \in D, i \in \{1,...,n+1\}\}$$

(initial upper bound; $\alpha_1 = \infty$ if $y^i \notin D \ \forall i$).

If $\alpha_1 \ne \infty$, then choose $x^1 \in \{y^i \in D: i \in \{1,...,n+1\}\}$ satisfying $cx^1 = \alpha_1$.

For each cone $M_i = M(F_i)$ $(i=1,...,n+1)$ solve the linear programming problem $LP(M_i,Q_1)$ where $Q_1 = P_1 \cap \{x: cx \le \alpha_1\}$, and $P_1 = T$.

Delete each $M_i \in \mathcal{M}_1$ for which $\mu^*(M_i) < 1$ (cf. Lemma X.11).

If $\{M_i \in \mathcal{M}_1: \mu^*(M_i) \geq 1\} = \emptyset$, then stop: the feasible set D of problem (CDC) is empty.

Otherwise, for each cone $M_i \in \mathcal{M}_1$ satisfying $\mu^*(M_i) \geq 1$, compute the lower bound

$$\beta(M_i) := \beta(M_i, Q_1)$$

(cf. Lemma X.11).

Set $\beta_1 := \min \{\beta(M_i): M_i \in \mathcal{M}_1, \mu^*(M_i) \geq 1\}$, and let $\tilde{x}^1$ be a point where β_1 is attained, i.e., $c\tilde{x}^1 = \beta_1$.

Iteration k=1,2,...:

At the beginning of Step k we have a polytope $P_k \supset D$, an upper bound α_k (possibly ∞) and, if a feasible point has been found, we have a point $x^k \in D$ satisfying $\alpha_k = f(x^k)$. Furthermore, we have a set $\mathcal{M}_k$ of cones generated from the initial partition $\mathcal{M}_1$ by deletion operations and subdivisions according to the rules stated below. Finally, for each cone $M \in \mathcal{M}_k$, a lower bound $\beta(M) \leq \min \{cx: x \in D \cap M\}$ is known, and we have the bound $\beta_k \leq \min \{cx: x \in D\}$ and a not necessarily feasible point $\tilde{x}^k$ associated to β_k, by $c\tilde{x}^k = \beta_k$.

k.1. Delete all $M \in \mathcal{M}_k$ satisfying

$$\beta(M) \geq \alpha_k .$$

Let $\mathcal{R}_k$ be the collection of remaining cones in $\mathcal{M}_k$. If $\mathcal{R}_k = \emptyset$, then stop: x^k is an optimal solution of problem (P) with optimal value $\beta_k = \alpha_k$.

k.2. Select a cone $M_k^* \in \mathcal{R}_k$ satisfying

$$\beta(M_k^*) = \min \{\beta(M): M \in \mathcal{R}_k\} .$$

Let $x^{*k} = Y(M_k^*)\lambda^*(M_k)$ be the point defined in (59) corresponding to the cone M_k^*.

If $x^{*k} \in D$, then set $P_{k+1} = P_k$ and go to k.4.

k.3. Determine the point w^k where the line segment $[0,x^{*k}]$ intersects the boundary $\partial\Omega$ of Ω. Compute a subgradient t^k of h at w^k, and define the affine functions

$$\ell_k(x) := t^k(x - w^k) .$$

Set

$$P_{k+1} = P_k \cap \{x \in \mathbb{R}^n : \ell_k(x) \le 0\} .$$

k.4. Subdivide M_k^* into a finite number of cones $M_{k,j}$ $(j \in J_k)$ by an exhaustive subdivision process and compute the point y^{*k} where the new (common) edge of the cones $M_{k,j}$ $(j \in J_k)$ intersects ∂C.

Set

$$\alpha'_{k+1} := \begin{cases} \min\ \{\alpha_k, cy^{*k}\} & \text{if } y^{*k} \in \Omega \\ \alpha_k & \text{if } y^{*k} \notin \Omega \end{cases} .$$

Let $\mathcal{M}_k^* = \{M_{k,j} \subset M_k^* : j \in J_k\}$ and set $\mathcal{N}_k = (\mathcal{R}_k \setminus \{M_k^*\}) \cup \mathcal{M}_k^*$.

For each cone $M \in \mathcal{N}_k$ let $y^i(M)$ denote the intersection points of its i–th edge with ∂C (i=1,...,n).

Delete $M \in \mathcal{N}_k$ if $cy^i(M) \ge \alpha'_{k+1}$ $\forall i=1,\dots,n$ (cf. (58)).

Let $\mathcal{N}'_k$ denote the set of remaining cones.

k.5. Set

$$Q_{k+1} = P_{k+1} \cap \{x : cx \le \alpha'_{k+1}\}$$

and, for each $M \in \mathcal{N}'_k$ solve the linear programming problem $LP(M,Q_{k+1})$ obtaining the optimal value $\mu^*(M)$.

Delete all $M \in \mathcal{N}'_k$ satisfying $\mu^*(M) < 1$. Let $\mathcal{M}_{k+1}$ denote the collection of cones in $\mathcal{N}'_k$ that are not deleted.

k.6. For all $M \in \mathcal{N}'_k$ determine the lower bound

$$\beta(M) = \max\{\beta(M_k^*), \beta(M,Q_{k+1})\}$$

(cf. Lemma X.11).

k.7. Set

$$\beta_{k+1} := \min\{\beta(M): M \in \mathcal{M}_{k+1}\},$$

and let $\tilde{x}^{k+1}$ be a point where β_{k+1} is attained, i.e., $c\tilde{x}^{k+1} = \beta_{k+1}$.

From α'_{k+1} and the new feasible points obtained in this iteration determine a current best feasible point x^{k+1} and the corresponding upper bound $\alpha_{k+1} = cx^{k+1}$.

Set $k \leftarrow k+1$ and go to the next iteration.

Convergence of the algorithm will be established by means of the general convergence theory developed in Chapter IV. According to Corollary IV.3 and Corollary IV.5, convergence in the sense that $\lim_{k\to\infty} \beta_k = \min\{cx: x \in D\}$ and every accumulation point of the sequence $\{\tilde{x}^k\}$ is an optimal solution of (CDC), is guaranteed if we show that any infinite decreasing sequence $\{M_q\}$ of successively refined partition cones M_q satisfies

$$\overline{M} \cap D \neq \emptyset, \ \beta(M_{q'}) \underset{q'\to\infty}{\longrightarrow} \min\{cx: x \in \overline{M} \cap D\},$$

where $\overline{M} = \bigcap_q M_q$ and $\{M_{q'}\}$ is an infinite subsequence of $\{M_q\}$.

Lemma X.12. *Assume that the algorithm is infinite. Then we have $x^* \in D \cap \partial\Omega$ for every accumulation point x^* of the sequence $\{x^{*k}\}$, where $x^{*k} = Y(M_k^*)\lambda^*(M_k)$ (cf. Step k.2.)*

Proof. From the convergence theory that is known for the outer approximation method defined in Step k.3 and the linear programs $LP(M,Q_k)$ it follows that $x^* \in \partial\Omega$ (cf. Chapter II). In order to prove $x^* \in D$, note that, using a standard argument on finiteness of the number of partition cones in each step, one can conclude that there exists a decreasing sequence $\{M_q\}$ of successively refined partition cones such that for the sequence $x^{*q} \longrightarrow x^*$ we have $[0,x^{*q}] \subset M_q$ and $[0,x^*] \subset \overline{M} := \bigcap_q M_q$.

Since the algorithm uses an exhaustive subdivision process, the limit $\overline{M}$ must be a ray emanating from 0 which intersects ∂C at a point y^*.

For $M = M_q$, let $y^{q,i}$, $z^{q,i}$ denote the quantities y^i, z^i introduced above. Let $\mu_q^* = \mu^*(M_q)$.

Suppose that $x^* \notin D$. Then, since $x^* \in \partial\Omega$, we must have $g(x^*) < 0$, i.e., x^* is an interior point of the open convex set C, and it is also a relative interior point of the line segment $[0,y^*]$. Hence, there exist a point $w \in [x^*,y^*]$ and a closed ball B around w such that $x^* \notin B$ and $B \subset C$. For sufficiently large q, we then must have $x^{*q} \notin B$, while the edges of M_q intersect B at points in the relative interior of $[0,y^{q,i}]$. It follows that x^* is contained in the open halfspace $\hat{H}_q^-$ generated by the hyperplanes H_q through $y^{q,i}$ $(i=1,...,n)$ and containing the origin. But this implies that we have $\mu_q^* < 1$, and this would have led to the deletion of M_q, a contradiction. ■

Lemma X.13. *Let $\{M_q\}$ be an infinite decreasing sequence of successively refined partition cones generated by the algorithm. Let x^* be an accumulation point of the corresponding sequence $\{x^{*q}\}$, and denote by y^* the intersection of the ray $\overline{M} := \lim_q M_q$ with ∂C. Then we have*

$$\overline{M} \cap D = [y^*, x^*] \ .$$

Proof. Clearly, $\{x^{*q}\}$ has accumulation points, since Ω and the initial polytope T are bounded sets, and $x^* \in \overline{M}$. In the proof of Lemma X.12 we showed that

$x^* \in \partial\Omega \cap D$. In particular, we have $g(x^*) \geq 0$, hence $y^* \in [0,x^*]$. This implies that $y^* \in \Omega$, since Ω is a convex set and $0 \in \Omega$. Therefore, $y^* \in \partial C \cap \Omega \subset D$. Again from the convexity of Ω and C it follows that the subset $[y^*,x^*]$ of $\overline{M}$ is contained in D, i.e., $[y^*,x^*] \subset \overline{M} \cap D$.

In a similar way, from the convexity of the sets involved it is easily seen that $x \in \overline{M} \setminus [y^*,x^*]$ implies that $x \notin D$. ■

Lemma X.14. *Let $\{M_q\}$ be an infinite decreasing sequence of successively refined partition cones generated by the algorithm, and let $\overline{M} = \lim_{q\to\infty} M_q$. Then there exists a subsequence $\{M_{q'}\}$ of $\{M_q\}$ such that*

$$\beta(M_{q'}) \xrightarrow[q'\to\infty]{} \min\{cx: x \in \overline{M} \cap D\} .$$

Proof. Consider the quantities

$$\beta(M_q,Q_q) = \min\{cz^{q,i}, cy^{q,i}: i=1,\dots,n\}$$

corresponding to M_q that we introduced in Lemma X.11.
Since M_q converges to the ray $\overline{M}$ and $y^{q,i} \in \partial C$ $(i=1,\dots,n)$, where ∂C is compact, we see that

$$y^{q,i} \xrightarrow[q]{} y^* \in \overline{M} \cap \partial C \quad (i=1,\dots,n) .$$

In the proof of Lemma X.13 we saw that every accumulation point x^* of $\{x^{*q}\}$ satisfies $x^* \in \overline{M} \cap \partial\Omega$, and $[y^*,x^*] = \overline{M} \cap D$.

Consider a subsequence $\{q'\}$ such that $x^{*q'} \xrightarrow[q']{} x^*$. We show (again passing to a subsequence if necessary; we also denote this subsequence by $\{q'\}$) that $z^{q',i} \xrightarrow[q']{} x^*$ $(i=1,\dots,n)$.

Recall that $z^{q,i} = \mu_q^* y^{q,i}$. Let y^q be the point where the line segment $[0,x^{*q}]$ intersects the hyperplane H_q through $y^{q,i}$ $(i=1,\dots,n)$. Since $y^q \in \text{conv}\,\{y^{q,i}: i=1,\dots,n\}$, we see from $y^{q,i} \xrightarrow[q]{} y^*$ that $y^q \xrightarrow[q]{} y^*$.
But the relations $x^{*q} = \mu_q^* y^q$, $x^{*q'} \xrightarrow[q']{} x^*$, $y^{q'} \xrightarrow[q']{} y^*$ and the boundedness of $\{\mu_q^*\}$

imply that (passing to a subsequence if necessary) we have $\mu^*_{q'} \xrightarrow[q']{} \mu^*$. It follows that

$$z^{q',i} \xrightarrow[q']{} \mu^* y^* = x^* \in \overline{M} \cap \partial\Omega$$

(cf. Lemma X.12).

From the continuity of the function cx we see that

$$\beta(M_{q'},Q_{q'}) \xrightarrow[q']{} \min\{cy^*,cx^*\} .$$

But $\min\{cy^*,cx^*\} = \min\{cx: x \in [y^*,x^*]\}$; and from Lemma X.13 it follows that

$$\beta(M_{q'},Q_{q'}) \xrightarrow[q']{} \min\{cx: x \in \overline{M} \cap D\} .$$

Finally, since $\beta(M_{q'},Q_{q'}) \le \beta(M_{q'}) \le \min\{cx: x \in M_{q'} \cap D\}$, we must also have

$$\beta(M_{q'}) \xrightarrow[q']{} \min\{cx: x \in \overline{M} \cap D\} .$$

■

Proposition X.12. *If Algorithm X.5 is infinite and $D \neq \emptyset$, then we have*

(i) $\beta := \lim_{k\to\infty} \beta_k = \min\{cx: x \in D\}$;

(ii) every accumulation point of the sequence $\{\bar{x}^k\}$ is an optimal solution of problem (CDC).

Proof. Since the selection of the cone to be subdivided in each iteration is bound improving and the preceding lemmas have shown that the lower bounding is strongly consistent (cf. Definition IV.7), the assertions follow from Corollary IV.3 and Corollary IV.5.

■

Extension.

Note that Algorithm X.5 can also be applied if the *objective function f(x) is a concave function.* The only modification that is necessary to cover this case is to omit the deletion rule (58) and the sets $\{x: cx \le \alpha_k\}$ whenever they occur. Since for concave f the corresponding set $\{x: f(x) \le \alpha_k\}$ cannot be handled by linear programming techniques, we replace $Q = P \cap \{x: cx \geq \alpha\}$ by $Q = P$, $M \cap D \cap \{x: cx \leq \alpha\}$ in Lemma X.11 by $M \cap D$, $Q_{k+1} = P_{k+1} \cap \{x: cx \leq \alpha'_{k+1}\}$ in Step k.5 by $Q_{k+1} = P_{k+1}$ etc. Then the concavity of f implies that all of the above lemmas and Proposition X.12 remain true when cx is replaced by f(x).

The following slight modification has improved the performance of the algorithm in all of the examples which we calculated. Instead of taking the point w defined in assumption (c) as both the vertex of all the cones generated by the procedure and the endpoint of the ray that determines the boundary point where the hyperplane constructed in Step k.3 supports the set Ω, we used two points w^0 and $\bar{w}^0$. The point w^0 has to satisfy the conditions

$$cw^0 \leq \min \{cx: h(x) \leq 0\},\ g(w^0) < 0\ ,$$

and serves as a vertex of the cones generated by the partitioning and subdivision process. The point $\bar{w}^0$ is an interior point of Ω satisfying $h(\bar{w}^0) < 0$, and it is used as endpoint of the rays that define the points where the hyperplanes constructed in Step k.3 support Ω. Note that w^0 can be found by solving the convex program

$$\begin{aligned} &\text{minimize } cx \\ &\text{s.t.} \quad h(x) \leq 0 \end{aligned}$$

After a coordinate transformation $w^0 \longrightarrow 0$ the algorithm can be run with the above modification and with $\{x: cx \leq \alpha_k\}$ replaced by $\{x: cw^0 \leq cx \leq \alpha_k\}$.

Example X.3.

Objective function: $f(x_1,x_2) = 0.5x_1 + 1.2x_2$

Constraints: $h(x_1,x_2) = \max \{(x_1-1)^2 + (x_2-1)^2 - 0.6,\ 1/(0.7x_1 + x_2)-5\}$,

$g(x_1,x_2) = x_1^2 + x_2^2 - 2.4$, $T = \{0 \leq x_1 \leq 5,\ 0 \leq x_2 \leq 5\}$.

The points w^0, $\bar{w}^0$ were chosen as $w^0 = (1.097729, 0.231592)$ with $f(w^0) = 0.826776$ and $\bar{w}^0 = (1.0,1.0)$.

The initiating step has led to $x^1 = (1.097729,1.093156)$, $\alpha_1 = f(x^1) = 1.860652$, $\beta_1 = 0.826776$.

In the first iteration the point $x^{*1} = (0,000000,1.461790)$ and the first supporting hyperplane $\ell_1(x_1,x_2) = 1.406x_1 + 0.650x - 0.443$ were determined.
After 16 iterations the algorithm stopped at an approximate optimal solution $x^* = (1.510419,0.417356)$ with $f(x^*) = 1.25603$ satisfying $\min f(D) \le f(x^*) \le \min f(D) + 10^{-6}$. The point x^* was found at iteration 13.

4. SOME SPECIAL D.C. PROBLEMS AND APPLICATIONS

Some special d.c. problems have already been treated in Chapter IX. In this section, we discuss *design centering problems* and *biconvex programming.*

4.1. The Design Centering Problem

Recall from Example I.5 that a design centering problem is defined as follows.

Let $K \subset \mathbb{R}^n$ be a *compact, convex set containing the origin in its interior.* Furthermore, let $M \subset \mathbb{R}^n$ be a nonempty, *compact* set. Then the problem of finding $x \in M$, $r \in \mathbb{R}_+$ satisfying

$$\max_{x,r} \{r\colon x + rK \subset M\} \tag{61}$$

is called the *design centering problem.*

Problems of the form (61) often arise in optimal engineering design (e.g., Polak and Vincentelly (1979), Vidigal and Director (1982), Polak (1982), Nguyen et al. (1985), Thoai (1987), Thach (1988)). For example, consider a fabrication process where the quality of a manufactured item is characterized by an n-dimensional parameter. An item is accepted if this parameter is contained in some region of acceptability M. Let x be the nominal value of this parameter, and let y be its actual value. Assume that for fixed x the probability

$$P(\|y-x\| \le r) = p(x,r)$$

that the deviation is no greater than r is monotonically increasing in r. Then for a given nominal value of x the production yield can be measured by the maximal value of $r = r(x)$ satisfying

$$\{y: \|y-x\| \le r\} \subset M .$$

In order to maximize the production yield, one should choose the nominal value $\bar{x}$ so that

$$r(\bar{x}) = \max \{r(x): x \in M\} . \tag{62}$$

Setting $K = \{z: \|z\| \le 1\}$ and $y = x + rz$, $z \in K$, we see that this is a design centering problem.

Another interesting application has been described in Nguyen et al. (1985). In the diamond industry, an important problem is to cut the largest diamond of a prescribed form inside a rough stone M. This form can often be described by a convex body K. Assume that the orientation of K is fixed, i.e., only translation and dilatation of K is allowed. Then, obviously, we are led to problem (61).

Note that, in (61), we may assume that int $M \neq \emptyset$, since otherwise (61) has the solution $r = 0$.

In many cases of interest, the set M is the intersection of a number of *convex and complementary convex sets*, i.e.,

$$M = C \cap D_1 \cap \dots \cap D_m , \tag{63}$$

where C is a closed convex set satisfying int $C \neq \emptyset$, and $D_i = \mathbb{R}^n \setminus C_i$ is the complement of an open convex subset C_i of $\mathbb{R}^n$ $(i=1,\dots,m)$. We show that in this case the design centering problem is a *d.c. programming problem* (cf. Thach (1988)).

Let

$$r_M(x) := \begin{cases} \sup\{r: x + rK \subset M\} & \text{if } x \in M \\ 0 & \text{if } x \notin M \end{cases} . \tag{64}$$

Note that max $\{r: x + rK \subset M\}$ exists if M is compact. The expression (64), however, is defined for arbitrary $M \subset \mathbb{R}^n$, and we will also consider unbounded closed sets M.

Obviously, if $M = \bigcap_{j\in J} M_j$, where $J \subset \mathbb{N}$, then

$$r_M(x) = \inf\{r_{M_j}(x): j \in J\} . \tag{65}$$

Lemma X.15. *Let H be a closed halfspace. Then $r_H(x)$ is affine on H.*

Proof. Consider

$$H = \{y: cy \le \alpha\} , c \in \mathbb{R}^n, c \neq 0, \alpha \in \mathbb{R}.$$

The programming problem

$$\max \{cz: z \in K\}$$

has a solution $\bar{z}$ satisfying $\bar{v} = c\bar{z} > 0$, because K is compact, $0 \in \text{int } K$, and $c \neq 0$.

Let $x \in H$, and define

$$\rho(x) = \frac{\alpha - cx}{\bar{v}} .$$

Obviously, $\rho(x) \ge 0$ and $x + \rho(x)K \subset H$ if $\rho(x) = 0$.

Let $\rho(x) > 0$ and $z \in K$. Then

$$c(x + \rho(x)z) = cx + \rho(x)cz \le cx + \rho(x)c\bar{z} = \alpha ;$$

hence $x + \rho(x)K \subset H$.

But $c(x + r\bar{z}) > c(x + \rho(x)\bar{z}) = \alpha$ whenever $r > \rho(x)$.

Therefore, $r_H(x) = \rho(x)$, i.e., we have

$$r_H(x) = \frac{\alpha - cx}{\bar{v}} \qquad \forall x \in H . \tag{66}$$

Note that $r_H(x) = \max\{0, \frac{1}{\bar{v}}(\alpha - cx)\} \quad \forall x \in \mathbb{R}^n$. ∎

Now consider a convex polyhedral set M, i.e.,

$$M = \{y: c^i y \le \alpha_i \quad (i=1,...,m)\} , \tag{67}$$

where $c^i \in \mathbb{R}^n \setminus \{0\}$, $\alpha_i \in \mathbb{R}$ $(i=1,...,m)$.

Then it follows from (65) and (65) that

$$r_M(x) = \min_{i=1,\ldots,m} \frac{\alpha_i - c^i x}{\bar{v}_i} \qquad \forall x \in M , \tag{68}$$

where

$$\bar{v}_i = \max\{c^i z: z \in K\} > 0 \quad (i=1,...,m) . \tag{69}$$

In this case, the design centering problem reduces to maximizing $r_M(x)$ over M. But since $x \notin M$ implies that $\alpha_i - c^i x < 0$ for at least one i, we may also maximize the expression (68) over $\mathbb{R}^n$.

Proposition X.13. *Let M be a polytope of the form (67). Then the design centering problem is the linear programming problem*

$$\begin{aligned} &\textit{maximize } t \\ &s.t. \quad \alpha_i - c^i x \ge t\bar{v}_i \quad (i = 1,...,m) \end{aligned} \tag{70}$$

where $\bar{v}_i$ *is defined by (69).*

Proof. Problem (70) with the additonal variable $t \in \mathbb{R}$ is equivalent to $\max\{r_M(x): x \in \mathbb{R}^n\}$ with $r_M(x)$ of the form (68). ∎

Since (69) must be evaluated, we see that the design centering problem for convex polyhedral M requires us to solve m convex programming problems (which reduce to linear programs when K is a polytope) and one linear program. The special case when K is the unit ball of an L_p–norm, $1 \leq p \leq \infty$, is investigated further in Shiau (1984).

Proposition X.14. *Let $M = C \cap D_1 \cap \ldots \cap D_m$, where C is a closed convex subset of $\mathbb{R}^n$, int $C \neq \emptyset$, and $D_i = \mathbb{R}^n \setminus C_i$ is the complement of an open convex set in $\mathbb{R}^n$ (i=1,...,m). Assume that int $M \neq \emptyset$. Then the design centering problem is equivalent to maximizing a d.c. function over M.*

Proof. a) It is well known that $C = \bigcap_{H \in \mathcal{H}} H$, where $\mathcal{H}$ is the family of closed halfspaces con– taining C. From Lemma X.15 and (65), we see that

$$r_C(x) = \inf_{H \in \mathcal{H}} r_H(x) \quad \forall x \in C\,, \tag{71}$$

where, for any $H \in \mathcal{H}$, $r_H(x)$ is the affine function (66). Let $H = \{y: cy \leq \alpha\}$, $\bar{v}(c) = \max\,\{cz: z \in K\}$. Then we have

$$r_C(x) = \inf_{H \in \mathcal{H}} \left\{\frac{\alpha - cx}{\bar{v}(c)}\right\}, \tag{72}$$

where the infimum is taken over all c, α for which $H \in \mathcal{H}$.

Note that $r_C(x)$ is finite for every x. To see this, let $x^0 \in C$. Then for all $H = \{y: cy \leq \alpha\} \in \mathcal{H}$, we have $cx^0 \leq \alpha$, and hence

$$\frac{\alpha - cx}{\bar{v}(c)} \geq \frac{c(x^0-x)}{\bar{v}(c)} \geq -\frac{\|c\|\ \|x^0-x\|}{\bar{v}(c)}\,. \tag{73}$$

Let

$$r_0 = \min\,\{\|y\|: y \in \partial K\} > 0\,, \tag{74}$$

where ∂K denotes the boundary of K. Then we see that

$$\overline{v}(c) \geq \max \{cy: \|y\| \leq r_0\} = \|c\| r_0 . \tag{75}$$

It follows from (73) that

$$\frac{\alpha - cx}{\overline{v}(c)} \geq - \frac{\|x^0 - x\|}{r_0} .$$

Hence, $r_C(x)$ is a *finite concave function.*

b) Now consider the set $D = \mathbb{R}^n \setminus \tilde{C}$, where $\tilde{C}$ is an open convex set. Then we have $\mathrm{cl}\, \tilde{C} = \bigcap_{H \in \mathscr{H}} H$, where $\mathscr{H}$ is the family of **all** halfspaces $H = \{y: cy \leq \alpha\}$ containing $\mathrm{cl}\, \tilde{C}$.

For any $H \in \mathscr{H}$, let $H' = \{y: cy \geq \alpha\}$ be the closed halfspace opposite H, and denote by $\mathscr{H}'$ the family of **all** halfspaces H' defined in this way.

Since $H' \subset D$ for every $H' \in \mathscr{H}'$, it follows that $r_D(x) \geq r_{H'}(x)$; hence,

$$r_D(x) \geq \sup_{H' \in \mathscr{H}'} r_{H'}(x) .$$

The converse inequality obviously holds for $r_D(x) = 0$, since $r_{H'}(x) \geq 0$. Assume that $r_D(x) = \delta > 0$. Then we see from the definition of $r_D(x)$ that there exists a closed ball B centered at $x \in D$ with radius δ satisfying $B \cap \tilde{C} = \emptyset$. Therefore, there is also a hyperplane separating $\tilde{C}$ and B. In other words, there is an $H' \in \mathscr{H}'$ such that $B \subset H'$, where H' was defined above. Then, we have $r_{H'}(x) \geq r_D(x)$, and hence

$$r_D(x) \leq \sup_{H' \in \mathscr{H}'} r_{H'}(x) . \tag{76}$$

It is easy to see that $r_D(x) = \sup_{H' \in \mathscr{H}'} r_{H'}(x)$ is finite everywhere.

c) Finally, since $M = \bigcap_{i=1}^{m} D_i \cap C$, by (65) we find that

$$r_M(x) = \min \{r_C(x), r_{D_i}(x): i=1,...,m\} . \tag{77}$$

This is a d.c. function as shown in Theorem I.7. ∎

Remark X.11. Part a) of the above proof showed that $r_C(x)$ is concave if C is convex, int $C \neq \emptyset$, i.e., in this case the design centering problem is a concave maximization problem.

When C and the collection of D_i (i=1,...,m) are given by a convex inequality and m reverse convex inequalities, then we are led to a d.c.–programming problem of the form discussed in the preceding section. Note, however, that the functions $r_C(x)$, $r_D(x)$ in the proof of Proposition X.14 can, by the same expression, be extended to the whole space $\mathbb{R}^n$. But if $x \notin M$, then $r_M(x) \leq 0$, whereas $r_M(x) > 0$ in int M. Therefore, (77) can be maximized over $\mathbb{R}^n$ instead over M.

Note that in part b) of the proof it is essential to consider all supporting halfspaces of cl $\tilde{C}$.

Example X.4. Let $D = \mathbb{R}^2 \setminus \tilde{C}$, where $\tilde{C} = \{(x_1,x_2)\colon x_1 + x_2 > 0, -x_1 + x_2 > 0\}$. Obviously, cl $\tilde{C}$ is defined by the two supporting halfspaces $x_1+x_2 \geq 0$, $-x_1+x_2 \geq 0$. But finding $r_D(x)$ for given K, x may require that one considers additional halfspaces that support cl $\tilde{C}$ at (0,0). In the case of K,x as in Fig. X.3, for example, it is the halfspace $x_2 \leq 0$ that determines $r_D(x)$.

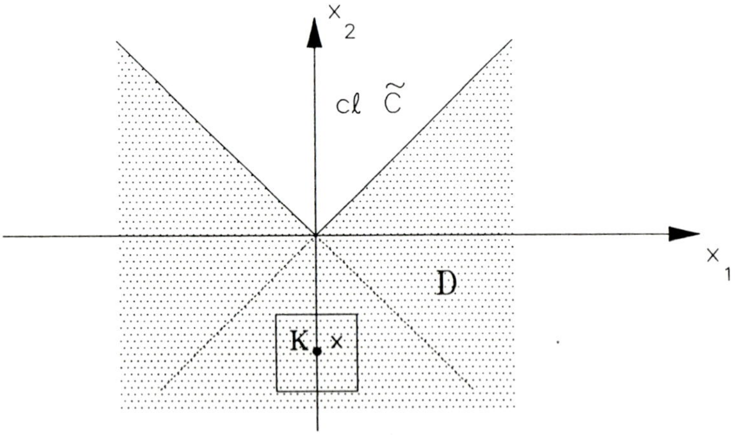

Fig. X.2. Design Centering and Supporting Halfspaces

The proof of Proposition X.14 shows that, under the assumptions there, the design centering problem can also be formulated as a semi–infinite optimization problem, i.e., a problem with an infinite number of linear constraints.

Knowledge of the d.c. structure is of little practical use if an explicit d.c. decomposition of $r_M(x)$ is not available. Such a d.c. representation is not known unless additional assumptions are made. To see how difficult this general d.c. problem is, let

$$x + rK = \{y \in \mathbb{R}^n: p(y-x) \le r\} , \tag{78}$$

where $p(z) = \inf \{\lambda > 0: z \in \lambda K\}$ is the *Minkowski functional of K* (cf. Thach (1988), Thach and Tuy (1988)). Then it is readily seen that

$$r_M(x) = \inf \{p(y-x): y \notin M\} . \tag{79}$$

Suppose that p(z) is given. (Note that $p(z) = \|z\|_N$ if K is the unit ball with respect to a norm $\|z\|_N$.) Then we have

$$r_{D_i}(x) = \inf \{p(y-x): y \in C_i\} , \tag{80}$$

which amounts to solving a convex programming problem, whereas

$$r_C(x) = \inf \{p(y-x): y \in \mathbb{R}^n \setminus C\} \tag{81}$$

requires minimizing a convex function over the complement of a convex set.

Since $r_M(x)$ is d.c. everywhere, it is also Lipschitzian on compact subsets of $\mathbb{R}^n$. A Lipschitz–constant can be found in the following way.

Proposition X.15. *Let M be given as in Proposition X.14. Assume that M is bounded and int $M \neq \emptyset$. Then $r_M(x)$ is Lipschitzian with Lipschitz–constant*

$$L = \frac{1}{r_0},$$

where $r_0 = \min \{\|x\|: x \in \partial K\}$.

Proof. Let $p(z)$ denote the Minkowski functional of K such that $K = \{z: p(z) \leq 1\}$. Then it is well known that for $y \in \mathbb{R}^n$ we have $p(y) = \frac{\|y\|}{\|\bar{y}\|}$, where $\|\cdot\|$ denotes the Euclidean norm, and where $\bar{y}$ is the intersection point of the ray $\{\rho y: \rho \geq K\}$ with the boundary ∂K. Therefore, for arbitrary $x^1, x^2 \in \mathbb{R}^n$, it follows that

$$p(x^2 - x^1) \leq \frac{1}{r_0} \|x^2 - x^1\| .$$

Using

$$p(y-x^1) \leq p(y-x^2) + p(x^2-x^1)$$

and

$$r_M(x) = \inf \{p(y-x): y \notin M\}$$

(cf. (79)), we see that

$$r_M(x^1) \leq r_M(x^2) + \frac{1}{r_0} \|x^2-x^1\| .$$

Interchanging x^1 and x^2, we conclude that

$$|r_M(x^1) - r_M(x^2)| \leq \frac{1}{r_0} \|x^2-x^1\| .$$ ■

Two algorithmic approaches to solve fairly general design centering problems have been proposed by Thach (1988) and Thach and Tuy (1988). Thach considers the case when M is defined as in Proposition X.12 and where $p(z) = (z(Az))^{1/2}$ with symmetric positive definite $(n \times n)$ matrix A. He reduces the design centering problem to concave minimization and presents a cutting plane method for solving it. However, there is a complicated, implicitly given function involved that requires that in each step of the algorithm, in addition to the calculation of new vertices generated by a cut, one minimizes a convex function over the complement of a convex set and solves several convex minimization problems. However, when C and K are polytopes, only linear programs have to be solved in this approach (cf. Thach (1988)).

Thach and Tuy (1988) treat the design centering problem in an even more general setting and develop an approach via so-called relief indicators which will be discussed in Section XI.4.

The first numerical tests (Boy (1988)) indicate, however, that – as expected from the complicated nature of the general problems considered – the practical impact of both methods is limited to very small problems.

4.2. The Diamond Cutting Problem

In the diamond cutting problem mentioned at the beginning of this section, the design centering problem can be assumed to have a polyhedral structure, i.e., K is a polytope and $M = C \cap D_1 \cap \ldots \cap D_m$, where C is a polytope and each D_i is the complement of an open convex polyhedral set C_i $(i=1,\ldots,m)$. Let

$$K = \{y: a^i y + \alpha_i \le 0 \, , \, i \in I\} \, , \tag{82}$$

$$C = \{y: b^j y + \beta_j \le 0 \, , \, j \in J\} \, , \tag{83}$$

and

$$C_i = \{y: c^{i,k} y + \gamma_{i,k} > 0, \, k \in K_i\} \, , \quad (i=1,\ldots,m) \, ,$$

where I, J, K_i are finite index sets; $a^i, b^j, c^{i,k} \in \mathbb{R}^n$; and $\alpha_i, \beta_j, \gamma_{i,k} \in \mathbb{R}$.

Then we have

$$D_i = \{y: \min_{k \in K_i} (c^{i,k} y + \gamma_{i,k}) \le 0\} \quad (i = 1,\ldots,m) \, , \tag{84}$$

and, according to Proposition X.14,

$$r_M(x) = \min \{r_C(x), r_{D_i}(x): i=1,\ldots,m\} \, . \tag{85}$$

It follows from Lemma X.15 (see also (68)) that

$$r_C(x) = \begin{cases} \min\limits_{j \in J} [-\frac{1}{\bar{v}_j}(\beta_j + b^j x)] , & x \in C \\ 0 & , x \notin C \end{cases} , \tag{86}$$

where $\bar{v}^i$ is the optimal value of the linear programming problem

$$\max \{b^j z: z \in K\} \quad (j \in J) .$$

Evaluating $r_{D_i}(x)$ $(i=1,...,m)$ by means of (80) amounts to solving m convex programming problems. This operation has to be repeated for each iteration point $x = x^k$ of the global optimization algorithm that is applied to solve the design centering problem. A computationally less expensive procedure for determining $r_{D_i}(x)$ can be derived under some additional assumptions that are reasonable, e.g., for the diamond cutting problem.

Consider the case $n = 3$, and replace D_i , C_i by D, $\tilde{C}$, respectively, i.e., we have $D = \mathbb{R}^3 \setminus \tilde{C}$, where $\tilde{C}$ is an open polyhedral convex set.

Assume that all of the extreme points, edges and facets of cl $\tilde{C}$ and of K are known.

Recall from part b) of the proof of Proposition X.14 that

$$r_D(x) = \sup_{H' \in \mathcal{H}'} r_{H'}(x) , \tag{87}$$

where $\mathcal{H}'$ is the set of closed halfspaces $H' = \{y: cy \geq \alpha\}$ determined by the supporting planes of cl $\tilde{C}$ which satisfy cl $\tilde{C} \subset \{y: cy \leq \alpha\}$. Observe that (87) is not useful for computing $r_D(x)$, because $\mathcal{H}'$ may contain an infinite number of elements. Following Nguyen and Strodiot (1988), we show that, by the above assumptions, one can find a *finite* subcollection $\mathcal{G}'$ of $\mathcal{H}'$ such that for all $x \in \mathbb{R}^3$

$$r_D(x) = \max_{H' \in \mathcal{G}'} r_{H'}(x) . \tag{88}$$

Then computing $r_D(x)$ amounts to using a formula similar to (86) (cf. also (66)) a finite number of times.

Proposition X.16. *If $\mathcal{G}'$ is a subcollection of $\mathcal{H}'$ such that*

$$\forall x \in D, \exists H'_0 \in \mathcal{G}' \text{ such that } x + r_D(x)K \subset H'_0 , \tag{89}$$

then for all $x \in \mathbb{R}^3$ we have

$$r_D(x) = \max_{H' \in \mathcal{G}'} r_{H'}(x) .$$

Proof. From the inclusion $\mathcal{G}' \subset \mathcal{H}'$ and (87) it follows that for all $x \in \mathbb{R}^3$ one has

$$0 \le \max_{H' \in \mathcal{G}'} r_{H'}(x) \le \sup_{H' \in \mathcal{H}'} r_{H'}(x) = r_D(x) . \tag{90}$$

But if $x \in D$, then by (86) we have $x + r_D(x)K \subset H'_0$ for some $H'_0 \in \mathcal{G}'$; hence, by the definition of $r_{H'}$,

$$r_D(x) \le r_{H'}(x) \le \max_{H' \in \mathcal{G}'} r_{H'}(x) .$$

Therefore, from (90) we see that $r_D(x) = \max_{H' \in \mathcal{G}'} r_{H'}(x)$, whenever $x \in D$.

On the other hand, when $x \notin D$, we have $r_D(x) = 0$ by definition, and thus, again using (87), we obtain the required equality. ■

Since cl $\tilde{C}$ is polyhedral in $\mathbb{R}^3$, there are three types of supporting planes for cl $\tilde{C}$: the ones containing a facet of cl $\tilde{C}$, the ones containing only a single extreme point of cl $\tilde{C}$ and the ones containing only one edge of cl $\tilde{C}$. Corresponding to this classification, $\mathcal{G}'$ will be constructed as the union of three *finite* subsets $\mathcal{G}_1'$, $\mathcal{G}_2'$ and $\mathcal{G}_3'$ of $\mathcal{H}'$.

$\mathcal{G}_1'$ will be generated by all of the supporting planes corresponding to the facets of cl $\tilde{C}$. Since cl $\tilde{C}$ has only a finite number of facets, the collection $\mathcal{G}_1'$ is finite. Let f be a facet of cl $\tilde{C}$, and let p_1 , p_2 , p_3 be three affinely independent points which characterize it. Suppose that they are numbered counter clockwise, when cl $\tilde{C}$ is viewed from outside. Then the halfspace $H' \in \mathscr{G}_1'$ which corresponds to the facet f has the representation

$$cy \geq \alpha ,$$

where c and α are defined by

$$c = (p_2-p_1) \times (p_3-p_1) \text{ and } \alpha = cp_1 .$$

Here "×" denotes the usual cross product in $\mathbb{R}^3$.

The set $\mathscr{G}_2'$ will be defined as follows. To each vertex p of cl $\tilde{C}$ and to each facet v of K we associate (if it exists) the plane

$$cy = \alpha$$

which is parallel to the facet v and passes through the vertex p but not through any other vertex of cl $\tilde{C}$. Moreover, it is required that

$$\text{cl } \tilde{C} \subset \{y: cy \leq \alpha\} \text{ and } K \subset \{y: cy \geq cq\} , \tag{91}$$

where q is any point of the facet v. Condition (91) ensures that $\{y: cy \geq \alpha\} \in \mathscr{H}'$ and that K is contained in the parallel halfspace $\{y: cy \geq cq\}$.

Computationally, $\mathscr{G}_2'$ can be obtained in the following way. Consider each couple (p,v), where p in a vertex of cl $\tilde{C}$ and v is a facet of K. Let q_1 , q_2 , q_3 be three affinely independent points which determine the facet v. Suppose that they are numbered counter clockwise when K is viewed from outside. Then

$$cy = \alpha$$

with $c = (q_3-q_1) \times (q_2-q_1)$, and $\alpha = cp$ represents a plane containing p which is parallel to v and satisfies $K \subset \{y \in \mathbb{R}^3: cy \geq cq\}$ $\forall q \in v$. The halfspace $\{y: cy \geq \alpha\}$ will be put into $\mathscr{G}_2'$ if cl $\tilde{C} \subset \{y: cy \leq \alpha\}$.

The latter condition can be verified by considering the edges of cl $\tilde{C}$ emanating from p.

Let each edge of cl $\tilde{C}$ emanating from p be represented by a point $\bar{p} \neq p$ on it. Since $\alpha = cp$, we have cl $\tilde{C} \subset \{y: cy \leq \alpha\}$ if $c\bar{p} \leq \alpha$ for of all these points (edges) $\bar{p}$. The collection $\mathcal{G}_2'$ is finite because there exist only a finite number of vertices p and facets v.

Finally, $\mathcal{G}_3'$ is defined in the following way. To each edge e of cl $\tilde{C}$ and to each edge w of K which is not parallel to e, we associate (if it exists) the plane

$$cy = \alpha$$

that contains the edge e and is parallel to the edge w. Then as in (91) we have

$$\text{cl } \tilde{C} \subset \{y: cy \leq \alpha\} \text{ and } K \subset \{y: cy \geq cq\} , \tag{92}$$

where q is now an arbitrary extreme point of w. Condition (92) means that $\{y: cy \geq \alpha\} \in \mathcal{H}'$, and that K is contained in the parallel halfspace $\{y: cy \geq cq\}$. Computationally, for each pair (e,w), where e is an edge of cl $\tilde{C}$ and w is an edge of K, one can proceed as follows. Let p_1 , p_2 be two points defining e, and let q_1 , q_2 be two points defining w. Set $E_{ew} = (p_1-p_2) \times (q_1-q_2)$. If $E_{ew} = 0$, then e and w are parallel to each other, and we must consider another pair of edges. Otherwise, we set $c = \pm E_{ew}$, where the plus or minus sign is determined in such a way that $cq_1 < 0$, and we set $\alpha = cp_1$. Note that $cq_1 < 0$ is always possible, because $0 \in \text{int } K$. Then the plane $cy = \alpha$ contains the edge e and is parallel to the edge w. Moreover, we have $0 \in \{y: cy \geq cq\}$ for any extreme point q of w. The halfspace $\{y: cy \geq \alpha\}$ will be put into $\mathcal{G}_3'$ if (92) is fulfilled, i.e., if $cp \leq \alpha$ for each point p which defines an edge of cl $\tilde{C}$ emanating from p_1 or p_2 , and if $cr \geq cq_1$ for each extreme point r of K.

Since there exist only a finite number of edges e and only a finite number of edges w, the collection $\mathcal{G}_3'$ is finite.

Finally, we set

$$\mathcal{G}' = \mathcal{G}_1' \cup \mathcal{G}_2' \cup \mathcal{G}_3' . \tag{90}$$

In order to prove that the collection $\mathcal{G}'$ satisfies condition (86) of Proposition X.16, we establish the following lemma (cf. Nguyen and Strodiot (1988)).

Lemma X.16. *Let $x \in D$, and consider the halfspaces $H' = \{y: cy \geq \alpha\} \subset \mathcal{H}'$ whose generating plane $P = \{y: cy = \alpha\}$ separates cl $\tilde{C}$ and $x + r_D(x)K$. Denote by $\mathcal{P}_x$ the set of these planes. Then the following assertions hold.*

(i) $\quad$ *cl $\tilde{C} \cap (x + r_D(x)K) \neq \emptyset$;*

(ii) $\quad$ *$\mathcal{P}_x \neq \emptyset$, and $(x + r_D(x)K) \cap$ cl $\tilde{C} \subset P \quad \forall P \in \mathcal{P}_x$;*

(iii) $\quad$ *if $P \in \mathcal{P}_x$ contains a relative interior point of a facet of cl $\tilde{C}$ (or of $x + r_D(x)K$, respectively), then P contains the whole facet;*

(iv) $\quad$ *if $P \in \mathcal{P}_x$ contains a relative interior point of an edge of cl $\tilde{C}$ (or of $x + r_D(x)K$, respectively), then P contains the whole edge;*

(v) $\quad$ *if $P \in \mathcal{P}_x$ contains an extreme point of cl $\tilde{C}$ and a facet of $x + r_D(x)K$, then we have $H' \in \mathcal{G}_2'$ for the halfspace H' corresponding to P;*

(vi) $\quad$ *if $P \in \mathcal{P}_x$ contains an edge of cl $\tilde{C}$ and an edge of $x + r_D(x)K$, then we have $H' \in \mathcal{G}_3'$ for the halfspace H' corresponding to P.*

Proof. In order to prove (i), we exhibit a point $y^* \in$ cl $\tilde{C} \cap (x + r_D(x)K)$. From (80) we see that $r_D(x) = \inf \{p(y-x): y \in \tilde{C}\}$. It follows, by the continuity of the Minkowski functional and the assumptions on D, that $r_D(x)$ is attained in cl $\tilde{C}$, i.e., there is a $y^* \in$ cl $\tilde{C}$ such that $p(y^*-x) = r_D(x)$. But the point y^* also belongs to $x+p(y^*-x)K$ because, by the well–known properties of p(z), we have $(y^*-x) / p(y^*-x) \in K$.

Since $x + r_D(x)K \subset D$ and $\tilde{C} \cap D = \emptyset$, assertion (ii) follows from (i) by a well–known classical theorem on separation of convex sets (cf., e.g., Rockafellar (1970)).

Finally, assertions (iii) and (iv) are straightforward because P is a plane separating cl $\tilde{C}$ and $x + r_D(x)K$, whereas (v) and (vi) are immediate consequences of the definition of $\mathcal{G}_2'$ and $\mathcal{G}_3'$. ■

The following proposition is also due to Nguyen and Strodiot (1988).

Proposition X.17. *The collection $\mathcal{G}'$ is finite and satisfies property (89) of Proposition X.16.*

Proof. Finiteness of $\mathcal{G}' = \mathcal{G}_1' \cup \mathcal{G}_2' \cup \mathcal{G}_3'$ has already been demonstrated above.

Let $x \in D$. First suppose that $x \in$ cl $\tilde{C}$. Then x belongs to a facet f of cl $\tilde{C}$. The supporting plane P of cl $\tilde{C}$ which contains f, determines a hyperplane $H' \in \mathcal{G}_1'$ which satisfies $x + r_D(x)K = \{x\} \subset P \subset H'$.

Next suppose that $x \notin$ cl $\tilde{C}$. Denote by Y^* the intersection of cl $\tilde{C}$ and $x + r_D(x)K$. From Lemma X.16 (i) and (ii) we know that $Y^* \neq \emptyset$ and $\mathcal{P}_x \neq \emptyset$. We consider several cases according to the dimension dim Y^* of Y^* (recall that dim Y^* is defined as the dimension of the affine hull of Y^*).

Case 1: dim $Y^* = 2$:

Let $P \in \mathcal{P}_x$. Then, by Lemma X.16 (ii), P contains a facet of cl $\tilde{C}$, and the halfspace $H' \in \mathcal{H}'$ generated by P belongs to $\mathcal{G}_1'$. Since $x + r_D(x)K \subset H'$, condition (89) is satisfied.

Case 2: dim $Y^* = 1$:

Since $x + r_D(x)K$ is bounded and dim $Y^* = 1$, we see that Y^* is a closed, bounded interval which does not reduce to a singleton. Consider the following three subcases.

Case 2.1: *Y^* is part of an edge of $x + r_D(x)K$ but is not contained in an edge of cl $\tilde{C}$.*

Then Y^* must contain a relative interior point of a facet of cl $\tilde{C}$. Let $P \in \mathcal{P}_x$. We see from Lemma X.16(ii) that, because $Y^* \in P$, it follows that P contains a relative interior point of a facet of cl $\tilde{C}$, and thus, by Lemma X.16 (iii), it contains the whole facet. Consequently, the halfspace $H' \in \mathcal{H}'$ generated by P again belongs to $\mathcal{G}_1'$, and condition (89) is implied by the inclusion $x + r_D(x)K \subset H'$.

Case 2.2: *Y^* is part of an edge of cl $\tilde{C}$ but is not contained in an edge of $x + r_D(x)K$.*

As in the case 2.1 above, we conclude that now each $P \in \mathcal{P}_x$ contains a facet of $x + r_D(x)K$. By Lemma X.16 (iv), we see that P also contains an edge of cl $\tilde{C}$. Hence, by Lemma X.16 (v), it follows that the halfspace $H' \in \mathcal{H}'$ generated by P belongs to $\mathcal{G}_2'$, and condition (89) holds because $x + r_D(x)K \subset H'$.

Case 2.3: *Y^* is part of an edge e of cl $\tilde{C}$ and part of an edge $\bar{w}$ of $x + r_D(x)K$.*

The edge $\bar{w}$ has the form $x + r_D(x)w$, where w is an edge of K. Let f_1 and f_2 be the two facets of cl $\tilde{C}$ such that $e = f_1 \cap f_2$, and let v_1 and v_2 be the two facets of K determining $w = v_1 \cap v_2$. Then, by Lemma X.16 (ii) and (iv), each $P \in \mathcal{P}_x$ contains the two colinear edges e and $x + r_D(x)w$. Therefore, among all planes $P \in \mathcal{P}_x$, there exists at least one plane P' which contains one of the four facets f_1, f_2, $x + r_D(x)v_1$, $x + r_D(x)v_2$. Let $H' \in \mathcal{H}'$ be the halfspace corresponding to P'. Then we have $H' \in \mathcal{G}_1'$ if $f_1 \in P'$ or $f_2 \in P'$. Otherwise, by Lemma X.16 (v), one has $H' \in \mathcal{G}_2'$. In any case, $x + r_D(x)K \subset H'$, and condition (89) is satisfied.

Case 3: dim $Y^* = 0$:

In this case, when Y^* is reduced to a singleton $Y^* = \{y^*\}$, we must consider six subcases according to the position of y^* with respect to cl $\tilde{C}$ and $x + r_D(x)K$.

Case 3.1: *y^* is a vertex of $x + r_D(x)K$ and belongs to the relative interior of a facet of cl $\tilde{C}$.*

Let $P \in \mathcal{P}_x$, and let $H' \in \mathcal{H}'$ be the corresponding halfspace. By Lemma X.16 (ii) and (iii), P contains a facet of cl $\tilde{C}$. Thus we have $H' \in \mathcal{G}_1'$ and, since

$x + r_D(x)K \subset H'$, condition (89) is satisfied.

Case 3.2: *y^* is a vertex of cl $\tilde{C}$ and belongs to the relative interior of a facet of $x + r_D(x)K$.*

Since $P \in \mathcal{P}_x$ and the corresponding halfspace $H' \in \mathcal{H}'$, we see from Lemma X.16 (ii) and (iii) that P contains a vertex of cl $\tilde{C}$ and a facet of $x + r_D(x)K$; hence, by Lemma X.16 (v), $H' \in \mathcal{G}_2'$. Since $x + r_D(x)K \subset H'$, it follows that condition (89) is satisfied.

Case 3.3: *y^* is a relative interior point of both an edge of cl $\tilde{C}$ and an edge of $x + r_D(x)K$.*

Since dim $Y^* = 0$ we see that these two edges cannot be collinear. Let $P \in \mathcal{P}_x$, and let $H' \in \mathcal{H}'$ be generated by P. Lemma X.16 (ii) and (iv) show that P contains an edge of cl $\tilde{C}$ and an edge of $x + r_D(x)K$ which are not collinear. Therefore, by Lemma X.16 (vi), $H' \in \mathcal{G}_3'$ and (86) is satisfied, because $x + r_D(x)K \subset H'$.

Case 3.4: *y^* is a vertex of $x + r_D(x)K$ and belongs to the relative interior of an edge e of cl $\tilde{C}$.*

Let f_1 and f_2 be the two facets of cl $\tilde{C}$ which determine e, i.e., $e = f_1 \cap f_2$. Then, by Lemma X.16 (ii) and (iv), each $P \in \mathcal{P}_x$ contains e. Therefore, there exists at least one plane $P' \in \mathcal{P}_x$ which contains either one of the faces f_1 and f_2 or an edge of $x + r_D(x)K$ emanating from x. (Observe that in the latter case this edge of $x + r_D(x)K$ cannot be collinear with e because $Y^* = \{y^*\}$ is a singleton.) If $f_1 \subset P'$ or $f_2 \subset P'$, then the halfspace $H' \in \mathcal{H}'$ generated by P' belongs to $\mathcal{G}_1'$; otherwise, by Lemma X.16 (vi), we have $H' \in \mathcal{G}_3'$. In any case $x + r_D(x)K \subset H'$, and (89) is satisfied.

Case 3.5: *y^* is a vertex of cl $\tilde{C}$ and belongs to the relative interior of an edge of $x + r_D(x)K$.*

The proof in this case is similar to Case 3.4.

Case 3.6: *y^* is a vertex of cl $\tilde{C}$ and a vertex of $x + r_D(x)K$.*

In this case, there exists at least one plane $P \in \mathcal{P}_x$ which contains either an edge of cl $\tilde{C}$ or an edge of $x + r_D(x)K$. Since both cases can be examined in a similar way, we suppose that there exist an edge e of cl $\tilde{C}$ and a plane $P \in \mathcal{P}_x$ satisfying $e \subset P$. Two possibilities can occur. If none of the edges of $x + r_D(x)K$ is collinear with e, then we can argue in the same way as in Case 3.4 to conclude that (89) holds.

Otherwise, there is an edge w of K such that $x + r_D(x)w$ is collinear with e. We can argue as in Case 2.3. Let f_1 and f_2 be the two facets of cl $\tilde{C}$ satisfying $e = f_1 \cap f_2$, and let v_1 and v_2 be the two facets of K determining $w = v_1 \cap v_2$. Then, among all of the planes $P \in \mathcal{P}_x$ containing e, there exists at least one plane P' which contains one of the four facets f_1, f_2, $x + r_D(x)v_1$, $x + r_D(x)v_2$. Let $H' \in \mathcal{H}'$ be the halfspace generated by P'. Then we have $H' \in \mathcal{G}_1'$ if $f_1 \in P'$ or $f_2 \in P'$. Otherwise, using Lemma X.16 (v), we see that $H' \in \mathcal{G}_2'$. In any case, $x + r_D(x)K \subset H'$, and condition (89) is satisfied. ■

Example X.5. An application of the above approach to the diamond cutting and dilatation problem is reported in Nguyen and Strodiot (1988). In all the tests discussed there, the reference diamond K has 9 vertices and 9 facets and the rough stone is a nonconvex polyhedron of the form $M = C \cap D_1$, $D_1 = \mathbb{R}^n \setminus C_1$ as described in (80), (81) with 9 facets and one "nonconvexity" D_1 determined by four planes. First, from the vertices, edges and facets of K and M the three finite collections $\mathcal{G}_1'$, $\mathcal{G}_2'$, $\mathcal{G}_3'$ were automatically computed. Then the design centering problem was solved by Algorithm X.2'. The following figure shows the reference diamond K, the rough stone M and the optimal diamond inside the rough stone for one of the test examples. In this example $\mathcal{G}_1'$ has 9 elements, $\mathcal{G}_2'$ has 1 element and $\mathcal{G}_3'$ has 3 elements. The optimal dilatation γ is 1.496.

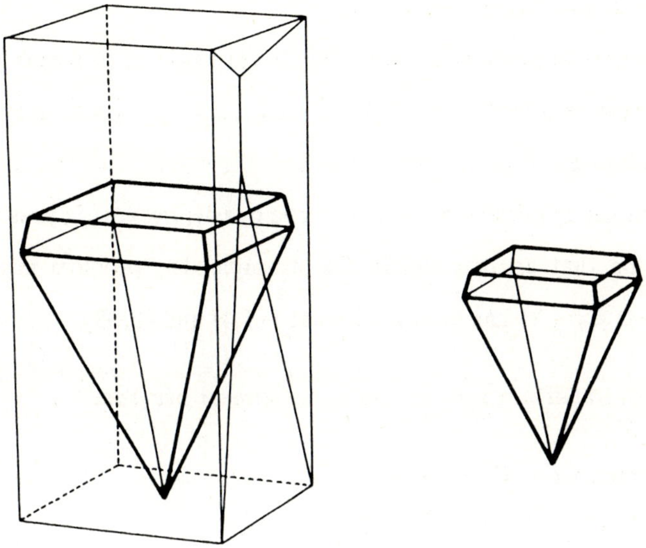

Fig. X.3. Diamond problem

4.3. Biconvex Programming and Related Problems

In this section, we consider a special *jointly constrained biconvex programming problem* namely

$$\text{(SBC)} \qquad \begin{array}{l} \text{minimize } F(x,y) = f(x) + xy + g(y) \\ \quad \text{s.t. } (x,y) \in K \cap R \end{array}, \tag{94}$$

where:

(a) $R = \{(x,y) \in \mathbb{R}^n \times \mathbb{R}^n : \underline{a} \le x \le \bar{a},\ \underline{b} \le y \le \bar{b}\}$ with

$\underline{a}, \bar{a}, \underline{b}, \bar{b} \in \mathbb{R}^n$, $\underline{a} < \bar{a}$, $\underline{b} < \bar{b}$;

(b) K is a closed convex set in $\mathbb{R}^{2n}$;

(c) f, g are real–valued convex functions on an open set containing R.

Problem (SBC) is a d.c. programming problem since $xy = \frac{1}{4}(\|x+y\|^2 - \|x-y\|^2)$ is a d.c. function (cf. Example I.6).

Some immediate extensions of problem (SBC) are obvious. For example, a term $x(Ay)$ can be transformed into xz if the linear constraint $z = Ay$ is included among the constraints defining K.

Note that, even though each term x_iy_i in xy is quasiconvex, it is possible to have proper local optima that are not global. For example, the problem $\min \{xy: (x,y) \in \mathbb{R}^2, -1 \le x \le 2, -2 \le y \le 3\}$ has local minima at (–1,3) and (2,–2).

An often treated important special case is the *bilinear program*

$$\text{(BLP)} \qquad \begin{array}{ll} \text{minimize} & px + x(Cy) + qy \\ \text{s.t.} & x \in X\,,\ y \in Y \end{array}, \qquad (95)$$

where $p, q \in \mathbb{R}^n$; C is a $(n \times n)$ matrix, and X and Y are polytopes in $\mathbb{R}^n$. This problem is treated in Section IX.1. In Section I.2.4, we showed that problem (BLP) is equivalent to a special concave minimization problem. From the form of this equivalent concave minimization problem (Section I.2.4) and the corresponding well–known property of concave functions it follows that (BLP) has an optimal solution $(\bar{x},\bar{y})$, where $\bar{x}$ is an extreme point of X and $\bar{y}$ is an extreme point of Y (cf. Theorem I.1 and Section I.2.4). As pointed out by Al–Khayyal and Falk (1983), this property is lost in the *jointly constrained* case.

Example X.5. The problem in $\mathbb{R}^2$

$$\begin{array}{ll} \text{minimize} & (-x + xy - y) \\ \text{s.t.} & -6x + 8y \le 3 \\ & 3x - y \le 3 \\ & 0 \le x,\ y \le 5 \end{array}$$

has an optimal solution at $(\frac{7}{6}, \frac{1}{2})$ which is not an extreme point of the feasible set. Moreover, no extreme point of the feasible set is a solution.

The *jointly constrained bilinear programming problem*, however, has an optimal solution on the boundary of a compact, convex feasible set. This can be shown in a somewhat more general context, by assuming that the objective function is biconcave in the sense of the following proposition (cf. Al–Khayyal and Falk (1983)).

Proposition X.18. *Let $F(x,y)$ be a biconcave real–valued function on a compact convex set $C \subset \mathbb{R}^n \times \mathbb{R}^n$, i.e., $F(x,\cdot)$ and $F(\cdot,y)$ are both concave on C. If $\min F(C)$ exists, then it is attained on the boundary ∂C of C.*

Proof. Assume that $\min F(C)$ is attained at an interior point $(\bar{x},\bar{y})$ of C. Let $(x^*,y^*) \in \partial C$ denote a point which minimizes the Euclidean distance between $(\bar{x},\bar{y})$ and ∂C, and let d denote this minimal distance.

Consider the point $(\tilde{x},\tilde{y}) = 2(\bar{x},\bar{y}) - (x^*,y^*)$, so that $(\bar{x},\bar{y})$ is the midpoint of the line segment $[(\tilde{x},\tilde{y}), (x^*,y^*)]$. Note that, by the definition of (x^*,y^*), we have $(\tilde{x},\tilde{y}) \in C$. Using the concavity of $F(x,\cdot)$ and $F(\cdot,y)$, respectively, we see that

$$
\begin{aligned}
F(\bar{x},\bar{y}) &= F(\tfrac{1}{2}(\tilde{x},\tilde{y}) + \tfrac{1}{2}(x^*,y^*)) \\
&\geq \tfrac{1}{2}F(\tilde{x}, \tfrac{1}{2}\tilde{y} + \tfrac{1}{2}y^*) + \tfrac{1}{2}F(x^*, \tfrac{1}{2}\tilde{y} + \tfrac{1}{2}y^*) \\
&\geq \tfrac{1}{4}[F(\tilde{x},\tilde{y}) + F(\tilde{x},y^*) + F(x^*,\tilde{y}) + F(x^*,y^*)].
\end{aligned}
$$

Note that the points $(\tilde{x},\tilde{y})$, $(\tilde{x},y^*)$, $(x^*,\tilde{y})$, (x^*,y^*) are contained in the ball $\{(x,y): \|(x,y) - (\bar{x},\bar{y})\| \leq d\}$, and hence in C.

Since $F(\bar{x},\bar{y}) = \min F(C)$ it follows that

$$F(\bar{x},\bar{y}) \leq \min \{F(\tilde{x},\tilde{y}), F(\tilde{x},y^*), F(x^*,\tilde{y}), F(x^*,y^*)]$$

which implies that

$$F(\bar{x},\bar{y}) \leq \tfrac{1}{4} [F(\tilde{x},\tilde{y}) + F(\tilde{x},y^*) + F(x^*,\tilde{y}) + F(x^*,y^*)] .$$

Again using the inequality $F(\bar{x},\bar{y}) = \min F(C)$, we then see that

$$F(\bar{x},\bar{y}) = F(x^*,y^*)$$

must hold. ■

Bilinear programming is of considerable interest because of its many applications. Recall, for example, that minimization of a (possibly indefinite) quadratic form over a polytope can be written as a bilinear problem. Another prominent example is the linear complementarity problem (cf. Chapter IX).

Several algorithms have been proposed for solving the bilinear programming problem, some of which are discussed in Section IX.1.

For the biconvex problem under consideration, observe that the methods in the preceding sections can easily be adapted to problem (SBC). Specifically, several branch and bound approaches are available. One of them is the method discussed in Section X.2 for a more general problem.

Another branch and bound algorithm is developed in Al–Khayyal and Falk (1983). Below we present a variant of this algorithm that differs from the original version in the subdivision rule which governs the refinement of partition sets and in the choice of iteration points.

Starting with $M_0 = R$, let all partition sets M be 2n–rectangles, and let the rectangles be refined by an exhaustive subdivision, e.g., by bisection. The procedure below will use bisection, but any other exhaustive subdivision will do as well (cf. also Chapter VII).

The lower bounds $\beta(M)$ will be determined by minimizing over $M \cap K$ a convex function $\phi_M(x,y)$ that underestimates F(x,y) on M. Let **$\varphi_M(x,y)$ denote the convex envelope of xy over M.** Then

$$\phi_M(x,y) = f(x) + \varphi_M(x,y) + g(y) \tag{96}$$

will be used.

The selection rule that determines the partition sets to be refined in the current iteration will be **bound improving**, as usual.

We begin by deriving a formula for $\varphi_M(x,y)$ (cf. Al–Khayyal and Falk (1983)).

Let $M = \{(x,y):= \underline{a} \le x \le \bar{a},\ \underline{b} \le y \le \bar{b}\}$ be a 2n–rectangle in $\mathbb{R}^{2n}$, and let $M_i = \{(x_i,y_i): \underline{a}_i \le x_i \le \bar{a}_i,\ \underline{b}_i \le y_i \le \bar{b}_i\}$, so that $M = M_1 \times M_2 \times \dots \times M_n$. Denote by $\varphi_{M_i}(x_i,y_i)$ the convex envelope of x_iy_i on M_i (i=1,...,n).

Proposition X.19. *(i)* $\varphi_M(x,y) = \sum_{i=1}^{n} \varphi_{M_i}(x_i,y_i)$,

(ii) *If* $f(x) = cx$ *and* $g(y) = dy$, $c,d \in \mathbb{R}^n$, *then* $cx + \varphi_M(x,y) + dy$ *is the convex envelope of* $F(x,y)$ *over M.*

Proof. Part (i) is an immediate consequence of Theorem IV.8, and part (ii) follows from Theorem IV.9 ■

Proposition X.20. $\varphi_{M_i}(x_i,y_i) = max\,\{\underline{b}_i x_i + \underline{a}_i y_i - \underline{a}_i\underline{b}_i,\ \bar{b}_i x_i + \bar{a}_i y_i - \bar{a}_i\bar{b}_i\}$.

Proof. We temporarily drop the subscripts in $\varphi_{M_i}, x_i, y_i, \underline{a}_i, \bar{a}_i, \underline{b}_i, \bar{b}_i$ and consider $\varphi_M(x,y)$, where x and y are now real variables rather than vectors.

Recall that the convex envelope of a function h on M may be equivalently defined as the pointwise supremum of all affine functions which underestimate h over M. Since $x - \underline{a} \ge 0$ and $y - \underline{b} \ge 0$, it follows after multiplication that

$$xy \ge \underline{b}x + \underline{a}y - \underline{a}\underline{b} ,$$

so that $\ell_1(x,y) = \underline{b}x + \underline{a}y - \underline{a}\underline{b}$ underestimates xy over M. Similarly, it follows from $\bar{a} - x \ge 0$ and $\bar{b} - y \ge 0$ that $\ell_2(x,y) = \bar{b}x + \bar{a}y - \bar{a}\bar{b}$ underestimates xy over M. Hence

$$\varphi(x,y) = \max\,\{\ell_1(x,y),\ \ell_2(x,y)\}$$

is a convex underestimating function for xy over M. In addition, a simple computation shows that $\varphi(x,y)$ agrees with xy at the four extreme points of M.

Let M_1, $M_2 \subset M$ be the closed triangle below and above the diagonal joining $(\underline{a},\bar{b})$ and $(\bar{a},\underline{b})$, respectively. Then it is easy to see that

$$\varphi(x,y) = \begin{cases} \ell_1(x,y) \,,\ (x,y) \in M_1 \\ \ell_2(x,y) \,,\ (x,y) \in M_2 \end{cases} . \tag{97}$$

If φ were not the convex envelope of xy over M, there would be a third affine function $\ell_3(x,y)$ underestimating xy over M such that

$$\varphi(\bar{x},\bar{y}) < \ell_3(\bar{x},\bar{y}) \text{ for some } (\bar{x},\bar{y}) \in M . \tag{98}$$

Suppose that $(\bar{x},\bar{y}) \in M_1$. Then $(\bar{x},\bar{y})$ is a unique convex combination of the three extreme points v^1,v^2,v^3 of M_1. Hence, for every affine function ℓ one has

$$\ell(\bar{x},\bar{y}) = \sum_{i=1}^{3} \lambda_i \ell(v^i)$$

with uniquely determined $\lambda_i \geq 0$ (i=1,...,3), $\sum_{i=1}^{3} \lambda_i = 1$. But since φ agrees with xy at these extreme points and ℓ_3 underestimates xy there, by (97) we must have

$$\varphi(\bar{x},\bar{y}) \doteq \sum_{i=1}^{3} \lambda_i \varphi(v^i) \geq \sum_{i=1}^{3} \lambda_i \ell_3(v^i),$$

contradicting (98). A similar argument holds when $(\bar{x},\bar{y}) \in M_2$. ■

Algorithm X.6

Step 0 (Initialization):

Set $\mathcal{M}_0 = \{M\}$, where M = R, and determine $\phi_M(x,y) = f(x) + \varphi_M(x,y) + g(y)$ according to Proposition X.19 (i) and Proposition X.20.

Solve the convex minimization problem

$$(P_M) \qquad \begin{array}{l} \text{minimize } \phi_M(x\,,\,y) \\ \text{s.t. } (x,y) \in M \cap K \end{array}$$

to determine $\beta_0 = \min \phi_M(M \cap K)$. Let S_M be the finite set of iteration points in $M \cap K$ obtained while solving (P_M). Set $\alpha_0 = \min F(S_M)$ and $(x^0,y^0) \in \operatorname{argmin} F(S_M)$.

If $\alpha_0 - \beta_0 = 0$ ($\le \varepsilon$), then stop. (x^0,y^0) is an (ε–)optimal solution.

Step k=1,2,...:

At the beginning of Step k we have the current partition $\mathcal{M}_{k-1}$ of a subset of R still of interest. Furthermore, for every $M \in \mathcal{M}_{k-1}$, we have $S_M \subset M \cap K$ and the bounds $\beta(M)$, $\alpha(M)$ satisfying $\beta(M) \le \min F(M \cap K) \le \alpha(M)$. Moreover, the current lower and upper bounds β_{k-1}, α_{k-1} satisfying $\beta_{k-1} \le \min F(K \cap R) \le \alpha_{k-1}$ are at hand, and we have a subset $\bar{\mathcal{M}}_{k-1}$ of $\mathcal{M}_{k-1}$ whose elements are the partition sets M such that $\beta_{k-1} = \beta(M)$. Finally, the current iteration point $(x^{k-1},y^{k-1}) \in K \cap R$ is the best feasible point obtained so far, i.e., one has $F(x^{k-1},y^{k-1}) = \alpha_{k-1}$.

k.1. Delete all $M \in \mathcal{M}_{k-1}$ satisfying $\beta(M) > \alpha_{k-1}$.

Let $\mathcal{R}_k$ be the collection of remaining members of $\mathcal{M}_{k-1}$.

k.2. Select a collection $\mathcal{P}_k \subset \mathcal{R}_k$ satisfying $\bar{\mathcal{M}}_{k-1} \subset \mathcal{P}_k$, and bisect each member of $\mathcal{P}_k$. Let $\mathcal{P}'_k$ be the collection of all new partition elements.

k.3. For each $M \in \mathcal{P}'_k$:

Determine $\phi_M(x,y)$ according to Proposition X.19 (i) and Proposition X.20. Solve the convex minimization problem (P_M). Delete M if the procedure for solving (P_M) detects $M \cap K = \emptyset$.

Let $\mathcal{M}'_k$ be the collection of all remaining members of $\mathcal{P}'_k$.

k.4. For each $M \in \mathcal{M}'_k$:

Let S_M be a finite set of iteration points in $M \cap K$ known so far (S_M contains iteration points obtained while solving (P_M) and best feasible points known from iteration k–1). Set $\alpha(M) = \min F(S_M)$, $\beta(M) = \min \phi_M(K \cap M)$.

k.5. Set $\mathcal{M}_k = (\mathcal{R}_k \setminus \mathcal{P}_k) \cup \mathcal{M}'_k$.

Compute

$$\alpha_k = \min \{\alpha(M): M \in \mathcal{M}_k\}, \ \beta_k = \min \{\beta(M): M \in \mathcal{M}_k\}.$$

Let $(x^k,y^k) \in K \cap R$ be such that $f(x^k,y^k) = \alpha_k$, and set $\bar{\mathscr{M}}_k = \{M \in \mathscr{M}_k: \beta_k = \beta(M)\}$.

If $\alpha_k - \beta_k = 0$ ($\leq \varepsilon$), then stop. (x^k,y^k) is an (ε–)optimal solution. Otherwise, go to Step k+1.

The following proposition shows convergence of the procedure ($\varepsilon = 0$).

Proposition X.21. *If Algorithm X.5 does not terminate after a finite number of iterations, then the sequence $\{x^k,y^k\}$ has accumulation points, and every accumulation point of $\{(x^k,y^k)\}$ is an optimal solution of problem (91) satisfying*

$$\lim_{k\to\infty} \beta_k = \min F(K \cap R) = \lim_{k\to\infty} \alpha_k = \lim F(x^k,y^k) . \tag{99}$$

Proof. Proposition X.21 can be derived from the general theory of branch and bound methods presented in Chapter IV.. We refer to Theorem IV.3 and Corollary IV.2.

Since the functions f, g are convex on an open set containing R, we have continuity of F on R. Obviously, because of the compactness of the feasible set $D = K \cap R$ we then see that $\{(x,y) \in D: F(x,y) \leq F(x^0,y^0)\}$ is bounded.

Recall from the discussion of Theorem IV.3 that for consistent bounding operations, bound improving selections are complete. Therefore, Proposition X.21 follows from Theorem IV.3 when consistency of the bounding operation is established. We show that any decreasing sequence $\{M_q\}$ of successively refined partition elements satisfies

$$\lim_{q\to\infty} (\alpha(M_q) - \beta(M_q)) = 0 , \tag{100}$$

and this implies consistency (cf. Section IV.2.).

Let $M_q = M_{x,q} \times M_{y,q}$, where $M_{x,q}$, $M_{y,q}$ denote the projection of M_q onto the x–space $\mathbb{R}^n$ and the y–space $\mathbb{R}^n$, respectively. Denote $h(x,y) = xy$. Recall from

Theorem IV.4 that for the convex envelope φ_{M_q} of h one has $\min \varphi_{M_q}(M_q) = \min h(M_q)$, i.e., the global minimum of h over M_q is equal to the global minimum of its convex envelope φ_{M_q} over M_q. Then, from the construction of the lower bounds $\beta(M_q)$ we see that

$$\min f(M_{x,q}) + \min h(M_q) + \min g(M_{y,q}) \le \beta(M_q) \le \alpha(M_q). \qquad (101)$$

But the subdivision is exhaustive, i.e., $M_q \xrightarrow[q]{} \{(\bar{x},\bar{y})\} \subset D$, and all of the infeasible partition sets are deleted. Recall that $\alpha(M_q) = F(\tilde{x}^q,\tilde{y}^q)$ for some $(\tilde{x}^q,\tilde{y}^q) \in D \cap M_q$. Using the continuity of f, h, g, it follows that as $q \longrightarrow \infty$ the left–hand side of (101) converges to $f(\bar{x}) + h(\bar{x},\bar{y}) + g(\bar{y}) = F(\bar{x},\bar{y})$. Since we also have $\alpha(M_q) = F(\tilde{x}^q,\tilde{y}^q) \xrightarrow[q]{} F(\bar{x},\bar{y})$, we see that condition (100) follows from (101) if we let $q \longrightarrow \infty$. ■

The original version of Al–Khayyal and Falk (1983) proceeds like Algorithm X.5 with two modifications.

The iteration point (x^k,y^k) is chosen to be the point, where the lower bound β_k is attained in the sense that

$$\beta_k = \min \{\beta(M) \colon M \in \mathcal{M}_k\} = \beta(\bar{M}) = f(x^k) + \varphi_M(x^k,y^k) + g(y^k)$$

(cf. the discussion in Section IV.2).

The subdivision of a partition element M involves setting up four new partition sets by choosing an index j satisfying

$$x_j^k y_j^k - \varphi_M(x_j^k y_j^k) = \max_{i=1,\ldots,n} \{x_i^k y_i^k - \varphi_M(x_i^k y_i^k)\}$$

and then splitting with two hyperplanes through (x_j^k,y_j^k) orthogonal to the (x_j,y_j)–plane, as illustrated in Fig. X.4 in the case k = 1 (cf. Chapter VII).

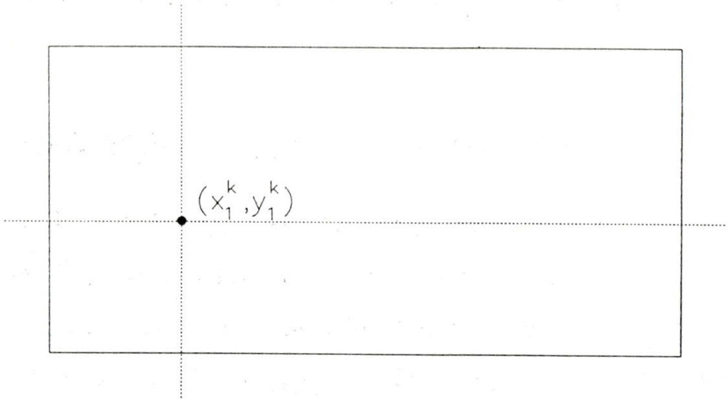

Fig. X.4. Splitting at Stage 1

CHAPTER XI

LIPSCHITZ AND CONTINUOUS OPTIMIZATION

In this chapter, we discuss global optimization problems where the functions involved are Lipschitz–continuous on certain subsets $M \subset \mathbb{R}^n$. Section 1 presents a brief introduction into the most often treated univariate case. Section 2 is devoted to branch and bound methods. First it is shown that the well–known univariate approaches can be interpreted as branch and bound methods. Then several extensions of univariate methods to the case of n dimensional problems with rectangular feasible sets are discussed. Finally, it is recalled from Chapter IV that very general Lipschitz optimization problems and also very general systems of equations and (or) inequalities can be solved by means of branch and bound techniques. As an example of Lipschitz optimization, the problem of minimizing a concave function subject to separable indefinite quadratic constraints is discussed in some detail.

In Section 3 it is shown that Lipschitz optimization problems can be transformed into equivalent special d.c. programs which can be solved by outer approximation techniques. This approach will then be generalized further to a "relief indicator" method.

1. BRIEF INTRODUCTION INTO THE GLOBAL MINIMIZATION OF UNIVARIATE LIPSCHITZ FUNCTIONS

1.1. Saw–Tooth Covers

Recall from Definition I.3 that a real–valued function f is called a ***Lipschitz function*** on a set $M \subset \mathbb{R}^n$ if there is a constant $L = L(f,M) > 0$ such that

$$|f(x) - f(y)| \leq L\|x - y\| \quad \forall x,y \in M . \tag{1}$$

In (1), $\|\cdot\|$ again denotes the Euclidean norm.

We first consider a ***univariate*** Lipschitz function defined on an interval [a,b]. We are interested in finding the global minimum of f over [a,b] and a point $x^* \in [a,b]$ such that

$$f^* := f(x^*) = \min \{f(x): x \in [a,b]\}$$

This problem will be denoted by

$$\text{(UL)} \qquad \begin{array}{l} \text{minimize } f(x) \\ \text{s.t.} \quad x \in [a,b] \end{array} . \tag{2}$$

It is not assumed that an analytic expression of f is known; i.e., f may be given by a so–called oracle.

This relatively simple problem (UL) is interesting because it arises in many applications and also because some algorithms for solving problem (UL) can easily be extended to the n–dimensional case.

Examples of applications are found in the optimization of the performance a system, which can in many cases be measured for given values of some parameter(s) even if the governing equations are unknown (e.g., Brooks (1958), Hanson et al. (1989) and references therein. Other examples are discussed in Pinter (1989).

Denote by $\Phi_L[a,b]$ the class of Lipschitz functions on $[a,b]$ having Lipschitz constant L. Then it is easy to see that no algorithm can solve (UL) for all $f \in \Phi_L[a,b]$ by using only a finite number of function evaluations (cf., Hansen et al. (1989)).

Theorem XI.1. *There is no algorithm for solving any problem (UL) in* $\Phi_L[a,b]$ *that uses only a finite number of function evaluations.*

Proof. Assume that there is such a finitely convergent algorithm A yielding a global minimizer x^* after k steps, i.e., we have $f(x^*) = \min_{i=1,\dots,k} f(x^i)$ $(k > 1)$.

Denote $X_k = \{x^1,\dots,x^k\}$, and let $f(X_k)$ be the set of corresponding function evaluations. Let $x^j \in X_k \setminus \{x^*\}$ be the evaluation point different from x^* which is closest to x^* on the left (if such a point does not exist, then a similar argument holds for the point in $X_k \setminus \{x^*\}$ closest to x^* on the right). Consider the function

$$f^1(x) = \begin{cases} f(x) & , x \in [a,x^j] \cup [x^*,b] \\ \max \{f(x^j) - L(x-x^j), f(x^*) - L(x^*-x)\} & , x \in [x^j,x^*] \end{cases}.$$

Obviously, we have $f^1 \in \Phi_L[a,b]$, and it is easy to see by a straightforward geometric argument that the global minimum of f^1 is attained at

$$\bar{x} = \frac{x^j + x^*}{2} + \frac{f(x^j) - f(x^*)}{2L}$$

with

$$\hat{f}(\bar{x}) = \frac{1}{2}(f(x^j) + f(x^*) - L(x^* - x^j)) < f(x^*) .$$

The strategy of algorithm A, however, depends only on L, X_k, $f(X_k)$ which coincide for f and f^1. Hence, we have

$$f(x^*) = \min_{i=1,\dots,k} f(x^i) = \min_{i=1,\dots,k} f^1(x^i) ,$$

and algorithm A concludes that x^* is also a global minimizer of f^1, which is a contradiction. ■

Frequently, instead of problem (UL), one investigates problem (UL_ε) which consists in finding a point $x^*_\varepsilon \in [a,b]$ such that

$$f^*_\varepsilon := f(x^*_\varepsilon) \leq f^* + \varepsilon \tag{3}$$

for a small $\varepsilon > 0$.

If is obvious that every problem (UL_ε) always can be solved by a finite algorithm. Evaluating f at the equidistant points

$$x^i = a + \frac{(2i - 1)\varepsilon}{L} \quad (i=1,...,k) , \tag{4}$$

where $k = \lceil \frac{L(b - a)}{2\varepsilon} \rceil$ is the smallest integer satisfying $k \geq \frac{L(b - a)}{2\varepsilon}$, yields a point satisfying (3) for any Lipschitz function $f \in \Phi_L[a,b]$.

Problems (UL) and (UL_ε) have been studied by several authors, of whom we mention Danilin (1971), Evtushenko (1971 and 1985), Piyavskii (1972), Shubert (1972), Strongin (1973 and 1978), Timonov (1977), Schoen (1982), Shepilov (1987), Pinter (1986 and 1988), Hansen et al. (1988 and 1989).

A number of procedures have also been designed to approximate the set

$$X^* = \{x^* \in [a,b]: f(x^*) = f^*\} \tag{5}$$

of all optimal solutions to (UL); see, for example, Basso (1982), Galperin (1985 and 1988), Pinter (1986 and 1988), Hansen et al. (1988).

An algorithm such as (4), where the evaluation points are chosen simultaneously, is frequently said to be *passive*, since the step size is predetermined and does not depend on the function values. Its counterpart is a *sequential* algorithm, in which the choice of new evaluation points depends on the information gathered at previous iterations.

For most functions f, the number of evaluation points required to solve (UL_ε) will be much smaller with a suitable sequential algorithm than with a passive algorithm. In the worst case, however, the number of evaluation points required by a passive and by a best possible sequential algorithm are the same (cf. Ivanov (1972), Archetti and Betró (1978), Sukharev (1985)). It can easily be seen from the following discussion that this case arises when f is a constant function over [a,b].

Given $X_k = \{x^1,...,x^k\}$, the corresponding set $f(X_k)$ of function values, and the Lipschitz constant L, it is natural to bound f* from below by a piecewise linear function with slope + L or − L that exploits the Lipschitz bounds given by

$$f(x) \geq f(x^i) - L|x - x^i| \quad \forall x \in [a,b] . \tag{6}$$

(cf. I.4.1).

Obviously, for a fixed set X_k, the best underestimating function using the above information is given by

$$F_k(x) = \max_{i=1,\ldots,k} \{f(x^i) - L|x - x^i|\} . \tag{7}$$

Because of its shape, a function F_k of the form (7) will be called a *saw−tooth cover of f*. Let X_k be ordered such that $a \leq y^1 \leq y^2 \leq \ldots \leq y^k \leq b$, where $\{y^1,...,y^k\} = \{x^1,...,x^k\}$. The restriction of F_k to the interval $[y^i,y^{i+1}]$ of two consecutive evaluation points is said to be the *tooth* on $[y^i,y^{i+1}]$. A straightforward simple calculation shows that the tooth on $[y^i,y^{i+1}]$ attains its minimal value (downward peak) at

$$x^{p,i} = \frac{y^i + y^{i+1}}{2} + \frac{f(y^i) - f(y^{i+1})}{2L} \tag{8}$$

with

$$F_k(x^{p,i}) = \frac{f(y^i) + f(y^{i+1})}{2} - L\,\frac{y^{i+1} - y^i}{2} \tag{9}$$

Since the number of necessary function evaluations for solving problem (UL_ε) measures the efficiency of a method, Danilin (1971) suggested studying the *minimum* number of evaluation points required to obtain a guaranteed solution of problem (UL_ε) (cf. also Hanson et al. (1988 and 1989)).

This can be done by constructing a *reference* saw–tooth cover

$$F_\beta(x) = \max_{i=1,\ldots,k_\beta} \{f(x^i) - L|x-x^i|\}$$

for solving (UL_ε) with a minimal number k_β of function evaluations. Such a reference cover is constructed with f* assumed to be known. It is, of course, designed not to solve problem (UL_ε) from the outset, but rather to give a reference number of necessary evaluation points in order to study the efficiency of other algorithms.

It is easy to see that a reference saw–tooth cover $F_\beta(t)$ can be obtained in the following way. Set $F_\beta(a) = f^* - \varepsilon$. The first evaluation point x^1 is then the intersection point of the line $(f^* - \varepsilon) + L(x - a)$ with the curve f(x). The next downward peak is at

$$x^{p,1} = x^1 + \frac{f(x^1) - (f^* - \varepsilon)}{L},$$

and it satisfies $F_\beta(x^{p,1}) = f^* - \varepsilon$.

Proceeding in this way, we construct a saw–tooth cover $F_\beta(x)$, for which the lowest value of a downward peak (with the possible exception of the last one) is $f^* - \varepsilon$ (Fig. XI.1).

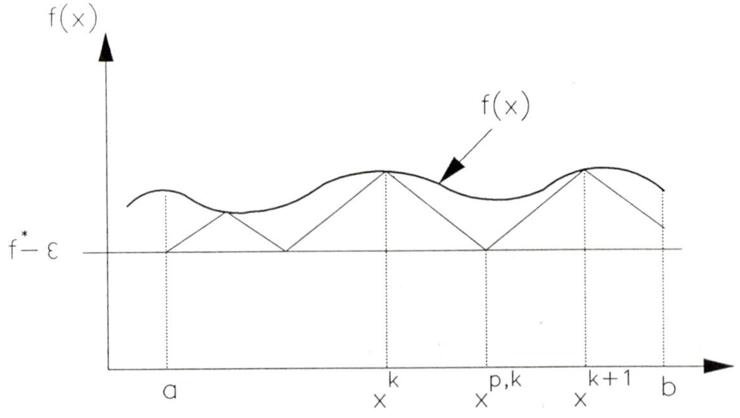

Fig. XI.1 Reference saw–tooth cover

Algorithm XI.1 (Reference saw–tooth cover).

Initialization:

Set $k = 1$, x^1 solution of the equation $f(x) = (f^* - \varepsilon) + L(x - a)$,

Reference saw–tooth cover:

Step k = 1,2,... :

Set $x^{p,k} = x^k + \frac{f(x^k) - (f^* - \varepsilon)}{L}$;

x^{k+1} solution of the equation $f(x) = (f^* - \varepsilon) + L(x - x^{p,k})$;

If $x^{k+1} \geq b$, then stop: $k_\beta = k$;

Otherwise, set $k = k+1$.

Go to Step k.

1.2. Algorithms for Solving the Univariate Lipschitz–Problem

Consider problem (UL_ε). Let x^k be the last evaluation point and let f_ε denote the current best known function value. We try to find x^{k+1} such that the step–size $x^{k+1} - x^k$ is maximal under the condition that, if $f(x^{k+1}) \geq f_\varepsilon$, then we have

$$f_\varepsilon - F_k(x^{p,k}) \leq \varepsilon . \tag{10}$$

Inserting (9) into (10), we obtain

$$x^{k+1} - x^k \leq \frac{1}{L} (2\varepsilon + f(x^k) + f(x^{k+1}) - 2f_\varepsilon) , \tag{11}$$

which, because of the condition $f(x^{k+1}) \geq f_\varepsilon$, leads to

$$x^{k+1} = x^k + \frac{1}{L} (2\varepsilon + f(x^k) - f_\varepsilon) . \tag{12}$$

This is essentially the procedure of Evtushenko (1971), (cf. also Hansen et al. (1989)).

Algorithm XI.2 (Evtushenko's saw–tooth cover).

Initialization:

Set $k = 1$, $x^1 = a + \frac{\varepsilon}{L}$, $x_\varepsilon = x^1$, $f_\varepsilon = f(x_\varepsilon)$;

Evtushenko's saw–tooth cover:

Step k = 1,2,... :

If $x^k > b$, then stop.

Otherwise, set

$$x^{k+1} = x^k + \frac{1}{L}(2\varepsilon + f(x^k) - f_\varepsilon).$$

If $f(x^{k+1}) < f_\varepsilon$, then set $f_\varepsilon = f(x^{k+1})$, $x_\varepsilon = x^{k+1}$;

Set $k = k+1$. Go to Step k.

Note that from the derivation of (12) it follows that, if $f(x^{k+1}) < f(x^k)$, then the downward peak $F(x^{p,k})$ differs from the new incumbent value $f(x^{k+1})$ by more than ε.

We see from (12) that in the worst case of a constant or a monotonically decreasing function f, we have $f(x^k) = f_\varepsilon$ for all k, hence $x^{k+1} - x^k = \frac{2\varepsilon}{L}$, which is the step–size of the passive algorithm (4)

The minimum number of evaluation points required by Evtushenko's algorithm is $1 + \lceil \log_2 (1 + \frac{L(b-a)}{2\varepsilon}) \rceil$. It is attained when f is an affine function on [a,b] with slope L. The efficiency of the procedure depends greatly on the position of the optimal solution x^*, and it tends to become worse when $x^* \longrightarrow b$. Its saw–tooth cover can differ considerably from the reference saw–tooth cover, particularly when x^* is far from the left bound a of the interval.

Evtushenko's algorithm was originally designed to globally optimize a multivariate Lipschitz function by repeatedly solving the univariate problems obtained by fixing all variables but one.

In contrast to Evtushenko's method, which is an *ordered sequential* algorithm, i.e., the evaluation points at successive iterations are increasing values of x belonging to [a,b], the algorithm of Piyavskii (1967 and 1972) constructs more and more refined saw–tooth covers of f in the following way. Starting with $x^1 = \frac{a+b}{2}$ and $f(x^1)$, the first saw–tooth cover

$$F_1 = f(x^1) - L|x - x^1|$$

is minimized over [a,b] in order to obtain its lowest downward peak at $x^2 \in \operatorname{argmin} F_1([a,b])$.

The function f is then evaluated at this "peak point" x^2, and the corresponding tooth is split into smaller teeth to obtain the next cover

$$F_2 = \max_{j=1,2} \{f(x^j) - L|x - x^j|\} ,$$

which in turn is minimized over [a,b], etc.

Thus, the procedure is governed by the formulas

$$F_k(x) = \max_{j=1,\ldots,k} \{f(x^j) - L|x - x^j|\} \tag{13}$$

(cf. (7)) and

$$x^{k+1} \in \operatorname{argmin} F_k([a,b]) .$$

Piyavskii's algorithm with various extensions seems to be the most often discussed approach for problems (UL) and (UL_ε). It was rediscovered by Shubert (1972) and Timonov (1977). Archetti and Betro (1978) discuss it in a general framework of sequential methods. Basso (1982) concentrates on convergence issues, and proposes some modifications in order to approximate the set X* of all optimal solutions. Pintér (1986 and 1986a) introduces five axioms which guarantee the convergence of certain global algorithms, and he shows that Piyavskii's algorithm, as well as others, satisfy them. Schoen (1982) proposes a variant of Piyavskii's approach that, instead of choosing the point of lowest downward peak to be the next evaluation point, selects the evaluation point of the passive strategy (4) that is closest to this peak point. Shen and Zhu (1987) discuss a simplified version in which at each iteration the new evaluation point is at the middle of a subinterval bounded by two consecutive previous evaluation points. Hansen et al. (1988 and 1989) present a thorough discussion of Piyavskii's univariate algorithm and related approaches which includes a theoretical study of the number of iterations which was initiated by Danilin (1971).

Multidimensional extensions where proposed by Piyavskii (1967 and 1972), Mayne and Polak (1984), Strigul (1985), Mladineo (1986), Pinter (1986, 1986a, and 1988), Skepilov (1987), Neferdov (1987) and Meewalla and Mayne (1988), and others.

Note that a direct extension of Piyavskii's method to the case of an n–rectangle D, where (13), (14) are replaced by

$$F_i(x) = \max_{j=1,\ldots,i} \{f(x^j) - L\|x - x^j\|\} \tag{13'}$$

and

$$x^{i+1} \in \operatorname{argmin} F_i(D) \tag{14'}$$

seems not to be very promising with respect to numerical efficiency for dimension $n > 2$ since (14') constitutes an increasingly difficult d.c. problem (cf. Section I.4).

In Horst and Tuy (1987) it is shown that Piyavskii's algorithm, as well as others, can be viewed as branch and bound algorithm. In this branch and bound reformulation, the subproblems correspond to a tooth of the current saw–tooth cover, and the selection is bound improving.

This way of viewing the procedure will allow one to delete unpromising teeth in order to reduce the need for memory space. We shall return to this branch and bound formulation in the more general framework of Section XI.2.

In the following algorithmic description of Piyavskii's method for solving (UL_ε), f_ε denotes the current best value of the function f, whereas F_ε denotes the minimal value of the current saw–tooth cover.

Algorithm XI.3 (Piyavskii's saw–tooth cover).

Initialization:

Set $k = 1$, $x^1 = \frac{a+b}{2}$, $x^\varepsilon = x^1$, $f_\varepsilon = f(x_\varepsilon)$,

$F_\varepsilon = f_\varepsilon - \frac{L(b-a)}{2}$, $F_1 = f(x^1) - L|x - x^1|$.

Piyavskii's saw–tooth cover.

Step k = 1,2,... :

If $f_\varepsilon - F_\varepsilon \le \varepsilon$, then stop.

Otherwise determine

$$x^{k+1} \in \operatorname{argmin} F_k([a,b]).$$

If $f(x^{k+1}) \le f_\varepsilon$, then set $f_\varepsilon = f(x^{k+1})$, $x_\varepsilon = x^{k+1}$.

Set $F_{k+1}(x) = \max_{i=1,\ldots,k+1} \{f(x^i) - L|x - x^i|\}$,

$$F_\varepsilon = \min F_{k+1}([a,b]),\ k = k+1.$$

Go to Step k.

Obviously, the second and the third evaluation points in Piyavskii's algorithm are a and b, where $x^2 = a$ implies $x^3 = b$ and $x^2 = b$ implies $x^3 = a$.

Now suppose that for $k \geq 3$ the first k evaluation points have already been generated and ordered so that we have $a = y^1 \le y^2 \le \dots \le y^k = b$, where $\{y^1,\dots,y^k\} = \{x^1,\dots,x^k\}$. Then, by (8), (9), we see that x^{k+1} is determined by

$$F_k(x^{k+1}) = \min_{i=1,\dots,k-1} \left\{ \frac{f(y^{i+1})+f(y^i)}{2} - L\frac{y^{i+1}-y^i}{2} \right\}.$$

Assume that the minimum is attained at i = j. Then we have

$$x^{k+1} = \frac{y^{j+1}+y^j}{2} + \frac{f(y^j) - f(y^{j+1})}{2L} \tag{15}$$

and

$$F_k(x^{k+1}) = \frac{f(y^{j+1})+f(y^j)}{2} - L\frac{y^{j+1}-y^j}{2}. \tag{16}$$

We show that Piyavskii's algorithm is *one–step optimal* in the sense explained below.

Let $f(x^1),f(x^2),\dots,f(x^{k-1})$ be the values of $f \in \Phi_L[a,b]$ at the first k–1 evaluation points of a saw–tooth cover algorithm for solving problem (UL) (or problem UL_ε). In addition, let $\Phi_{k-1}(f) \subset \Phi_L[a,b]$ denote the set of all Lipschitz functions (with Lipschitz constant L) which coincide with f(x) at these points. For all $\varphi \in \Phi_{k-1}(f)$, denote $\varphi^* = \min \varphi[a,b]$. Consider the saw–tooth cover $\overline{\varphi}_k(x)$ of $\varphi(x)$ at iteration k which, given $\overline{\varphi}_{k-1}(x)$, is determined by the choice of the evaluation point x^k. We are interested in making the choice in such a way that the error

$$\Delta_k(\varphi) := \min\{\varphi(x^1),\varphi(x^2),\dots,\varphi(x^k)\} - \varphi^* \tag{17}$$

is *minimized* in the *worst case.* In other words, we attempt to minimize the quantity

$$\Delta_k := \sup\{\Delta_k(\varphi) \colon \varphi \in \Phi_{k-1}(f)\} \tag{18}$$

("*optimality in one step*", cf. Sukharev (1985) for a related definition).

Proposition XI.1. *Piyavskii's algorithm is optimal in one step in the sense of (17), (18).*

Proof. Since for all $\varphi \in \Phi_{k-1}(f)$ we have $f(x^i) = \varphi(x^i)$ $(i=1,...,k-1)$, the previous saw–tooth covers $\overline{\varphi}_{k-1}(x)$ coincide for all $\varphi \in \Phi_k(f)$, i.e., one has $\overline{\varphi}_{k-1}(x) = F_{k-1}(x)$.

By the construction of $F_{k-1}(x)$, we have

$$\varphi(x) \geq F_{k-1}(x) \quad \forall x \in [a,b],\ \varphi \in \Phi_{k-1}(f)$$

and

$$F_{k-1}(x) \in \Phi_{k-1}(f).$$

It follows from the construction of F_{k-1} that the maximal error after k–1 iterations occurs when $\varphi(x) = F_{k-1}(x)$, i.e., we have

$$\Delta_{k-1} = \min \{f(x^1),...,f(x^{k-1})\} - \min F_{k-1}[a,b] \tag{19}$$

which is attained at the Piyavskii evaluation point $x^{p,k}$.

In order to investigate the worst case error in the next iteration, first note that there exist functions $\varphi \in \Phi_{k-1}(f)$ satisfying

$$\varphi(x) \geq \min_{i=1,\ldots,k-1} \varphi(x^i) = \min_{i=1,\ldots,k-1} f(x^i) \quad \forall x \in [a,b]\ . \tag{20}$$

Furthermore, let a_{k-1}, $b_{k-1} \in \{x^1,...,x^{k-1}\}$, $a_{k-1} < b_{k-1}$, be the nearest evaluation points to the left and to the right of $x^{p,k}$, respectively, i.e., we have

$$x^{p,k} = \frac{b_{k-1} + a_{k-1}}{2} + \frac{f(a_{k-1}) - f(b_{k-1})}{2L}$$

(cf. (15), (16), where a_{k-1}, b_{k-1} correspond to y^j, y^{j+1}).

Let x^k denote the next evaluation point and consider the two cases $x^k \notin [a_{k-1}, b_{k-1}]$ and $x^k \in [a_{k-1},b_{k-1}]$. Denote

$$M_{k-1} = \min F_{k-1}[a,b]$$

and let $\hat{f}_{k-1} = \min \{f(x^1),...,f(x^{k-1})\}$.

Suppose that $x^k \notin [a_{k-1}, b_{k-1}]$. Then for all $\varphi \in \Phi_{k-1}(f)$ satisfying (20), we have

$$\Delta_k(\varphi) = \Delta_{k-1} = \hat{f}_{k-1} - M_{k-1} \,.$$

(cf. (17)). In this case, the maximal error is not improved by the choice of x^k and

$$\Delta_k = \hat{f}_{k-1} - M_{k-1} \,. \tag{21}$$

Now suppose that $x^k \in [a_{k-1}, b_{k-1}]$. In this case, we have $x^k \neq a_{k-1}$ and $x^k \neq b_{k-1}$, since a_{k-1}, b_{k-1} are previous evaluation points, and the optimal error reduction is obtained by setting $x^k = x^{p,k}$. ■

The algorithms for solving problem (UL_ε) discussed so far regard the Lipschitz constant L as known a priori. We would like to mention that Strongin (1973 and 1978) proposes an algorithm that, instead of L, uses an estimate of L which is a multiple of the greatest absolute value of the slopes of the lines joining successive evaluation points. Convergence to a global minimum ($\varepsilon = 0$) can be guaranteed whenever this estimate is a sufficiently large upper bound for the Lipschitz constant L (for details, see Strongin (1978), Hansen et al. (1989)).

2. BRANCH AND BOUND ALGORITHMS

In this section, the branch and bound concept developed in Chapter IV will be applied to certain Lipschitz optimization problems. We begin with an interpretation of Piyavskii's univariate algorithm as a branch and bound procedure. Then the case of an n–dimensional rectangular feasible set D is considered, where a generalization of Piyavskii's univariate algorithm and an axiomatic approach are discussed in the branch and bound framework. Finally, it is recalled from Chapter IV that branch and bound methods can be designed for minimizing Lipschitz functions over convex sets, over intersections of a convex set with finitely many complements of convex sets, and over sets defined by a finite number of Lipschitz inequalities. The resulting approach will be applied to global optimization problems with indefinite separable quadratic constraints.

2.1. Branch and Bound Interpretation of Piyavskii's Univariate Algorithm

Let the first k ($k \geq 3$) evaluation points of Piyavskii's algorithm for solving problem (UL) be ordered in such a way that we have $a = y^1 \leq y^2 \leq \ldots \leq y^k = b$, where $\{y^1,\ldots,y^k\} = \{x^1,\ldots,x^k\}$. Obviously, the intervals

$$M_{k,i} = [y^i, y^{i+1}] \quad (i=1,\ldots,k-1) \, ,$$

define a partition of $[a,b]$, and

$$\beta(M_{k,i}) = \frac{f(y^{i+1}) + f(y^i)}{2} - L \frac{y^{i+1} - y^i}{2} \quad (i=1,\ldots,k-1) \tag{22}$$

constitute the associated lower bounds (cf. (15) and (16)). Changing notation slightly, let $x^{k,i} \in \{y^i, y^{i+1}\}$ satisfy $f(x^{k,i}) = \min \{f(y^i), f(y^{i+1})\}$;

set

$$\alpha(M_{k,i}) = f(x^{k,i}) \quad (i=1,\ldots,k-1) \, , \tag{23}$$

and let the current iteration point $x^k \in \{x^{k,i}: i=1,\ldots,k-1\}$ be such that

$$\alpha_k := f(x^k) := \min \{f(x^{k,i}): i=1,\ldots,k-1\}. \tag{24}$$

Subdivide an interval $M_{k,j}$ satisfying

$$\beta_k = \beta(M_{k,j}) = \min \{\beta(M_{k,i}): i=1,\ldots,k-1\} \tag{25}$$

into the two intervals $[y^j, z^k]$, $[z^k, y^{j+1}]$, where

$$z^k = \frac{y^{j+1} + y^j}{2} + \frac{f(y^j) - f(y^{j+1})}{2L} \tag{26}$$

is the next evaluation point (cf. (15)).

Rearranging the evaluation points $y^1,...,y^k,z^k$ in the order of increasing values to obtain $a = y^1 \le y^2 \le ... \le y^{k+1} = b$, $\{y^1,...,y^k,z^k\} = \{y^1,...,y^{k+1}\}$, we see that all of the ingredients of a BB procedure as described in Section IV.1 are at hand. In this way, Piyavskii's algorithm can be interpreted as a branch and bound method, in which deletion operations can be used to reduce memory space compared to the version presented in XI.1.2.

Proposition XI.2. *Consider Piyavskii's algorithm for solving problem (UL) in its branch and bound interpretation described above. Then we have*

$$\lim_{k\to\infty} \beta_k = \min f([a,b]) = \lim_{k\to\infty} \alpha_k ,$$

and every accumulation point of $\{x^k\}$ *is an optimal solution of problem (UL).*

Proof. We refer to the convergence theory of the BB procedure presented in Chapter IV, and we verify the assumptions of Theorem IV.3 and Corollary IV.3.

The selection is obviously *bound improving* and the set $\{x \in [a,b]: f(x) \le f(x^1)\}$ is compact by continuity of f. It remains to verify *consistency* of the bounding operation.

Let $\{M_{k_q}\}$ be a nested sequence of partition intervals generated by the algorithm. The sequence $\alpha(M_{k_q})$ of associated upper bounds is nonincreasing and bounded from below by $\min f([a,b])$; the sequence $\beta(M_{k_q})$ of associated lower bounds is nondecreasing and bounded from above by $\min f([a,b])$. Hence, we have the existence of limits α,β satisfying

$$\alpha = \lim_{q\to\infty} \alpha_{k_q} \ge \min f([a,b]) \ge \lim_{q\to\infty} \beta_{k_q} = \beta . \tag{27}$$

Now consider the sequence $\{z^k\}$ of evaluation points z^k generated at each Step k. Since M_{k_q} is subdivided in iteration k_{q+1}, we have

$$z^{k_{q+1}} \in M_{k_q} \text{ and } \alpha(M_{k_{q+1}}) \leq f(z^{k_{q+1}}) . \tag{28}$$

Let $\bar{z}$ be an accumulation point of the subsequence $\{z^{k_{q+1}}\}$ of $\{z^k\}$. If follows from (28) and the continuity of f, that

$$\alpha \leq f(\bar{z}) \tag{29}$$

holds.

Now suppose that there is an $\varepsilon > 0$ such that

$$f(\bar{z}) > \beta + \varepsilon . \tag{30}$$

By definition of $\bar{z}$, there is a subsequence $\{z^{k_{q'+1}}\} \subset \{z^{k_{q+1}}\}$ and a $q_\varepsilon \in \mathbb{N}$ such that, for $q' > q_\varepsilon$

$$|z^{k_{q'+1}} - \bar{z}| < \frac{\varepsilon}{2L} . \tag{31}$$

It follows from the Lipschitz–continuity of f and from (30) that

$$f(z^{k_{q'+1}}) \geq f(\bar{z}) - L|z^{k_{q'+1}} - \bar{z}| > \beta + \frac{\varepsilon}{2} \quad \forall q' > q_\varepsilon .$$

Hence, for $x \in [a,b]$ satisfying

$$|x - z^{k_{q'+1}}| < \frac{\varepsilon}{2L} , \quad q' > q_\varepsilon ,$$

we have

$$f(z^{k_{q'+1}}) - L|x - z^{k_{q'+1}}| > \beta . \tag{32}$$

However, by construction of BB procedures, we have the inequality

$$\beta_k \leq \beta \quad \forall k ,$$

and hence, from (32), it follows that

$$z^k \notin \{x: |x - z^{k_{q'+1}}| < \frac{\varepsilon}{2L}, q' > q_\varepsilon\} \quad \forall k .$$

This contradicts the assumption that $\bar{z}$ is an accumulation point of $\{z^k\}$. Therefore, there is no $\varepsilon > 0$ such that (30) holds, and this (by using (29)) implies that

$$\alpha \leq f(\bar{z}) \leq \beta .$$

Considering (27), we finally see that there must hold

$$\alpha = \min f([a,b]) = \beta ,$$

from which consistency follows. ■

In the preceding discussion of Lipschitz optimization it was always assumed that the Lipschitz constant L is used. It is obvious, however, that, for a given Lipschitz function, there are infinitely many Lipschitz constants, i.e., if L is a Lipschitz constant, then all numbers L' > L are Lipschitz constants as well. Let f be a Lipschitz function on [a,b] and let

$$L = \inf \{L': L' \text{ is Lipschitz constant of f on } [a,b]\} .$$

Then, in practice, we often know only some L' > L. Assuming this and applying Piyavskii's algorithm with L' > L instead of L, Proposition XI.2 can also be derived along the very simple lines presented in Section IV.4.5.

To see this, consider the subdivision of the interval $[y^j,y^{j+1}]$ into two subintervals $[y^j,z^k]$, $[z^k,y^{j+1}]$ as described by (26). From (26), it follows that

$$\max \{y^{j+1} - z^k, z^k - y^j\} \leq \frac{y^{j+1} - y^j}{2} + \frac{|f(y^{j+1}) - f(y^j)|}{2L'} ,$$

and, using the Lipschitz continuity with constant L,

$$\max \{y^{j+1}-z^k, z^k-y^j\} \leq \frac{1}{2} (1 + \frac{L}{L'}) (y^{j+1}-y^j) . \tag{33}$$

Considering a nested sequence $\{M_q\}$ of successively refined partition intervals M_q with length $\delta(M_q)$, we see that

$$\delta(M_{q+1}) \le \gamma\delta(M_q) , \tag{34}$$

where $\gamma = \frac{1}{2}(1 + \frac{L}{L'}) < 1$. This establishes the exhaustiveness of the subdivision procedure.

Consistency is then an obvious consequence of (22), (23), and the continuity of f.

2.2. Branch and Bound Methods for Minimizing a Lipschitz Function over an n–dimensional Rectangle

Now let the feasible set D be an n–dimensional interval, i.e., there are vectors $a,b \in \mathbb{R}^n$, $a < b$, such that

$$D := \{x \in \mathbb{R}^n : a \le x \le b\} , \tag{35}$$

where the inequalities are understood with respect to the componentwise ordering of $\mathbb{R}^n$. Let the objective function f be Lipschitzian on D with Lipschitz constant L and consider the global optimization problem

$$\underset{x \in D}{\text{minimize}}\ f(x) . \tag{36}$$

As shown in Section XI.1.2, Piyavskii's univariate algorithm can easily be formulated for the case of problem (36), and one obtains a corresponding convergence result as a straightforward n–dimensional extension of Proposition XI.2 (cf. Horst and Tuy (1987)). Since, however, the computational effort in solving the corresponding subproblems is enormous in dimension $n \ge 2$, we prefer to present branch and bound extensions of Piyavskii's approach that essentially apply the univariate version to the main diagonal of rectangular partition sets.

Let $L' > L$ be an upper bound for the optimal Lipschitz constant L. Denote by a_M, b_M the lower left and upper right vertex of an n–rectangle M, respectively, i.e., we have

$$M = \{x \in \mathbb{R}^n: a_M \le x \le b_M\} .$$

Algorithm XI.4 (Prototype Diagonal Extension of Piyavskii's saw–tooth cover).

Step 0 (Initialization):

Set $M_0 = D, \mathcal{M}_0 = \{M_0\},$

$$\alpha_0 = \min \{f(a), f(b)\}, x^0 \in \{a,b\} \text{ such that } \alpha_0 = f(x^0) , \tag{37}$$

$$\beta_0 = \max \{f(a), f(b) - L'\|b - a\|\} . \tag{38}$$

Go to Step 1.

Step k (k=1,2,...) :

At the beginning of Step k we have the current rectangular partition $\mathcal{M}_{k-1}$ of a subset of $M_0 = D$ which is still of interest, and for every $M \in \mathcal{M}_{k-1}$ we have bounds $\beta(M)$, $\alpha(M)$ satisfying

$$\beta(M) \le \min f(M) \le \alpha(M) .$$

Moreover, we have the current lower and upper bounds β_{k-1}, α_{k-1} satisfying

$$\beta_{k-1} \le \min f(D) \le \alpha_{k-1}$$

and a point x^{k-1} such that $f(x^{k-1}) = \alpha_{k-1}$.

k.1. Delete all $M \in \mathcal{M}_{k-1}$ satisfying

$$\beta(M) \ge \alpha_{k-1} .$$

Let $\mathcal{R}_k$ be the collection of remaining partition sets in $\mathcal{M}_{k-1}$.

k.2. Select

$$\mathcal{P}_k \subset \mathcal{R}_k$$

satisfying

$$\mathcal{P}_k \cap \operatorname{argmin} \{\beta(M): M \in \mathcal{R}_k\} \neq \emptyset . \tag{39}$$

For each $\tilde{M} \in \mathcal{P}_k$ choose

$$x_{\tilde{M}} = \frac{1}{2}(a_{\tilde{M}} + b_{\tilde{M}}) + \frac{f(a_{\tilde{M}}) - f(b_{\tilde{M}})}{2L' \| b_{\tilde{M}} - a_{\tilde{M}} \|} (b_{\tilde{M}} - a_{\tilde{M}}) , \tag{40}$$

and subdivide $\tilde{M}$ into two n–dimensional subintervals using the hyperplane which contains $x_{\tilde{M}}$ and is orthogonal to one of the longest edges of $\tilde{M}$.

Let $\mathcal{M}_k'$ be the collection of new partition elements. For each $M' \in \mathcal{M}_k'$ denote by $\tilde{M}' \in \mathcal{P}_k$ the rectangle whose subdivision generated M'.

k.3. For each $M' \in \mathcal{M}_k'$ set

$$\alpha(M') = \min \{f(a_{M'}), f(b_{M'}), f(x_{\tilde{M}'})\} , \tag{41}$$

$$\beta(M') = \max \{\beta(\tilde{M}'), \max\{f(a_{M'}),f(b_{M'}),f(x_{\tilde{M}'})\} - L'\|b_{M'} - a_{M'}\|\} . \tag{42}$$

k.4. Set $\mathcal{M}_k = (\mathcal{R}_k \setminus \mathcal{P}_k) \cup \mathcal{M}_k'$.

Compute

$$\alpha_k = \min \{\alpha(M): M \in \mathcal{M}_k\} ,$$

$$\beta_k = \min \{\beta(M): M \in \mathcal{M}_k\} .$$

Let $x^k \in D$ be such that $f(x^k) = \alpha_k$.

k.5. If $\alpha_k - \beta_k = 0$ ($\le \varepsilon$), then stop. Otherwise, go to Step k+1.

Proposition XI.3. *Consider Algorithm XI.4 (with $\varepsilon = 0$). Then we have*

$$\lim_{k\to\infty} \beta_k = \min f(D) = \lim_{k\to\infty} \alpha_k ,$$

and every accumulation point x^* of $\{x^k\}$ satisfies $f(x^*) = \min f(D)$.

Proof. Proposition XI.3 readily follows from Theorem IV.3, Corollary IV.3 and Proposition IV.3 if the subdivision procedure is exhaustive. But from (33), (34), we see that the length of the diagonal $[a_M, b_M]$ is reduced by a factor $\gamma = \frac{1}{2}(1 + \frac{L}{L'}) < 1$.

Exhaustiveness then follows (cf. also the proof of Proposition IV.2). ■

Remarks XI.1. (i) Many variants of this algorithm are possible.

For example, recall from Chapter IV that, if we choose a subset $V'(M) \supset \{a_M, b_M\}$ of the vertices of a partition set M, the lower and upper bounds may be determined by

$$\alpha(M) = \min f(V'(M))$$

and

$$\beta(M) = \max f(V'(M)) - L'\|b_M - a_M\| ,$$

respectively. Moreover, any exhaustive subdivision process will yield a convergent procedure as long as one of these bounds or any uniformly better bound is used.

(ii) In practical computation, an interval M will not be subdivided further when its diameter $\|b_M - a_M\|$ is less than a fixed parameter $\delta > 0$.

(iii) Whenever adaptive estimates L(M) of the Lipschitz constant of f over current intervals are available, then, of course, these should be used instead of L'.

(iv) When the branch–and–bound algorithm is stopped at an iteration point x^k because we have $\alpha_k - \beta_k < \varepsilon$ or $\|b_M - a_M\| < \delta$ for all sets M which are still of interest, then, of course, a local search starting from x^k could lead to an improved approximation of a global minimum.

A so–called "axiomatic" approach to problem (36) has been proposed by Pintér (1986 and 1986a). This approach uses typical branch and bound elements, such as partitions of the feasible n–interval D into finite sets of n–intervals and refinement by subdivision of selected partition elements. The selection of partition elements is governed by a selector function that takes into account the size of a given partition interval as well as the objective function values at its vertices. Each selected n–interval M is subdivided into 2^n subintervals using all of the hyperplanes through selected interior points of M that are parallel to the facets of M.

However, the current iteration point in Pintér's approach is not necessarily the best feasible point obtained so far. Moreover, Pintér's method does not make use of lower bounds, and hence it does not provide estimates of the quality of a current iteration point or a deletion rule to remove partition elements not of interest.

In Horst and Tuy (1987) it was shown, however, that Pintér's approach can readily be modified, improved and generalized by viewing it within the framework of branch and bound methods discussed in Chapter IV. A simplified and slightly generalized version of the presentation in Horst and Tuy (1987) follows.

Consider Algorithm XI.4 with the following modifications:

(i) Upper and lower bounds are determined on the complete vertex set V(M) of a partition interval M (cf. Remark XI.1.(i)).

Following Pintér (1986 and 1986(a)), the vertex set V(M) can be described by an $(n \times 2^n)$ matrix X(M) whose columns are the lexicographically ordered vertices of M. The corresponding values of the objective function f at the vertices of M define a 2^n vector z(M).

Using this notation, in Step k.3 of Algorithm XI.4 we replace the bounds (41), (42) by

$$\alpha(M') = \min \{z_j(M'): j=1,\dots,2^n\} \tag{41'}$$

and

$$\beta(M') = \max \{\beta(\tilde{M}'), \max\{z_i(M'):i=1,\dots,2^n\} - L'\|b_{M'}-a_{M'}\|\} . \tag{42'}$$

(ii) Rule (39) in Step k.2 of Algorithm XI.4, which selects the n–intervals to be subdivided further is replaced by the following procedure.

Let R(X(M),z(M)) be a *suitable* real–valued function of the vertices of M (represented by the matrix X(M)) and of the vector z(M) of the objective function values at these vertices.

A function R(X(M),z(M)) is suitable if it satisfies the requirements R.2 – R.5 listed below. Given such a suitable (selector) function, select

$$\mathcal{P}_k = \{\tilde{M} \in \mathcal{R}_k: R(X(\tilde{M}),z(\tilde{M})) = \max\{R(X(M),z(M)):M \in \mathcal{R}_k\} \ .(39')$$

Subdivide each $\tilde{M} \in \mathcal{P}_k$ into r n–intervals, $2 \le r \le 2^n$, using hyperplanes parallel to certain facets of $\tilde{M}$ that pass through a chosen interior point of $\tilde{M}$. Any subdivision of that kind is admitted as long as the following requirement R.1 is fullfilled.

R.1. The subdivision is exhaustive.

The requirements for a suitable selector function R(X(M),z(M)) are as follows.

R.2. R(X(M), z(M)) is continuous in (X(M), z(M)) and, for every decreasing sequence $\{M_q\}$ of n–intervals $M_q \subset D$, the limit $\lim_{q\to\infty} R(X(M_q), z(M_q))$ exists and is continuous. If $\{M_q\}$ converges to a singleton $\{\bar{x}\}$, then

$$R(\bar{X},\bar{z})) = \lim_{q\to\infty} R(X(M_q), z(M_q)) \ , \tag{43}$$

holds. In (43), $\bar{X}$ is the $(n\times 2^n)$–matrix having 2^n identical columns $\bar{x} \in \mathbb{R}^n$, and $\bar{z}$ is the vector of 2^n identical components $f(\bar{x})$.

R.3. R(X(M), z(M)) is translation–invariant with respect to M, i.e., for an arbitrary vector $c \in \mathbb{R}^n$ satisfying $M + c \in D$, we have

$$R(X(M), z(M)) = R(X(M + c), z(M)) \ . \tag{44}$$

R.4. R(X(M), z(M)) is strictly monotonically decreasing in z(M), i.e., for an arbitrary $d \in \mathbb{R}^n$*,* $d \neq 0$*,* $d \geq 0$ *(componentwise), we have*

$$R(X(M), z(M)) < R(X(M), z(M) - d)\,. \qquad (45)$$

R.5. If M is an n–interval and $\bar{x} \in M$*, then*

$$R(\bar{X}, \bar{z}) < R(X(M), z(M))\,, \qquad (46)$$

where $\bar{X}$*,* $\bar{z}$ *are defined as in R.2.*

Example XI.1. Consider the branch–and–bound interpretation of Piyavskii's univariate algorithm as discussed in Section XI.2.1. Suppose that a Lipschitz constant L' > L is used, where L denotes the infimum taken over all Lipschitz constants of f on D = [a,b]. Then exhaustiveness of the subdivision (26) has been demonstrated in Section XI.2.1., i.e., the requirement R.1. holds.

Let

$$R(X(M), z(M)) = L'\,\frac{x^2(M)-x^1(M)}{2} - \frac{z_2(M)+z_1(M)}{2}\,, \qquad 47)$$

where $M = \{x \in \mathbb{R}: x^1(M) \leq x \leq x^2(M)\}$ and $z_1(M) = f(x^1(M))$, $z_2(M) = f(x^2(M))$. Note that $(-R(x(M), z(M))$ describes Piyavskii's lower bound (cf. (22), where slightly different notation is used).

The function R(X(M), z(M)) defined by (47) is obviously continuous in its arguments $x^i(M)$, $z_i(M)$ (i=1,2), and from the continuity of f we see that requirement R.2 is satisfied.

Requirements R.3 and R.4 obviously hold.

Finally, in order to verify R.5, let $\bar{x} \in M$, $\bar{z} = f(\bar{x})$. Then

$$R(\bar{X}, \bar{z}) = -\bar{z} = -\tfrac{1}{2}(\bar{z} + \bar{z})$$

$$\leq -\tfrac{1}{2}\,[z_1(M) - L(\bar{x} - x^1(M))] - \tfrac{1}{2}\,[z_2(M) - L(x^2(M) - \bar{x})]$$

$$= L \frac{x^2(M) - x^1(M)}{2} - \frac{z_2(M) + z_1(M)}{2}$$

$$< L' \frac{x^2(M) - x^1(M)}{2} - \frac{z_2(M) + z_1(M)}{2} = R(X(M), z(M)) .$$

Proposition XI.4. *Suppose that in the above branch and bound interpretation of Pintér's method the requirements R.1 – R.5 are satisfied. Then, if the algorithm is infinite, we have*

$$\lim_{k\to\infty} \alpha_k = \min f(D) ,$$

and every accumulation point x^* of $\{x^k\}$ satisfies $f(x^*) = \min f(D)$.

Proof. We show that the assumptions of Theorem IV.2 and Corollary IV.2 are satisfied.

Since every Lipschitz function is continuous, and hence $\{x \in D: f(x) \le f(x^0)\}$ is compact because of the compactness of D, it remains to verify that the bounding operation is consistent and the selection operation is complete.

Requirement R.1 (exhaustiveness of the subdivision process) implies that every nested sequence of n–intervals $\{M_{k_q}\}$ generated by the algorithm converges to a singleton $\{\bar{x}\}$; hence we have

$$\lim_{q\to\infty} \delta(M_{k_q}) = 0 ,$$

where $\delta(M_{k_q})$ denotes the diameter $\|b_{M_{k_q}} - a_{M_{k_q}}\|$ of M_{k_q}. Using (41'), (42') and the continuity of f, we see that

$$\lim_{q\to\infty} \alpha(M_{k_q}) = f(\bar{x}) \text{ and } \lim_{q\to\infty} \beta(M_{k_q}) \ge f(\bar{x}) ;$$

hence $\lim_{q\to\infty} \beta(M_{k_q}) = f(\bar{x})$, because $\alpha(M_{k_q}) \ge \beta(M_{k_q})$ $\forall q$.
This implies consistency of the bounding operation.

To verify completeness, let $\tilde{x}$ be an arbitrary point of

$$M \in \bigcup_{p=1}^{\infty} \bigcap_{k=p}^{\infty} \mathscr{R}_k ;$$

i.e., there is a $k_0 \in \mathbb{N}$ such that M is not subdivided further if $k > k_0$. We have to show that $\inf f(M \cap D) = \inf f(M) \geq \alpha$ where $\alpha = \lim_{k \to \infty} \alpha_k$.

Let M be represented by the $n \times 2^n$ matrix $X = X(M)$ of its vertices, and let $z = z(M)$ be the corresponding vector of the values of f at the vertices of M. Consider any nested sequence $\{M_{k_q}\}$ of intervals where M_{k_q} is subdivided in Step $k_q > k_0$. In order to simplify notation, set $M_q = M_{k_q}$, let X_q be the matrix representation of M_q, and let z^q be the associated vector of the function values at the vertex set $V(M_q)$ of M_q.

The selection rule (39') implies that

$$R(X_q, z^q) \geq R(X,z) \quad \forall q \in \mathbb{N}, \tag{48}$$

and because of R.1 there is a point $\bar{x} \in D$ such that $\lim_{q \to \infty} M_q = \{\bar{x}\}$.

Let $\bar{X}$, $\bar{z}$ be the quantities associated to $\bar{x}$ defined as above, and take the limit as $q \longrightarrow \infty$ in (48). Then R.2 and R.5 yield

$$R(\bar{X},\bar{z}) = \lim_{q \to \infty} R(X_q, z^q) \geq \lim_{q \to \infty} R(X,z) = R(X,z) > R(\tilde{X}, \tilde{z}) \tag{49}$$

where $\tilde{X}$, $\tilde{z}$ correspond to $\tilde{x}$ as specified above.

Using R.3, from (43) we obtain

$$R(\bar{X}, \bar{z}) = R(\tilde{X}, \bar{z}) > R(\tilde{X}, \tilde{z}), \tag{50}$$

which, by R.4, implies that $\bar{z} < \tilde{z}$, i.e.,

$$f(\bar{x}) < f(\tilde{x}). \tag{51}$$

Now consider the sequence of points $\bar{x}^q \in M_q$ satisfying $f(\bar{x}^q) = \alpha(M_q)$. Since

$M_q \underset{q\to\infty}{\longrightarrow} \{\bar{x}\}$, $\alpha(M_q) \geq \alpha$, and since f is continuous it follows that

$$\alpha \leq \lim_{q\to\infty} \alpha(M_q) = f(\bar{x}) ;$$

and hence, by (51), $\alpha < f(\tilde{x})$.

Since $\tilde{x}$ is an arbitrary point of M, it follows that inf $f(M) \geq \alpha$, i.e., the selection process is complete. ■

A class of functions R(X(M), z(M)) that satisfies R.2 – R.5 is proposed in Pintér (1986a).

Let $x^j(M)$ $(j=1,...,2^n)$ denote the 2^n lexicographically ordered vertices of M and consider selector functions of the form

$$R(X(M),z(M)) = R_1\Big(\sum_{i=1}^{n} (x_i^{2^n}(M)-x_i^1(M))\Big) + R_2\Big(\frac{1}{2^n}\sum_{j=1}^{2^n} z_j(M)\Big) , \quad (52)$$

where R_1: $\mathbb{R}_+ \longrightarrow \mathbb{R}$ is *strictly monotonically increasing* and satisfies $R_1(0) = 0$, whereas R_2: $\mathbb{R} \longrightarrow \mathbb{R}$ is *strictly monotonically decreasing*. Furthermore, it is assumed that R_1 and R_2 are *continuous* in their respective domains.

Recall that $z_j(M) = f(x^j(M))$ $(j=1,...,2^n)$. The function R_1 depends only on the "lower left" vertex x^1 and the "upper right" vertex x^{2^n}, whereas R_2 takes into account the function values at all of the vertices of M.

It is easily seen that under the above assumptions the requirements R.2, R.3, R.4 are satisfied.

In order to meet requirement R.5, suppose in addition that R_2 is *Lipschitzian* with Lipschitz constant $L(R_2)$ and that

$$R_1(y) \geq L(R_2)\cdot L\cdot y \quad \forall y \in \mathbb{R}_+ , \qquad (53)$$

where L is the Lipschitz constant of f.

Let $\bar{x} \in M$, and consider $\bar{x}$ as an n–interval $\overline{M}$ with coinciding vertices $x^j(\overline{M}) = \bar{x}$

$(j=1,\ldots,2^n)$. Since $R_1(0) = 0$ and $z_j(\overline{M}) = f(\overline{x})$ $(j=1,\ldots,2^n)$, we see that

$$R(\overline{X}, \overline{z}) = R(X(\overline{M}), z(\overline{M})) = R_2(f(\overline{x})) .$$

By (53), the inequality $R(\overline{X}, \overline{z}) < R(X(M), z(M))$ of R.5 then follows from the following chain of relations:

$$R_2(f(\overline{x})) - R_2\left(\frac{1}{2^n}\sum_{j=1}^{2^n} z_j(M)\right) \leq L(R_2)\, |f(\overline{x}) - \frac{1}{2^n}\sum_{j=1}^{2^n} z_j(M)|$$

$$= \frac{1}{2^n} L(R_2)\, |\sum_{j=1}^{2^n} f(\overline{x}) - z_j(M))| \leq \frac{1}{2^n} L(R_2)\cdot L\cdot \sum_{j=1}^{2^n} \|\overline{x} - x^j(M)\|$$

$$< \frac{1}{2^n} L(R_2)\cdot L\cdot \sum_{j=1}^{2^n} \|x^{2^n}(M) - x^1(M)\| = L(R_2)\, L\, \|x^{2^n}(M) - x^1(M)\|$$

$$< L(R_2)\cdot L\cdot \sum_{i=1}^{n} (x_i^{2^n}(M) - x_i^1(M)) \leq R_1\left(\sum_{i=1}^{n} (x_i^{2^n}(M) - x_i^1(M)\right).$$

Example XI.2. An example of functions R_1 , R_2 satisfying the above assumption is given by

$$R_1 = c\cdot \sum_{i=1}^{n} (x_i^{2^n}(M) - x_i^1(M)) , \ (c \in \mathbb{R}_+ , c > L) ,$$

and

$$R_2 = \exp\left(-\frac{1}{2^n} \sum_{j=1}^{2^n} (z_j(M) - \beta_0)\right) ,$$

where β_0 is a lower bound for min f(D). Then $L(R_2)<1$ and $R_1(y) = cy \geq L(R_2)\cdot L\cdot y$ $\forall y \in \mathbb{R}_+$.

2.3. Branch and Bound Methods for Solving Lipschitz Optimization Problems with General Constraints

The purpose of this section is to recall certain classes of general multiextremal global optimization problems

$$\begin{aligned} &\text{minimize } f(x) \\ &\text{s.t. } x \in D \end{aligned} \tag{54}$$

and the corresponding branch and bound procedures that were already treated in Chapter IV.

Let the *compact* feasible set D and the objective function f: $\mathbb{R}^n \longrightarrow \mathbb{R}$ belong to one of the following classes:

Feasible set D:

(D_1) – *robust, convex and defined by a finite number of convex constraints;*

(D_2) – *robust, intersection of a convex set with finitely many complements of convex sets and defined by a finite number of convex and reverse convex constraints;*

(D_3) – *defined by a finite number of Lipschitzian inequalities.*

Objective Function f:

(f_1) – convex ,

(f_2) – concave,

(f_3) – d.c.,

(f_4) – Lipschitzian .

In Chapter IV we saw that a convergent branch–and–bound procedure can be developed for every optimization problem (54) where D and f belong to one of the above classes, respectively. One has to apply the prototype BB procedure of Section IV.1 with

- an appropriate *choice of partition sets* (e.g., simplices, rectangles, cones (cf. Section IV.3));
- an *exhaustive subdivision process* (cf. Section IV.3, Chapter VII);
- *bound improving selection* (cf. Definition IV.6);
- an appropriate *bounding operation* (cf. Section IV.4);
- an appropriate "*deletion by infeasibility*" rule (cf. Section IV.5).

We expect the reader to be able to formulate a convergent branch and bound procedure following the lines of the discussion in Chapter IV for each of the resulting problem classes (cf. also Horst (1988 and 1989)).

It is also recalled from Section I.4.2 that broad classes of *systems of equalities and (or) inequalities* can be solved by transforming them into an equivalent optimization problem out of the above classes (for some numerical results, see Horst and Thoai (1988)).

2.4. Global Optimization of Concave Functions Subject to Separable Quadratic Constraints

The bounds provided by the general methods referred to in the preceding section can often be improved for problems with additional structure. In this section as an example we consider problems of the following form:

$$\text{minimize } f(x)$$

$$\text{s.t. } g_i(x) = \sum_{k=1}^{n} (\tfrac{1}{2} p_{ik} x_k^2 + q_{ik} x_k + r_{ik}) \le 0 \quad (i=1,...,m), \tag{55}$$

$$\underline{m}_k \le x_k \le \overline{m}_k \quad (k=1,...,n) ,$$

where p_{ik} , q_{ik} , r_{ik}, $\underline{m}_k$, $\overline{m}_k$ (i=1,...,n; k=1,...,m) are given real numbers, and f(x) is a real valued *concave* function defined on an open convex set containing the rectangle $R := \{x \in \mathbb{R}^n : \underline{m} \le x \le \overline{m}\}$, where $\underline{m} = (\underline{m}_1,...,\underline{m}_n)^T$, $\overline{m} = (\overline{m}_1,...,\overline{m}_n)^T$.

Denote by D the feasible set of problem (55).

Note that several other problems of importance can be included under problem (55). For example, the problem of globally minimizing a separable possibly indefinite quadratic form subject to separable quadratic constraints, i.e., the problem

$$\text{minimize } f_0(x) = \sum_{k=1}^{n} (\tfrac{1}{2} p_{0k} x_k^2 + q_{0k} x_k + r_{0k}) , \tag{56}$$

$$\text{s.t. } x \in D ,$$

where p_{0k} , q_{0k} , $r_{0k} \in \mathbb{R}$ (k=1,...,n), is equivalent to

$$\begin{array}{l} \text{minimize } t \\ \text{s.t. } x \in D, f_0(x) \le t \end{array} \tag{56'}$$

with the additional variable $t \in \mathbb{R}$.

Practical problems that can be formulated as in (55) include location problems, production planning, and minimization of chance–constrained risks (cf., e.g., Phan (1982)). Quadratic programs with quadratic constraints of the form (56) also arise from some VLSI chip design problems (e.g., Maling et al. (1982)).

A direct application of the branch and bound methods developed in Chapter IV would probably use exhaustive rectangular partitions, lower bounding by means of vertex minima (cf. Section IV.4.5, Example IV.2) and the "Lipschitzian" deletion–

by–infeasibility rule (DR3) (cf. Section IV.5.).

A Lipschitz constant L(M) for a separable indefinite quadratic function $\sum_{k=1}^{n} (\frac{1}{2} p_k x_k^2 + q_k x_k + r_k)$ on a rectangle $M = \{x \in \mathbb{R}^n: a_k \leq x_k \leq b_k, k=1,...,n\}$ was derived in Section I.4.1:

$$L(M) = [\sum_{k \in I_1} (p_k a_k + q_k)^2 + \sum_{k \in I_2} (p_k b_k + q_k)^2]^{1/2}, \tag{57}$$

where

$$I_1 = \left\{k: -\frac{q_k}{p_k} \geq \frac{a_k + b_k}{2}\right\},$$

$$I_2 = \left\{k: -\frac{q_k}{p_k} < \frac{a_k + b_k}{2}\right\}.$$

In this way we are led to the following algorithm.

Algorithm XI.5.

Step 0 (Initialization):

Let $M_0 = R$, choose a finite set $S_{M_0} \subset D$ (S_{M_0} possibly empty), and determine $\beta(M_0) = \min f(V(M))$ (where V(M) denotes the vertex set of M), $\alpha_0 = \min f(S_{M_0})$ ($\alpha_0 = \infty$ if $S_{M_0} = \emptyset$).

Let $\mathcal{M}_0 = \{M_0\}$, and $\beta_0 = \beta(M_0)$.

If $\alpha_0 < \infty$, then choose $x^0 \in \operatorname{argmin} f(S_{M_0})$ (i.e., $f(x^0) - \alpha_0$).

If $\alpha_0 - \beta_0 = 0$ (or, in practice $\leq \varepsilon$, where $\varepsilon > 0$), then stop: $\alpha_0 = \beta_0 = \min f(D)$ ($\alpha_0 - \beta_0 \leq \varepsilon$, x^0 is an ε–approximate solution). Otherwise, set r = 1 and go to Step r.

Step r = 1,2,... :

At the beginning of Step r we have the current rectangular partition $\mathcal{M}_{r-1}$ of a subset of M_0 still under consideration. Furthermore, for every $M \in \mathcal{M}_{r-1}$ we have

$S_M \subset M \cap D$ and bounds $\beta(M)$, $\alpha(M)$ satisfying

$$\beta(M) \le \min f(M) \le \alpha(M) .$$

Moreover, we have the current lower and upper bounds β_{r-1} , α_{r-1} satisfying

$$\beta_{r-1} \le \min f(D) \le \alpha_{r-1} .$$

Finally, if $\alpha_{r-1} < \infty$, then we have a point $x^{r-1} \in D$ satisfying $f(x^{r-1}) = \alpha_{r-1}$ (the best feasible point obtained so far).

r.1. Delete all $M \in \mathscr{M}_{r-1}$ satisfying $\beta(M) \ge \alpha_{r-1}$.

Let $\mathscr{R}_r$ be the collection of remaining rectangles in the portion $\mathscr{M}_{r-1}$.

r.2. Select a nonempty collection of sets $\mathscr{P}_r \subset \mathscr{R}_r$ satisfying

$$\operatorname{argmin} \{\beta(M): M \in \mathscr{M}_{r-1}\} \subset \mathscr{P}_r$$

and subdivide every member of $\mathscr{P}_r$ by bisection (or any other exhaustive or normal subdivision yielding rectangular partitions). Let $\mathscr{P}'_r$ be the collection of all new partition elements.

r.3. Remove any $M \in \mathscr{P}'_r$ for which there is an $i \in \{1,...,m\}$ satisfying

$$\max \{g_i(x): x \in V(M)\} - L_i(M)\, \delta(M) > 0 , \tag{58}$$

where $L_i(M)$ is a Lipschitz constant for g_i over M given as in (57), and where $\delta(M)$ is the diameter of M (rule (DR3); note that V(M) can be replaced by any nonempty subset V'(M) of V(M)).

Let $\mathscr{M}'_r$ be the collection of all remaining members of $\mathscr{P}'_r$.

r.4. Assign to each $M \in \mathscr{M}'_r$ the set $S_M \subset M \cap D$ of feasible points in M known so far and the bounds

$$\beta(M) = \min f(V(M)),\ \alpha(M) = \min f(S_M) \quad (\alpha(M) = \infty \text{ if } S_M = \emptyset)$$

r.5. Set $\mathcal{M}_r = (\mathcal{R}_r \setminus \mathcal{P}_r) \cup \mathcal{M}_r'$. Compute

$$\alpha_r = \inf \{\alpha(M): M \in \mathcal{M}_r\}, \beta_r = \min \{\beta(M): M \in \mathcal{M}_r\}.$$

If $\alpha_r < \infty$, then let $x^r \in D$ be such that $f(x^r) = \alpha_r$.

r.6. If $\alpha_r - \beta_r = 0$ ($\leq \varepsilon$), then stop: x^r is an (ε–approximate) optimal solution. Otherwise go to Step r+1.

From the theory developed in Chapter IV we know that

$$\beta := \lim_{r\to\infty} \beta_r = \min f(D).$$

Moreover, if $S_M \neq \emptyset$ for all partition sets M, then

$$\beta = \lim_{r\to\infty} \beta_r = \min f(D) = \lim_{r\to\infty} \alpha_r =: \alpha,$$

and every accumulation point of the sequence $\{x^r\}$ solves problem (55).
If not enough feasible points can be obtained such that $S_M \neq \emptyset$ for all partition elements M, then, as discussed in Chapter IV, one may consider the iteration sequence $\{\tilde{x}^r\}$ defined by $f(\tilde{x}^r) = \beta_r$. Although $\tilde{x}^r$ is not necessarily feasible for problem (49), we know that every accumulation point of $\{\tilde{x}^r\}$ solves problem (55).

The general algorithm uses only simple calculations to make decisions on partitioning, deleting and bounding. However, the lower bounds $\beta(M) = \min f(V(M))$ are weak, and a closer examination leads to procedures that allow improved bounding, at the expense of having to solve additional subproblems. Moreover, a mechanism needs to be devised for identifying points in $S_M \subset M \cap D$. Following Al–Khayyal, Horst and Pardalos (1988), we next attempt to obtain better bounds than $\beta(M)$ using only linear programming calculations. In the process, at times, we will also be able to identify when $M \cap D = \emptyset$ or possibly uncover feasible points of $M \cap D \neq \emptyset$ for inclusion in S_M.

Let

$$G = \{x: g_i(x) = \sum_{k=1}^{n} g_{ik}(x_k) \le 0 \quad (i=1,...,m)\},$$

where $g_{ik}(x_k) = \frac{1}{2} p_{ik}x_k^2 + q_{ik}x_k + r_{ik}$ (i=1,...,m; k=1,...,n).
Note that for each partition element M we have $M \cap D = M \cap G$ since $M_0 \subset R$ and $D = R \cap G$.

Linearization of the constraints in G:

We begin with a simple linearization of the constraints in G. Let $M = \{x \in \mathbb{R}^n: a \le x \le b\}$ and let $\varphi_{M,h}(x)$ denote the *convex envelope* of a function h over M. Then we know from Theorem IV.8 that

$$\varphi_{M,g_i}(x) = \sum_{k=1}^{n} \varphi_{M_k,g_{ik}}(x_k) \le g_i(x) , \tag{53}$$

where

$$M_k = \{x_k: a_k \le x_k \le b_k\} \quad (k=1,...,n) .$$

Each $g_{ik}(x_k)$ can be linearized according to one of the following three cases:

Case 1: $p_{ik} = 0$. Then $g_{ik}(x_k)$ is linear.

Case 2: $p_{ik} < 0$. Then $g_{ik}(x_k)$ is concave and

$$\varphi_{M_k,g_{ik}}(x_k) = \alpha_{ik}x_k + \beta_{ik} ,$$

where

$$\alpha_{ik} = [g_{ik}(b_k) - g_{ik}(a_k)] \,/\, (b_k - a_k), \ \beta_{ik} = g_{ik}(a_k) - \alpha_{ik}a_k$$

(cf. the remark after Theorem IV.7).

In this case we can replace $g_{ik}(x_k)$ by $\ell_{ik}^{(0)}(x_k) = \alpha_{ik}x_k + \beta_{ik}$.

Case 3: $p_{ik} > 0$. Then g_{ik} is convex. Compute $\hat{x}_k = -q_{ik}/p_{ik}$ which minimizes g_{ik}.

If $g_{ik}(\hat{x}_k) \geq 0$, then replace $g_{ik}(x_k)$ by the constant $g_{ik}(\hat{x}_k)$ in the constraint $g_i(x) \leq 0$. (This effectively enlarges the region which is feasible for that constraint.) Otherwise, continue.

Case 3a: If $\hat{x}_k < a_k$, then compute

$$\rho_{ik}^{(1)} = \frac{1}{p_{ik}} [- q_{ik} + (q_{ik}^2 - 2p_{ik}r_{ik})^{1/2}] .$$

($\rho_{ik}^{(1)}$ is the zero of g_{ik} to the right of $\hat{x}_k$). Replace $g_{ik}(x_k)$ by the linear support of its graph at $\rho_{ik}^{(1)}$. This is given by

$$\ell_{ik}^{(1)}(x_k) = \alpha_{ik}x_k + \beta_{ik} \leq g_{ik}(x_k) ,$$

where $\alpha_{ik} = p_{ik}\rho_{ik}^{(1)} + q_{ik} > 0$ and $\beta_{ik} = - \alpha_{ik}\rho_{ik}^{(1)}$.

Case 3b: If $\hat{x}_k > b_k$, then replace $g_{ik}(x_k)$ by the linear support of its graph at

$$\rho_{ik}^{(2)} = \frac{1}{p_{ik}} [- q_{ik} - (q_{ik}^2 - 2p_{ik}r_{ik})^{1/2}] ;$$

namely,

$$\ell_{ik}^{(2)}(x_k) = \alpha_{ik}x_k + \beta_k \leq g_{ik}(x_k) ,$$

where $\alpha_{ik} = p_{ik}\rho_{ik}^{(2)} + q_{ik} < 0$ and $\beta_{ik} = - \alpha_{ik}\rho_{ik}^{(2)}$ ($\rho_{ik}^{(2)}$ is the zero of g_{ik} to the left of $\hat{x}_k$).

Case 3c: If $a_k \leq \hat{x}_k \leq b_k$, then compute $\rho_{ik}^{(1)}$ and $\rho_{ik}^{(2)}$, as above, and replace $g_{ik}(x_k)$ by the maximum of the supports at $\rho_{ik}^{(1)}$ and $\rho_{ik}^{(2)}$,

$$\max \{\ell_{ik}^{(1)}(x_k) , \ell_{ik}^{(2)}(x_k)\} .$$

Let t_i be the number of terms in $g_i(x)$ that fall into Case 3c. That is, t_i is the cardinality of the index set

$$K_i = \{k: p_{ik} > 0, g_{ik}(\hat{x}_k) \leq 0 , a_k \leq \hat{x}_k \leq b_k , k \in \{1,...,n\}\} .$$

Setting

$$\ell_{ik}(x_k) = \begin{cases} g_{ik}(x_k) & \text{if Case 1} \\ \ell_{ik}^{(0)}(x_k) & \text{if Case 2} \\ g_{ik}(\hat{x}_k) & \text{if Case 3} \\ \ell_{ik}^{(1)}(x_k) & \text{if Case 3a} \\ \ell_{ik}^{(2)}(x_k) & \text{if Case 3b} \end{cases} ,$$

we have

$$\ell_i(x) := \sum_{k \notin K_i} \ell_{ik}(x_k) + \sum_{k \in K_i} \max\{\ell_{ik}^{(1)}(x_k), \ell_{ik}^{(2)}(x_k)\} \le g_i(x) ,$$

and $\ell_i(x)$ is a linear underestimate of $g_i(x)$ when $K_i = \emptyset$, and it is piecewise linear and convex, otherwise.

In particular, the region defined by $\ell_i(x) \le 0$ is equivalent to the polyhedral set defined by

$$\sum_{k \notin K_i} \ell_{ik}(x_k) + \sum_{k \in K_i} z_{ik} \le 0 ,$$

$$\ell_{ik}^{(1)}(x_k) \le z_{ik} \quad (k \in K_i) ,$$

$$\ell_{ik}^{(2)}(x_k) \le z_{ik} \quad (k \in K_i) ,$$

which involves t_i new variables z_{ik} $(k \in K_i)$ and $2t_i$ additional constraints.

Performing the above linearization for every constraint i, we let L(G) denote the resulting polyhedral set, and we let $L_x(G)$ denote its projection onto the n–dimensional space of x. It is clear that

$$L_x(G) \supset \text{conv}\,(G) \supset G .$$

It is also clear that, if $(\bar{x}, \bar{z})$ minimizes f(x) subject to $(x,z) \in L(G)$ and $x \in M$, then $\bar{x}$ minimizes f(x) over $L_x(G) \cap M$. If $L_x(G) \cap M = \emptyset$, then $G \cap M = \emptyset$ and $D \cap M = \emptyset$.

Therefore, if we solve the linearly constrained concave minimization problem

$$\begin{aligned} &\text{minimize } f(x) \\ &\text{s.t. } (x,z) \in L(G),\ x \in R \end{aligned} \tag{60}$$

by one of the methods developed in Part B, we obtain an initial lower bound for min f(D). (A refined linearization technique will be discussed below.)

Solving the problem

$$\begin{aligned} &\text{minimize } f(x) \\ &\text{s.t. } (x,z) \in L(G),\ x \in M \end{aligned} \tag{61}$$

would lead to a better lower bound than min f(V(M)). However, since f is concave (and is not assumed to have any additional exploitable structure), too much effort is required to solve (61). Instead, we seek an $\hat{x} \in L_x(G) \cap M$ that satisfies $\min f(V(M)) \le f(\hat{x}) \le f(\bar{x})$. This can be done by using any of the algorithms discussed in Part B of this book. We run one of these algorithms on problem (61) until a point $(\hat{x},\hat{z}) \in L(G)$, $\hat{x} \in M$ satisfying $\min f(V(M)) \le f(\hat{x})$ is found. If $\hat{x} \in G$, then $S_M \neq \emptyset$.

For problem (56) in the form (56'), however, problems (60) and (61) are linear programs that can be solved easily.

Bilinear constraint approach:

An alternative linearization of G can be derived from the bilinear constraints that arise from the quadratic constraints in the following way.

Each constraint function can be written as

$$g_i(x) = \sum_{k=1}^{n} \left(\tfrac{1}{2} p_{ik} x_k^2 + q_{ik} x_k + r_{ik}\right) = \sum_{k=1}^{n} \left(x_k y_{ik} + r_{ik}\right),$$

where $y_{ik} = \frac{1}{2} p_{ik} x_k + q_{ik}$ (i=1,...,m; k=1,...,n).

From Theorem IV.8 and Proposition X.20 we also have that the convex envelope of each g_i over Ω_i can be expressed by

$$\varphi_{\Omega_i,g_i}(x) = \sum_{k=1}^{n} (\varphi_{\Omega_{ik}}(x_k y_{ik}) + r_{ik}) , \qquad (62)$$

where $\varphi_{\Omega_{ik}}(x_k y_{ik})$ denotes the convex envelope of $x_k y_k$ over the set Ω_{ik} and

$$\Omega_i = \Omega_{i1} \times \Omega_{i2} \times \dots \times \Omega_{in} ,$$

$$\Omega_{ik} = \{(x_k, y_{ik}): a_k \le x_k \le b_k , c_{ik} \le y_{ik} \le d_{ik}\} ,$$

$$c_{ik} = \frac{1}{2} \min \{p_{ik} a_i , p_{ik} b_k\} + q_{ik} ,$$

$$d_{ik} = \frac{1}{2} \max \{p_{ik} a_i , p_{ik} b_k\} + q_{ik} ,$$

$$\varphi_{\Omega_{ik}}(x_k y_{ik}) = \max \{\ell_{ik}^{(1)}(x_k, y_{ik}), \ell_{ik}^{(2)}(x_k, y_{ik})\} ,$$

$$\ell_{ik}^{(1)}(x_k, y_{ik}) = c_{ik} x_k + a_k y_{ik} - a_k c_{ik} ,$$

$$\ell_{ik}^{(2)}(x_k, y_{ik}) = d_{ik} x_k + b_k y_{ik} - b_k d_{ik} .$$

Let $y_i = (y_{i1},\dots,y_{in})^T$. Since $\varphi_{\Omega_i,g_i}(x)$ is piecewise linear and convex, the set defined by

$$\varphi_{\Omega_i,g_i}(x) \le 0 ,$$

$$y_{ik} = \frac{1}{2} p_{ik} x_k + q_{ik} \quad (k=1,\dots,n) ,$$

$$(x, y_i) \in \Omega_i ,$$

is equivalent to a polyhedral set whose description involves n additional variables $(z_{i1},\dots,z_{in})$ and 2n additional linear constraints:

$$\sum_{k=1}^{n} z_{ik} \le 0 ,$$

$$y_{ik} = \frac{1}{2} p_{ik} x_k + q_{ik} \quad (k=1,\dots,n) ,$$

$$\ell_{ik}^{(1)}(x_k,y_k) + r_{ik} \le z_{ik} \quad (k=1,...,n) ,$$

$$\ell_{ik}^{(2)}(x_k,y_k) + r_{ik} \le z_{ik} \quad (k=1,...,n) ,$$

$$(x,y_i) \in \Omega_i .$$

This is done for each constraint i=1,...,m to yield a polyhedral set P(G) whose projection onto $\mathbb{R}^n$ (the space of x–variables), denoted by $P_x(G)$, contains the convex hull of G; that is we have

$$G \subseteq \text{conv}\,(G) \subseteq P_x(G) .$$

Hence,

$$\min \{f(x): (x,y,z) \in P(G), x \in M\} \le \min \{f(x): x \in G \cap M\} .$$

The discussion at the end of the preceding subsection ("Linearisation of the constraints") can now be followed with L(G) replaced by P(G) and the vector (x,z) replaced by (x,y,z).

Piecewise linear approximation:

Piecewise linear approximation of separable functions belongs to the folklore of mathematical programming. Recently this approach has been studied in the context of separable (quadratic) concave problems subject to linear constraints (e.g., Rosen and Pardalos (1986), Pardalos and Rosen (1987)). A detailed discussion is given in Chapter IX. An extension of these techniques to problems with an indefinite quadratic objective function is presented in Pardalos, Glick and Rosen (1987).

Let $g: [a,b] \longrightarrow \mathbb{R}$ be a continuous *univariate* function on the interval $[a,b] \subset \mathbb{R}$. Choose a fixed grid of points by partitioning [a,b] into r subintervals of length

$$h = \frac{b - a}{r}$$

and determine the piecewise linear function $\ell(x)$ that linearly interpolates g(x) at the grid points

$$x_j = a + jh \quad (j=0,\dots,r) .$$

Replacing a constraint $g(x) \leq 0$ by $\ell(x) \leq 0$ leads to linear constraints and additional zero–one integer variables having a special structure. To see this, one writes $x \in [a,b]$ in the form

$$x = a + \sum_{j=1}^{r} h\omega_j$$

where

$$0 \leq \omega_j \leq 1 \quad (j=1,\dots,r) \tag{63}$$

$$\omega_{j+1} \leq z_j \leq \omega_j , \quad z_j \in \{0,1\} \quad (j=1,\dots,r-1) .$$

The last constraints in (63) imply that the vector $z = (z_1,\dots,z_{r-1})$ of zero–one variables must have the form $(1,1,\dots,1,0,0,\dots0)$, i.e., whenever $z_j = 0$ for some j, one has $z_{j+1} = \dots = z_{r-1} = 0$. Hence, z takes only r possible values, instead of 2^{r-1} values as in the case of a general zero–one vector with r–1 components. Under the transformation (63), $\ell(x)$ can be written in the form

$$\begin{aligned} \ell(x) &= \ell(a) + \sum_{j=1}^{r} \omega_j(\ell(a + jh) - \ell(a + (j-1)h) \\ &= g(a) + \sum_{j=1}^{r} \omega_j(g(a + jh) - g(a + (j-1)h). \end{aligned} \tag{64}$$

Replacing each term g_{ik} in the constraints of (55) by its corresponding piecewise linear approximation in the form (63), (64), one obtains an approximation for G which is described by mixed integer linear constraints. Then, in particular, problem (56), (56') can be approximated by a mixed integer linear program.

Since the constraints are quadratic, it is easy to derive bounds on the maximum interpolation error.

3. OUTER APPROXIMATION

Consider the problem

$$(P) \qquad \begin{array}{ll} \text{minimize } f(x) & \\ \text{s.t. } g_i(x) \le 0 & (i=1,\dots,m) \end{array}, \tag{65}$$

where f, g_i: $\mathbb{R}^n \longrightarrow \mathbb{R}$ are *Lipschitz functions* (i=1,...,m). Suppose that the feasible set

$$D = \{x \in \mathbb{R}^n: g_i(x) \le 0 \quad (i=1,\dots,m)\} \tag{66}$$

is nonempty and *compact*, and suppose that a real number r > 0 is known which satisfies

$$D = \{x \in \mathbb{R}^n: \|x\| \le r\} , \tag{67}$$

i.e., we know a ball of radius r containing D ($\|\cdot\|$ denotes the Euclidean norm).

Moreover, it is assumed that *Lipschitz constants* L_i of g_i on the ball $\|x\| \le r$ (i=1,...,m) are known.

We attempt to apply to (65) an outer approximation method of the type discussed in Chapter II. Recall from Section II.1 that in each step of an outer approximation method a subproblem of the form

$$(Q_k) \qquad \begin{array}{l} \text{minimize } f(x) \\ \text{s.t. } x \in D_k \end{array} \tag{68}$$

must be solved, where D_k is a relaxation of D and

$$D_{k+1} = D_k \cap \{x: \ell_k(x) \le 0\} \tag{69}$$

with a suitable function ℓ_k: $\mathbb{R}^n \longrightarrow \mathbb{R}$ satisfying

$$\ell_k(x) \le 0 \quad \forall x \in D , \tag{70}$$

$$\ell_k(x^k) > 0 \tag{71}$$

(x^k denotes a solution of (Q_k)).

A direct application of such an outer approximation scheme does not seem to be promising, because the subproblems (Q_k) are still Lipschitz optimization problems, and, moreover, it will be difficult to find suitable functions ℓ_k such that $\{x: \ell_k(x) = 0\}$ separates x^k from D in the sense of (70), (71). Convexity does not seem to be present in problem (65), and this makes it difficult to apply outer approximation methods.

Therefore, we shall first transform problem (65) into an equivalent program where convexity is present. Specifically, problem (65) will be converted into a problem of globally minimizing *a concave* (in fact, even linear) *function subject to a convex and a reverse convex constraint.*

This idea and the following outer approximation method are due to Thach and Tuy (1987).

First we note that in (65) one may always assume that the objective function f(x) is concave (even linear). Indeed, in the general case of a nonconcave function f, it would suffice to write the problem as

$$\begin{array}{ll} \text{minimize} & t \\ \text{s.t.} & f(x) \le t \,,\; g_i(x) \le 0 \quad (i=1,...,m) \end{array}$$

which involves the additional variable t and the additional constraint $f(x) \le t$.

Now, in the space $\mathbb{R}^{n+1}$ let us consider the hemisphere

$$S = \{u \in \mathbb{R}^{n+1}: \|u\|^2 = \sum_{i=1}^{n+1} u_i^2 = r^2 \,,\; u_{n+1} \ge 0\} \,,$$

whose projection onto $\mathbb{R}^n$ is just the ball $B:= \{x \in \mathbb{R}^n: \|x\| \le r\}$ introduced above. Using the projection $\pi: \mathbb{R}^{n+1} \rightarrow \mathbb{R}^n$ defined by $u = (u_1,...,u_{n+1}) \rightarrow \pi(u) = (u_1,...,u_n)$ we can establish an obvious homeomorphism between the hemisphere S in $\mathbb{R}^{n+1}$ and the ball B in $\mathbb{R}^n$.

Let

$$\varphi(u) = f(\pi(u)), \;\; \varphi_i(u) = g_i(\pi(u)) \;\; (i=1,...,m) \,. \tag{72}$$

Then we rewrite problem (P) in the form

$$(P_S) \qquad \begin{aligned} &\text{minimize } \varphi(u) \\ &\text{s.t. } \varphi_i(u) \le 0 \qquad (i=1,...,m), \\ &\|u\| = r, \ u_{n+1} \ge 0. \end{aligned} \tag{73}$$

In fact, if $\bar{x}$ solves (P), then $\bar{u}$ with $u_{n+1} = \sqrt{r^2 - \|\bar{x}\|^2}$, $\pi(\bar{u}) = \bar{x}$ (i.e., $\bar{u}_i = \bar{x}_i$ (i=1,...,n)) solves (P_S). Conversely, if $\bar{u}$ solves (P_S), then $\bar{x} = \pi(\bar{u})$ solves (P).

Since π is a linear mapping, the function φ is still concave, while the functions φ_i (i=1,...,m) are still Lipschitzian on S with the same Lipschitz constants L_i as g_i (on B).

At first glance, problem (P_S) seems to be more complicated than the original problem (P). However, the following proposition shows that the feasible set of (73) can be expressed as the difference of two convex sets.

Let C denote the feasible set of problem (P_S), i.e.,

$$\begin{aligned} C &= \{u \in \mathbb{R}^{n+1}: \varphi_i(u) \le 0 \ \ (i=1,...,m), \ \|u\| = r, \ u_{n+1} \ge 0\} \\ &= \{u \in S: \varphi_i(u) \le 0 \ \ (i=1,...,m)\}. \end{aligned} \tag{74}$$

Proposition XI.5. *The feasible set C of problem (P_S) is a difference of two convex sets:*

$$C = \Omega \setminus G, \tag{75}$$

where Ω is the (closed) convex hull of C and G is the (open) ball of radius r in $\mathbb{R}^{n+1}$, i.e.,

$$\Omega = conv\, C, \quad G = \{u \in \mathbb{R}^{n+1}: \|u\| < r\}. \tag{76}$$

Proof. Clearly, we have $C \subset \Omega \setminus G$.

To prove the converse inclusion, consider an arbitrary point $u \in \Omega \setminus G$. Then u can be written as $u = \sum_{i\in I} \lambda_i u^i$, with $\sum_{i\in I} \lambda_i = 1$, $\lambda_i > 0$, $u^i \in C$ $(i \in I)$. Suppose that

$|I| > 1$. Then we would have

$$\|u\| < \sum_{i \in I} \lambda_i \|u^i\| = \sum_{i \in I} \lambda_i r = r$$

(since the norm $\|\cdot\|$ is strictly convex), i.e., $u \in G$. Therefore, $u \notin G$ implies $|I| = 1$, and hence $u \in C$. ■

Note that Proposition XI.5 simply expresses the observation that, when one forms the convex hull of a closed subset of a sphere in $\mathbb{R}^{n+1}$ (with respect to the Euclidean norm), one only needs to add to this set strictly interior points of the ball which determines the sphere.

A related observation is that, given a closed subset M of the hemisphere S in $\mathbb{R}^{n+1}$ and a point $u^0 \in S \setminus M$, there always exists a hyperplane H in $\mathbb{R}^{n+1}$ which strictly separates u^0 from M: take the hyperplane which supports the ball $\|u\| \leq r$ at u^0, and move it parallel to itself a sufficiently small distance towards the interior of the ball. In particular, a separating hyperplane can easily be constructed for $M = C$, where the functions φ_i are Lipschitzian functions with Lipschitz constants L_i.

Proposition XI.6. *For every $u \in S$ define*

$$h(u) := \frac{1}{2} \max_{i=1,\ldots,m} \left[\frac{\varphi_i^+(u)}{L_i}\right]^2 \tag{77}$$

where $\varphi_i^+(u) = \max\{0, \varphi_i(u)\}$. If $u^0 \in S \setminus C$, then the affine function

$$\ell(u) = u^0 u - r^2 + h(u^0) \tag{78}$$

strictly separates u^0 from C, i.e., one has

$$\ell(u^0) > 0, \quad \ell(u) \leq 0 \quad \forall u \in C. \tag{79}$$

Proof. Since $u^0 \notin C$, we have $\varphi_i(u^0) > 0$ for at least one $i \in \{1,\ldots,m\}$. Therefore, $h(u^0) > 0$, and since $\|u^0\| = r$, it follows that

$$\ell(u^0) = \|u^0\|^2 - r^2 + h(u^0) = h(u^0) > 0 .$$

On the other hand, $u \in C$ implies $\varphi_i^+(u) = 0$ $(i=1,...,m)$.

It follows that for any $u \in C$ we have

$$\varphi_i(u^0) \leq \varphi_i(u) + L_i\|u^0 - u\| \leq L_i\|u^0 - u\| \quad (i=1,...,m).$$

Therefore, if i^* denotes an index such that

$$h(u^0) = \frac{1}{2} \max_{i=1,...,m} \left[\frac{\varphi_i^+(u^0)}{L_i}\right]^2 = \frac{1}{2}\left[\frac{\varphi_{i^*}^+(u^0)}{L_{i^*}}\right]^2 = \frac{1}{2}\left[\frac{\varphi_{i^*}(u^0)}{L_{i^*}}\right]^2 ,$$

then

$$h(u^0) \leq \frac{1}{2}\left[\frac{L_{i^*}\|u^0 - u\|}{L_{i^*}}\right]^2 = \frac{1}{2}\|u^0 - u\|^2 . \tag{80}$$

But, since $u^0u - r^2 = u^0u - u^0u^0 = u^0(u - u^0)$, it follows from (78) and (80) that

$$\begin{aligned} \ell(u) &\leq u^0(u - u^0) + \frac{1}{2}\|u^0 - u\|^2 \\ &= (u^0 + \frac{1}{2}(u - u^0))\,(u - u^0) = \frac{1}{2}(u + u^0)\,(u - u^0) \\ &= \frac{1}{2}(\|u\|^2 - \|u^0\|^2) = \frac{1}{2}(r^2 - r^2) = 0 . \end{aligned}$$

This completes the proof. ■

As a consequence of the above results we see that problem (P_S) is actually a special d.c. programming problem, namely, it is a problem of minimizing a concave function over the intersection of the convex set Ω with the complement of the convex set $\{u \in \mathbb{R}^{n+1}: \|u\| < r\}$ (cf. Proposition XI.5). Therefore, an outer approximation method, such as those discussed in Chapter X, could be applied. A difficulty which then arises is that the convex set Ω is not defined explicitly by a set of finitely many convex constraints.

However, for any $u^0 \in S \setminus C$, Proposition XI.6 allows us to construct a hyperplane which strictly separates u^0 from C, as required by outer approximation methods.

Before describing such a method we first prove a result which is very similar to Proposition IX.11.

Proposition XI.7. *Consider a polytope P in* $\mathbb{R}^{n+1}$ *and the problem*

$$\begin{aligned} &\text{minimize } \varphi(u) \\ &\text{s.t. } u \in P \text{ , } \|u\| \geq r \end{aligned} \tag{81}$$

where $\varphi(u)$ *is a concave function on* $\mathbb{R}^{n+1}$. *If this problem is solvable, then there is always an optimal solution of the problem which is a vertex of P or lies on the intersection of the surface* $\|x\| = r$ *with an edge of P.*

Proof. For every $w \in P$, let F_w denote the face of P containing w of smallest dimension.

Suppose that problem (81) has an optimal solution in the region $\|u\| > r$ and let w be such an optimal solution with minimal F_w.

We first show that in this case $\dim F_w = 0$, i.e., w must be a vertex of P. Indeed, if $\dim F_w \geq 1$, then there exists a line whose intersection with $F_w \cap \{u: \|u\| \geq r\}$ is a segment $[w',w'']$ containing w in its relative interior. Because of the concavity of $\varphi(u)$, we must have $\varphi(w') = \varphi(w'') = \varphi(w)$. Moreover, if $\|w'\| > r$, then $\dim F_{w'} < \dim F_w$, and this contradicts the minimality of $\dim F_w$. Therefore, $\|w'\| = r$, and similarly $\|w''\| = r$. But then we must have $[w', w''] \subset \{u: \|u\| \leq r\}$, contradicting the assumption that $\|w\| > r$.

Now consider the case when all of the optimal solutions of (73) lie on the surface $\|u\| = r$, and let w be an optimal solution. If $\dim F_w > 1$, then the tangent hyperplane to the sphere $\|x\| = r$ at the point w and $F_w \cap \{u: \|u\| \geq r\}$ would have a common line segment $[w', w'']$ which contains w in its relative interior. Because of the concavity of φ, we must have $\varphi(w') = \varphi(w'') = \varphi(w)$, i.e., w' and w'' are also optimal solutions. But since $\|w'\| > r$ and $\|w''\| > r$, this contradicts the hypothesis. Therefore, $\dim F_w \leq 1$, i.e., w lies on an edge of P. ■

Algorithm XI.6:

Initialization:

Select a polytope D_1 with known vertex set V_1 satisfying

$$\Omega \subset D_1 \subset \{u: \|u\| \leq r, u_{n+1} \geq 0\} .$$

Compute the set $\tilde{V}_1$ of all points w such that w is either a vertex of D_1 satisfying $\|w\| \geq r$ or else the intersection of an edge of D_1 with the sphere $\|u\| = r$.

Set $k = 1$.

Step 1. Compute

$$w^k \in \operatorname{argmin} \{\varphi(w): w \in \tilde{V}_k\} \tag{82}$$

and the point u^k where the vertical line through w^k meets the hemisphere $S = \{u: \|u\| = r, u_{n+1} \geq 0\}$.

If $\varphi_i(u^k) \leq 0$ $(i=1,\ldots,m)$, i.e., $u^k \in C$, then stop: u^k is an optimal solution of (P_S).

Otherwise go to Step 2.

Step 2. Define the additional linear constraint

$$\ell_k(u) = u^k u - r^2 + h(u^2) \leq 0 \tag{83}$$

and set

$$D_{k+1} = D_k \cap \{u: \ell_k(u) \leq 0\} .$$

Compute the vertex set V_{k+1} of D_{k+1} and the set $\tilde{V}_{k+1}$ of all points w such that w is either a vertex of D_{k+1} satisfying $\|w\| \geq r$ or else the intersection of an edge of D_{k+1} with the sphere $\|u\| = r$.

Set $k \leftarrow k+1$ and return to Step 1.

Remark XI.2. The sets V_{k+1} , $\tilde{V}_{k+1}$ can be obtained from V_k, $\tilde{V}_k$ using the methods discussed in Section II.4.2.

Remark XI.3. The constraint (83) corresponds to a quadratic cut in the original variables $(x_1,\ldots,x_n)$ of the form

$$x^k x + [(r^2 - \|x^k\|^2)(r^2 - \|x\|^2]^{1/2} - r^2 + h(x^k) \leq 0 .$$

The convergence of Algorithm XI.6 follows from the general theory of outer approximation methods that was discussed in Chapter II.

Proposition XI.8. *If Algorithm XI.6 is infinite, then every accumulation point of* $\{u^k\}$ *is an optimal solution of problem* (P_S).

Proof. We refer to Theorem II.1, and we verify that the assumptions of this theorem are satisfied.

The conditions $\ell_k(u^k) > 0$ and $\ell_k(u) \leq 0 \;\; \forall u \in C$ follow from Proposition XI.6.

Since the functions $\varphi_i(u)$ are Lipschitzian, and hence continuous, we have continuity of $\ell_k(u)$.

Now consider a subsequence $\{u^{k_q}\} \subset \{u^k\}$ satisfying $u^{k_q} \to \bar{u}$. Clearly

$$\ell_{k_q}(u^{k_q}) = u^{k_q}u - r^2 + h(u^{k_q}) \xrightarrow[q\to\infty]{} \ell(u) := \bar{u}u - r^2 + h(\bar{u}) .$$

Moreover, since $\|u^{k_q}\| = r$, $u^{k_q}_{n+1} \geq 0 \;\; \forall q$, we have $\|\bar{u}\| = r$, $\bar{u}_{n+1} \geq 0$. Therefore, using $h(u) \geq 0 \;\; \forall u$, we see that $\ell(\bar{u}) = \|\bar{u}\|^2 - r^2 + h(\bar{u}) = 0$ implies $h(\bar{u}) = 0$. But $h(\bar{u}) = 0$ is equivalent to $\bar{u} \in C$.

Thus, the assumptions of Theorem II.1 are satisfied, and Proposition XI.8 follows from Theorem II.1. ∎

Example XI.3. Consider the feasible set $D \subset \mathbb{R}^2$ defined by

$$g_1(x) = (x_1^2 + x_2^2)^{1/2} - 5 \leq 0 ,$$

$$g_2(x) = 0.125 \sin 0.625\, x_1 + 0.0391\, x_2 \leq 0 ,$$

which is contained in the rectangle $-5 \leq x_1 \leq 5$, $-5 \leq x_2 \leq 3.2$.

Consider the ball B of radius 1 around $(-1,0)$. Let $a = (-1.5,0.5)$ and for every x define

$$p(x) = \max \{t \in \mathbb{R}: t(x - a) \in B\} .$$

It is easily seen that p(x) is a convex function having several local extrema over A (actually p(x) is equal to the value of the gauge of $B - a$ at the point $x - a$). We want to find the global maximum of p(x), or equivalently, the global minimum of $f(x) = -p(x)$ over D. By an easy computation we see that $f(a) = 0$ and

$$f(x) = \frac{2x_1^2 + 2x_2^2 + 6x_1 - 2x_2 + 5}{(3x_1^2 + 3x_2^2 - 2x_1x_2 + 10x_1 - 6x_2 + 9)^{1/2} + x_1 - x_2 + 2}$$

for $x \neq a$. The Lipschitz constants of g_1, g_2 are $L_1 = 1$, $L_2 = 0.0875$.

We can choose $r = 100$ and the starting polytope

$$D_1 = \{(x_1, x_2, x_3): -5 \leq x_1 \leq 5, -5 \leq x_2 \leq 3.2, 0 \leq x_3 \leq 100\} .$$

After 13 iterations the algorithm finds a solution with $\bar{x} = (-3.2142, 3.0533)$, $p(\bar{x}) = 10.361$, $\max \{g_1(\bar{x}), g_2(\bar{x})\} = 0.0061$.

The intermediary results are shown in the table below, where N_k denotes the number of vertices of the current polytope and $\varepsilon_k = \max \{g_1(x^k), g_2(x^k)\}$.

k	N_k	$x^k = (x_1^k, x_2^k)$	$p(x^k)$	ϵ_k	$\lambda(x^k)$
1	8	$(-5, 3.2)$	15.005	0.9363	0.9868
2	10	$(-5, 1.7956)$	11.927	0.3126	0.3026
3	12	$(-3.5964, 3.2)$	11.606	0.0276	0.0496
4	14	$(-3.6321, 2.8871)$	10.915	0.0171	0.0191
5	16	$(-3.2816, 3.2)$	10.886	0.0141	0.0131
6	18	$(-3.6891, 2.7)$	10.596	0.0128	0.0106
7	20	$(-3.3231, 3.0439)$	10.585	0.0096	0.0060
8	22	$(-3.4406, 2.9265)$	10.562	0.0098	0.0062
9	24	$(-3.1201, 3.2)$	10.524	0.0089	0.0052
10	26	$(-5, 1.0178)$	10.435	0.1026	0.0927
11	28	$(-4.2534, 2.0118)$	10.423	0.0205	0.0275
12	30	$(-2.748, 2.57)$	10.420	0.0107	0.0074
13	32	$(-3.2142, 3.0533)$	10.361	0.0061	0.0024

Table XI.1.

4. THE RELIEF INDICATOR METHOD

The conceptual method that follows, which is essentially due to Thach and Tuy (1990), is intended to get beyond local optimality in general global optimization and to provide solution procedures for certain important special problem classes.

Consider the problem

$$\text{(P)} \qquad \begin{array}{ll} \text{minimize} & f(x) \\ \text{s.t.} & x \in D \end{array} , \tag{84}$$

where $f: \mathbb{R}^n \to \mathbb{R}$ is *continuous* and $D \subset \mathbb{R}^n$ is *compact*. Moreover, it will be assumed that

$$\inf\{f(x): x \in D\} = \inf\{f(x): x \in \text{int } D\} . \tag{85}$$

Assumption (85) is fulfilled, for example, if D is robust (cf. Definition I.1).

The purpose of our development is to associate to f, D and to every $\alpha \in \mathbb{R}$ a d.c. function $\varphi_\alpha(x)$ such that $\bar{x}$ is a global minimizer of f over D if and only if

$$0 = \min\{\varphi_{\bar{\alpha}}(x): x \in \mathbb{R}^n\} , \tag{86}$$

where $\bar{\alpha} = f(\bar{x})$. Based on this optimality criterion, a method will be derived to handle problem (P) in the sense mentioned above.

4.1. Separators for f on D

The function $\varphi_\alpha(x)$ will be defined by means of a separator of f on D in the following sense.

Let $\bar{\mathbb{R}} = \mathbb{R} \cup \{\infty\}$, and for each $\alpha \in \bar{\mathbb{R}}$ consider the level sets

$$D_\alpha := \{x \in D: f(x) < \alpha\} ,$$

$$\bar{D}_\alpha := \{x \in D: f(x) \le \alpha\} .$$

Let $d_A(x) := \inf \{\|x - y\|: y \in A\}$ denote the distance from $x \in \mathbb{R}^n$ to a set $A \subset \mathbb{R}^n$ (with the usual convention that $d_A(x) = +\infty$ if A is empty).

Definition XI.1. *A real valued function $r(\alpha,x)$ defined on $\overline{\mathbb{R}}\times\mathbb{R}^n$ is called a separator for the function $f(x)$ on the set D if it satisfies the following conditions:*

(i) $0 \le r(\alpha,x) \le d_{\overline{D}_\alpha}(x)$ *for every* $\alpha \in \overline{\mathbb{R}}$, $x \in \mathbb{R}^n$;

(ii) for each fixed $\alpha \in \mathbb{R}$, $x^k \underset{k\to\infty}{\longrightarrow} \tilde{x} \notin \overline{D}_\alpha$ *implies that*

$$\underline{\lim}_{k\to\infty} r(\alpha,x^k) > 0;$$

(iii) $r(\alpha,x)$ *is monotonically nonincreasing in* α, *i.e.,* $\alpha \ge \alpha'$ *implies that* $r(\alpha,x) \le r(\alpha',x) \ \forall x \in \mathbb{R}^n$.

Note that the notion of a separator for f on D is related to but is different from the notion of a separator (for a set) as introduced in Definition II.1

Example XI.4. The distance function $d_{\overline{D}_\alpha}(x)$ is a separator for f on D.

Example XI.5. Let $D = \{x \in \mathbb{R}^n: g_i(x) \le 0 \quad (i=1,...,m)\}$ with $g_i: \mathbb{R}^n \to \mathbb{R}$ $(i=1,...,m)$. Suppose that f is (L,μ)–Hölder continuous and g_i is (L_i,μ_i)–Hölder continuous $(i=1,...,m)$, i.e., for all $x,y \in \mathbb{R}^n$ one has

$$|f(x) - f(y)| \le L\|x - y\|^{\mu}, \tag{87}$$

$$g_i(x) - g_i(y)| \le L_i\|x - y\|^{\mu_i} \quad (i=1,...,m) \tag{88}$$

with $L, L_i > 0$; $\mu, \mu_i \in (0,1]$ $(i=1,...,m)$.

Then

$$r(\alpha,x) = \max \{[\max (0, \frac{f(x)-\alpha}{L})]^{1/\mu},$$

$$[\max (0, \frac{g_i(x)}{L_i})]^{1/\mu_i} \quad (i=1,...,m)\} \tag{89}$$

is a separator for f on D.

We verify the conditions (i), (ii), (iii) of Definition XI.1:

(i) Let $\alpha \in \overline{\mathbb{R}}$, $x \in \mathbb{R}^n$, $y \in \overline{D}_\alpha$. Then it follows from (87) and (88) that

$$\|x - y\| \geq \max \{ |\frac{f(x)-f(y)}{L}|^{1/\mu}, |\frac{g_i(x)-g_i(y)}{L_i}|^{1/\mu_i} \quad (i=1,\dots,m)\} .$$

But since

$$|f(x) - f(y)| \geq \max \{0, f(x) - f(y)\} \geq \max \{0, f(x) - \alpha\}$$

and

$$|g_i(x) - g_i(y)| \geq \max \{0, g_i(x) - g_i(y)\} \geq \max \{0, g_i(x)\} \quad (i=1,\dots,m) ,$$

we have

$$\|x - y\| \geq r(\alpha,x) \quad \forall x \in \mathbb{R}^n, y \in \overline{D}_\alpha ;$$

and hence

$$r(\alpha,x) \leq \inf \{\|x - y\|: y \in \overline{D}_\alpha\} = d_{\overline{D}_\alpha}(x) .$$

The conditions (ii) and (iii) in Definition XI.1 obviously hold for the function $r(\alpha,x)$ given in (89).

Example XI.6. Let $f: \mathbb{R} \to \mathbb{R}$ be twice continuously differentiable with bounded second derivative, i.e., there is a constant $M > 0$ such that

$$|f''(x)| \leq M \quad \forall x \in \mathbb{R} .$$

For every $\alpha \in \overline{\mathbb{R}}$, $x \in \mathbb{R}$ set

$$\rho(\alpha,x) := \begin{cases} 0 & \text{if } f(x) \leq \alpha \\ -\frac{1}{M}|f'(x)| + (|f'(x)|^2 + 2M(f(x)-\alpha))^{1/2} & \text{if } f(x) > \alpha \end{cases} \tag{90}$$

Then a separator for f on the ball $D = \{x \in \mathbb{R}: |x| \leq c\}$ is given by

$$r(\alpha,x) = \max \{\rho(\alpha,x),\ |x| - c\} . \tag{91}$$

Note that the second expression in (90) describes the unique positive solution $t = \rho(\alpha,x)$ of the equation

$$\frac{M}{2} t^2 + |f'(x)| t = f(x) - \alpha .$$

Conditions (ii) and (iii) in Definition XI.1 are again obviously satisfied by (91). To demonstrate (i), it suffices to show that $|y| \le c$ and $f(y) \le \alpha$ imply that $r(\alpha,x) \le |x-y|$. But, since $|y| \le c$ implies that $|x| - c \le |x| - |y| \le |x-y|$, we need only to show that $f(y) \le \alpha$ implies that $\rho(\alpha,x) \le |x-y|$. This can be seen using Taylor's formula

$$|f(y) - f(x) - f'(x)(x-y)| \le \frac{M}{2} |y-x|^2 .$$

From this it follows that

$$|f(y) - f(x)| \le \frac{M}{2} |y-x|^2 + |f'(x)|\ |y-x| ,$$

and hence

$$\frac{M}{2} |y-x|^2 + |f'(x)|\ |y-x| \ge f(x) - f(y) . \tag{92}$$

Now let $f(y) \le \alpha$. Then from (92) we have that

$$\frac{M}{2} |y-x|^2 + |f'(x)|\ |y-x| \ge f(x) - \alpha .$$

But $\frac{M}{2} t^2 + |f'(x)|\ t$ is monotonically increasing in $t > 0$ (for x fixed). Therefore, it follows from the definition of $\rho(\alpha,x)$ that $\rho(\alpha,x) \le |x-y|$.

Note that, by the same arguments, Example XI.6 can be extended to the case where f is defined on $\mathbb{R}^n$ with bounded Hessian $f''(x) := \nabla^2 f(x)$ and $f'(x) := \nabla f(x)$.

4.2. A Global Optimality Criterion

Suppose that a separator $r(\alpha,x)$ for the function f on the set D is available, and for every $\alpha \in \overline{\mathbb{R}}$ define the function

$$h(\alpha,x):= \sup_{v \notin D_\alpha} \{2vx + r^2(\alpha,v) - \|v\|^2\} \tag{93}$$

(with the usual convention that $\sup \emptyset = -\infty$, $\inf \emptyset = +\infty$). Clearly, for fixed α, $h(\alpha,x)$ is the pointwise supremum of a family of affine functions, and hence is convex (it is a so-called closed convex function, cf. Rockafellar (1970)).

Now consider the d.c. function

$$\varphi_\alpha(x):= h(\alpha,x) - \|x\|^2 . \tag{94}$$

Lemma XI.1. *We have*

$$\varphi_\alpha(x) > 0 \quad \text{if } x \notin \overline{D}_\alpha , \tag{95}$$

$$\varphi_\alpha(x) = -\inf\{\|x - v\|^2 : v \notin D_\alpha\} \quad \text{if } x \in \overline{D}_\alpha . \tag{96}$$

Proof. If $x \notin \overline{D}_\alpha$, then it follows from Definition XI.1 (ii) that $r(\alpha,x) > 0$, and hence, by (93),

$$\varphi_\alpha(x) = \sup_{v \notin D_\alpha} \{2vx + r^2(\alpha,v) - \|v\|^2\} - \|x\|^2$$

$$\geq 2xx + r^2(\alpha,x) - 2\|x\|^2 = r^2(\alpha,x) > 0 .$$

This proves (95).

In order to prove (96), we observe that from (93) it follows that

$$\varphi_\alpha(x) = \sup_{v \notin D_\alpha} \{2vx + r^2(\alpha,v) - \|v\|^2\} - \|x\|^2$$

$$= \sup_{v \notin D_\alpha} \{r^2(\alpha,v) - \|x-v\|^2\}$$

$$\geq \sup_{v \notin D_\alpha} \{-\|x-v\|^2\} = - \inf_{v \notin D_\alpha} \{\|x-v\|^2\} . \tag{97}$$

Now consider an arbitrary point $x \in \overline{D}_\alpha$. If $f(x) = \alpha$, then it follows that

$$\varphi_\alpha(x) \geq - \inf_{v \notin D_\alpha} \|x - v\|^2 \geq -\|x - x\|^2 = 0 . \tag{98}$$

This implies that $- \inf_{v \notin D_\alpha} \|x - v\|^2 = 0$.

On the other hand, from Definition XI.1 (i) we know that

$$r^2(\alpha,v) \leq d^2_{D_\alpha}(v) \leq \|x - v\|^2 \quad \forall v \in \mathbb{R}^n .$$

It follows that

$$\varphi_\alpha(x) = \sup_{v \notin D_\alpha} \{-\|x - v\|^2 + r^2(\alpha,v)\} \leq 0 ,$$

and hence by (98)

$$\varphi_\alpha(x) = 0 .$$

Therefore, (96) holds for $x \in \overline{D}_\alpha$ satisfying $f(x) = \alpha$.

If $f(x) < \alpha$, then to each point $v \notin D_\alpha$ we associate a point $z(v)$ on the intersection of the line segment $[x,v]$ with the boundary ∂D_α of D_α. Such a point $z(v)$ exists, because $x \in D_\alpha$ while $v \notin D_\alpha$.

Since $z(v) \in [x,v]$, $z(v) \in \overline{D}_\alpha$, and because of Definition XI.1 (i), we have

$$\|x - v\| = \|x - z(v)\| + \|v - z(v)\| \geq \|x - z(v)\| + r(\alpha,v) .$$

It follows that

$$-\|x - z(v)\|^2 \geq r^2(\alpha,v) - \|x - v\|^2 ,$$

and hence

$$-\inf_{v \notin D_\alpha} \{\|x-v\|^2\} = \sup_{v \notin D_\alpha} \{-\|x-v\|^2\} = \sup \{-\|x-v\|^2 : v \in \partial D_\alpha\}$$

$$\geq \sup_{v \notin D_\alpha} \{r^2(\alpha,v) - \|x-v\|^2\} = \varphi_\alpha(x) . \qquad (99)$$

Finally, from (97) and (99) it follows that we must have (96) for $x \in D_\alpha$. ∎

Corollary XI.1. *For every $\alpha \in \overline{\mathbb{R}}$ satisfying $\overline{D}_\alpha \neq \emptyset$ we have*

$$\inf_{x \in \mathbb{R}^n} \varphi_\alpha(x) \leq 0 .$$

Furthermore, for every $x \in D$ and $\alpha = f(x)$ we have

$$\varphi_\alpha(x) = 0 .$$

Proof. Corollary XI.1 is an immediate consequence of Lemma XI.1. ∎

Theorem XI.2. *Let $\tilde{x} \in D$ be a feasible point of problem (P), and let $\tilde{\alpha} = f(\tilde{x})$. Consider the function $\varphi_\alpha(x)$ defined in (93), (94).*

(i) If

$$\inf_{x \in \mathbb{R}^n} \varphi_{\tilde{\alpha}}(x) < 0 , \qquad (100)$$

then for any x satisfying $\varphi_{\tilde{\alpha}}(x) < 0$ we have $x \in D$, $f(x) < \tilde{\alpha}$ (i.e., $\tilde{x}$ is a better feasible point than $\tilde{x}$).

(ii) If $\tilde{x}$ is a (globally) optimal solution of problem (P), then

$$\min_{x \in \mathbb{R}^n} \varphi_{\tilde{\alpha}}(x) = 0 . \qquad (101)$$

(iii) If problem (P) is regular in the sense of (85), i.e., if

$$inf\{f(x): x \in D\} = inf\{f(x): x \in int\ D\} ,$$

then any $\tilde{x}$ *satisfying (101) is a (globally) optimal solution of problem (P).*

Proof. Because of Lemma XI.1, every $x \in \mathbb{R}^n$ satisfying $\varphi_{\tilde{\alpha}}(x) < 0$ must be in $D_{\tilde{\alpha}}$. This proves (i) by the definition of $D_{\tilde{\alpha}}$.

Using the assertion (i) just proved and Corollary XI.1, we see that condition (101) is necessary for the global optimality of $\tilde{x}$, i.e., we have (ii).

In order to prove (iii), suppose that $\tilde{x}$ satisfies (101) but is not a globally optimal solution of (P). Then, using the regularity condition (85) we see that there exists a point $x' \in int\ D$ satisfying $f(x') < \tilde{\alpha}$. Because of the continuity of f it follows that $x' \in int\ D_{\tilde{\alpha}}$. By Lemma XI.1 this implies that

$$\inf_{x \in \mathbb{R}^n} \varphi_{\tilde{\alpha}}(x) \le \varphi_{\tilde{\alpha}}(x') = - \inf_{v \notin D_{\tilde{\alpha}}} \|x' - v\| < 0 ,$$

i.e., $\tilde{x}$ does not satisfy (101). This contradiction proves (iii). ■

A slightly modified version of the above results is the following corollary.

Corollary XI.2. *If* $\tilde{\alpha} = min\ f(D)$ *is the optimal objective function value of problem (P), then* $\tilde{\alpha}$ *satisfies (101), and every optimal solution of (P) is a global minimizer of* $\varphi_{\tilde{\alpha}}(x)$ *over* $\mathbb{R}^n$.

Conversely, if the regularity condition (85) is fulfilled and $\tilde{\alpha}$ *satisfies (101), then* $\tilde{\alpha}$ *is the optimal objective function value of problem (P), and every global minimizer of* $\varphi_{\tilde{\alpha}}(x)$ *over* $\mathbb{R}^n$ *is an optimal solution of (P).*

Proof. The first assertion follows from Corollary XI.1 and Theorem XI.2.

In order to prove the second assertion, assume that the regularity condition (85) is fulfilled. Let $\tilde{\alpha}$ satisfy (101) and let $\bar{x}$ be a global minimizer of $\varphi_{\tilde{\alpha}}(x)$ over $\mathbb{R}^n$. Then $\varphi_{\tilde{\alpha}}(\bar{x}) = 0$, and from Lemma XI.1 we have $\bar{x} \in D$, $f(\bar{x}) \le \tilde{\alpha}$.
Let $\bar{\alpha} = f(\bar{x})$. Since $\bar{\alpha} \le \tilde{\alpha}$, it follows from Definition XI. (iii) that $r(\bar{\alpha},x) \ge r(\tilde{\alpha},x)$,

and hence

$$\varphi_{\bar{\alpha}}(x) \geq \varphi_{\tilde{\alpha}}(x) \geq 0 \qquad \forall x \in \mathbb{R}^n .$$

Using the first part of Corollary XI.1, we deduce that $\inf_{x \in \mathbb{R}^n} \varphi_{\bar{\alpha}}(x) = 0$. Therefore, by Theorem XI.2 (iii) we conclude that $\bar{x}$ is an optimal solution of problem (P).

Furthermore, it follows from the regularity condition (85) that we cannot have $\bar{\alpha} < \tilde{\alpha}$, because in that case there would exist a point $x' \in \text{int } D$ satisfying $f(x') < \tilde{\alpha}$. But, by Lemma XI.1, this would imply that $\varphi_{\tilde{\alpha}}(x') = -\inf \{\|x' - v\|^2 : v \notin D_{\tilde{\alpha}}\} < 0$, which contradicts the hypothesis that $\tilde{\alpha}$ satisfies (101). Therefore, we must have $\tilde{\alpha} = \bar{\alpha}$, i.e., $\bar{\alpha}$ is the optimal objective function value of (P). ∎

4.3. The Relief Indicator Method

The properties of the function $\varphi_\alpha(x)$ presented in the preceding section suggest interpreting $\varphi(x)$ as a sort of generalized gradient or relief indicator. Following Thach and Tuy (1990), we shall call $\varphi_\alpha(x)$ a *relief indicator function for f on D.* (A slightly more general notion of relief indicator is discussed in Thach and Tuy (1989)).

We have seen that under the regularity assumption (85)

$$\min_{x \in \mathbb{R}^n} \varphi_\alpha(x) = 0 \text{ if and only if } \alpha = \min f(D) .$$

This suggests replacing problem (P) by the *parametric unconstrained d.c. minimization problem:*

$$\textit{find } \alpha \textit{ such that } 0 = \inf_{x \in \mathbb{R}^n} \varphi_\alpha(x). \tag{102}$$

Suppose that $D \neq \emptyset$. A straightforward conceptual iterative method to solve problem (102) is as follows:

Start with an arbitrary feasible point x^1.

At iteration k, solve the auxiliary problem

(P_k) minimize $\varphi_{\alpha_k}(x)$ over $\mathbb{R}^n$,

where $\alpha_k = f(x^k)$. Let x^{k+1} denote an optimal solution of (P_k). If $\varphi_{\alpha_k}(x^{k+1}) = 0$, then stop: x^k is an optimal solution of (P). Otherwise, go to iteration k+1 with $\alpha_{k+1} = f(x^{k+1})$.

Proposition XI.9. *Let problem (P) be regular in the sense of (85). If the above iterative procedure is infinite, then every accumulation point $\bar{x}$ of the sequence $\{x^k\}$ is an optimal solution of (P).*

Proof. We first note that problem (P_1) has an optimal solution x^2 with $\varphi_{\alpha_1}(x^2) \le 0$. To see this, recall from Corollary XI.1 that $\inf_{x\in\mathbb{R}^n} \varphi_{\alpha_1}(x) \le 0$ since $\bar{D}_{\alpha_1} \neq \emptyset$ (because $x^1 \in D$). Moreover, from Lemma XI.1 we know that $\varphi_{\alpha_1}(x) > 0$ if $x \notin \bar{D}_{\alpha_1}$. It follows that $\inf_{x\in\mathbb{R}^n} \varphi_{\alpha_1}(x) = \inf_{x\in D} \varphi_{\alpha_1}(x)$. But $\varphi_{\alpha_1}(x)$ is lower semicontinuous, since $h(\alpha,x)$ is lower semicontinuous (cf. Rockafellar (1970)), and from the compactness of D we see that $\min_{x\in D} \varphi_{\alpha_1}(x)$ exists.

We have $\varphi_{\alpha_1}(x^2) < 0$, since otherwise the above procedure would terminate at x^2. It follows that $x^2 \in D$, $f(x^2) < f(x^1)$. By induction, it is then easily seen that we have $x^k \in D$, $f(x^{k+1}) < f(x^k)$ for all k. Therefore, the sequence $\{\alpha_k\}$, $\alpha_k = f(x^k)$, is monotonically decreasing. It is also bounded from below by min f(D), which exists because of the inclusion $x^1 \in D$, the continuity of f and the compactness of D. It follows that $\bar{\alpha} := \lim_{k\to\infty} \alpha_k$ exists.

Let $\bar{x} = \lim_{q\to\infty} x^{k_q}$ be an accumulation point of $\{x^k\}$. Then we have $\bar{\alpha} = f(\bar{x})$.

Now, for every $r < k_q$ one has

$$\varphi_{\alpha_r}(x^{k_q}) \le \varphi_{\alpha_{q}-1}(x^{k_q}) \le \varphi_{\alpha_{q}-1}(x) \le \varphi_{\bar{\alpha}}(x) \quad \forall x \in \mathbb{R}^n . \tag{103}$$

The first and last inequalities in (103) follow from Definition XI.1 (iii) and the definition of $\varphi_\alpha(x)$. The second inequality in (103) holds since $\varphi_{\alpha_q-1}(x^{k_q}) = \min_{x \in \mathbb{R}^n} \varphi_{\alpha_q-1}(x)$. Keeping α_r fixed and letting $q \longrightarrow \infty$ in (103), we see from (103) and the lower semicontinuity of $\varphi_{\alpha_r}(x)$ that

$$\varphi_{\alpha_r}(\bar{x}) \le \varphi_{\bar{\alpha}}(x) \quad \forall x \in \mathbb{R}^n . \tag{104}$$

Letting $r \longrightarrow \infty$ in (104), by virtue of Corollary XI.1 we obtain

$$0 \le \varphi_{\bar{\alpha}}(x) \quad \forall x \in \mathbb{R}^n .$$

Since $\varphi_{\bar{\alpha}}(\bar{x}) = 0$, it follows from this inequality that $0 = \min_{x \in \mathbb{R}^n} \varphi_{\bar{\alpha}}(x)$. But this is equivalent to $f(\bar{x}) = \min f(D)$ by Theorem XI.2. ∎

The implementation of the above iterative method requires that we investigate two issues. The first matter concerns the solution of the subproblems (P_k). These problems in general might not be solvable by a finite procedure. Therefore, one should replace problem (P_k) by a suitable approximation (Q_k) which can be solved by a finite algorithm.

The second matter regards the computation of the initial feasible point x^1. In addition, the implicit definition of $r_\alpha(x)$ by means of (93), (94) necessitates suitable approximations.

One possibility for handling these matters is as follows.

Replace problem (P_k) by a relaxed problem of the form

$$(Q_k) \qquad \begin{array}{l} \text{minimize } (h_k(x) - \|x\|^2) \\ \text{s.t. } \ x \in S \end{array} , \tag{105}$$

where S is a suitable polytope and $h_k(x)$ is a suitable polyhedral convex function that underestimates $h(\alpha_k,x)$. Since $\min \{\varphi_{\alpha_k}(x): x \in \mathbb{R}^n\}$ is attained in D, it suffices to choose any polytope S that encloses D. Moreover, the form (93) of $h(\alpha,x)$ suggests that we consider

$$h_k(x) = \sup_{i=1,\ldots,k} \{2x^i x + r^2(\alpha_i,x^i) - \|x^i\|^2\} , \tag{106}$$

where α_i is the smallest value of f(x) at all feasible points evaluated until iteration i, and where x^{i+1} $(i \geq 1)$ is an optimal solution of problem (Q_i).

By the definition of the α_i in (106), we must have $\alpha_i \leq f(x^i)$ whenever $x^i \in D$. It follows that $x^i \notin D_{\alpha_i}$, and hence $x^i \notin D_{\alpha_k}$ for i=1,...,k. Since $r(\alpha_i,x) \leq r(\alpha_k,x)$ for all $x \in \mathbb{R}^n$, i=1,...,k (Definition XI.1 (iii)), we see from (93) and (106) that for all x

$$h_k(x) \leq \sup_{v \notin D_{\alpha_k}} \{2vx + r^2(\alpha_k,v) - \|v\|^2\} = h(\alpha_k,x),$$

i.e., the functions $h_k(x)$ defined by (106) underestimate $h(\alpha_k,x)$.

Moreover, the functions $h_k(x)$ are polyhedral convex functions since

$$h_k(x) = \max \{\ell_i(x): i=1,\ldots,k\}$$

with

$$\ell_i(x) = 2x^i x + r^2(\alpha_i,x^i) - \|x^i\|^2 . \tag{107}$$

It follows that problem (Q_k) is equivalent to

$$(\overline{Q}_k) \qquad \begin{array}{ll} \text{minimize} & (t - \|x\|^2) \\ \text{s.t.} & x \in S,\ \ell_i(x) \leq t \quad (i=1,\ldots,k) \end{array} . \tag{108}$$

In this way we have replaced the unconstrained d.c. problem (P_k) by the linearly constrained concave minimization problem $(\overline{Q}_k)$ with quadratic objective function. Several finite algorithms to solve problem $(\overline{Q}_k)$ were discussed in Part B of this book.

Since $h_k(x) \leq h(\alpha_k,x)$ for all $x \in S$, it follows from Lemma XI.1 that the optimal objective function value of $(\overline{Q}_k)$ is nonpositive. Moreover, if this value is zero, then we must have

$$0 = \min_{x \in \mathbb{R}^n} \varphi_{\alpha_k}(x) ,$$

and it follows from Theorem XI.2 that $\alpha_k = \min f(D)$ and every point $\bar{x}^k \in D$ satisfying $f(x^k) = \alpha_k$ solves (P).

However, if $h_k(x^{k+1}) - \|x^{k+1}\|^2 < 0$, then it is not guaranteed that $f(x^{k+1}) < \alpha_k$. Therefore, we set $\alpha_{k+1} = \alpha_k$ if $x^{k+1} \notin D$. When the solution x^{k+1} of $(\overline{Q}_k)$ is feasible, then we can run a local optimization procedure that starts with x^{k+1} and yields a feasible point $\bar{x}^{k+1}$ satisfying $f(\bar{x}^{k+1}) \le f(x^{k+1})$. In this case we set $\alpha_{k+1} = \min \{\alpha_k, f(\bar{x}^{k+1})\}$.

Algorithm XI.6 (approximate relief indicator method).

Initialization:

Construct a polytope $S \supset D$ and choose $x^1 \in S$ (if available, then choose $x^1 \in D$). Set $\alpha_0 = \infty$ if no feasible point is known, and set α_0 equal to the minimal value of f at known feasible points, if feasible points are known.

Iteration k=1,2,... :

k.1.: If $x^k \in D$, then, using a local optimization procedure that starts with x^k, find a point $\bar{x}^k \in D$ satisfying $f(\bar{x}^k) < f(x^k)$, and set $\alpha_k = \min \{\alpha_{k-1}, f(\bar{x}^k)\}$. If $x^k \notin D$, then set $\alpha_k = \alpha_{k-1}$.

Denote by $\tilde{x}^k$ the best feasible point known so far, i.e., we have $f(\tilde{x}^k) = \alpha_k$.

k.2.: Set

$$\ell_k(x) = 2x^k x + r^2(\alpha_k, x^k) - \|x^k\|^2$$

and solve the relaxed problem

$$(\overline{Q}_k) \qquad \begin{array}{l} \text{minimize } (t - \|x\|^2) \\ \text{s.t. } x \in S,\ \ell_i(x) \le t \quad (i=1,\dots,k) \end{array}$$

Let (x^{k+1}, t^{k+1}) be an optimal solution of $(\overline{Q}_k)$.

If

$$t^{k+1} - \|x^{k+1}\|^2 = 0,$$

then stop: $\tilde{x}^k$ is an optimal solution of (P), and $\alpha_k = f(\tilde{x}^k) = \min f(D)$. Otherwise ($t^{k+1} - \|x^{k+1}\|^2 < 0$), go to iteration k+1.

Remark XI.4. Let D_k denote the feasible set of $(\overline{Q}_k)$, i.e.,

$$D_k = \{(x,t): x \in S, \ell_i(x) \le t \quad (i=1,...,k)\} ,$$

and let V_k denote the vertex set of D_k. Then we know that

$$\min \{t - \|x^2\|: (x,t) \in D_k\} = \min \{t - \|x^2\|: (x,t) \in V_k\} .$$

Since $D_{k+1} = D_k \cap \{(x,t): \ell_k(x) \le t\}$, we have the same situation as with the outer approximation methods discussed in Chapter II, and V_{k+1} can be determined from V_k by one of the methods presented in Chapter II.

Before proving convergence of Algorithm XI.6, we state the following result.

Proposition XI.10. *Let problem (P) be regular in the sense of (85), and let the feasible set D of (P) be nonempty. Assume that we have a separator $r(\alpha,x)$ for f on D which is continuous in $\mathbb{R}\times\mathbb{R}^n$. Then Algorithm XI.6 either terminates after a finite number of iterations with an optimal solution $\tilde{x}^k$ of (P), or else it generates an infinite sequence $\{x^k\}$, each accumulation point of which is an optimal solution of (P).*

Proof. If the algorithm terminates after a finite number of iterations, then, by construction, it terminates with an optimal solution of (P).

In order to prove the second assertion, first note that the sequence $\{\alpha_k\}$ generated by the algorithm is nonincreasing and is bounded from below by min f(D). Therefore, $\bar{\alpha} := \lim_{k\to\infty} \alpha_k$ exists. We show that $\bar{\alpha} = \min f(D)$, and that every accumulation point $\bar{x}$ of $\{x^k\}$ satisfies $\bar{x} \in D$ and $f(\bar{x}) = \bar{\alpha}$.

We use the general convergence results for outer approximation methods (cf. Theorem II.1) in the following form:

Consider the nonempty set

$$A = \{(x,t) \in \mathbb{R}^n \times \mathbb{R}: x \in \overline{D}_{\bar{\alpha}},\ \|x\|^2 \le t\}$$

and the sequence $\{(x^k,t^k)\}$ generated by Algorithm X.6. This sequence is bounded, because $\{x^k\} \subset S$ and $\ell_1(x^k) \le h_{k-1}(x^k) \le t^k < \|x^k\|^2$.

Then we know from the outer approximation theory (cf. Theorem II.1) that every accumulation point $(\bar{x},\bar{t})$ belongs to A if the functions $\tilde{\ell}_k(x,t) := \ell_k(x) - t$ (k=1,2,...) satisfy the following conditions:

(1) $\tilde{\ell}_k(x^k,t^k) > 0$;

(2) $\tilde{\ell}_k(x,t) \le 0 \quad \forall (x,t) \in A$;

(3) $\tilde{\ell}_k(x,t)$ is continuous, and for every convergent subsequence $\{(x^{k_q},t^{k_q})\} \subset \{(x^k,t^k)\}$ satisfying $(x^{k_q},t^{k_q}) \xrightarrow[(q\to\infty)]{} (\tilde{x},\tilde{t})$ we have

$$\tilde{\ell}_{k_q}(x^{k_q},t^{k_q}) \xrightarrow[(q\to\infty)]{} \tilde{\ell}(\tilde{x},\tilde{t}), \text{ where } \tilde{\ell}: \mathbb{R}^{n+1} \to \mathbb{R};$$

(4) $\tilde{\ell}(\tilde{x},\tilde{t}) = 0$ implies that $(\tilde{x},\tilde{t}) \in A$.

We show that the conditions (1) – (4) are satisfied.

(1): We have

$$\tilde{\ell}_k(x^k,t^k) = 2x^kx^k + r^2(\alpha_k,x^k) - \|x^k\|^2 - t^k = \|x^k\|^2 - t^k + r^2(\alpha_k,x^k)$$

$$\ge \|x^k\|^2 - t^k > 0,$$

where the first inequality is obvious and the second inequality follows from the assumption that the algorithm is infinite (cf., Step k.2).

(2): Let $(x,t) \in A$. Then

$$\tilde{\ell}_k(x,t) = 2x^kx + r^2(\alpha_k,x^k) - \|x^k\|^2 - t$$

$$\le 2x^k x + r^2(\alpha_k,x^k) - \|x^k\|^2 - \|x\|^2$$

$$= r^2(\alpha_k,x^k) - \|x^k - x\|^2 \le r^2(\overline{\alpha},x^k) - d^2_{\overline{D}_{\overline{\alpha}}}(x^k) \le 0.$$

Here the last two inequalities follow from Definition XI (iii) and (i), respectively.

(3): The affine functions $\ell_k(x,t)$ are continuous. Consider a subsequence (x^{k_q},t^{k_q}) satisfying $(x^{k_q},t^{k_q}) \xrightarrow[(q\to\infty)]{} (\tilde{x},\tilde{t})$. Then we see from the continuity of $r(\alpha,x)$ that

$$\ell_{k_q}(x^{k_q},t^{k_q}) \xrightarrow[(q\to\infty)]{} \|\tilde{x}\|^2 + r^2(\overline{\alpha},\tilde{x}) - \tilde{t} = \overline{\ell}(\tilde{x},\tilde{t}),$$

where $\overline{\ell}(x,t) = \|x\|^2 + r^2(\overline{\alpha},x) - t$.

(4): Let $\overline{\ell}(\tilde{x},\tilde{t}) = \|\tilde{x}\|^2 + r^2(\overline{\alpha},\tilde{x}) - \tilde{t} = 0$. Since the algorithm is infinite, i.e., we have $\|x^{k_q}\|^2 > t^{k_q}$ $\forall q$, it follows that $\|\tilde{x}\|^2 - \tilde{t} \ge 0$, and hence, because $\overline{\ell}(\tilde{x},\tilde{t}) = 0$, we have $r^2(\overline{\alpha},\tilde{x}) \le 0$. This is possibly only if $r(\overline{\alpha},\tilde{x}) = 0$, and hence $\|\tilde{x}\|^2 = \tilde{t}$. But from Definition XI.1 (ii) we see that then we must have $\tilde{x} \in \overline{D}_{\overline{\alpha}}$, and hence $(\tilde{x},\tilde{t}) \in A$.

Therefore, by Theorem II.1, every accumulation point $(\overline{x},\overline{t})$ of $\{(x^k,t^k)\}$ satisfies $(\overline{x},\overline{t}) \in A$, i.e.

$$\overline{x} \in \overline{D}_{\overline{\alpha}} \text{ and } \|\overline{x}\|^2 \le \overline{t} . \tag{109}$$

Now we show that the optimality condition of Theorem XI.2 (resp. Corollary XI.2) is satisfied.

Since $\overline{x} \in \overline{D}_{\overline{\alpha}}$ it follows by Lemma XI.1 that

$$\varphi_{\overline{\alpha}}(\overline{x}) = h(\overline{\alpha},\overline{x}) - \|\overline{x}\|^2 = -\inf\{\|x - v\|^2 : v \notin D_{\overline{\alpha}}\} \le 0 . \tag{110}$$

Let $\{(x^{k_s},t^{k_s})\}$ be a subsequence converging to $(\overline{x},\overline{t})$. Then (110) implies that

$$0 \ge h(\bar{\alpha},\bar{x}) - \|\bar{x}\|^2 \ge \min_{x \in S} \{h(\bar{\alpha},x) - \|x\|^2\}$$

$$\ge \min_{x \in S} \{h(\alpha_{k_s-1},x) - \|x\|^2\} \ge \min_{x \in S} \{h_{k_s-1}(x) - \|x\|^2\}$$

$$= h_{k_s-1}(x^{k_s}) - \|x^{k_s}\|^2 = t^{k_s} - \|x^{k_s}\|^2 . \qquad (111)$$

Now let $s \longrightarrow \infty$ in (111) and observe that $\|\bar{x}\|^2 \le \bar{t}$ (cf. 109). We obtain

$$0 \ge h(\bar{\alpha},\bar{x}) - \|\bar{x}\|^2 \ge \min_{x \in S} \{h(\bar{\alpha},x) - \|x\|\} \ge 0 .$$

But since $S \supset \bar{D}_{\bar{\alpha}}$, this implies, because of Lemma XI.1, that

$$0 = h(\bar{\alpha},\bar{x}) - \|\bar{x}\|^2 = \min_{x \in \mathbb{R}^n} \{h(\bar{\alpha},x) - \|x\|^2\}.$$

The assertion follows from Theorem XI.2. ■

For a version of Proposition XI.10 that does not require the continuity of $r(\alpha,x)$, see Thach and Tuy (1990).

Example XI.7. Consider the problem

$$\text{minimize } f(x) = (x_1)^2 + (x_2)^2 - \cos 18x_1 - \cos 18x_2$$

$$\text{s.t. } [(x_1 - 0.5)^2 + (x_2 - 0.415331)^2]^{1/2} \ge 0.65 ,$$

$$0 \le x_1 \le 1, 0 \le x_2 \le 1 .$$

Let $g(x_1,x_2) = -[(x_1 - 0.5)^2 + (x_2 - 0.415331)^2]^{1/2} + 0.65$ and let $S = \{(x_1,x_2) \in \mathbb{R}^2: 0 \le x_1 \le 1, 0 \le x_2 \le 1\}$.

The functions g and f are Lipschitz continuous on S with Lipschitz constants 28.3 for f and 1.0 for g.

According to Example XI.6 we choose the separator

$$r(\alpha,x) = \max \{0, \frac{f(x)-\alpha}{28.3}, g(x)\} .$$

With the stopping criterion inf $(\overline{Q}_k) > -10^{-4}$ the algorithm terminates after 16 iteratons with $\overline{x} = (0,1)$, $f(\overline{x}) = -0.6603168$. No local methods were applied, i.e., we used $x^k = \tilde{x}^k$.

The intermediate results are given in Table XI.2 .

Iter.	x^k	$f(x)^k$	$\tilde{x}^k$	α_k	$inf(\bar{Q}_k)$	$\lvert V_k \rvert$
1	(0, 1)	−0.6603	(0, 1)	−0.6603	−0.1693	4
2	(1, 1)	0.6793	—	—	−0.1693	6
3	(0.1693, 1)	1.3639	—	—	−0.0286	8
4	(0.8284, 1)	1.7252	—	—	−0.0271	10
5	(0, 0.8551)	0.6822	—	—	−0.0209	12
6	(1, 0.8532)	2.0074	—	—	−0.0192	14
7	(0.0695, 0.9353)	0.9939	—	—	−0.0090	16
8	(0.9284, 0.9493)	2.4908	—	—	0.0054	18
9	(0, 0.9567)	−0.0271	—	—	−0.0018	18
10	(0.0402, 1)	−0.4080	—	—	−0.0016	20
11	(0.0041, 0.9145)	0.5678	—	—	−0.0012	22
12	(0.0899, 1)	0.3967	—	—	−0.0011	24
13	(0.0191, 0.9841)	−0.3948	—	—	−0.0006	26
14	(0.0445, 0.8794)	1.0731	—	—	−0.0003	28
15	(0.0006, 0.9841)	−0.4537	—	—	−0.0002	30
16	(0.0138, 1)	−0.6293	—	—	−0.0001	32

Table XI.2

REFERENCES

ABRHAM, J., and BUIE, R.N. (1975), *A Note on Nonconcave Continuous Programming.* Zeitschrift für Operations Research, Serie A, 3, 107–114.

ADAMS, W.P. and SHERALI, H.D. (1986), *A Tight Linearization and an Algorithm for zero–one Quadratic Programming Problems.* Management Science, 32, 1274–1290.

AFENTAKIS, P, GARISH, B. and KARMARKAR, U. (1984), *Computationally efficient optimal solutions to the lot–sizing problem in multistage assembly systems,* Management Science, 30, 222–239.

AL–KHAYYAL, F.A. (1983), *Jointly Constrained Bilinear Programming and Related Problems.* Industrial and Systems Engineering Report Series No. J–83–3, The Georgia Institute of Technology,Atlanta, GA.

AL–KHAYYAL, F.A. (1985), *Necessary and Sufficient Conditions for the Existence of Complementarity Solutions and Characterizations of the Matrix* Q *and* Q_0. Report PDRC 85–05, School of Industrial and Systems Engineering, Georgia Institute of Technology, Atlanta.

AL–KHAYYAL, F.A. (1986), *Further Simplified Characterizations of Linear Complementarity Problems Solvable as Linear Programs.* Report PDRC 86–08, School of Industrial and Systems Engineering, Georgia Institute of Technology, Atlanta.

AL–KHAYYAL, F.A. (1986 a), *Linear, Quadratic and Bilinear Programming Approaches to the Linear Complementarity Problem.* European Journal of Operational Research, 24, 216–227.

AL–KHAYYAL, F.A. (1987), *An Implicit Enumeration Procedure for the General Linear Complementarity Problem.* Mathematical Programming Study, 31, 1–20.

AL–KHAYYAL, F.A. (1989), *Jointly Constrained Bilinear Programs and Related Problems: An Overview,* to appear in Computers and Mathematics with Applications.

AL–KHAYYAL, F.A. and FALK, J.E. (1983), *Jointly constrained biconvex programming.* Mathematics of Operations Research, 8, 273–286.

AL–KHAYYAL, F.A., HORST, R., and PARDALOS, P. (1989), *Global Optimization of Concave Functions Subject to Separable Quadratic Constraints and of All–Quadratic Separable Problems,* submitted.

ALLGOWER, E.L. and GEORG, K. (1980), *Simplicial and Continuation Methods for Approximating Fixed Points and Solutions to Systems of Equations.* SIAM Review, 22, 28–85.

ALLGOWER, E.L. and GEORG, K. (1983), *Predictor–corrector and Simplicial Methods for Approximating Fixed Points and Zero Points of Nonlinear Mappings.* In: Bachem, A., Grötschel, M. and Korte, B. (editors), Mathematical Programming: The State of the Art. Springer, Berlin

ALTMANN, M. (1968), *Bilinear Programming.* Bull. Acad. Polon. Sci. Ser. Sci. Math. Astronom. Phys., 16, 741–745.

ANEJA, Y. P., AGGARWAL, V. and NAIR, K. P. (1984), *On a Class of Quadratic Programs.* European Journal of Operations Research, 18, 62–72.

ARCHETTI, F. and BETRO, B. (1978), *A Priori Analysis of Deterministic Strategies for Global Optimization Problems.* In: L.C.W. Dixon and G.P. Szegö eds. Towards Global Optimization 2, North Holland.

ARCHETTI, F. and BETRO, B. (1979), *A Probabalistic Algorithm for Global Optimization.* Calcolo, 16, 335–343

ARCHETTI, F. and BETRO, B. (1980), *Stochastic Models and Global Optimization.* Bolletino della Unione Matematica Italiana, 5, 17–A, 295–301.

ARSOVE, G. (1953), *Functions Representable as the Difference of Subharmonic Functions.* Transactions of the Amererican Mathematical Society, 75, 327–365.

ASPLUND, E. (1973), *Differentiability of the Metric Projection in Finite Dimensional Euclidean Space.* Proceedings of the American Mathematical Society, 38, 218–219.

AUBIN, J.P. and EKELAND, I. (1976), *Estimates of the Duality Gap in Nonconvex Optimization.* Mathematics of Operations Research, 1, 225–245.

AVRIEL, M. (1973), *Methods for Solving Signomial and Reverse Convex Programming Problems.* In: Avriel et al. (editors), Optimization an Design, Prentice–Hall Inc., Englewood Cliffs, N.J. 307–320.

AVRIEL, M. (1976), *Nonlinear Programming: Analysis and Methods.* Prentice Hall Inc. Englewood Cliffs, N.J.

AVRIEL, M. and WILLIAMS, A.C. (1970), *Complementary Geometric Programming.* SIAM Journal of Applied Mathematics, 19, 125–141.

AVRIEL, M. and ZANG, I. (1981), *Generalized Arcwise Connected Functions and Characterizations of Local–Global Minimum Properties.* In: Schaible, S. and Ziemba, W.T. (editors) Generalized Concavity in Optimization and Economics, Academic Press, New York.

BALAS, E. (1968), *A Note on the Branch and Bound Principle.* Operations Research 16, 442–445.

BALAS, E. (1971), *Intersection Cuts – a New Type of Cutting Planes for Integer Programming.* Operations Research, 19, 19–39.

BALAS, E. (1972), *Integer Programming and Convex Analysis: Intersection Cuts from Outer Polars.* Mathematical Programming, 2, 330–382.

BALAS, E. (1975), *Disjunctive Programming: Cutting Planes from Logical Conditions.* In "Nonlinear Programming, 2", Academic Press, Inc., New York, San Francisco, London, 279–312.

BALAS, E. (1975 a), *Nonconvex Quadratic Programming via Generalized Polars.* SIAM Journal on Applied Mathematics, 28, 335–349.

BALAS, E. (1979), *Disjunctive Programming.* Annals of Discrete Mathematics, 5, 3–51.

BALAS, E., and BURDET, C.A. (1973), *Maximizing a Convex Quadratic Function Subject to Linear Constraints.* Management Science Research Report No. 299, Carnegie–Mellon University (Pittsburg, P.A.).

BALI, S. (1973), *Minimization of a concave function on a bounded convex polyhedron.* Ph.D. Dissertation, UCLA.

BALINSKI, M.L. (1961), *An Algorithm for Finding All Vertices of Convex Polyhedral Sets.* SIAM Journal, 9, 72–88.

BAN, V.T. (1982), *A finite Algorithm for Globally Minimizing a Concave Function Under Linear Constraints and its Applications.* Preprint, Institute of Mathematics, Hanoi.

BANSAL, P.P. and JACOBSEN, S.E. (1975), *Characterization of Local Solutions for a Class of Nonconvex Programs.* Journal of Optimzation Theory and Applications, 15, 549–564.

BANSAL, P.P. and JACOBSEN, S.E. (1975 a), *An Algorithm for Optimizing Network Flow Capacity under Economies of Scale.* Journal of Optimization Theory and Applications, 15, 565–586.

BARD, J.F. and FALK, J.E. (1982), *A Separable Programming Approach to the Linear Complementarity Problem.* Computation and Operations Research, 9, 153–159.

BARON, D.P. (1972), *Quadratic Programming with Quadratic Constraints.* Naval Research Logistics Quarterly, 19, 253–260.

BARR, R.S., GLOVER, F. and KLINGMAN, D (1981), *A New Optimization Method for Large Scale Fixed Charge Transportation Problems.* Operations Research, 29, 3, 448–463.

BASSO, P. (1982), *Iterative methods for the locallization of the global maximum.* SIAM Journal on Numerical Analysis, 19, 781–792.

BASSO, P. (1985), *Optimal Search for the Global Maximum of Functions with Bounded Seminorm.* SIAM Journal on Numerical Analysis, 22, 888–903.

BAZARAA, M.S. (1973), *Geometry and Resolution of Duality Gaps.* Naval Research Logistics, 20, 357–366.

BAZARAA, M.S. and SHETTY, C.M. (1979), *Nonlinear Programming: Theory and Algorithms.* John Wiley and Sons, New York.

BAZARAA, M.S. and SHERALI, H.D. (1982), *On the Use of Exact and Heuristic Cutting Plane Methods for the Quadratic Assignment Problem.* Journal Operational Society, 33, 991–1003.

BEALE, E.M.L. (1980), *Branch and Bound Methods for Numerical Optimization of Nonconvex Functions.* In: Compstat 1980, Physica, Vienna, 11–20.

BEALE, E.M.L. and FORREST, J.J.H. (1978), *Global Optimization as an Extension of Integer Programming.* In: Towards Global Optimization 2, Dixon, L.C.W. and Jzego, editors, North Holland, Amsterdam, 131–149.

BENACER, R. (1986), *Contribution à l'étude des algorithmes de l'optimisation non convexe et non differentiable.* Thèse, Université Scientifique, Technologique et Médicale de Grenoble.

BENACER, R. and TAO, P.D. (1986), *Global Maximization of a Nondefinite Quadratic Function over a Convex Polyhedron.* In: Hiriart–Urruty, editor, FERMAT DAYS 1985: Mathematics for Optimization, North–Holland, Amsterdam, 65–76.

BENDERS, J.F. (1962), *Partitioning Procedures for Solving Mixed–Variables Programming Problems.* Numerische Mathematik, 4, 238–252.

BEN NAHIA, K. (1986), *Autour de la biconvexité en optimisation.* Thèse, Université Paul Sabatier de Toulouse.

BENSON, H.P. (1982), *On the Convergence of Two Branch and Bound Algorithms for Nonconvex Programming Problems.* Journal of Optimization Theory and Applications, 36, 129–134.

BENSON, H.P. (1982 a), *Algorithms for Parametric Nonconvex Programming.* Journal of Optimization Theory and Applications, 38, 319–340.

BENSON, H.P. (1985), *A Finite Algorithm for Concave Minimization Over a Polyhedron.* Naval Research Logistics Quarterly, 32, 165–177.

BENSON, H.P. (1990), *Separable Concave Minimization via Partial Outer Approximation and Branch and Bound,* to appear in Operations Research Letters.

BENSON, H.P. and ERENGUC, S. (1988), *Using Convex Envelopes to Solve the Interactive Fixed Charge Linear Programming Problem.* Journal of Optimization Theory and Applications 59, 223–246.

BENSON, H.P. and ERENGUC, S. (1988 a), *An Algorithm for Concave Integer Minimization Over a Polyhedron.* Discussion Paper No. 137, Center for Econometrics and Decision Sciences, University of Florida, to appear in Naval Research Logistics Quarterly.

BENSON, H.P. and HORST, R. (1988), *A Branch and Bound – Outer Approximation Algorithm for Concave Minimization Over a Convex Set.* Discussion Paper No. 139, Center for Econometrics and Decision Sciences, University of Florida, to appear in Journal Computers and Mathematics with Applications.

BERGE, C. (1958), *Théorie des Graphes et ses Applications.* Dunod Paris.

BHATIA, H.L., (1981), *Indefinite Quadratic Solid Transportation Problem.* Journal of Information and Optimization, 2, 297–303.

BITRAN, G.R., MAGNANTI, T.L. and YANASSE, H.H. (1984), *Approximation Methods for the Uncapacitated Dynamic Lot Size Problem.* Management Science, 30, 1121–1140.

BITTNER (1970), *Some Representation Theorems for Functions and Sets and their Applications to Nonlinear Programming.* Numerische Mathematik, 16, 32–51.

BLUM, E. and OETTLI, W. (1975), *Mathematische Optimierung.* Springer–Verlag, Berlin.

BOD, P. (1970), *Solution of a Fixed–Charge Linear Programming Problem.* In: Proceedings of Princeton Symposium on Mathematica Programming. Princeton University Press, 367–375.

BODLAENDER, H.L., GRITZMANN, P., KLEE, V. and VAN LEEUWEN, J. (1989), *Computational Complexity of Norm–Maximization.* Preprint 528, Institute for Mathematics and its Application, University of Minnesota.

BOHRINGER, M.C. (1973), *Linear Programs with Additional Reverse Convex Constraints.* Ph.D. Dissertation, UCLA.

BOY, M. (1988), *The Design Centering Problem.* Diplom–Thesis. Department of Mathematics, University of Trier.

BRACKEN, J. and McCORMICK, G.P. (1968), *Selected Applications of Nonlinear Programming,* J. Wiley, New York.

BRAESS, D. (1986), *Nonlinear Approximation Theory.* Springer–Verlag, Berlin.

BRANIN, F.H. (1972), *Widely Convergent Method for Finding Multiple Solutions of Simultaneous Nonlinear Equations.* IBM J. Res. Dev. 504–522.

BRENT, R.P. (1973), *Algorithms for Minimization Without Derivatives.* Prentice–Hall, Englewood Cliffs.

BROOKS, S.H. (1958), *Discussion of Random Methods for Locating Surface Maxima.* Operations Research, 6, 244–251.

BROTCHI, J.F. (1971), *A Model for National Development.* Management Science, 18, B14 – B18.

BULATOV, V.P. (1977), *Embedding Methods in Optimization Problems.* Nauka, Novosibirsk (in Russian).

BULATOV, V.P. (1987), *Methods for Solving Multiextremal Problems (Global Search).* In: Methods of Numerical Analysis and Optimization, ed. B.A. Beltiukov and V.P. Bulatov, Nauka, Novosibirsk, 133–157 (in Russian).

BULATOV, V.P. and KASINKAYA, L.I. (1982), *Some Methods of Concave Minimization on a Convex Polyhedron and Their Applications.* In: Methods of Optimization and their Applications, Nauka, Novosibirsk, 71–80 (in Russian).

BURDET, C.A. (1973), *Polaroids: A New Tool in Nonconvex and in Integer Programming.* Naval Research Logistics Quarterly, 20, 13–24.

BURDET, C.A. (1973), *Enumerative Cuts I.* Operations Research, 21, 61–89.

BURDET, C.A. (1977), *Convex and Polaroid Extensions.* Naval Research Logistics Quarterly, 26, 67–82.

BURDET, C.A. (1977a), *Elements of a Theory in Nonconvex Programming.* Naval Research Logistics Quarterly, 24, 47–66.

CABOT, A.V. (1974), *Variations on a Cutting Plane Method for Solving Concave Minimization Problems with Linear Constraints..* Naval Research Logistics, 21, 265–274.

CABOT, A.V. and ERENGUC, S.S. (1984), *Some Branch and Bound Procedures for Fixed–Cost Transportation Problems.* Naval Research Logistics Quarterly, 31, 129–138.

CABOT, A.V. and ERENGUC, S.S. (1986), *A Branch and Bound Algorithm for Solving a Class of Nonlinear Integer Programming Problems.* Naval Research Logistics, 33, 559–567.

CABOT, A.V. and FRANCIS, R.L. (1970), *Solving Nonconvex Quadratic Minimization Problems by Ranking the Extreme Points.* Operations Research, 18, 82–86.

CANDLER, W. and TOWNSLEY, R.J. (1964), *The Maximization of a Quadratic Function of Variables subject to Linear Inequalities.* Management Science, 10, 515–523.

CARILLO, M.J. (1977), *A Relaxation Algorithm for the Minimization of a Quasiconcave Function on a Convex Polyhedron.* Mathematical Programming, 13, 69–80.

CARVAJAL–MORENO, R. (1972), *Minimization of Concave Functions Subject to Linear Constraints.* Operations Research Center Report ORC 72–3, University of Berkely, California.

CHENEY, E.W. and GOLDSTEIN, A.A. (1959), *Newton's Method for Convex Programming and Tchebycheff Approximation.* Numerische Mathematik, 1, 253–268.

CHENG, Y.C. (1984), *On the Gradient Projection Method for Solving the Nonsymmetric Linear Complementarity Problem.* Journal of Optimization Theory and Applications, 43, 527–541.

CHEW, S.H. and ZHENQ, Q. (1988), Integral Global Optimization, Lecture Notes in Economics and Mathematical Systems, 289, Springer–Verlag, Berlin.

CIRINA, M. (1983), Recent Progress in Complementarity. Paper presented at TIMS/ORSA Joint National Meeting, Chicago, IL.

CIRINA, M. (1985), *A Class of Nonlinear Programming Test Problems.* Working Paper, Dipartimento di Informatica, Torino, Italy.

CIRINA, M. (1986), *A Finite Algorithm for Global Quadratic Minimization.* Working Paper, Dip. di Informat. Torino, Italy.

COHEN, J.W. (1975), *Plastic–Elastic–Torsion, Optimal Stopping and Free Boundaries.* Journal of Engineering Mathematics, 9, 219–226.

COOPER, M. (1981), *Survey of Methods for Pure Nonlinear Integer Programming.* Management Science, 27, 353–361.

COTTLE, R.C. and DANTZIG, G.B. (1968), *Complementarity Pivot Theory of Mathematical Programming.* Linear Algebra and its Applications, 1, 103–125.

COTTLE, R. and MYLANDER, W.C. (1970), *Ritter's Cutting Plane Method for Nonconvex Quadratic Programming.* In: J. Abadie (editor), Integer and Nonlinear Programming, North–Holland, Amsterdam.

COTTLE, R. and PANG, J.S. (1978), *On Solving Linear Complementarity Problems as Linear Programs.* Mathematical Programming Study, 7, 88–107.

CZOCHRALSKA, I. (1982), *Bilinear Programming.* Zastosow. Mat., 17, 495–514.

CZOCHRALSKA, I. (1982 a), *The Method of Bilinear Programming for Nonconvex Quadratic Problems.* Zastosow. Mat., 17, 515–523.

DAJANI, J.S. and HASIT, Y. (1974), *Capital Cost Minimization of Drainage Networks.* Journal Environ. Eng. Div., 100, 325–337.

DANILIN, Y.M. (1971), *Estimation of the Efficiency of an Absolute–Minimum–Finding Algorithm.* USSR Computational Mathematics and Mathematical Physics, 11, 261–267.

DANTZIG, G.B. and WOLFE, P. (1960), *Decomposition Principle for Linear Programs.* Operations Research, 8, 101–111.

DENNIS, J.E. and SCHNABEL, R.B. (1983), *Numerical Methods for Nonlinear Equations and Unconstrained Optimization.* Prentice–Hall, Englewood Cliffs, New Jersey.

DEWDER, D.R. (1967), *An Approximate Algorithm for the Fixed Charge Problem.* Naval Research Logistics Quarterly, 14, 101–113.

DIENER, I. (1987), *On the Global Convergence of Path–Following Methods to Determine all Solutions to a System of Nonlinear Equations.* Mathematical Programming, 39, 181–188.

DINH ZUNG (1987), *Best Linear Methods of Approximation for Classes of Periodic Function of Several Variables.* Matematicheskie Zametki 41, 646–653 (in Russian).

DIXON, L.C.W. (1978), *Global Optima Without Convexity.* In: Greenberg, H. (editor), Design and Implementation Optimization Software, Sijthoff and Noordhoff, Alphen aan den Rijn, 449–479.

DIXON, L.C.W., and SZEGO, G.P. (editors) (1975), *Towards Global Optimization,* Volume I. North–Holland, Amsterdam.

DIXON, L.C.W., and SZEGO, G.P. (editors) (1978), *Towards Global Optimization,* Volume II. North–Holland, Amsterdam.

DUONG, P.C. (1987), *Finding the Global Extremum of a Polynomial Function.* in: Essays on Nonlinear Analysis and Optimization Problems, Institute of Mathematics, Hanoi (Vietnam), 111–120.

DUTTON, R., HINMAN, G. and MILLHAM, C.B. (1974), *The Optimal Location of Nuclear–Power Facilities in the Pacific Northwest.* Operations Research, 22, 478–487.

DYER, M.E. (1983), *The Complexity of Vertex Enumeration Methods.* Mathematics of Operations Research, 8, 381–402.

DYER, M.E. and PROLL, L.G. (1977), *An Algorithm for Determining All Extreme Points of a Convex Polytope.* Mathematical Programming, 12, 81–96.

DYER, M.E. and PROLL, L.G. (1982), *An Improved Vertex Enumeration Algorithm.* European Journal of Operational Research, 9, 359–368.

EAVES, B.C. AND ZANGWILL, W.I. (1971), *Generalized Cutting Plane Algorithms.* SIAM Journal on Control, 9, 529–542.

ECKER, J.G. and NIEMI, R.D. (1975), *A Dual Method for Quadratic Programs with Quadratic Constraints.* SIAM Journal Applied Mathematics, 28, 568–576.

ELLAIA, R. (1984), *Contribution à l'Analyse et l'Optimisation de Différences de Fonctions Convexes.* Thèse du 3ème Cycle, Université Paul Sabatier, Toulouse.

ELSHAFEI, A.N. (1975), *An Approach to Locational Analysis.* Operational Research Quarterly, 26, 167–181.

EMELICHEV, V.A. and KOVALEV, M.M. (1970), *Solving Certain Concave Programming Problems by Successive Approximation I.* Izvestya Akademii Nauk BSSR, 6, 27–34 (in Russian).

ERENGUC, S.S. (1988), *Multiproduct Dynamic Lot–Sizing Model with Coordinated Replenishment.* Naval Research Logistics, 35, 1–22.

ERENGUC, S.S. and BENSON, H.P. (1986), *The Interactive Fixed Charge Linear Programming Problem.* Naval Research Logistics, 33, 157–177.

ERENGUC, S.S. and BENSON, H.P. (1987), *An Algorithm for Indefinite Integer Quadratic Programming.* Discussion Paper 134, Center for Econometrics and Decision Sciences, University of Florida.

ERENGUC, S.S. and BENSON, H.P. (1987 a), *Concave Integer Minimizations Over a Compact, Convex Set.* Working Paper No. 135, Center for Econometrics and Decision Science, University of Florida.

ERICKSON, R.E., MONMA, C.L. and VEINOTT, A.F. (1987), *Send–and–Split Method for Minimum–Concave–Cost Network Flows.* Mathematics of Operations Research, 12, 634–663.

EVTUSHENKO, Y.G. (1971), *Numerical Methods for Finding the Global Extremum of a Function.* USSR Computational Mathematics and Mathematical Physics, 11, 38–54.

EVTUSHENKO, Y.G. (1985), *Numerical Optimization Techniques.* Translation Series in Mathematics and Engineering, Optimization Software Inc. Publication Division, New York

EVTUSHENKO, Y.G. (1987), *Bisection Method for Global Optimization of Functions of Many Variables.* Technicheskaya Kibernetika, 1, 119–127 (in Russian).

FALK, J.E. (1967), *Lagrange Multipliers and Nonlinear Programming.* J. Mathematical Analysis and Applications, 19, 141–159.

FALK, J.E. (1969), *Lagrange Multipliers and Nonconvex Programs.* SIAM Journal on Control, 7, 534–545.

FALK, J.E. (1972), *An Algorithm for Locating Approximate Global Solutions of Nonconvex, Separable Problems.* Working Paper Serial T–262, Program in Logistics, The George Washington University.

FALK, J.E. (1973), *A Linear Max–Min Problem.* Mathematical Programming, 5, 169–188.

FALK, J.E. (1973 a), *Conditions for Global Optimality in Nonlinear Programming.* Operations Research, 21, 337–340.

FALK, J.E. (1974), *Sharper Bounds on Nonconvex Programs.* Operations Research, 22, 410–412.

FALK, J.E., BRACKEN, J. and McGILL, J.T. (1974), *The Equivalence of Two Mathematical Programs with Optimization Problems in the Constraints.* Operations Research, 22, 1102–1104.

FALK, J.E. and HOFFMAN, K.L. (1976), *A Successive Underestimation Method for Concave Minimization Problems.* Mathematics of Operations Research, 1, 251–259.

FALK, J.E. and HOFFMAN, K.L. (1986), *Concave Minimization via Collapsing Polytopes.* Operations Research, 34, 919–929.

FALK, J.E. and SOLAND, R.M. (1969), *An Algorithm for Separable Nonconvex Programming Problems.* Management Science, 15, 550–569.

FEDOROV, V.V. (editor) (1985), *Problems of Cybernetics, Models and Methods in Global Optimization.* USSR Academy of Sciences, Moscow (in Russian).

FENCHEL, W. (1949), *On Conjugate Convex Functions.* Canadian Journal of Mathematics, 1, 73–77.

FENCHEL, W. (1951), *Convex Cones, Sets and Functions.* Mimeographed Lecture Notes, Princeton University.

FERLAND, A.J. (1975), *On the Maximum of a Quasi–Convex Quadratic Function on a Polyhedral Convex Set.* SIAM Journal on Applied Mathematics, 29, 409–415.

FLORIAN, M. (1986), *Nonlinear Cost Network Models in Transportation Analysis.* Mathematical Programming Study, 26, 167–196.

FLORIAN, M. and ROBILLARD, P. (1971), *An Implicit Enumeration Algorithm for the Concave Cost Network Flow Problem.* Management Science, 18, 184–193.

FLORIAN, M., ROSSIN, M.A. and de WERRA, D. (1971), *A Property of Minimum Concave Cost Flows in Capacitated Networks.* Canadian Journal of Operations Research, 9, 293–304.

FORGO, F. (1972), *Cutting Plane Methods for Solving Nonconvex Quadratic Problems.* Acta Cybernetica, 1, 171–192.

FORGO, F. (1988), *Nonconvex Programming.* Akademiai Kiado, Budapest.

FORSTER, W. (editor) (1980), *Numerical Solution of Highly Nonlinear Problems.* North–Holland, Amsterdam.

FRIEZE, A.M. (1974), *A Bilinear Programming Formulation of the 3–dimensional Assignment Problem.* Mathematical Programming, 7, 376–379.

FUJIWARA O. and KHANG, D.B. (1988), *Optimal Design of Looped Water Distribution Networks via Concave Minimization.* 13th International Symposium on Mathematical Programming 1988.

FUKUSHIMA, M. (1983), *An Outer Approximation Algorithm for Solving General Convex Programs.* Operations Research, 31, 101–113.

FUKUSHIMA, M. (1984), *On the Convergence of a Class of Outer Approximation Algorithms for Convex Programs.* Journal of Computational and Applied Mathematics, 10, 147–156.

FÜLÖP, J. (1988), *A Finite Procedure to Generate Feasible Points for the Extreme Point Mathematical Programming Problem*, to appear in European Journal of Operational Research.

FÜLÖP, J. (1988 a), *A Finite Cutting Plane Method for Solving Linear Programs with an Additional Reverse Convex Constraint.* Working Paper, Computation and Automation Institute, Budapest.

GAL, T. (1975), *Zur Identifikation Redundanter Nebenbedingungen in Linearen Programmen.* Zeitschrift für Operations Research, 19, 19–28.

GAL, T. (1985), *On the Structure of the Set Bases of a Degenerate Point.* Journal of Optimization Theory and Applications, 45, 577–589.

GAL, T., KRUSE, H.J. and ZÖRNIG, P. (1988), *Survey of Solved and Open Problems in the Degeneracy Phenomenon.* Mathematical Programming B, 42, 125–133.

GALLO, G., SANDI, C. and SODINI, C. (1980), *An Algorithm for the Min Concave Cost Flow Problem,* European Journal of Operational Research, 4, 249–255.

GALLO, G. and SODINI, C. (1979), *Adjacent Extreme Flows and Application to Min Concave Cost Flow Problem.* Networks, 9, 95–122.

GALLO, G. and SODINI, C. (1979 a), *Concave Cost Minimization on Networks.* European Journal of Operations Research, 3, 239–249.

GALLO, G. and ÜLKUCÜ, A. (1977), *Bilinear Programming: An Exact Algorithm.* Mathematical Programming, 12, 173–194.

GALPERIN, E.A. (1985), *The Cubic Algorithm.* Journal of Mathematical Analysis and Applications, 112, 635–640.

GALPERIN, E.A. (1988), *Precision, Complexity, and Computational Schemes of the Cubic Algorithm.* Journal of Optimization Theory and Applications, 57, 223–238.

GALPERIN, E.A. and ZHENG, Q. (1987), *Nonlinear Observation via Global Optimization Methods: Measure Theory Approach.* Journal of Optimization Theory and Applications, 54, 1, 63–92.

GANSHIN, G.S. (1976), *Function Maximization.* USSR Computational Mathematics and Mathematical Physics, 16, 26–36.

GANSHIN, G.S. (1976 a), *Simplest Sequential Search Algorithm for the Largest Value of a Twice–Differentiable Function.* USSR Computational Mathematics and Mathematical Physics, 16, 508–509.

GANSHIN, G.S. (1977), *Optimal Passive Algorithms for Evaluating the Maximum of a Function in an Interval.* USSR Computational Mathematics and Mathematical Physics, 17, 8–17.

GARCIA, C.B. and ZANGWILL, W.I. (1981), *Pathways to Solutions, Fixed Points and Equilibra.* Prentice–Hall, Englewood Cliffs, N.J.

GAREY, M.P., JOHNSON, D.S. and STOCKMEYER, L. (1976), *Some Simplified NP–Complete Problems.* Theoretical Computer Science, 1, 237–267.

GAREY, M.P. and JOHNSON, D.S. (1979), *Computers and Intractability: A Guide to the Theory of NP–Completeness.* Freeman, San Francisco.

GASANOV, I.I. and RIKUN, A.D. (1984), *On Necessary and Sufficient Conditions for Uniextremality in Nonconvex Mathematical Programming Problems.* Soviet Mathematics Doklady, 30, 457–459.

GASANOV, I.I. and RIKUN, A.D. (1985), *The Necessary and Sufficient Conditions for Single Extremality in Nonconvex Problems of Mathematical Programming,* USSR Computational Mathematics and Mathematical Physics, 25, 105–113.

GEOFFRION, A.M. (1971), *Duality in Nonlinear Programming: A Simplified Applications–Oriented Development.* SIAM Reviews, 13, 1–37.

GEOFFRION, A.M. (1972), *Generalized Benders Decomposition.* Journal of Optimization Theory and Applications, 10, 237–260.

GIANNESSI, F., JURINA, L. and MAIER, G. (1979), *Optimal Excavation Profile for a Pipeline Freely Resting on the Sea Floor.* Engineering Structures, 1, 81–91.

GIANNESSI, F. and NICCOLUCCI, F. (1976), *Connections between Nonlinear and Integer Programming Problems.* Istituto Nazionale di Alta Mathematica, Symposia Mathematica, 19, 161–176.

GINSBERG, W. (1973), *Concavity and Quasiconcavity in Economics.* Journal of Econ. Theory, 6, 596–605.

GLOVER, F. (1972), *Cut Search Methods in Integer Programming.* Mathematical Programming, 3, 86–100.

GLOVER, F. (1973), *Convexity Cuts and Cut Search.* Operations Research, 21, 123–124.

GLOVER, F. (1973 a), *Concave Programming Applied to a Special Class of 0–1 Integer Programs.* Operations Research, 21, 135–140.

GLOVER, F. (1974), *Polyhedral Convexity Cuts and Negative Edge Extensions.* Zeitschrift für Operations Research, 18, 181–186.

GLOVER, F. (1975), *Polyhedral Annexation in Mixed Integer and Combinatorial Programming.* Mathematical Programming, 8, 161–188.

GLOVER, F. (1975a), *Surrogate Constraint Duality in Mathematical Programming.* Operations Research, 23, 434–451.

GLOVER, F. and KLINGMAN, D. (1973), *Concave Programming Applied to a Special Class of 0–1 Integer Programs.* Operations Research, 21, 135–140.

GLOVER, F. and WOLSEY, E. (1973), *Further Reduction of Zero–One Polynomial Programming Problems to Zero–One Linear Programming Problems.* Operations Research, 21, 141–161.

GLOVER, F. and WOLSEY, E. (1974), *Converting the 0–1 Polynomial Programming Problem to a Linear 0–1 Program.* Operations Research, 22, 180–182.

GOMORY, R.E. (1958), *Outline of an Algorithm for Integer Solutions to Linear Programs.* Bulletin of the Americal Mathematical Society, 64, 275–278.

GOMORY, R.E. (1960), *Solving Linear Programming Problems in Integers.* In: Combinatorial Analysis (R. Bellman and M. Hall eds.), Providence, 211–215.

GONZAGA, C. and POLAK, E. (1979), *On Constraint Dropping Schemes and Optimality Functions for a Class of Outer Approximation Algorithms.* SIAM Journal on Control and Optimization, 17, 477–493.

GOULD, F.J. (1969), *Extension of Lagrange Multipliers in Nonlinear Programming.* SIAM Journal on Applied Mathematics, 17, 1280–1297.

GOULD, F.J. and TOLLE, J.W. (1983), *Complementary Pivoting on a Pseudomanifold Structure with Applications in the Decision Sciences.* Heldermann, Berlin.

GRAVES, S.T. and ORLIN, J.B. (1985), *A Minimum–Cost Dynamic Network Flow Problem with an Application to Lot–Sizing.* Networks, 15, 59–71.

GRAVES, G.W. and WHINSTON, A.B. (1970), *An Algorithm for the Quadratic Assignment Problem.* Management Science, 17, 453–471.

GRAY, P. (1971), *Exact Solution of the Fixed Charge Transportation Problem.* Operations Research, 19, 1529–1538.

GREENBERG, H.J. (1973), *Bounding Nonconvex Programs by Conjugates.* Operations Research, 21, 346–347.

GRIEWANK, A.O. (1981), *Generalized Descent for Global Optimization.* Journal of Optimization Theory and Applications, 34, 11–39.

GRITZMANN, P. and KLEE, V. (1988), *On the 0–1–Maximization of Positive Definite Quadratic Forms.* Discussion Paper, submitted.

GROCH, A., VIDIGAL, L. and DIRECTOR, S. (1985), *A New Global Optimization Method for Electronic Circuit Design.* IEEE Transactions on Circuits and Systems, CAS–32, 160–170.

GROTTE, J.H. (1975), *A Nontarget Damage Minimizer which achieves Targeting Objectives: Model Description and Users Guide.* Working Paper WP–8, Institute of Defense Analysis, Arlington, Virginia.

GUISWITE, G.M. and PARDALOS, P.M. (1989), *On the Complexity of Minimum Concave–Cost Network Flow Problems.* Discussion Paper, Department of Computer Science, Pennsylvania State University.

GUPTA, A.K. and SHARMA, J.K. (1983), *A Generalized Simplex Technique for Solving Quadratic Programming Problems.* Indian Journal of Technology, 21, 198–201.

GURLITZ, T.R. (1985), *Algorithms for Reverse Convex Programs.* Ph.D. Dissertation, UCLA.

HAMAMI, M. (1982), *Finitely Convergent Tuy–Type Algorithms for Concave Minimization.* Ph.D. Dissertation, UCLA.

HAMAMI, M. and JACOBSEN, S.E. (1988), *Exhaustive Nondegenerate Conical Processes for Concave Minimization on Convex Polytope.* Mathematics of Operations Research, 13, 479–487.

HAMMER, P.L. (1965), *Some Network Flow Problems with Pseudo–Boolean Programming.* Operations Research, 13, 388–399.

HANSEN, E. (1979), *Global Optimization Using Interval Analysis – The one–dimensional Case.* Journal of Optimization Theory and its Applications, 29, 331–344.

HANSEN, E. (1980), *Global Optimization using Interval Analysis – The Multidimensional Case.* Numerische Mathematik, 34, 247–270.

HANSEN, P. (1979), *Methods of Nonlinear 0–1 Programming.* Annals of Discrete Mathematics, Discrete Optimization II, 5, 53–70.

HANSEN, P., JAURMARD, B. and LU, S.H. (1988), *On the Number of Iterations of Piyavskii's Global Optimization Algorithm,* to appear in Mathematics of Operations Research.

HANSEN, P., JAUMARD, B and LU, S.H. (1988a), *A Framework for Algorithms in Globally Optimal Design.* Research Report G–88–17, HEC–GERAD, University of Montréal.

HANSEN, P., JAUMARD, B and LU, S.H. (1989), *Global Optimization of Univariate Lipschitz Functions: I. Survey and Properties.* Research Report G–89–10, HEC–GERAD, University of Montréal

HANSEN, P., JAUMARD, B and LU, S.H. (1989a), *Global Optimization of Univariate Lipschitz Functions: Part II. New Algorithms and Computational Comparison.* Cahier du GERAD, Montréal

HANSEN, P., JAUMARD, B and LU, S.H. (1990), *Global Minimization of Univariate Functions by Sequential Polynomial Approximation,* to appear in International Journal of Computer Mathematics.

HARMAN, J.K. (1973), *Some Experience in Global Optimization.* Naval Research Logistics Quarterly, 20, 569–576.

HARTMAN, P. (1959), *On Functions Representable as a Difference of Convex Functions.* Pacific Journal of Mathematics, 9, 707–713.

HASHAM, A. and SACK, J.–R. (1987), *Bounds for Min–Max Heaps.* BIT, 27, 315–323.

HERON, B. and SERMANGE, M. (1982), *Nonconvex Methods for Computing Free Boundary Equilibra of Axially Symmetric Plasmas.* Applied Mathematics and Optimization, 8, 351–382.

HEYDEN, L. Van der (1980), *A Variable Dimension Algorithm for the Linear Complementarity Problem.* Mathematical Programming, 19, 123–130.

HILLESTAD, R.J. (1975), *Optimization Problems Subject to a Budget Constraint with Economies of Scale.* Operations Research, 23, 1091–1098.

HILLESTAD, R.J. and JACOBSEN, S.E. (1980), *Reverse Convex Programming.* Applied Mathematics and Optimization, 6, 63–78.

HILLESTAD, R.J. and JACOBSEN, S.E. (1980a), *Linear Programms with an Additional Reverse Convex Constraint.* Applied Mathematics and Optimization, 6, 257–269.

HIRIART–URRUTY, J.B. (1985), *Generalized Differentiability, Duality and Optimization for ProblemsDealing with Differences of Convex Functions.* Lecture Notes in Economics and Mathematical Systems, 256, 37–69, Springer–Verlag, Berlin.

HIRIART–URRUTY, J.B. (1986), *When is a point x satisfying* $\Delta f(x) = 0$ *a global minimum of f.* Amer. Math. Monthly, 93, 556–558.

HOFFMAN, K.L. (1981), *A Method for Globally Minimizing Concave Functions over Convex Sets.* Mathematical Programming, 20, 22–32.

HOGAN, W.W. (1973), *Applications of a General Convergence Theory for Outer Approximation Algorithms.* Mathematical Programming, 5, 151–168.

HORST, R. (1976), *An Algorithm for Nonconvex Programming Problems.* Mathematical Programming, 10, 312–321.

HORST, R. (1976 a), *On the Characterization of Affine Hull–Functional.* Zeitschrift für Angewandte Mathematik und Mechanik, 3, 347–348 (in German).

HORST, R. (1976 b), *A New Branch and Bound Approach for Concave Minimization Problems.* Lecture Notes in Computer Science 41, 330–337, Springer–Verlag, Berlin.

HORST, R. (1978), *A New Approach for Separable Nonconvex Minimization Problems Including a Method for Finding the Global Minimum of a Function of a Single Variable.* Proceedings in Operations Research, 7, 39–47, Physica, Heidelberg.

HORST, R. (1979), *Nichtlineare Optimierung.* Carl Hanser–Verlag, München.

HORST, R. (1980), *A Note on the Convergence of an Algorithm for Nonconvex Programming Problems.* Mathematical Programming, 19, 237–238.

HORST, R. (1980 a), *A Note on the Duality Gap in Nonconvex Optimization and a very simple Procedure for Bild Evaluation Type Problems.* European Journal of Operational Research, 5, 205–210.

HORST, R. (1982), *A Note on Functions whose Local Minima are Global.* Journal of Optimization Theory and Applications, 36, 457–463.

HORST, R. (1984), *On the Convexification of Nonconvex Programming Problems.* European Journal of Operational Research, 15, 382–392.

HORST, R. (1984 a), *Global Minimization in Arcwise Connected Spaces.* Journal of Mathematical Analysis and Applications, 104, 481–483.

HORST, R. (1984 b), *On the Global Minimization of Concave Functions: Introduction and Survey.* Operations Research Spektrum, 6, 195–205.

HORST, R. (1986), *A General Class of Branch and Bound Methods in Global Optimization with some New Approaches for Concave Minimization.* Journal of Optimization Theory and Applications, 51, 271–291.

HORST, R. (1987), *On Solving Lipschitzian Optimization Problems.* in: Essays on Nonlinear Analysis and Optimization Problems, National Center for Scientific Research, Hanoi, 73–88.

HORST, R. (1987 a), *Outer Cut Methods in Global Optimization.* Lecture Notes in Economics and Mathematical Systems, 304, 28–40.

HORST, R. (1988), *Deterministic Global Optimization with Partition Sets whose Feasibility is not Known. Application to Concave Minimization, Reverse Convex Constraints, D.C.–Programming and Lipschitzian Optimization.* Journal of Optimization Theory and Application, 58, 11–37.

HORST, R. (1989), *On Consistency of Bounding Operations in Deterministic Global Optimization.* Journal of Optimization Theory and Applications, 61, 143–146.

HORST, R. (1990), *Deterministic Global Optimization: Some Recent Advances and New Fields of Application*, to appear in Naval Research Logistics.

HORST, R. and DIEN, L.V. (1987), *A Solution Concept for a very General Class of Decision Problems.* In: Opitz and Rauhut (editors): Mathematik und Ökonomie, Springer 1987, 143–153.

HORST, R., PHONG, T.Q. and THOAI, N.V. (1989), *On Solving General Reverse Convex Programming Problems by a Sequence of Linear Programs and Line Searches*, submitted.

HORST, R., PHONG, T.Q., THOAI, N.V., TUY, H. and VRIES, J. de (1990), *On Solving a D.C. Programming Problem by a Sequence of Linear Programs.* Preprint, Department of Mathematics, University of Trier, submitted.

HORST, R. and THACH, P.T. (1988), *A Topological Property of Limes–Arcwise Strictly Quasiconvex Functions.* Journal of Mathematical Analysis and Applications, 134, 426–430.

HORST, R. and THOAI, N.V. (1988), *Branch–and–Bound Methods for Solving Systems of Lipschitzian Equations and Inequalities.* Journal of Optimization Theory and Applications, 58, 139–146.

HORST, R. and THOAI, N.V. (1989), *Modification, Implementation and Comparison of Three Algorithms for Globally Solving Linearly Constrained Concave Minimization Problems.* Computing, 42, 271–289.

HORST, R. and THOAI, N.V. (1989 a), *Conical Algorithms for the Global Minimization of Linearly Constrained Large–Scale Concave Minimization Problems*, submitted.

HORST, R., THOAI, N.V., and BENSON, H.P. (1988), *Concave Minimization via Conical Partitions and Polyhedral Outer Approximation.* Discussion Paper No. 140, Center for Econometrics and Decision Sciences, University of Florida, to appear in Mathematical Programming.

HORST, R., THOAI, N.V. and TUY, H. (1987), *Outer Approximation by Polyhedral Convex Sets.* Operations Research Spektrum, 9, 153–159.

HORST, R., THOAI, N.V. and TUY, H. (1989), *On an Outer Approximation Concept in Global Optimization.* Optimization, 20, 255–264.

HORST, R., THOAI, N.V. and de VRIES, J. (1988), *On Finding New Vertices and Redundant Constraints in Cutting Plane Algorithms for Global Optimization.* Operations Research Letters, 7, 85–90.

HORST, R. and TUY, H. (1987), *On the Convergence of Global Methods in Multiextremal Optimization.* Journal of Optimization Theory and Applications, 54, 253–271.

IBARAKI, T. (1971), *Complementarity Programming.* Operations Research, 19, 1523–1528.

ISTOMIN, L.A. (1977), *A Modification of Tuy's Method for Minimizing a Concave Function Over a Polytope.* USSR Computational Mathematics and Mathematical Physics, 17, 1582–1592 (in Russian).

IVANOV, V.V. (1972), *On Optimal Algorithms of Minimization in the Class of Functions with the Lipschitz Condition.* Information Processing, 71, 1324–1327.

JACOBSEN, S.E. (1981), *Convergence of a Tuy–Type–Algorithm for Concave Minimization Subject to Linear Inequality Constraints.* Applied Mathematics and Optimization, 7, 1–9.

JACOBSEN, S.E. and TORABI, M. (1978), *A Global Minimization Algorithm for a Class of One–Dimensional Functions.* Journal of Mathematical Analysis and Applications, 62, 310–324.

JENSEN, P.A. and BARNES, J.W. (1980), *Network Flow Programming.* John Wiley, New York.

JEROSLOW, R.G. (1976), *Cutting Planes for Complementarity Constraints.* SIAM Journal on Control and Optimization, 16, 56–62.

JEROSLOW, R.G. (1977), *Cutting Plane Theory: Disjunctive Methods.* Annals of Discrete Mathematics, 1, 293–330.

JONES, A.P. and SOLAND, R.M. (1969), *A Branch–and–Bound Algorithm for Multi–Level Fixed–Charge Problems.* Management Science, 16, 67–76.

KALANTARI, B. (1984), *Large Scale Global Minimization of Linearly Constrained Concave Quadratic Functions and Related Problems.* Ph.D. Thesis, Computer Sci. Dept., University of Minnesota.

KALANTARI, B. (1986), *Quadratic Functions with Exponential Number of Local Maxima.* Operations Research Letters, 5, 47–49.

KALANTARI, B. and ROSEN, J.B. (1986), *Construction of Large–Scale Global Minimum Concave Test Problems.* Journal of Optimization Theory and Applications, 48, 303–313.

KALANTARI, B. and ROSEN, J.B. (1987), *An Algorithm for Global Minimization of Linearly Constrained Concave Quadratic Functions.* Mathematics of Operations Research, 12, 544–561.

KALL, P. (1986), *Approximation to Optimization Problems: An Elementary Review.* Mathematics of Operations Research, 11, 9–17.

KAO, E. (1979), *A Multi–Product Lot–Size Model with Individual and Joint Set–Up Costs.* Operations Research, 27, 279–289.

KEARFOTT, R.B. (1987), *Abstract Generalized Bisection and a Cost Bound.* Mathematics of Computation, 49, 187–202.

KEDE, G. and WATANABE, H. (1983), *Optimization Techniques for IC Layout and Compaction.* Proceedings IEEE Intern. Conf. in Computer Design: KSI in Computers, 709–713.

KELLEY, J.E. (1960), *The Cutting–Plane Method for Solving Convex Programs.* Journal SIAM, 8, 703–712.

KHACHATUROV, V. and UTKIN, S. (1988), *Solving Multiextremal Concave Programming Problems by Combinatorial Approximation Method.* Preprint, Computer Center of the Academy of Sciences, Moscow (in Russian).

KHANG, D.B. and FUJIWARA, O. (1989), *A New Algorithm to Find All Vertices of a Polytope.* Operations Research Letters, 8, 261–264.

KIEFER, J. (1957), *Optimum Search and Approximation Methods Under Minimum Regularity Assumptions.* SIAM Journal, 5, 105–136.

KIWIEL, K.C. (1985), *Methods of Descent for Nondifferentiable Optimization.* Lecture Notes in Mathematics, 1133, Springer–Verlag, Berlin.

KLEIBOHM, K. (1967), *Remarks on the Non–Convex Programming Problem.* Unternehmensforschung, 2, 49–60 (in German).

KLEITMAN, D.J. and CLAUS, A. (1972), *A Large Scale Multicommodity Flow Problem: Telepack. Proceedings Symp. Comput. Commun. Networks and Teletraffic,* Polytechn. Institute New York 22, 335–338.

KOEHLER, G., WHINSTON, A.B. and WRIGHT, G.P. (1975), *Optimization over Leontiev Substitution Systems.* North Holland, Amsterdam.

KONNO, H. (1971), *Bilinear Programming: Part I. An Algorithm for Solving Bilinear Programs.* Technical Report No. 71–9, Operations Research House, Stanford University.

KONNO, H. (1971 a), *Bilinear Programming: Part II. Applications of Bilinear Programming.* Technical Report No. 71–10, Operations Research House, Stanford University, Stanford, CA.

KONNO, H. (1973), *Minimum Concave Series Production Systems with Deterministic Demand–Backlogging.* Journal of the Operations Research Society of Japan, 16, 246–253.

KONNO, H. (1976), *A Cutting–Plane Algorithm for Solving Bilinear Programs.* Mathematical Programming, 11, 14–27.

KONNO, H. (1976 a), *Maximization of a Convex Quadratic Function under to Linear Constraints.* Mathematical Programming, 11, 117–127.

KONNO, H. (1980), *Maximizing a Convex Quadratic Function over a Hypercube.* Journal of the Operations Research Society of Japan, 23, 171–189.

KONNO, H. (1988), *Minimum Concave Cost Production System: A Further Generalization of Multi–Echelon Model.* Mathematical Programming, 41, 185–193..

KONNO, H., (1981), *An Algorithm for Solving Binlinear Knapsack Problems.* Journal of Operations Research Society of Japan, 24, 360–373.

KOUGH, P.L. (1979), *The Indefinite Quadratic Programming Problem.* Operations Research, 27, 516–533.

KRIPFGANZ, A. and SCHULZE, R. (1987), *Piecewise affine function as a difference of two convex functions.* Optimization, 18, 23–29.

KRUSE, H.–J. (1986), *Degeneracy Graphs and the Neighbourhood Problem.* Lecture Notes in Economics and Mathematical Systems 260, Springer–Verlag, Berlin.

LANDIS, E.M. (1951), *On Functions Representable as the Difference of two Convex Functions.* Dokl. Akad. Nauk SSSR, 80, 9–11.

LAWLER, E.L. (1963), *The Quadratic Assignment Problem.* Management Science, 9, 586–699.

LAWLER, E.L. (1966), *Branch and Bound Methods: A Survey.* Operations Research, 14, 669–179.

LAZIMY, R. (1982), *Mixed–Integer Quadratic Programming.* Mathematical Programming, 22, 332–349.

LAZIMY, R. (1985), *Improved Algorithm for Mixed–Integer Quadratic Programs and a Computational Study.* Mathematical Programming, 32, 100–113.

LEBEDEV, V.Y. (1982), *A Method of Solution of the Problem of Maximization of a Positive Definite Quadratic Form on a Polyhedron.* USSR Computational Mathematics and Mathematical Physics, 22, 1344–1351 (in Russian).

LEICHTWEISS, K. (1980), *Konvexe Mengen.* VEB Deutscher Verlag der Wissen–schaften, Berlin.

LEMKE, C.E. (1965), *Bimatrix Equilibrium Points and Mathematical Programming.* Management Science, 4, 681–689.

LEVITIN, E.S. and POLYAK, B.T. (1966), *Constrained Minimization Methods.* USSR Computational Mathematics and Mathematics Physics, 6, 1–50.

LEVY, A.V. and MONTALVO, A. (1985), *The Tunneling Algorithm for the Global Minimization of Functions.* SIAM Journal on Scientific and Statistical Computing, 6, 15–29.

LOVE, F.C. (1973), *A Facilities in Series Inventory Model with Nested Schedules.* Management Science, 18, 327–338.

LUENBERGER, D.G. (1969), *Optimization by Vector Space Methods.* John Wiley, New York.

LÜTHI, H.J. (1976), *Komplementaritäts– und Fixpunktalgorithmen in der Mathematischen Programmierung, Spieltheorie und Ökonomie.* Lecture Notes in Economics and Mathematical Systems 129, Springer–Verlag, Berlin.

MAJTHAY, A. and WHINSTON, A. (1974), *Quasiconcave Minimization Subject to Linear Constraints.* Discrete Mathematics, 9, 35–59.

MAHJOUB, Z. (1983), *Contribution a L'Etude de l'Optimisation des Reseaux MaillΘs.* Thèse de Doctorat d'Etat. Institut Nationale Polytechnique de Toulouse.

MALING, K., MUELLER, S.H. and HELLER, W.R. (1982), *On Finding Most Optimal Rectangular Package Plans.* Proceeding of the 19th Design Automation Conference, 663–670.

MANAS, M. and NEDOMA, J. (1968), *Finding all Vertices of a Convex Polyhedron.* Numerische Mathematik, 12, 226–229.

MANGASARIAN, O.L. (1964), *Equilibrum Points of Bimatrix Games.* SIAM Journal, 12, 778–780.

MANGASARIAN, O.L. (1969), *Nonlinear Programming.* McGraw–Hill, New–York.

MANGASARIAN, O.L. (1976), *Linear Complementarity Problems Solvable by a Single Linear Program.* Mathematical Programming, 10, 263–270.

MANGASARIAN, O.L. (1978), *Characterization of Linear Complementarity Problems as Linear Programs.* Mathematical Programming Study, 7, 74–87.

MANGASARIAN, O.L. (1979), *Simplified Characterizations of Linear Complementarity Problems Solvable as Linear Programs.* Mathematics of Operations Research, 4, 268–273.

MANGASARIAN, O.L. and SHIAU, T.H. (1985), *A Variable Complexity Norm Maximization Problem.* Technical Summary Report University of Wisconsin, Madison.

MANGASARIAN, O.L. and STONE, H. (1964), *Two Person Nonzero–Sum Games and Quadratic Programming.* Journal Mathematical Analysis and Applications, 9, 345–355.

MANNE, A.S. (1964), *Plant Location under Economies of Scale – Decentralization and Computation.* Management Science, 11, 213–235.

MARTOS, B. (1967), *Quasiconvexity and Quasimonotonicity in Nonlinear Programming,* Studia Scientiarum Mathematicarum Hungaria, 2, 265–273.

MARTOS, B. (1971), *Quadratic Programming with a Quasiconvex Objective Function.* Operations Research, 19, 87–97.

MARTOS, B. (1975), *Nonlinear Programming,* North Holland Publishing Company, Amsterdam.

MATTHEIS, T.H. (1973), *An Algorithm for Determining Irrelevant Constraints and all Vertices in Systems of Linear Inequalities.* Operations Research, 21, 247–260.

MATTHEIS, T.H. and RUBIN, D.S. (1980), *A Survey and Comparison of Methods for Finding all Vertices of Convex Polyhedral Sets.* Mathematics of Operations Research, 5, 167–184.

MAYNE, D.Q. AND POLAK, E. (1984), *Outer Approximation Algorithm for Nondifferentiable Optimization Problems.* Journal Optimization Theory and Applications, 42, 19–30.

MAYNE, D.Q. AND POLAK, E. (1986), *Algorithms for Optimization Problems with Exclusion Constraints.* Journal of Optimization Theory and Applications, 51, 453–473.

McCORMICK, G.P. (1972), *Attempts to Calculate Global Solutions of Problems that may have Local Minima.* In: Numerical Methods for Nonlinear Optimization, F. Lootsma, Ed., Academic, London and New York, 209–221.

McCORMICK, G.P. (1973), *Algorithmic and Computational Methods in Engineering Design.* Computers and Structures, 3, 1241–1249.

McCORMICK, G.P. (1976), *Computability of Global Solutions to Factorable Nonconvex Programs: Part I – Convex Underestimating Problems.* Mathematical Programming, 10, 147–175.

McCORMICK, G.P. (1980), *Locating an Isolated Global Minimizer of a Constrained Nonconvex Program.* Mathematics of Operations Research, 5, 435–443.

McCORMICK, G.P. (1981), *Finding the Global Minimum of a Function of one Variable Using the Methods of Constant Signed Higher Order Derivatives.* In: Mangasarian, O.L., Meyer, R.R., Robinson, S.M. (eds.), Nonlinear Programming 4, Proceedings Symposium Madison, July 14–16, Academic Press, New York, 223–243.

McCORMICK, G.P. (1983), *Nonlinear Programming: Theory, Algorithms and Applications.* John Wiley & Sons, New York.

MEEWELLA, C.C. and MAYNE, D.Q. (1988), *An Algorithm for Global Optimization of Lipschitz Continuous Functions.* Journal of Optimization Theory and Applications, 57, 307–322.

MEYER, R.R. (1970), *The Validity of a Family of Optimization Methods.* SIAM Journal on Control, 8, 41–54.

MEYER, R.R. (1983), *Computational Aspects of Two–Segment Separable Programming.* Mathematical Programming, 26, 21–39.

MILLS, H. (1960), *Equilibrum Points in Finite Games.* SIAM Journal, 8, 397–402.

MINOUX, M. (1986), *Mathematical Programming.* John Wiley, 1986.

MINOUX, M. and SERREAULT, J.Y. (1981), *Synthèse Optimale d'un Réseau de Télécommunications avec Constraintes de Securité.* Annales des Télécommunications, 36, 211–230.

MITTEN, L.G. (1970), *Branch and Bound Methods: General Formulation and Properties.* Operations Research, 18, 24–34.

MLADINEO, R.H. (1986), *An Algorithm for Finding the Global Maximum of a Multimodal, Multivariate Function.* Mathematical Programming, 34, 188–200.

MOORE, R.E. (1979), *Methods and Applications of Interval Analysis.* SIAM Studies in Applied Mathematics, W.F. Ames (editor), Philadelphia.

MOORE, R.E. (1988) (ed.), *Reliability in Computing: The Role of Interval Methods.* Academic Press, New York.

MUELLER, R.K. (1970), *A Method for Solving the Indefinite Quadratic Programming Problem.* Management Science, 16, 333–339.

MUKHAMEDIEV, B.M. (1982), *Approximate Methods for Solving Concave Programming Problems.* USSR Computational Mathematics and Mathematical Physics, 22, 238–245 (in Russian).

MURTY, K.G. (1969), *Solving the Fixed–Charge Problem by Ranking the Extreme Points.* Operations Research, 16, 268–279.

MURTY, K.G. (1974), *Note on a Bard–Type Scheme for Solving the Complementarity Problem.* Operations Research, 11, 123–130.

MURTY, K.G. (1988), *Linear Complementarity, Linear and Nonlinear Programming.* Heldermann Verlag, Berlin.

MURTY, K.G. and KABADI, S.N. (1987), *Some NP–Complete Problems in Quadratic and Nonlinear Programming.* Mathematical Programming, 39, 117–130.

MUU, L.D. (1985), *A Convergent Algorithm for Solving Linear Programs with an Additional Reverse Convex Constraint.* Kybernetika, 21, 428–435.

MUU, L.D. (1989), *An Algorithm for Solving Convex Programs with an Additional Convex–Concave Constraint.* Manuskripte der Fakultät für Mathematik und Informatik der Universität Mannheim, 112.

MUU, L.D. and OETTLI, W. (1989), *An Algorithm for Indefinite Quadratic Programming with Convex Constraints.* Preprint, Fakultät für Mathematik und Informatik, Mannheim.

NEFERDOV, V.N. (1987), *The Search for the Global Maximum of a Function of Several Variables on a Set Defined by Constraints of Inequality Type.* USSR Computational Mathematics and Mathematical Physics, 27, 23–32.

NEMHAUSER, G.L. and WOLSEY, L.A. (1988), *Integer and Combinatorial Optimization.* John Wiley & Sons, New York.

NETZER, D. and PASSY, W. (1975), *A Note on the Maximum of Quasiconvex Functions,* Journal of Optimization Theory and Applications, 16, 565–569.

NGHIA, N.D. and HIEU, N.D. (1986), *A Method for Solving Reverse Convex Programming Problems.* Acta Mathematica Vietnamica, 11, 241–252.

NGUYEN, V.H. and STRODIOT, J.J. (1988), *Computing a Global Optimal Solution to a Design Centering Problem.* Technical Report 88/12, Facultés Universitaires de Namur, Namur, Belgium.

NGUYEN, V.H., STRODIOT, J.J. and THOAI, N.V. (1985), *On an Optimum Shape Design Problem.* Technical Report 85/5. Department of Mathematics, Facultes Universitaires de Namur.

OWEN, G. (1973), *Cutting Planes for Programs with Disjunctive Constraints.* Journal of Optimization Theory and Applications, 11, 49–55.

PARDALOS, P.M. (1985), *Integer and Separable Programming Techniques for Large–Scale Global Optimization Problems.* Ph.D. Thesis. Computer Science Department, Univ. Minnesota, Minneapolis, M.N.

PARDALOS, P.M. (1986), *Aspects of Parallel Computation in Global Optimization.* Proc. of the 24th Annual Allerton Conference on Communication, Control and Computing, 812–821.

PARDALOS, P.M. (1986 a), *An Algorithm for a Class of Nonlinear Fractional Problems Using Ranking of the Vertices.* BIT, 26, 392–395.

PARDALOS, P.M. (1987), *Generation of Large–Scale Quadratic Programs for Use as Global Optimization Test Problems.* ACM Transactions on Mathematical Software, 13, No. 2, 133–137.

PARDALOS, P.M. (1987 a), *Objective Function Approximation in Nonconvex Programming.* Proceedings of the 18th Modeling and Simulation Conference, 1605–1610.

PARDALOS, P.M. (1988), *Quadratic Programming Defined on a Convex Hull of Points.* BIT, 28, 323–328.

PARDALOS, P.M. (1988 a), *Enumerative Techniques for Solving some Nonconvex Global Optimization Problems.* Operations Research Spektrum, 10, 29–35.

PARDALOS, P.M. (1988 b), *Linear Complementarity Problems Solvable by Integer Programming.* Optimization, 19, 467–474.

PARDALOS, P.M. (1990), *Global Optimization Algorithms for Linearly Constrained Indefinite Quadratic Problems*, to appear in Journal of Computers and Mathematics with Applications.

PARDALOS, P.M., GLICK, J.H. and ROSEN, J.B. (1987), *Global Minimization of Indefinite Quadratic Problems.* Computing, 39, 281–291.

PARDALOS, P.M. and GUPTA, S. (1988), *A Note on a Quadratic Formulation for Linear Complementarity Problems.* Journal of Optimization Theory and Applications, 57, 197–202.

PARDALOS, P.M. and KOVOOR, N. (1987), *An Algorithm for Singly Constrained Quadratic Programs.* Technical Report, Computer Sci. Dept., Penn. State Univ.

PARDALOS, P.M. and RODGERS, G.P. (1989), *A Branch and Bound Algorithm for the Maximum Clique Problem*, to appear in Mathematical Programming.

PARDALOS, P.M. and ROSEN, J.B. (1986), *Methods for Global Concave Minimization: A Bibliographic Survey.* SIAM Review, 28, 367–379.

PARDALOS, P.M. and ROSEN, J.B. (1987), *Constrained Global Optimization: Algorithms and Applications.* Lecture Notes in Computer Science, 268, Springer–Verlag, Berlin.

PARDALOS, P.M. and ROSEN, J.B. (1988), *Global Optimization Approach to the Linear Complementarity Problem.* SIAM Journal on Scientific and Statistical Computing, 9, 341–353.

PARDALOS, P.M. and SCHNITGER, G. (1987), *Checking Local Optimality in Constrained Quadratic Programming is NP–hard.* Operations Research Letters, 7, 33–35.

PARIKH, S.C. (1976), *Approximate Cutting Planes in Nonlinear Programming.* Mathematical Programming, 11, 184–198.

PEVNYI, A.V. (1982), *On Optimal Search Strategies for the Maximum of a Function with Bounded Highest Derivative.* USSR Computational Mathematics and Mathematical Physics, 22, 38–44.

PFORR, E.A. and GUNTHER, R.H. (1984), *Die Normalform des Quadratischen Optimierungsproblems und die Pol–Polaren Theorie.* Mathematische Operationsforschung und Statistik, Ser. Optimization, 15, 41–55.

PHAN, H. (1982), *Quadratically Constrained Quadratic Programming: Some Applications and a Method of Solution.* Zeitschrift für Operations Research, 26, 105–119.

PHAM, D. T. (1984), *Algorithmes de calcul du maximum d'une forme quadratique sur la boule unite de la norme du maximum.* Numerische Mathematik, 45, 163–183.

PHAM, D. T. and SOUAD, E.B. (1986), *Algorithms for Solving a Class of Nonconvex Optimization Problems. Method of Subgradients.*in: Hiriart–Urruty (editor), Fermat Days 1985: Mathematics for Optimization, 249–272.

PHILLIPS, A.T. and ROSEN, J.B. (1987), *A Parallel Algorithm for Constrained Concave Quadratic Global Minimization: Computational Aspects.* Technical Report 87, 101, Super–Computer Institute, University of Minnesota.

PINCUS, M. (1968), *A Closed Form Solution of Certain Programming Problems.* Operations Research, 16, 690–694.

PINTER, J. (1983), *A Unified Approach to Globally Convergent One–Dimensional Optimization Algorithms.* Technical Report IAMI 83–5, Istituto per le Applicazioni della Matematica e dell' Informatica CNR, Milan.

PINTER, J. (1986), *Extended Univariate Algorithms for n–dimensional Global Optimization,* Computing, 36, 91–103.

PINTER, J. (1986 a), *Globally Convergent Methods for n–dimensional Multiextremal Optimization.* Optimization, 17, 187–202.

PINTER, J. (1986 b), *Global Optimization on Convex Sets.* Operations Research Spektrum, 8, 197–202.

PINTER, J. (1987), *Convergence Qualification of Partition Algorithms in Global Optimization.* Research Report 87–61, Department of Mathematics and Informatics, Delft University of Technology.

PINTER, J. (1988), *Branch–and–Bound Algorithm for Solving Global Optimization Problems with Lipschitzian Structure.* Optimization, 19, 101–110.

PINTER, J. (1989), *Adaptive Partition Methods for Global Optimization: Theory and Some Applications.* Research Report, VITUKI, submitted.

PINTER, J., SZABO, J. and SOMLYODY, L. (1986), *Multiextremal Optimization for Calibrating Water Resources Models.* Environmental Software, 1, 98–105.

PIYAVSKII, S.A. (1967), *An Algorithm for Finding the absolute Minimum of a Function.* Theory of Optimal Solutions, No. 2, Kiev, IK AN USSR, 13–24, (in Russian).

PIYAVSKII, S.A. (1972), *An Algorithm for Finding the Absolute Extremum of a Funtion.*USSR Computational Mathematics and Mathematical Physics, 12, 57–67.

POLAK, E. (1982), *An Implementable Algorithm for the Optimal Design Centering, Tolerancing and Tuning Problem.* Journal of Optimization Theory and Applications, 37, 45–67.

POLAK, E. (1987), *On the Mathematical Foundations of Nondifferentiable Optimization in Engineering Design.* SIAM Review, 29, 21–89.

POLAK, E. and VINCENTELLI, A.S. (1979), *Theoretical and Computational Aspect of the Optimal Design Centering, Tolerancing and Turning Problem.* IEEE Transactions Circuits and Systems, CAS–26, 795–813.

POLJAK, B.T. (1969), *Minimization of Nonsmooth Functionals.* USSR Computational Mathematics and Mathematical Physics, 9, 14–29.

RAGHAVACHARI, M. (1969), *On Connections between Zero–One Integer Programming and Concave Programming under Linear Constraints.* Operations Research, 17, 680–684.

RAMARO, B. and SHETTY, C.M. (1984), *Application of Disjunctive Programming to the Linear Complementarity Problem.* Naval Research Logistics, 31, 589–600.

RATSCHEK, H. (1985), *Inclusion Functions and Global Optimization.* Mathematical Programming, 33, 300–317.

RATSCHEK, H. and ROKNE, J. (1984), *Computer Methods for the Range of Functions.* Ellis Horwood Series Mathematics and its Applications, Wiley, New York.

RATSCHEK, H. and ROKNE, J. (1988), *New Computer Methods for Global Optimization.* Ellis Horwood, Chichester.

REEVES, G.R. (1975), *Global Minimization in Nonconvex All–Quadratic Programming.* Management Science, 22, 76–86.

RINNOOY KAN, A.H.G. and TIMMER, G.T. (1987), *Stochastic Global Optimization Methods. Part I: Clustering Methods.* Mathematical Programming, 39, 27–56.

RINNOOY KAN, A.H.G. and TIMMER, G.T. (1987), *Stochastic Global Optimization Methods. Part II: Multi–Level Methods.* Mathematical Programming, 39, 57–78.

RITTER, K. (1965), *Stationary Points of Quadratic Maximum Problems.* Zeitschrift für Wahrscheinlichkeitstheorie und Verwandte Gebiete, 4, 149–158.

RITTER, K. (1966), *A Method for Solving Maximum–Problems with a Nonconcave Quadratic Objective Function.* Zeitschrift für Wahrscheinlichkeitstheorie und Verwandte Gebiete, 4, 340–351.

ROCKAFELLAR, R.T. (1970), *Convex Analysis.* Princeton University Press, Princeton, N.J.

ROCKAFELLAR, R.T. (1974), *Augmented Lagrange Multiplier Functions and Duality in Nonconvex Programming.* SIAM Journal on Control, 12, 268–285.

ROCKAFELLAR, R.T. (1981), *Favorable Classes of Lipschitz Continuous Functions in Subgradient Optimization.* in: Nurminskii (editor): Progress in Nondifferentiable Optimization, Pergamon Press.

ROESSLER, M. (1971), *Eine Methode zur Berechnung des Optimalen Produktionsprogramms bei konkaven Kosten.* Unternehmensforschung, 15, 103–111.

ROKNE, J. (1977), *Bounds for an Interval Polynomial.* Computing, 18, 225–240.

ROSEN, J.B. (1960), *The Gradient Projection Method for Nonlinear Programming, I: Linear Constraints.* SIAM Journal on Applied Mathematics, 8, 181–217.

ROSEN, J.B. (1966), *Iterative Solution of Nonlinear Optimal Control Problems.* SIAM Journal on Control, 4, 223–244.

ROSEN, J.B. (1983), *Global Minimization of a Linearly Constrained Concave Function by Partition of Feasible Domain.* Mathematics of Operations Research, 8, 215–230.

ROSEN, J.B. (1983 a), *Parametric Global Minimization for Large Scale Problems.* Technical Report 83–11 (revised), Computer Sci. Dept. Univ. of Minnesota.

ROSEN, J.B. (1984), *Performance of Approximate Algorithms for Global Minimization.* Math. Progr. Study, 22, 231–236.

ROSEN, J.B. (1984 a), Computational Solution of Large–Scale Constrained Global Minimization Problems. Numerical Optimization 1984. (P.T. Boggs, R.H. Byrd, R.B. Schnabel, Eds.). SIAM, Phil., 263–271.

ROSEN, J.B. and PARDALOS, P.M. (1986), *Global Minimization of Large–Scale Constrained Concave Quadratic Problems by Separable Programming.* Mathematical Programming, 34, 163–174.

RUBINSTEIN, R. and SAMORODNITSKY, G. (1986), *Optimal Coverage of Convex Regions.* Journal of Optimization Theory and Applications, 51, 321–343.

SAHNI, S. (1974), *Computationally Related Problems.* SIAM J. on Computing, 3, 262–279.

SCHOCH, M. (1984), *Über die Äquivalenz der allgemeinen quadratischen Optimierungsaufgabe zu einer linearen parametrischen komplementären Optimierungsaufgabe.* Mathematische Operationsforschung und Statistik, Ser. Optimization, 16, 211–216.

SCHOEN, F. (1982), *On a Sequential Search Strategy in Global Optimization Problems.* Calcolo, 19, 321–334.

SEN, S. (1987), *A Cone Splitting Algorithm for Reverse Convex Programming.* Preprint, University of Arizona, Tucson.

SEN, S. and SHERALI, H.D. (1985), *On the Convergence of Cutting Plane Algorithms for a Class of Nonconvex Mathematical Programs.* Mathematical Programming, 31, 42–56.

SEN, S. and SHERALI, H.D. (1985 a), *A Branch–and–Bound Algorithm for Extreme Point Mathematical Programming Problems.* Discrete Applied Mathematics, 11, 265–280.

SEN, S. and SHERALI, H.D. (1986), *Facet Inequalities from Simple Disjunctions in Cutting Plane Theory.* Mathematical Programming, 34, 72–83.

SEN, S. and SHERALI, H.D. (1987), *Nondifferentiable Reverse Convex Programs and Facial Convexity Cuts via a Disjunctive Characterization.* Mathematical Programming, 37, 169–183.

SEN, S. and WHITESON, A. (1985), *A Cone Splitting Algorithm for Reverse Convex Programming.* Proceedings of IEEE Conference on Systems, Man, and Cybernetics, Tucson, Az. 656–660.

SHEN, Z. and ZHU, Y. (1987), *An Interval Version of Shubert's Iterative Method for the Localization of the Global Maximum.* Computing 38, 275–280.

SHEPIVOV, M.A. (1987), *Determination of the Roots and of the Global Extremum of a Lipschitz Function,* 23, 233–238.

SHERALI, H.D. and SHETTY, C.M. (1980), *Optimization with Disjunctive Constraints.* Lecture Notes in Economics and Mathematical Systems, 181, Springer–Verlag, Berlin.

SHERALI, H.D. and SHETTY, C.M. (1980 a), *A Finitely Convergent Algorithm for Bilinear Programming Problems Using Polar and Disjunctive Face Cuts.* Mathematical Programming, 19, 14–31.

SHERALI, H.D. and SHETTY, C.M. (1980 b), *Deep Cuts in Disjunctive Programming.* Naval Research Logistics, 27, 453–476.

SHIAU, T.H. (1984), *Finding the Largest ℓ^p–ball in a Polyhedral Set.* Technical Summary Report, University of Wisconsin.

SHOR, N.Z. (1987), *Quadratic Optimization Problems.* Technicheskaya Kibernetika, 1, 128–139 (in Russian).

SHUBERT, B.O. (1972), *A Sequential Method Seeking the Global Maximum of a Function,* SIAM Journal on Numerical Analysis, 9, 379–388.

SINGER, I. (1980), *Minimization of Continuous Convex Functionals on Complements of Convex Subsets of Locally Convex Spaces.* Mathematische Operationsforschung und Statistik, Serie Optimization, 11, 221–234.

SOLAND, R.M. (1971), *An Algorithm for Separable Nonconvex Programming Problems II: Nonconvex Constraints.* Management Science, 17, 759–773.

SOLAND, R.M. (1974), *Optimal Facility Location with Concave Costs.* Operations Research, 22, 373–382.

STEINBERG, D.I. (1970), *The Fixed Charge Problem.* Naval Research Logistics, 17, 217–235.

STONE, R. E. (1981), *Geometric Aspects of the Linear Complementarity Problem.* Technical Report SOL81–6, Department of Operations Research, Stanford University.

STRIGUL, O.I. (1985), *Search for a Global Extremum in a Certain Subclass of Functions with the Lipschitz Condition.* Cybernetics, 21, 193–195.

STRONGIN, R.G. (1973), *On the Convergence of an Algorithm for Finding a Global Extremum.* Engineering Cybernetics, 11, 549–555.

STRONGIN, R.G. (1978), *Numerical Methods for Multiextremal Problems.* Nauka, Moscow (in Russian).

STRONGIN, R.G. (1984), *Numerical Methods for Multiextremal Nonlinear Programming Problems with Nonconvex Constraints.* In: Demyanov, V.F. and Pallaschke, D. (editors), Non–Differentiable Optimization: Motivations and Applications. Proceedings IIASA Workshop, Sopron 1984, IIASA, Laxenburg.

STUART, E.L., PHILLIPS, A.T. and ROSEN, J.B. (1988), *Fast Approximate Solution of Large–Scale Constrained Global Optimization Problems.* Technical Report No. 88–9, Computer Science Department, University of Minnesota.

SUKHAREV, A.G. (1971), *Best Sequential Search Strategies for Finding an Extremum.* USSR Compurations Mathematics and Mathematical Physics, 12, 35–50.

SUKHAREV, A.G. (1985), *The Equality of Errors in Classes of Passive and Sequential Algorithms.* USSR Computational Mathematics and Mathematical Physics, 25, 193–195.

SUNG, Y.Y. and ROSEN, J.B. (1982), *Global Minimum Test Problem Construction.* Mathematical Programming, 24, 353–355.

SWARUP, K. (1966), *Quadratic Programming.* Cahiers du Centre d'Etudes de Récherche Opérationelle, 8, 223–234.

TAHA, H.A. (1972), *A Balasian–Based Algorithm for zero–one Polynomial Programming.* Management Science, 18, 328–343.

TAHA, H.A. (1973), *Concave Minimization over a Convex Polyhedron.* Naval Research Logistics, 20, 533–548.

TAM, B.T. and BAN, V.T. (1985), *Minimization of a Concave Function Under Linear Constraints.* Ekonomika i Matematicheskie Metody, 21, 709–714 (in Russian).

TAMMER, K. (1976), *Möglichkeiten zur Anwendung der Erkenntnisse der parametrischen Optimierung für die Lösung indefiniter quadratischer Optimierungsprobleme.* Mathematische Operationsforschung und Statistik, Serie Optimization, 7, 206–222.

THACH, P.T. (1985), *Convex Programs with Several Additional Reverse Convex Constraints.* Acta Mathematica Vietnamica, 10, 35–57.

THACH, P.T. (1987), *D.C. Sets, D.C. Functions and Systems of Equations.* Preprint, Institute of Mathematics, Hanoi.

THACH, P.T. (1987 a), *Concave Minimization Under Nonconvex Constraints with Special Structure.* in: Essays on Nonlinear Analysis and Optimization, Institute of Mathematics, Hanoi, 121–139.

THACH, P.T. (1988), *The Design Centering Problem as a D.C. Programming Problem.* Mathematical Programming, 41, 229–248.

THACH, P.T. (1988 a), *Quasiconjugates of Functions and Duality Correspondence Between Quasiconcave Minimization Under a Reverse Convex Constraint and Quasiconvex Maximization Under a Convex Constraint.* Preprint, Institute of Mathematics, Hanoi.

THACH, P.T. (1988 b), *A Decomposition Method for the Min Concave Cost Flow Problem with a Special Structure.* Preprint, Institute of Mathematics, Hanoi.

THACH, P.T. (1989), *Convex Minimization Under Lipschitz Constraints*, submitted.

THACH, P.T., THOAI, N.V. and TUY, H. (1987), *Design Centering Problem with Lipschitzian Structure.* Preprint, Institute of Mathematics, Hanoi, submitted.

THACH, P.T. and TUY, H. (1987). *Global Optimization under Lipschitzian Constraints.* Japan Journal of Applied Mathematics, 4, 205–217.

THACH, P.T. and TUY, H. (1988), *Parametric Approach to a Class of Nonconvex Global Optimization Problems.* Optimization, 19, 3–11.

THACH, P.T. and TUY, H. (1989), *Remarks on a New Approach to Global Optimization: The Relief Indicator Method.* Proceedings of the International Seminar on Optimization Methods and Applications, Irkutsk.

THACH, P.T. and TUY, H. (1990), *The Relief Indicator Method for Constrained Global Optimization*, to appear in Naval Research Logistics.

THIEU, T.V. (1980), *Relationship between Bilinear Programming and Concave Programming.* Acta Mathematica Vietnamica, 2, 106–113.

THIEU, T.V. (1987), *Solving the Lay–Out Planning Problem with Concave Cost.* in: Essays in Nonlinear Analysis and Optimization, Institute of Mathematics, Hanoi, 101–110.

THIEU, T.V. (1988), *A Note on the Solution of Bilinear Programming Problem by Reduction to Concave Minimization.* Mathematical Programming, 41, 249–260.

THIEU, T.V., TAM, B.T. and BAN, V.T. (1983), *An Outer Approximation Method for Globally Minimizing a Concave Function over a Compact Convex Set.* Acta Mathematica Vietnamica, 8, 21–40.

THOAI, N.V. (1981), Anwendung des Erweiterungsprinzips zu Lösung konkaver Optimierungsaufgaben. Mathematische Operationsforschung und Statistik, Serie Optimization, 11, 45–51.

THOAI, N.V. (1984), *Verfahren zur Lösung konkaver Optimierungsaufgaben.* Seminarbericht 63 der Humboldt Universität Berlin, Sektion Mathematik.

THOAI, N.V. (1987), *On Canonical D.C. Programs and Applications.* in: Essays on Nonlinear Analysis and Optimization Problems, Institute of Mathematics, Hanoi, 88–100.

THOAI, N.V. (1988), *A Modified Version of Tuy's Method for Solving D.C. Programming Problems.* Optimization, 19, 665–674.

THOAI, N.V. and de VRIES, J. (1988), Numerical Experiments on Concave Minimization Algorithms. Preprint, Department of Mathematics, University of Trier.

THOAI, N.V. and TUY, H. (1980), *Convergent Algorithms for Minimizing a Concave Function.* Mathematics of Operations Research, 5, 556–566.

THOAI, N.V. and TUY, H. (1983), *Solving the Linear Complementarity Problem through Concave Programming.* USSR Computational Mathematics and Mathematical Physics, 23, 602–608.

THUONG, T.V. and TUY, H. (1984), *A Finite Algorithm for Solving Linear Programs with an Additional Reverse Convex Constraint.* Lecture Notes in Economics and Mathematical Systems, 225, Springer, 291–302.

TICHONOV, A.N. (1980), *On a Reciprocity Principle.* Soviet Mathematics Doklady, 22, 100–103.

TIMONOV, L.N. (1977), *An Algorithm for Search of a Global Extremum.* Engineering Cybernetics, 15, 38–44.

TÖRN, A. and ZILINSKAS, A. (1989), *Global Optimization.* Lecture Notes in Computer Science, 350, Springer–Verlag, Berlin.

TOLAND, J.F. (1978), *Duality in Nonconvex Optimization.* Journal Mathematical Analysis and Applications, 66, 399–415.

TOLAND, J.F. (1979), *A Duality Principle for Nonconvex Optimisation and the Calculus of Variations.* Archive of Rational Mechanics and Analysis, 71, 41–61.

TOMLIN, J.A. (1978), *Robust Implementation of Lemke's Method for the Linear Complementarity Problem.* Mathematical Programming Study, 7, 55–60.

TOPKIS, D.M. (1970), *Cutting Plane Methods without Nested Constraint Sets.* Operations Research, 18, 404–413.

TUY, H. (1964), *Concave Programming under Linear Constraints.* Soviet Mathematics, 5, 1437–1440.

TUY, H. (1981), *Conical Algorithm for Solving a Class of Complementarity Problems.* Acta Mathematica Vietnamica, 6, 3–17.

TUY, H. (1982), *Global Maximization of a Convex Function over a Closed, Convex, Non necessarily Bounded Set.* Cahiers de Mathématiques de la Décision, 8223, Université Paris IX–Dauphine.

TUY, H. (1983), *On Outer Approximation Methods for Solving Concave Minimization Problems.* Acta Mathematica Vietnamica, 8, 3–34.

TUY, H. (1983 a), *Global Minimization of a Concave Function Subject to Mixed Linear and Reverse Convex Constraints.* IFIP Conference on "Recent Advances in System Modeling and Optimization", Hanoi.

TUY, H. (1984), *Global Minimization of a Difference of Two Convex Functions.* Proceedings of the VII Symposium on Operations Research, Karlsruhe, August 1983, Lecture Notes In Economics and Mathematical Systems, 226, Springer–Verlag, Berlin, 98–108.

TUY, H. (1985), *Concave Minimization Under Linear Constraints with Special Structure.* Optimization, 16, 335–352.

TUY, H. (1986), *A General Deterministic Approach to Global Optimization via D.C. Programming.* In: Hiriart–Urruty, J.B. (editor), Fermat Days 1985: Mathematics for Optimization, Elsevier, Amsterdam, 137–162.

TUY, H. (1986 a), *On a Polyhedral Annexation Method for Concave Minimization.* Preprint Institute of Mathematics, Hanoi.

TUY, H. (1987), *Global Minimization of a Difference of two Convex Functions.* Mathematical Programming Study, 30, 150–182.

TUY, H. (1987 a), *Convex Programs with an Additional Reverse Convex Constraint.* Journal of Optimization Theory and Applications, 52, 463–485.

TUY, H. (1987 b), *Normal Conical Algorithm for Concave Minimization Over Polytopes.* Preprint 87.20, Institute of Mathematics, Hanoi.

TUY, H. (1987 c), *An Implicit Space Covering Method with Applications to Fixed Point and Global Optimization Problems.* Acta Mathematica, 12, 162–170.

TUY, H. (1988), *On Large Scale Concave Minimization Problems with Relatively Few Nonlinear Variables.* Preprint, Institute of Mathematics, Hanoi.

TUY, H. (1988 a), *On Cutting Methods for Concave Minimization.* Preprint, Institute of Mathematics, Hanoi.

TUY, H. (1988 b), *Polyhedral Underestimation Method for Concave Minimization.* Preprint, Institute of Mathematics, Hanoi.

TUY, H. (1989), *Computing Fixed Points by Global Optimization Methods.* Proceedings of the International Conference on Fixed Point Theory and Applications. Marseille–Luminy, to appear.

TUY, H. (1989 a), *On D.C. Programming and General Nonconvex Global Optimization Problems.* Preprint, Institute of Mathematics, Hanoi.

TUY, H. (1989 b), *The Geometric Complementarity Problem and Transcending Stationarity in Global Optimization.* Preprint, Institute of Mathematics, Hanoi.

TUY, H. (1990), *Concave Minimization and Constrained Nonsmooth Global Optimization: A Survey,* to appear in Optimization.

TUY, H. and AL–KHAYYAL, F. (1988), *Concave Minimization by Piecewise Affine Approximation.* Preprint, Institute of Mathematics, Hanoi.

TUY, H. and HORST, R. (1988), *Convergence and Restart in Branch and Bound Algorithms for Global Optimization. Application to Concave Minimization and DC–Optimization Problems.* Mathematical Programming, 41, 161–183.

TUY, H., KHACHATUROV, V. and UTKIN, S. (1987), *A Class of Exhaustive Cone Splitting Procedures in Conical Algorithms for Concave Minimization.* Optimization, 18, 791–807.

TUY, H. and THAI, N.Q. (1983), *Minimizing a Concave Function Over a Compact Convex Set.* Acta Mathematica Vietnamica, 8, 12–20.

TUY, H., THIEU, T.V. and THAI, N.Q. (1985), *A Conical Algorithm for Globally Minimizing a Concave Function Over a Closed Convex Set.* Mathematics of Operations Research, 10, 498–515.

TUY, H. and THUONG, N.V. (1985), *Minimizing a Convex Function over the Complement of a Convex Set.* Methods of Operations Research, 49, 85–89.

TUY, H. and THUONG, N.V. (1988), *On the Global Minimization of a Convex Function under General Nonconvex Constraints.* Applied Mathematics and Optimization, 18, 119–142.

UEING, U. (1972), *A Combinatorial Method to Compute the Global Solution of Certain Non–Convex Optimization Problems.* in: Lootsma, F.A. (editor), Numerical Methods for Nonlinear Optimization, Academic Press, 223–230.

UTKIN, S., KHACHATUROV, V. and TUY, H. (1988), *A New Exhaustive Cone Splitting Procedure for Concave Minimization.* USSR Computational Mathematics and Mathematical Physics, 7, 992–999 (in Russian).

VAISH, H. and SHETTY, C.M. (1976), *The Bilinear Programming Problem.* Naval Research Logistics Quarterly, 23, 303–309.

VAISH, H. and SHETTY, C.M. (1977), *A Cutting Plane Algorithm for the Bilinear Programming Problem.* Naval Research Logistics Quarterly, 24, 83–94.

VAN der HEYDEN, L. (1980), *A Variable Dimension Algorithm for the Linear Complementarity Problem.* Mathematical Programming, 19, 328–346.

VASIL'EV, N.S. (1985), *Minimum Search in Concave Problems Using the Sufficient Condition for a Global Extremum.* USSR Computational Mathematics and Mathematical Physics, 25, 123–129 (in Russian).

VASIL'EV, S.B. and GANSHIN, G.S. (1982), *Sequential Search Algorithm for the Largest Value of a Twice Differentiable Function.* USSR Computational Mathematics and Mathematical Physics, 31, 312–316.

VEINOTT, A.F. (1967), *The Supporting Hyperplane Method for Unimodal Programming.* Operations Research, 15, 147–152.

VEINOTT, A.F. (1969), *Minimum Concave Cost Solution of Leontiev Substitution Models of Multi–Facility Inventory Systems.* Operations Research, 14, 486–507.

VERGIS, A., STEIGLITZ, K. and DICKINSON, B. (1986), *The Complexity of Analog Computation.* Mathematics and Computers in Simulation, 28, 91–113.

VIDIGAL, L.M. and DIRECTOR, S.W. (1982), *A Design Centering Algorithm for Nonconvex Regions of Acceptability.* IEEE Transactions on Computer–Aided–Design of Integrated Circuits and Systems, CAD–I, 13–24.

WATANABE, H. (1984), *IC Layout Generation and Compaction Using Mathematical Programming.* Ph.D.–Thesis, Computer Sc. Dept. Univ. of Rochester.

WINGO, D.R. (1985), *Globally Minimizing Polynomials Without Evaluating Derivatives.* International Journal of Computer Mathematics, 17, 287–294.

WOLFE, D. (1970), *Convergence Theory in Nonlinear Programming.* In: Integer and Nonlinear Programming, J. Abadie, Ed., North–Holland, Amsterdam.

WOOD, G.R. (1985), *Multidimensional Bisection and Global Minimization.* Research Report, University of Canterbury, New Zealand.

YAGED, B. (1971), *Minimum Cost Routing for Static Network Models.* Networks, 1, 139–172.

YOUNG, R.D. (1971), *Hypercylindrically deduced Cuts in Zero–One Integer Programs.* Operations Research, 19, 1393–1405.

ZALEESKY, A.B. (1980), *Non–Convexity of Admissible Areas and Optimization of Economic Decisions.* Ekonomika i Matematitcheskie Metody, XVI, 1069–1081 (in Russian).

ZALIZNYAK, N.F. and LIGUN, A.A. (1978), *Optimal Strategies for Seeking the Global Maximum of a Function.* USSR Computational Mathematics and Mathematical Physics, 18, 31–38.

ZANG, I. and AVRIEL, M. (1975), *On Functions whose Local Minima are Global,* Journal of Optimization Theory and Applications, 16, 183–190.

ZANG, I., CHOO, E.W. and AVRIEL, M. (1976), *A Note on Functions whose Local Minima are Global,* Journal of Optimization Theory and Applications, 18, 556–559.

ZANGWILL, W.I. (1966), *A Deterministic Multi–Product Multi–Facility Production and Inventory Model.* Operations Research, 14, 486–507.

ZANGWILL, W.I. (1968), *Minimum Concave Cost Flows in Certain Networks.* Management Science, 14, 429–450.

ZANGWILL, W.I. (1969), *A Backlogging Model and a Multi–Echelon Model of a Dynamic Economic Lot Size Production System – A Network Approach.* Management Science, 15, 506–527.

ZANGWILL, W.I. (1985), *Setup Cost Reduction in Series Facility Production.* Discussion Paper, Graduate School of Business, University of Chicago.

ZHIROV, V.S. (1985), *Searching for a Global Extremum of a Polynomial on a Parallelepiped.* USSR of Computational Mathematics and Mathematical Physics, 25, 163–180 (in Russian).

ZIELINSKI, R. and NEUMANN, P. (1983), *Stochastische Verfahren zur Suche nach dem Minimum einer Funktion.* Mathematical Research, 16, Akademie–Verlag, Berlin.

ZILINSKAS, A. (1981), *Two Algorithms for one–dimensional Multimodal Minimization.* Optimization, 12, 53–63.

ZILINSKAS, A. (1982), *Axiomatic Approach to Statistical Models, Applications and Their Use in Multimodal Optimization Theory.* Mathematical Programming, 22, 104–116.

ZILINSKAS, A. (1986), *Global Optimization: Axiomatics of Statistical Models, Algorithms and their Applications,* Mokshar Publishers, Vilnius (in Russian).

ZWART, P.B. (1973), *Nonlinear Programming: Counterexamples to two Global Optimization Algorithms.* Operations Research, 21, 1260–1266.

ZWART, P.B. (1974), *Global Maximization of a Convex Function with Linear Inequality Constraints.* Operations Research, 22, 602–609.

NOTATION

$\mathbb{N}$	set of natural numbers
$\mathbb{R}$	set of real numbers
$\overline{\mathbb{R}}$	set of extended real numbers ($\overline{\mathbb{R}} = \mathbb{R} \cup \{+\infty, -\infty\}$)
$\mathbb{R}_+$	set of real n–vectors
$\lfloor \alpha \rfloor$	lower integer part of α
$\lceil \alpha \rceil$	upper integer part of α
$M \subset N$	M (not necessarily strict) subset of N
$M \setminus N$	difference of sets M and N
$M - N$	algebraic difference of sets M and N
$M + N$	sum of sets M and N
$\|M\|$	cardinality of set M
lin M	linear hull of set M
aff M	affine hull of set M
conv M	convex hull of set M
cone M	conical hull of set M
$\delta(M) = d(M)$	diameter of M
xy	inner product of vectors x,y
I	identity matrix
I_n	(n×n) identity matrix

$\mathrm{diag}(\alpha) =$	
$\mathrm{diag}(\alpha_1,\ldots,\alpha_n)$	diagonal matrix with entries $\alpha_1,\ldots,\alpha_n$ (where $\alpha=(\alpha_1,\ldots,\alpha_n)$
A^T	transpose of matrix A
A^{-1}	inverse of matrix A
det A	determinant of A
$x \leq y$	$x_i \leq y_i$ for all i (where $x = (x_1,\ldots,x_n)$, $y = (y_1,\ldots,y_n)$)
$Ax = b$	system of linear equalities
$Ax \leq b$	system of linear inequalities
$Q = (z^1,\ldots,z^n)$	matrix of columns $z^1,\ldots,z^n$
con(Q)	cone spanned by the columns of the matrix Q
$\mathrm{conv}\,[x^0,\ldots,x^n] =$	
$[x^0,\ldots,x^n]$	simplex spanned by its n+1 affinely independent vertices $x^0,\ldots,x^n$
polyhedron,	
convex polyhedral set	set of solutions of a system of linear inequalities
polytope	bounded polyhedron
vert(P), V(P)	vertex set of polyhedron (polytope) P
extd(P), U(P)	set of extreme directions of a polyhedron P
R(K)	recession cone of convex set K
G = (V,A)	directed graph
epi (f)	epigraph of f
hypo (f)	hypograph of f
dom (f)	effective domain of a function f: $\mathbb{R}^n \to \overline{\mathbb{R}}$
$\nabla f(x)$	gradient of f at x
$\partial f(x)$	subdifferential of f at x
$\lVert\cdot\rVert$	Euclidean norm
$N(x,\varepsilon)$	open ball centered at x with radius ε
int M	interior of set M

cl M = $\overline{M}$	closure of set M
∂M	boundary of set M
$\pi(x)$	projection of a point x onto a specified set
min f(M)	global minimum of function f over set M
argmin f(M)	set of global minimizers of f over M
(BCP)	basic concave programming problem (=linearly constrained concave minimization problem)
(BLP)	bilinear programming problem
(CCP)	concave complementarity problem
(CDC)	canonical dc programming problem
(CF)	minimum concave cost flow problem
(CP)	concave minimization problem (concave programming problem)
(LCP)	linear complementarity problem
(LP)	linear programming problem
(LRCP)	linear program with an additional reverse convex constraint
(MIP)	mixed integer programming problem
(PCP)	parametric concave programming problem
(QCP)	quadratic concave programming problem
(SBC)	special biconvex programming problem
(SCP)	separable concave programming problem
(SUCF)	single source uncapacitated minimum concave cost flow problem
(UCF)	uncapacitated minimum concave cost flow problem
(UL)	univariate Lipschitz optimization problem
NCS	normal conical subdivision
NRS	normal rectangular subdivision
NSS	normal simplicial subdivision

INDEX

P

Q

R

S

T

U

V

W, X, Y, Z